MW01356649

VOLUME ONE HUNDRED AND FIFTY EIGHT

ADVANCES IN CANCER RESEARCH

Epigenetic Regulation of Cancer in Response to Chemotherapy

VOLUME ONE HUNDRED AND FIFTY EIGHT

ADVANCES IN CANCER RESEARCH

Epigenetic Regulation of Cancer in Response to Chemotherapy

Edited by

JOSEPH W. LANDRY
Department of Human and Molecular Genetics,
VCU Institute of Molecular Medicine (VIMM);
VCU Massey Cancer Center, Virginia Commonwealth University, School of Medicine, Richmond, VA, United States

SWADESH K. DAS
Department of Human and Molecular Genetics,
VCU Institute of Molecular Medicine (VIMM);
VCU Massey Cancer Center, Virginia Commonwealth University, School of Medicine, Richmond, VA, United States

PAUL B. FISHER
Professor and Chairman,
Department of Human and Molecular Genetics,
VCU Institute of Molecular Medicine (VIMM);
VCU Massey Cancer Center, Virginia Commonwealth University, School of Medicine, Richmond, VA, United States

Academic Press is an imprint of Elsevier
50 Hampshire Street, 5th Floor, Cambridge, MA 02139, United States
525 B Street, Suite 1650, San Diego, CA 92101, United States
The Boulevard, Langford Lane, Kidlington, Oxford OX5 1GB, United Kingdom
125 London Wall, London, EC2Y 5AS, United Kingdom

First edition 2023

Notices
Knowledge and best practice in this field are constantly changing. As new research and experience broaden our understanding, changes in research methods, professional practices, or medical treatment may become necessary.

Practitioners and researchers must always rely on their own experience and knowledge in evaluating and using any information, methods, compounds, or experiments described herein. In using such information or methods they should be mindful of their own safety and the safety of others, including parties for whom they have a professional responsibility.

To the fullest extent of the law, neither the Publisher nor the authors, contributors, or editors, assume any liability for any injury and/or damage to persons or property as a matter of products liability, negligence or otherwise, or from any use or operation of any methods, products, instructions, or ideas contained in the material herein.

ISBN: 978-0-443-19418-4
ISSN: 0065-230X

For information on all Academic Press publications
visit our website at https://www.elsevier.com/books-and-journals

Publisher: Zoe Kruze
Developmental Editor: Naiza Ermin Mendoza
Production Project Manager: James Selvam
Cover Designer: Miles Hitchen

Typeset by STRAIVE, India

Contents

Yanqiu Xie
School of Pharmacy, Key Laboratory of Molecular Pharmacology and Drug Evaluation (Yantai University), Ministry of Education, Collaborative Innovation Center of Advanced Drug Delivery System and Biotech Drugs in Universities of Shandong, Yantai University, Yantai; Drug Discovery and Design Center, State Key Laboratory of Drug Research, Shanghai Institute of Materia Medica, Chinese Academy of Sciences, Shanghai, China

Jin-Shu Yang
MOE Laboratory of Biosystem Homeostasis and Protection, College of Life Sciences, Zhejiang University, Hangzhou, Zhejiang, China

Wei-Jun Yang
MOE Laboratory of Biosystem Homeostasis and Protection, College of Life Sciences, Zhejiang University, Hangzhou, Zhejiang, China

Yao-Shun Yang
MOE Laboratory of Biosystem Homeostasis and Protection, College of Life Sciences, Zhejiang University, Hangzhou, Zhejiang, China

Manzoor Ahmad Yatoo
Division of Basic Sciences and Humanities, FoA, SKUAST-Kashmir, Sopore-193201, India

Kehao Zhao
School of Pharmacy, Key Laboratory of Molecular Pharmacology and Drug Evaluation (Yantai University), Ministry of Education, Collaborative Innovation Center of Advanced Drug Delivery System and Biotech Drugs in Universities of Shandong, Yantai University, Yantai, China

Lauren May
Department of Human and Molecular Genetics, VCU Institute of Molecular Medicine, Massey Cancer Center, Virginia Commonwealth University School of Medicine, Richmond, VA, United States

Paul C. Moore
Pfizer Centers for Therapeutic Innovation, San Francisco, CA, United States

Sudeshna Mukherjee
Department of Biological Sciences, Birla Institute of Technology and Science (BITS), Pilani, Rajasthan, India

Zainab Rentia
The George Washington University Cancer Center (GWCC), Washington, DC, United States

Basit Amin Shah
Directorate of Forensic Science Laboratory, Bemina, Srinagar, Jammu & Kashmir, India

Manjulata Singh
Department of Human and Molecular Genetics, VCU Institute of Molecular Medicine, Massey Cancer Center, Virginia Commonwealth University School of Medicine, Richmond, VA, United States

Ashish Vaidyanathan
Department of Human and Molecular Genetics, VCU Institute of Molecular Medicine, Massey Cancer Center, Virginia Commonwealth University School of Medicine, Richmond, VA, United States

Reddick R. Walker
The George Washington University Cancer Center (GWCC); Department of Microbiology, Immunology & Tropical Medicine, The George Washington University, Washington, DC, United States

Christopher R. Wood
MOE Laboratory of Biosystem Homeostasis and Protection, College of Life Sciences, Zhejiang University, Hangzhou, Zhejiang, China

Wen-Tao Wu
MOE Laboratory of Biosystem Homeostasis and Protection, College of Life Sciences, Zhejiang University, Hangzhou, Zhejiang, China

Xinyu Wu
School of Pharmacy, Key Laboratory of Molecular Pharmacology and Drug Evaluation (Yantai University), Ministry of Education, Collaborative Innovation Center of Advanced Drug Delivery System and Biotech Drugs in Universities of Shandong, Yantai University, Yantai; Drug Discovery and Design Center, State Key Laboratory of Drug Research, Shanghai Institute of Materia Medica, Chinese Academy of Sciences, Shanghai, China

Yongmei Xi
The Women's Hospital, and Institute of Genetics, Zhejiang University School of Medicine, Zhejiang Provincial Key Laboratory of Genetic & Developmental Disorders, Hangzhou, Zhejiang, China

Paul Dent
Department of Biochemistry and Molecular Biology, Massey Cancer Center, Virginia Commonwealth University, Richmond, VA, United States

Luni Emdad
Department of Human and Molecular Genetics; VCU Institute of Molecular Medicine; VCU Massey Cancer Center, Virginia Commonwealth University, School of Medicine, Richmond, VA, United States

Paul B. Fisher
Department of Human and Molecular Genetics; VCU Institute of Molecular Medicine; VCU Massey Cancer Center, Virginia Commonwealth University, School of Medicine, Richmond, VA, United States

Shabir Ahmad Ganai
Division of Basic Sciences and Humanities, FoA, SKUAST-Kashmir, Sopore-193201, India

Gordon Ginder
Department of Internal Medicine, Massey Cancer Center, Virginia Commonwealth University, Richmond, VA, United States

Steven Grant
Department of Human and Molecular Genetics, VCU Institute of Molecular Medicine, Massey Cancer Center, Virginia Commonwealth University School of Medicine; Department of Internal Medicine; Department of Biochemistry and Molecular Biology, Massey Cancer Center, Virginia Commonwealth University; Department of Microbiology & Immunology, Virginia Commonwealth University School of Medicine, Massey Cancer Center, Richmond, Richmond, VA, United States

Kurt W. Henderson
Pfizer Centers for Therapeutic Innovation, San Francisco, CA, United States

Amit Kumar
Department of Human and Molecular Genetics; VCU Institute of Molecular Medicine, Virginia Commonwealth University, School of Medicine, Richmond, VA, United States

Joseph Landry
Department of Human and Molecular Genetics, VCU Institute of Molecular Medicine, Massey Cancer Center, Virginia Commonwealth University School of Medicine, Richmond, VA, United States

Jing Lu
School of Pharmacy, Key Laboratory of Molecular Pharmacology and Drug Evaluation (Yantai University), Ministry of Education, Collaborative Innovation Center of Advanced Drug Delivery System and Biotech Drugs in Universities of Shandong, Yantai University, Yantai, China

Nilanjana Mani
Department of Biological Sciences, Birla Institute of Technology and Science (BITS), Pilani, Rajasthan, India

Contributors

Christiana O. Appiah
Department of Human and Molecular Genetics, VCU Institute of Molecular Medicine, Massey Cancer Center, Virginia Commonwealth University School of Medicine; Wright Center for Clinical and Translational Research, Virginia Commonwealth University, Richmond, VA, United States

Manny D. Bacolod
Department of Microbiology and Immunology, Weill Cornell Medicine, New York, NY, United States

Ishita Bakshi
Department of Human and Molecular Genetics, VCU Institute of Molecular Medicine, Massey Cancer Center, Virginia Commonwealth University School of Medicine, Richmond, VA, United States

Francis Barany
Department of Microbiology and Immunology, Weill Cornell Medicine, New York, NY, United States

Harry Bear
Department of Surgery; Department of Microbiology & Immunology, Virginia Commonwealth University School of Medicine, Massey Cancer Center, Richmond, Richmond, VA, United States

Katherine B. Chiappinelli
The George Washington University Cancer Center (GWCC); Department of Microbiology, Immunology & Tropical Medicine, The George Washington University, Washington, DC, United States

Rajdeep Chowdhury
Department of Biological Sciences, Birla Institute of Technology and Science (BITS), Pilani, Rajasthan, India

Shibasish Chowdhury
Department of Biological Sciences, Birla Institute of Technology and Science (BITS), Pilani, Rajasthan, India

Marie Classon
Pfizer Centers for Therapeutic Innovation, San Francisco, CA, United States

Ankita Daiya
Department of Biological Sciences, Birla Institute of Technology and Science (BITS), Pilani, Rajasthan, India

Swadesh K. Das
Department of Human and Molecular Genetics; VCU Institute of Molecular Medicine; VCU Massey Cancer Center, Virginia Commonwealth University, School of Medicine, Richmond, VA, United States

Preface

Cancer treatments have evolved and improved significantly over the last 100 years. Over this time frame, therapies (combining surgery, chemotherapy, and/or radiation) have been conceived and validated which have resulted in significant increases in 5-year survival rates, and in some forms of cancer now being considered essentially curable (early-stage prostate, thyroid, testicular, melanoma, breast). However, despite improvements in cancer treatment, certain cancers, particularly those in advanced stages, remain challenging to treat. Traditionally, chemotherapy and radiation are the standard of care used to treat these advanced cancers, with significantly reduced 5-year survival rates. Relapse, months to years after initial treatment, is a significant contributor to the reduced survival rates for these advanced cancers. How relapse occurs remains a significant clinical problem and it will continue to be a major focus of the cancer research community.

After several decades of intensive research, it is now appreciated that cancer relapse after initial treatment is multifaceted and can occur by many different resistance and tolerance mechanisms. Resistance mechanisms can result from cancer cells acquiring an ability to render therapies ineffective (i.e., cellular quiescence), whereas tolerance mechanisms allow improved recovery from therapy-induced damage after exposure (i.e., cytoprotective forms of autophagy). The underlying molecular basis of these mechanisms can be either genetic, through changes in the DNA sequence, or epigenetic, which results in changes in gene expression. A distinct difference between these two mechanisms is that genetic changes are acquired through mutations, which are permanent, whereas epigenetic changes endure through cellular division, and in many (but not all) cases are reversible.

Over the last several decades, our understanding of epigenetic regulatory mechanisms in both normal development and disease states, especially cancer, has expanded. Many of these mechanisms operate at the level of chromatin to regulate gene expression (i.e., histone-modifying enzymes), or can operate to regulate mRNA transcript stability (i.e., miRNAs). At the most basic level, epigenetic changes alter gene expression to change cellular responses. In the context of acquired resistance, cancer cells can acquire epigenetic changes that render them more resistant to therapy exposure or can make them better able to recover postexposure. Because epigenetic mechanisms are reversible (in the majority of cases), and are created by enzymes with active sites, these changes can in theory be targeted by small molecules to restore the original ground state of gene expression. With this strategy in mind, small molecules that target epigenetic targets (epi-drugs) can be used to alter the ability of cancer cells to be sensitized to or recover from therapy exposure. A wide variety of epigenetic regulators has been targeted successfully either genetically or pharmacologically (when small molecule inhibitors exist) in preclinical studies to alter the sensitivity or resistance to therapy exposure. Some of these strategies have been proven to have efficacy in a clinic setting.

In this thematic volume of *Advances in Cancer Research* entitled "Epigenetic Regulation of Cancer in Response to Chemotherapy," we asked nine leaders in the field of cancer epigenetics to share their views on how epigenetics can alter the cancer cell response to therapy, and how epigenetics can be targeted to improve responses to therapy.

Chapter 1 by Moore, Henderson, and Classon provides an overview of the many epigenetic mechanisms that contribute to drug tolerance. In this chapter, the authors present the hallmarks of cancer drug tolerance and how these hallmarks are regulated epigenetically. In Chapter 2, Walker, Rentia, and Chiappinelli extend the discussion by including elements of the anti-tumor immune response, and describing how epigenetics can contribute to immune evasion. These additions are timely, as it is now well accepted that the immune system responds to therapy-treated cancer cells and the nature of this response has significant impacts on cancer progression. These two chapters in combination provide a nice overview of cancer epigenetics and how it relates to the many facets of drug resistance and tolerance.

The discussion continues in Chapter 3 by Kumar, Emdad, Fisher, and Das, where the authors contribute a comprehensive survey of the small molecules available to target epigenetic regulators. Effective small molecule regulators are essential to any clinical success targeting epigenetics to prevent drug tolerance. Chapter 3 provides a useful overview of where several classes of these molecules stand in the developmental pipeline.

Histone deacetylase (HDAC) inhibitors are one of the more effective epi-drugs and have achieved approval from the U.S. Food and Drug Administration (FDA) to treat several cancers. Chapter 4 by Ganai, Shah, and Yatoo describes the current status of HDAC utilization for the treatment of lung cancers.

In order to achieve research improvements in the field of cancer epigenetics, new and more tractable model systems must be developed. In Chapter 5, Wood, Wu, Yang, Yang, Xi, and Yang introduce the brine shrimp as a model for the study of cancer stem cells and cancer cell dormancy. The authors describe some of the parallels that this novel model systems have to humans, and how it may be superior as a model to study drug tolerance by cancer stem cells.

Chapter 6, provided by Bacolod, Fisher, and Barany, showcases a bioinformatic method for using CpG methylation to predict sensitivity of cancer cell lines to therapy exposure. This tool, and those which are similar, could provide oncologists with the ability to predict the sensitivity of a tumor to a variety of therapies.

Chapter 7, by Mani, Daiya, Chowdhury, Mukherjee, and Chowdhury, describes how tumor cells adapt their epigenetics to achieve tolerance to therapy exposure. These mechanisms are prime therapeutic targets for preventing therapy-induced dormancy, and prevent the potential for relapse. These discussions then lead into the Chapter 8 by Appiah, Singh, May, Bakshi, Vaidyanathan, Dent, Ginder, Grant, Bear, and Landry, which describes roles for epigenetic regulation in promoting cancer cell recovery from therapy exposure. In this chapter, the authors propose that recovery from therapy exposure is based in epigenetics, and that by targeting these epigenetics, recovery can be prevented. Finally, Chapter 9 by Wu, Xie, Zhao, and Lu examines roles for the superelongation complex in cancer cell biology.

As editors, we hope that ACR Volume 158 will provide a platform for showcasing the important roles that epigenetics has in regulating the response of cancer cells to therapy exposure. We hope that this collection of work will stimulate discussion on this topic, possibly leading to discoveries that are at the interface between cancer epigenetic and cancer therapies, with enduring and transformative impacts on cancer research and cancer treatments.

Joseph W. Landry, PhD
Associate Professor, Department of Human and Molecular Genetics
Member, VCU Institute of Molecular Medicine (VIMM)
Member, VCU Massey Cancer Center
Virginia Commonwealth University, School of Medicine
Richmond, VA 23239, United States

Swadesh K. Das, PhD
Associate Professor, Department of Human and Molecular Genetics
Member, VCU Institute of Molecular Medicine (VIMM)
Member, VCU Massey Cancer Center
Virginia Commonwealth University, School of Medicine
Richmond, VA 23239, United States

Paul B. Fisher, MPh, PhD, FNAI
Professor and Chair, Department of Human and Molecular Genetics
Director, VCU Institute of Molecular Medicine (VIMM)
Member, VCU Massey Cancer Center
Virginia Commonwealth University, School of Medicine
Richmond, VA 23239, United States

CHAPTER ONE

The epigenome and the many facets of cancer drug tolerance

Paul C. Moore, Kurt W. Henderson, and Marie Classon*
Pfizer Centers for Therapeutic Innovation, San Francisco, CA, United States
*Corresponding author: e-mail address: marie.classon@pfizer.com

Contents

Abstract

The use of chemotherapeutic agents and the development of new cancer therapies over the past few decades has consequently led to the emergence of myriad therapeutic resistance mechanisms. Once thought to be explicitly driven by genetics, the coupling of reversible sensitivity and absence of pre-existing mutations in some tumors opened the way for discovery of drug-tolerant persisters (DTPs): slow-cycling subpopulations of tumor cells that exhibit reversible sensitivity to therapy. These cells confer multi-drug tolerance, to targeted and chemotherapies alike, until the residual disease can establish a stable, drug-resistant state. The DTP state can exploit a multitude of distinct, yet interlaced, mechanisms to survive otherwise lethal drug exposures. Here, we categorize these multi-faceted defense mechanisms into unique Hallmarks of Cancer Drug Tolerance. At the highest level, these are comprised of *heterogeneity, signaling plasticity, differentiation, proliferation/metabolism, stress management, genomic integrity, crosstalk with the tumor microenvironment, immune escape*, and *epigenetic regulatory mechanisms*. Of these, epigenetics was both one of the first proposed means of non-genetic resistance and one of the first discovered. As we describe in this review, epigenetic regulatory factors are involved in most facets of DTP biology, positioning this hallmark as an overarching mediator of drug tolerance and a potential avenue to novel therapies.

Advances in Cancer Research, Volume 158
ISSN 0065-230X
https://doi.org/10.1016/bs.acr.2022.12.002

1. Principles of non-genetic drug tolerance

Resistance to cancer therapy can manifest as primary or secondary resistance, characterized by non-response or initial response followed by acquired resistance and relapse (Cabanos & Hata, 2021; De Conti, Dias, & Bernards, 2021). Advances in radiotherapy and traditional chemotherapy as well as the advent of targeted and immune therapies have improved clinical outcomes. However, as cancer treatments have evolved, so too have secondary resistance mechanisms.

One perspective is that acquired resistance to targeted agents is driven by a pre-existing subpopulation of cells harboring one or more genetic mutations (Fig. 1A). Therapy leads to eradication of the bulk tumor and Darwinian selection and enrichment of the mutant subpopulation, which relapses into a drug-resistant population (Cabanos & Hata, 2021). Classical examples of this mechanism are well documented in resistance of non-small cell lung cancers (NSCLC) to targeted therapies with receptor tyrosine kinase (RTK) inhibitors of the epidermal growth factor receptor (EGFR). Namely, a T790M mutation in EGFR or amplification of MET, another RTK, each result in reactivation of MAPK signaling and establishment of drug resistance (Engelman & Janne, 2008). Following these initial discoveries, a multitude of genetic mechanisms driving resistance to targeted agents have been identified. There may be similar alterations in tumor cells which dampen/alter the response to chemotherapeutic or other agents. However, relapsed tumors do not always carry a clear genetic mutation (Engelman & Janne, 2008) and, critically, it was observed that resistance to therapy could be reversible after drug holiday (Cara & Tannock, 2001; Chevallier et al., 2008; Hashimoto et al., 2006; Kaminski et al., 2005; Naing & Kurzrock, 2010; Oh, Ban, Kim, & Kim, 2012; Taverna, Voegeli, Trojan, Olie, & von Rohr, 2012; Yano et al., 2005). The fact that refractory tumors can still respond to their original treatments, following a period of drug withdrawal, paved the way for investigations into non-genetic mechanisms of drug resistance (Fig. 1B).

Early research into bacterial response to antibiotics and viral infection established proof-of-concept mechanisms for non-genetic mechanisms of resistance (Bigger, 1944; Hobby, Meyer, & Chaffee, 1942; Luria & Delbruck, 1943). Bacteria that survive the initial onslaught, termed "bacterial persisters," display a consistent set of traits: they arise from a heterogeneous population containing a rare, stochastically arising, low- or non-proliferating subpopulation that can be re-sensitized upon withdrawal of treatment

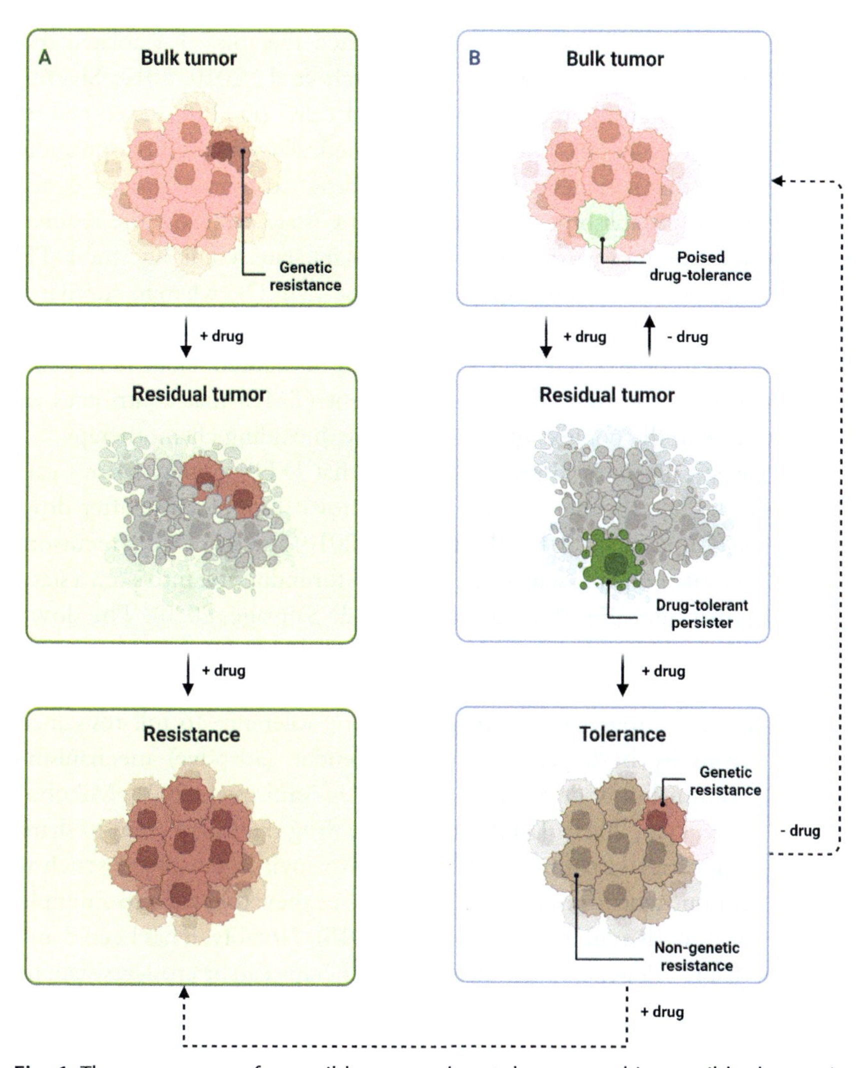

Fig. 1 The emergence of reversible cancer drug tolerance and irreversible drug resistance. The squares in column A illustrate a scenario where pre-existing mutations, or other alterations, result in irreversible resistance. Column B describes the emergence of a drug-tolerant population from a potentially poised, stochastically arising cell population. The drug-tolerant cells can transition to reversibly drug-tolerant cells or develop into an irreversible drug-resistant state. A combination of these two states can evolve within a tumor.

(Fisher, Gollan, & Helaine, 2017). However, sustained treatment of the remaining "bacterial persister" population can give rise to irreversible resistance, requiring a distinction between reversible (tolerant) and stable (resistant) states (Levin-Reisman et al., 2017). This same phenomenon was first observed in cultured cancer cell lines exposed to targeted therapies that ablated the majority of the tumor cell populations (Roesch et al., 2010, 2013; Sharma et al., 2010). The term "drug-tolerant persister cells" (DTPs), was coined to characterize this analogous, stochastic, non-genetically tolerant subpopulation of cancer cells that survives otherwise lethal drug exposures (Sharma et al., 2010) (Fig. 1B). Establishment of this "dormant" state can allow the residual tumor to progress to a permanently drug resistant state. Clinically, the DTP state is associated with minimal residual disease (MRD), wherein surviving tumor cells can eventually progress from tolerance to full resistance and relapse (De Conti et al., 2021). There is still much to be learned about DTPs and MRD as well as the tumor microenvironment (TME) that contributes to patient relapse in the context of cancer therapy, including chemotherapy.

Mechanistically, it is important to note that DTP subpopulations can reestablish the heterogeneity of the bulk tumor cell population after drug withdrawal (Gupta et al., 2011; Sharma et al., 2010) and that DTP precursors may pre-exist in a capable state or be "poised" for induction into such a state after drug treatment (Fig. 1B) (Boumahdi & de Sauvage, 2020). The slow- or non-cycling DTP populations have been classified as dormant, quiescent, or senescent (Saleh, Tyutyunyk-Massey, & Gewirtz, 2019; Vallette et al., 2019); and their progression from non-genetic tolerance to full resistance can be driven by both genetic and non-genetic (adaptive) mechanisms (Fig. 1B) (Boumahdi & de Sauvage, 2020; De Conti et al., 2021; Mikubo, Inoue, Liu, & Tsao, 2021). DTPs that survive drug treatment seem to share some characteristics with tumor cells subjected to environmental stress such as hypoxia and nutrient deprivation, particularly in their tolerance to multiple therapies/stresses (Ravindran Menon et al., 2015). Notably, it has been demonstrated that DTPs generated under targeted therapy can be cross-resistant to chemotherapy (Nilsson et al., 2020; Sharma et al., 2010), establishing that mechanisms that drive drug tolerance under one type of therapy may be broadly applicable to others.

While still in its infancy, the field has accumulated some degree of detail surrounding the molecular and cellular drivers of drug tolerance. This has led to numerous so-called hallmarks of drug tolerance in cancer, each of which is intricately connected to others: *heterogeneity, signaling plasticity, differentiation, proliferation/metabolism, stress management, genomic integrity, crosstalk with the tumor microenvironment, immune escape,* and *regulatory epigenetic mechanisms.* While a hallmark on its own, epigenetic regulation can also be seen as an overarching regulator or mediator of other facets of drug tolerance (Fig. 2). The goal of this review is to outline various epigenetic mechanisms

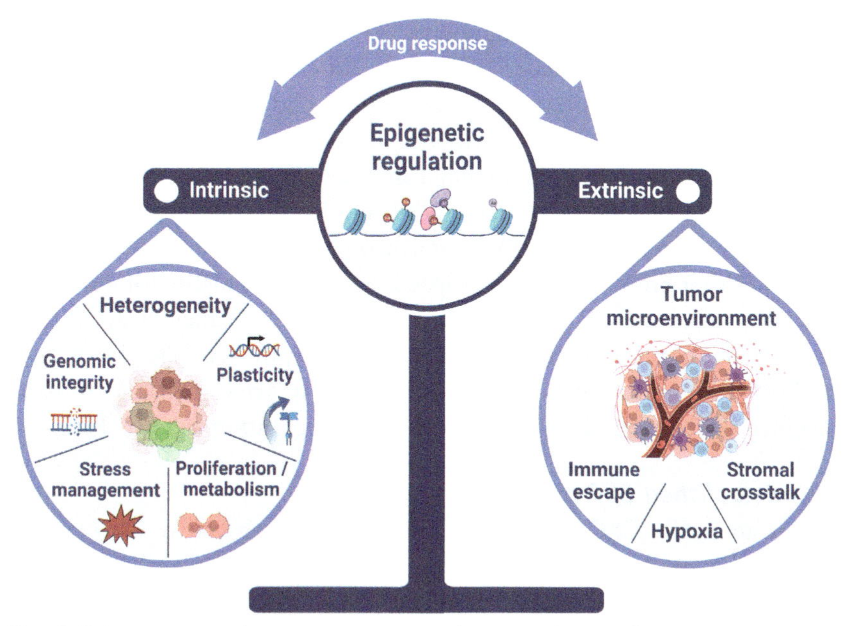

Fig. 2 Epigenetic regulation as an overarching mediator of tumor-intrinsic and -extrinsic hallmarks of cancer drug tolerance/resistance.

that contribute to drug tolerance, framing them in the context of the other hallmarks. Here, we first provide a high-level overview of the of the hallmarks of drug tolerance in cancer.

2. The hallmarks of cancer drug tolerance

2.1 Heterogeneity

2.1.1 DTP subpopulation dynamics

Once it was established that a rare, seemingly stochastic, population of cancer cells can give rise to DTPs, several hypotheses arose to describe this state before and after drug exposure. The DTP state could simply pre-exist, arise upon drug exposure, or fluctuate stochastically in the population and be stabilized upon treatment (Boumahdi & de Sauvage, 2020). As the tools to study intratumoral heterogeneity evolved from multi-region sequencing (Gerlinger et al., 2012) to single-molecule FISH (Shaffer et al., 2017), barcoding (Hata et al., 2016; Kurppa et al., 2020) and single-cell RNA sequencing (Hong et al., 2019; Maynard et al., 2020), so did our understanding of DTPs. Some of this work demonstrated that non-genetic heterogeneity and, importantly, rare DTP subpopulations can be re-established by expansion of isolated clones (Gupta et al., 2011; Ramirez et al., 2016; Shaffer et al., 2017), lending credence to the stochastic model of tolerance and a mechanism for re-sensitization upon drug holiday. Additionally, such experiments provided concrete evidence that cells with pre-existing genetic and non-genetic means of resistance can occur simultaneously within the same population (Hata et al., 2016). While not fully resolved, further experiments provide evidence that DTP precursors within the stochastic population may either pre-exist in a fully capable state or be "poised" for induction after drug treatment (Fig. 1B) (Hong et al., 2019; Shaffer et al., 2017). Co-existence of these precursor types appears to create heterogeneity within the DTP subpopulation, allowing one tumor to have multiple paths to tolerance (Hata et al., 2016; Kurppa et al., 2020; Ramirez et al., 2016).

2.1.2 Established DTP markers

Cancer stem cells (CSC) are tumor-initiating cells with known markers in a broad range of different cancer types (Makena, Ranjan, Thirumala, & Reddy, 2020). CSCs, which are inherently plastic and associated with therapeutic resistance, may be precursors of DTPs (Bai, Ni, Beretov, Graham, & Li, 2018; Makena et al., 2020). These cells are commonly characterized by markers such as ALDH, CD44, CD133 (Dhanyamraju, Schell, Amin, & Robertson, 2022) and the Wnt target LGR5 (Sanchez-Danes et al., 2018), as well as ABCG2, a drug efflux pump identified as a breast cancer CSC marker (Leccia et al., 2014). Other characterized markers of this state include increased expression of the histone demethylase KDM5B (Roesch et al., 2010, 2013) and the receptor tyrosine kinase AXL (Shaffer et al., 2017). Such markers can co-exist in the same population, or within the same cells, as the classic CSC markers (Dhanyamraju et al., 2022). There are also examples where DTPs lack CSC markers or typical properties of stemness (Farge et al., 2017). Furthermore, other types of subpopulations may serve as DTPs and determine treatment response, as exemplified by more mesenchymal CSC subsets in glioblastoma and breast cancer (Jin et al., 2017; Luo et al., 2018).

2.2 Plasticity: Signaling

To withstand lethal drug treatment and enable relapse, drug-tolerant cancer cells exhibit a great degree of plasticity, including the ability to rewire signaling pathways to either override or altogether circumvent the effects of targeted therapies. In this context, much less is known about "re-wiring" when cancer cells are treated with DNA damaging agents.

2.2.1 Signaling bypass

The idea of signaling bypass originated in studies of drug tolerance/resistance to targeted therapies, where efficacy of inhibitors of targets such as mutant EGFR and BRAF is blunted by reactivation of MAPK and PI3K/AKT signaling pathways (Engelman & Janne, 2008; Niederst & Engelman, 2013; Shi et al., 2014). This concept was extended by demonstrations that many RTK ligands, with pathways typically converging on MAPK/PI3K/AKT, could confer resistance to various targeted treatments in the absence of pre-existing mutations (Wilson et al., 2012). In DTPs, the same effect is achieved by non-genetic means: EGFR inhibition in NSCLC can be rescued by FGFR (Raoof et al., 2019) and other receptors that signal through pathways converging on ERK and AKT (Shah et al., 2019); tolerance to BRAF inhibition in colorectal cancer (CRC) and melanoma is dependent on maintaining MEK signaling (Corcoran et al., 2018, 2012; Prahallad et al., 2012; Sun et al., 2014); and inhibition of mutant KRAS can be thwarted by a variety of mechanisms, including autocrine signaling through IGF1 (Rajbhandari, Lin, Wehde, Triplett, & Wagner, 2017) and SHP2-mediated signaling of various RTKs to wild-type Ras (Ryan et al., 2020). Interestingly, increases in IGF signaling can also increase the number of DTPs in the context of chemotherapy (Sharma et al., 2010).

2.2.2 Other compensatory pathways

In addition to the aforementioned modes of bypass, DTPs commonly activate a number of other signaling pathways to compensate for targeted inhibition. In melanoma, resistance to treatment with BRAF or MEK inhibitors can activate the AXL or TGFβ pathways, with dependence on signaling illustrated using inhibitors against AXL to enhance drug sensitivity, or induction of TGFβ to manufacture resistance (Boshuizen et al., 2018; Sun et al., 2014). The TGFβ pathway, which is commonly associated with EMT, has also been suggested to promote signaling bypass via upregulation of EGFR (Sun et al., 2014). AXL and TGFβ are similarly implicated in NSCLC tolerance to EGFR inhibitors (Taniguchi et al., 2019; Yao et al., 2010; Zhang et al., 2016). NSCLC has additionally been shown to tolerate treatment by signaling through mTOR (Song, Hosono, et al., 2018) and direct activation of NFκB signaling by inhibitor-bound EGFR (Blakely et al., 2015). Signaling compensation also occurs in more distinct pathways, as exemplified by Wnt-driven tolerance to endocrine therapy in prostate cancer, and to Hedgehog inhibition in basal cell carcinoma (Rajan et al., 2014;

Sanchez-Danes et al., 2018). The ultimate implications of signaling through these alternate pathways are discussed in the following sections.

2.3 Plasticity: Differentiation

Signaling plasticty can lead to a shift in the transcriptional landscape and "differentiation," including changes in stemness, epithelial-mesenchymal transistion (EMT), and other forms of transdifferentiation (Boumahdi & de Sauvage, 2020; Cabanos & Hata, 2021).

2.3.1 Stemness

As discussed in relation to heterogeneity, cancer stem cells, or cells that exhibit characteristics of stemness, play a role in drug tolerance. This has been shown to be mediated by a myriad of mechanisms including the dynamic interplay between signaling and transcriptional programs. For example, the YAP-TEAD transcriptional machinery has been suggested to influence stemness and tolerance by acting as a master regulator of SOX family transcription factors (Shaffer et al., 2017; Wang et al., 2019), directors of cell fate including stem cell pluripotency (Takahashi & Yamanaka, 2006). Upregulation of SOX2-SOX9 promotes stemness and dormancy in tolerant subpopulations (Domenici et al., 2019; Wang et al., 2019; Yuan et al., 2018), while suppression of another SOX family member, SOX10, can de-differentate tumor cells and drive signaling plasticity in response to targeted therapies (Shaffer et al., 2017; Sun et al., 2014). SOX-mediated tolerance may go through Wnt/β-catenin signaling (Domenici et al., 2019; Malladi et al., 2016), which is broadly associated with stemness and tolerance to targeted and chemotherapies (Milanovic et al., 2018; Sanchez-Danes et al., 2018). Additional studies have proposed a role for NOTCH signaling pathways in CSC self-renewal and drug tolerance, including through stabilization of β-catenin (Arasada et al., 2018; Sansone et al., 2016).

2.3.2 Epithelial-mesenchymal transition

Transdifferentiation of epithelial cancers to a mesenchymal state, known as EMT, is a well-described process associated with tumor progression, metastasis, and most relevantly, stemness (Singh & Settleman, 2010; Wilson, Weinberg, Lees, & Guen, 2020). This process is commonly propagated through TGFβ and Wnt/β-catenin, both of which are strongly associated with the stable or induced presence of mesenchymal markers in stem-like DTPs that are tolerant of targeted and chemotherapies (Li et al., 2019; Liu et al., 2020; Sun et al., 2012; Yao et al., 2010; Yu et al., 2017).

EMT-mediated drug tolerance is also linked to a number of other factors associated with DTP hallmarks such as signaling plasticity (AXL, FGFR, Src/FAK) and stemness (YAP-TEAD) (Kurppa et al., 2020; Nilsson et al., 2020; Raoof et al., 2019; Terry et al., 2019; Wilson et al., 2014; Zhang et al., 2012).

2.3.3 Neuroendocrine transdifferentiation

Upon treatment, some tumors have been observed to switch to a state resembling tumors derived from neuroendocrine (NE) tissues. This path to drug tolerance is most well studied in prostate cancer (PCa) and NSCLC (Beltran et al., 2019; Oser, Niederst, Sequist, & Engelman, 2015). NE differentiation of PCa (NEPC) requires loss of TP53 and RB and is commonly, but not firmly, associated with stemness, EMT, and lineage factors such as SOX2. Similarly, it is marked by a transition from dependence on androgen receptor (AR) signaling to AR "indifference" and finally AR independence, which may or may not be fully reversible (Beltran et al., 2019; Bishop et al., 2017). Interestingly, NEPC also display chemotherapeutic tolerance (Chang et al., 2014). In NSCLC tumors, NE differentiation results in NSCLC conversion to small cell lung cancer (SCLC). This form of transdifferentiation is relatively rare among NSCLC, with tumors driven by mutant EGFR being more susceptible (Ferrer et al., 2019) and, like NEPC, requires loss of TP53 and RB (SCLC) (Lee, Lee, et al., 2017; Niederst et al., 2015).

2.3.4 Melanoma differentiation

The microphthalmia-associated transcription factor (MITF), a master regulator of melanocyte development, displays functional divergence to permit tolerance of melanoma to BRAF inhibition (Arozarena & Wellbrock, 2019). For example, elevated expression of MITF has been shown to create a reversible, drug-tolerant state that shows some similarities to neural crest stem cells (Johannessen et al., 2013; Levy, Khaled, & Fisher, 2006; Smith et al., 2016). However, melanoma cells that display low MITF levels have been suggested to form a distinct drug-tolerant sub-population under the same mode of treatment (Konieczkowski et al., 2014; Muller et al., 2014).

2.4 Proliferation/metabolism

2.4.1 Restriction of proliferation

Although canonically classified as slow-cycling, drug-tolerant populations have also been described as dormant, senescent or quiescent, but always with a common theme of restricted proliferation that diminishes reliance on pathways and processes beset by targeted and chemotherapies (De Angelis, Francescangeli, La Torre, & Zeuner, 2019; De Conti et al., 2021). Evidence has also been presented suggesting that persister cells exist in distinct non-cycling states such as diapause, a reversible, stress-induced state of embryonic development that is mirrored in some tumors under chemotherapy (De Conti et al., 2021), and senescence, an "irreversible" state of growth arrest that can be induced by targeted and chemotherapies (Saleh et al., 2019). In DTPs, the line between these states can be blurred, creating a spectrum of proliferative states (De Conti et al., 2021). Ultimately, these cells must undergo rewiring to escape their non- or slow-cycling state and acquire full drug resistance (De Angelis et al., 2019; Xue et al., 2019).

2.4.2 Metabolic switch

In alignment with the energetic demand of proliferative restriction, the drug-tolerant state has been shown to adopt a variety of metabolic programs. The most commonly observed metabolic switch is to mitochondrial oxidative phosphorylation (OXPHOS) (De Angelis et al., 2019). These observations stemmed from the hypothesis that highly proliferative cells, which are dependent on aerobic glycosylation, would require a switch to OXPHOS to gain chemoresistance (De Angelis et al., 2019). It is still unclear how universal the requirement for the switch to OXPHOS is in DTPs, and whether some features of OXPHOS pre-exist in a subpopulation prior to drug exposure. For example, an increase in OXPHOS does not always supplant glycolysis, and some DTPs have been observed to favor the latter (Aldonza et al., 2020; Karki, Angardi, Mier, & Orman, 2021). On the other hand, it has also been shown that tumor cells can show metabolic plasticity, readily switching between glycolysis and OXPHOS as necessary (Elgendy et al., 2019). Regardless, it has been shown that targeting of mitochondrial energetics can reduce DTP survival (De Conti et al., 2021). DTPs also rely on fatty acid oxididation (FAO) and autophagy for nutrient acquisition. It is noteworthy that some studies have shown that upregulation of these pathways can be integrated into the metabolic glycolysis-OXPHOS switch through the loss of mTOR signaling, and individual targeting of these pathways has been shown to reduce the number of drug-tolerant cells (De Conti et al., 2021; Kwon, Kim, Jung, Kim, & Jeoung, 2019; Lue et al., 2017).

2.4.3 Stress management

At its core, drug tolerance is a survival mechanism that most likely derives from pathways that evolved to counteract and respond to environmental stress. "Stress" comes in many flavors including hypoxia, nutrient deprivation, and changes in the tumor microenvironment. Compared to normal cells, cancer cells can tolerate elevated stress, allowing them to survive and proliferate under challenging endogenous and environmental conditions. As illustrated here, the stress response also takes on many forms in drug-tolerant cells.

2.4.4 Oxidative stress

One consequence of the metabolic switch to OXPHOS and FAO is accumulation of toxic metabolites. The most prominent example of this is production of reactive oxygen species (ROS), which have been found to be elevated in DTPs that rely on these metabolic pathways (Farge et al., 2017). DTPs, or more broadly CSCs and dormant cells, can counter ROS accumulation acutely through expression of ALDHs, alteration of DNA damage respsone (DDR) pathways, or more bluntly through clearance of damaged mitochondria via autophagy (Raha et al., 2014; Skvortsov, Debbage, Lukas, & Skvortsova, 2015; Vera-Ramirez, Vodnala, Nini, Hunter, & Green, 2018). Additionally, activation of the NRF2 antioxidant response, which upregulates thioredoxin and glutathione pathways, has been demonstrated in chemotolerant DTPs (Oshimori, Oristian, & Fuchs, 2015). Toxic lipid peroxidases resulting from FAO can also be neutralized by such responses, with the glutathione peroxidase 4 (GPX4) enzyme being of particular importance to DTP survival (Hangauer et al., 2017; Viswanathan et al., 2017).

2.4.5 Drug metabolism

Direct metabolism and removal of xenobiotics is one of the earliest observed mechanisms of chemoresistance. While this can occur in a variety of ways, it typically follows a three-phase process of redox/hydrolysis, conjugation, and transport/efflux across the plasma membrane (Iyanagi, 2007). Chemotoxic stress can lead to upregulation of Phase I, II and III drug metabolism machinery through the NRF2 pathway, and has been linked to glutathione-mediated metabolism of cisplatin in drug-tolerant cells (Bai, Chen, Hou, Huang, & Jin, 2016; Oshimori et al., 2015). Of the Phase III enzymes, or multi-drug resistance pumps, members of the ATP-binding cassette (ABC) family are commonly associated with chemoresistance and tolerance. One family member may serve as a marker of breast cancer stem cell populations, and several others have been shown to be upregulated in response to expression of EMT markers and OXPHOS-mediated stress in DTPs (Bai et al., 2018; Matassa et al., 2016; Saxena, Stephens, Pathak, & Rangarajan, 2011). In an interesting twist, drug efflux pumps have also been shown to contribute to oxidative stress (Szakacs et al., 2014). However, while initially seen as the most important mechanism of drug resistance, it is not clear how crucial drug efflux actually is in the context of cancer cell drug tolerance.

2.4.6 Proteostasis networks

Cells operate two major networks centered around protein folding homeostasis: the cytosolic Heat Shock Response (HSR) and the endoplasmic reticulum (ER)-associated Unfolded Protein Response (UPR). Activation, and reliance, on these pathways depends on integration of environmental factors with the metabolic demands of the cell. The ER chaperone GRP78/BiP plays a central role in DTPs, as it is associated with dormancy/stemness and, through the PERK arm of the UPR, mediates tolerance to both targeted and chemotherapies (Dauer et al., 2019; Ma et al., 2014; Ranganathan, Zhang, Adam, & Aguirre-Ghiso, 2006). GRP78 has also been implicated in mediating ROS status, while upregulation of HSR and UPR target genes has been linked to elevated mitochondrial activity after chemotherapy, highlighting a broader role for proteostasis networks in resolving oxidative stress (Dauer et al., 2019; Dobson et al., 2020). ATF4, which integrates signaling from PERK and other stress pathways, is a key driver of autophagy in DTPs and has been shown to be activated by ERK kinases during signaling bypass of BRAF/MEK inibition (Ma et al., 2014; Moeckel et al., 2019; Ojha et al., 2019). Lastly, the IRE1-XBP1 axis of the UPR, while not explicitly linked to drug tolerance, has been suggested to play a role in tumor relapse following cycles of chemotherapy (Madden, Logue, Healy, Manie, & Samali, 2019).

2.4.7 Suppression of apoptosis

An additional "safety net" that can be utilized to promote DTP survival is suppression of apoptosis. The apoptotic response is, in part, controlled by a balance between anti- and pro-apopotic members of the BCL-2 family. To tip the balance in favor of survival, drug-tolerant cells can directly regulate expression of these proteins. In this context, it has been shown that suppression of pro-apoptotic (BH3-only) proteins such as BIM, BMF and PUMA is observed in response to various treatment types and is commonly linked to EMT (Kurppa et al., 2020; Song, Niederst, et al., 2018; Wu et al., 2005). Conversely, anti-apoptotic proteins, most commonly MCL-1 and

BCL-XL, can be upregulated in response to a range of cancer therapies (Hata et al., 2016; Song, Hosono, et al., 2018; Terai et al., 2018). In addition to its normal anti-apopotic role, MCL-1 has also been suggested to directly mediate tolerance through induction of OXPHOS (Lee, Giltnane, et al., 2017). The pro-apoptotic factor BCL2A1 has also been shown to have a unique role, as it is directly regulated by MITF and associated with melonama transdifferentiation (Haq et al., 2013).

2.5 Genomic integrity

In the context of chemotherapy, many studies have shown that inhibition of components of the DNA damage response (DDR) sensitize tumor cells to treatment (De Conti et al., 2021). This is particularly noted in cancer stem cells, which use crosstalk between DDR and other survival pathways to mitigate otherwise lethal DNA damage (Skvortsov et al., 2015). In addition, drug-tolerant cells may use shifting genomic integrity to their advantage.

2.5.1 Adaptive mutability

A defining characterisitic of DTPs is that they seemingly rely on non-genetic mechanisms to survive therapy-induced stress, exemplified by studies where drug removal re-generates a drug responsive population. However, as DTPs transition into a drug-tolerant expanded persister (DTEP) population, they can acquire mutations that provide a path to stable resistance (Ramirez et al., 2016). This can occur in response to both targeted and chemotherapies, exemplified by de novo generation of the inhibitor-resistant EGFR T790M mutation in NSCLC and mutations in ovarian cancer that restore BRCA2 function in response to cisplatin treatment (Hata et al., 2016; Sakai et al., 2008). This gain of specific, resistance-conferring mutations occurs in the context of a global increase in mutation rates, termed adaptive mutability (Cabanos & Hata, 2021). Although borrowed from bacteria, proof-of-concept for this mechanism has been observed in DTPs that downregulate genes involved in high-fidelity mismatch repair and homologous recombination, while upregulating error-prone polymerases (Russo et al., 2019).

2.5.2 Chromosomal instability

Chromosomal instability has also been demonstrated to confer multidrug resistance by promoting genomic heterogeneity across the population (Lee, Endesfelder, et al., 2011). This can occur due to replication stress, which is elevated in cancer stem cells but balanced by changes in DDR to prevent lethality (Manic et al., 2018). DTPs have also been shown to depend on upregulation of spindle assembly checkpoint proteins, such as PLK1 and Aurora Kinase A (AURKA), which can result in mitotic errors and chromosomal instability (Nilsson et al., 2020; Shah et al., 2019). Additionally, recent observations have identified altered regulation of repetitive elements, including activation of retrotransposons and generation of R-loops, in drug-tolerant cell survival (Kermi, Lau, Asadi Shahmirzadi, & Classon, 2022). This concept, and its relationship to genomic integrity, will be discussed in more detail in the context of epigenetics.

2.6 Tumor microenvironment

In addition to intrinsic mechanisms of tolerance, the tumor microenvironment (TME) plays a pivotal role in the creation and maintenance of DTPs. Because crosstalk between the tumor cells and their environment is a complicated web of interactions, we've selected some prominent examples to illustrate how the TME reinforces drug tolerance.

2.6.1 Hypoxia

The combination of rapid tumor growth and errant vascular structure contributes to a hostile environment commonly marked by poor oxygen (hypoxia) and nutrient availability. These factors not only influence tumor progression, but also therapeutic sensitivity (Gillies, Verduzco, & Gatenby, 2012). Hypoxia has been specifically tied to tolerance of targeted and chemotherapies, signaling through hypoxia-inducible factors (HIFs) to affect cellular dormancy and pro-survival signaling pathways (Endo et al., 2017; Fluegen et al., 2017; Mendez-Blanco, Fondevila, Garcia-Palomo, Gonzalez-Gallego, & Mauriz, 2018; Qin et al., 2016). Hypoxia also plays an important role in tumor differentiation, establishing stemness and promoting EMT (Liu, Kumar, Martin, Yang, & Xu, 2011; Mohlin, Wigerup, Jogi, & Pahlman, 2017).

2.6.2 Stromal crosstalk

As outlined above, soluble factors can alter signaling and transcriptional programs to establish drug-tolerant states. Many of the factors used by the tumor in an autocrine fashion, including TGFβ, Wnt ligands, and inflammatory cytokines, are also secreted by stromal cells within the TME (Kaur et al., 2016; Matassa et al., 2016; Oshimori et al., 2015; Saha et al., 2016; Sun et al., 2012; Vermeulen et al., 2010). Notably, stromal cells damaged by chemotherapy have been shown to secrete such factors and promote tolerance in tumor cells (Sun et al., 2012).

Hepatocyte growth factor (HGF), which activates MAPK/PI3K/AKT signaling through its receptor MET, promotes tumor intrinsic drug tolerance driven by signaling bypass, stemness, and EMT (Straussman et al., 2012; Todaro et al., 2014; Wang et al., 2009; Wilson et al., 2012; Yi et al., 2018). HGF is commonly secreted by cancer-associated fibroblasts (CAFs), which can broadly promote tolerance by remodeling the extracellular matrix and secreting numerous growth factors that can promote the DTP state (Hirata et al., 2015; Meador & Hata, 2020; Zhang et al., 2016). Stromal cells can also promote tolerance through physical interaction, as seen between osteoblasts, a component of the bone marrow niche, and metastatic, refractory prostate cancer cells (Kim et al., 2013).

In turn, tumor cells also "signal" to the stroma to reinforce their own drug-tolerant state. Tumor cells have been observed directly stimulating paracrine signaling from stromal cells (Sharma et al., 2016) and recruiting cells to the local microenvironment. Recruitment of tumor associated macrophages (TAMs), for example, can influence tumor cell lineage plasticity (Lee, Kwon, et al., 2011; Smith et al., 2014).

2.7 Immune escape

The immune system, another facet of the TME, presents unique challenges to tumor cells. The progression of various immune therapies, including immune checkpoint blockade (ICB) and adoptive T cell transfer therapies (ACT), have greatly improved clinical outcomes, alone and in combination with chemotherapies (Paz-Ares et al., 2018). However, these advancements have also uncovered new forms of tolerance.

One key predictor of immunogenicity, and response to ICB therapy, is tumor mutational burden (Ribas & Wolchok, 2018). Mutation rate is possibly associated with an increase in tumor specific neoantigens, which are presented by the host cell and recognized by the immune system. Tumors treated with combined chemotherapy and ICB, while displaying an increase in neoantigens, exhibited an ability to evade immune surveillance (McGranahan et al., 2016). While this could be explained by downregulation of antigen-presenting machinery, other work has demonstrated that an increase in mutational burden can manifest as a form of tumor heterogeneity, giving some cells a selective advantage (Gejman et al., 2018; Wolf et al., 2019). As such, adaptive mutability may also double as a strategy for immune evasion.

Stemness may also play a role in immune evasion. Cancer stem cells may be innately immune resistant, as seen in squamous cell carcinoma CSCs that express CD80 to block CTLA4-mediated T cell activity (Miao et al., 2019). It has also been shown that low levels of Type I interferon (IFN) signaling, which normally increases antigen presentation and immunogenecity, can induce stemness and promote immune evasion in chemotherapy treated cells (Galluzzi & Kroemer, 2022; Wang et al., 2017). Similarly, signaling through IFNγ is typically associated with enhanced immunogenecity, but has also been shown to promote dormancy and immune evasion in cancer stem cells (Liu et al., 2017). "Immunotherapy persisters" have also been characterized: stem- and mesenchymal-like cells that survive PD-1 blockade (Sehgal et al., 2021).

Another strategy for immune evasion is to suppress expression of factors that mediate immune detection. Melanomas that resist ACT switch to a dedifferentiated or stem-like state characterized by loss of the T cell-specific antigen (Landsberg et al., 2012; Mehta et al., 2018). In patients with B-progenitor acute lymphoblastic leukemia, residual cells capable of driving relapse were linked to downregulation of antigen processing and T/B cell receptor pathways (Dobson et al., 2020). A variety of dormant/latent tumor populations with stem-like or tumor-initiating capacity have developed unique methods to achieve similar goals. This includes downregulation of MHC-I by activation of the UPR and suppression of Natural Killer (NK) cell ligands by SOX2-SOX9 transcriptional programs (Malladi et al., 2016; Pommier et al., 2018).

Taken together, these hallmarks present a plethora of interconnected mechanisms that may contribute to therapy response and relapse in patients.

3. Epigenetics in drug tolerance

In its broadest sense, epigenetics is everything that makes otherwise genetically identical cells different; an integration of extrinsic and intrinsic signaling, transcriptional and translational events that determine cellular phenotype. In this section, we apply the term epigenetics to more narrowly describe factors that orchestrate a large web of reversible alterations to chromatin, resulting in global changes in cellular function and identity. This includes enzymes that methylate and demethylate DNA (DNMTs), enzymes that modify histone tails (including histone acetylases, HATs; histone methyltransferases, HMTs; histone deacetylases, HDACs; histone lysine demethylases, KDMs), proteins that "read" chromatin modifications, factors that may scaffold epigenetic enzymes, and those that actively remodel chromatin structure. Additionally, enzymes and RNA species that can reversibly alter transcript stability and translation fall under this epigenetic classification, including RNA methyltransferases/demethylases, microRNAs (miRNAs) and long non-coding RNAs (lncRNAs) (Allis & Jenuwein, 2016).

As with drug resistance, genetic mutations were once a monolith for our understanding of tumorigenesis, tumor progression, and metastasis. However, it is now evident that epigenetics plays a pivotal role in cancer evolution and metastasis, simultaneously providing the plasticity and rigidity required for tumor development (Flavahan, Gaskell, & Bernstein, 2017; Sandoval & Esteller, 2012). The combined absence of genetic drivers and reversible drug sensitivity of some refractory tumors implied that epigenetics may also play a key role in drug resistance (Engelman & Janne, 2008; Yano et al., 2005). In fact, evidence of such epigenetic mechanisms predated a formal characterization of non-genetic, drug-tolerant cancer cell populations, particularly with observations suggesting that DNA and histone methylation states may be correlated with resistant subpopulations (Glasspool, Teodoridis, & Brown, 2006; Varambally et al., 2002). Epigenetic regulation was also instrumental at the inception of the DTP concept, with the H3K4 demethylases KDM5A and KDM5B maintaining rare, slow-cycling subpopulations of cells that confer reversible, multi-drug tolerance (Roesch et al., 2010, 2013; Sharma et al., 2010). Moreover, the efficacy of HDAC inhibitors in ablating such DTPs provided the first evidence that epigenetic blockade could be a potent weapon against drug tolerance (Sharma et al., 2010).

Following on these initial findings, various epigenetic factors have been linked to the survival of DTPs, and the DTP state has been suggested to be defined in part by a repressed chromatin state (Guler et al., 2017). Slow-cycling, reversibly tolerant subpopulations have been shown to harbor alterations in epigenetic regulatory mechanisms including histone and DNA modifications and pathways such as RNA-mediated transcriptional repression (Banelli et al., 2015; Dalvi et al., 2017; Liau et al., 2017; Lim et al., 2017; Puig et al., 2018; Ravindran Menon et al., 2015; Sahu et al., 2016; Sharma et al., 2010). As such, epigenetics collectively stands as its own hallmark of drug tolerance. Additionally, epigenetic-mediated tolerance involves a dynamic interplay with the other hallmarks, hinting at a role for epigenetics as a mediator of intrinsic and extrinsic paths to resistance.

3.1 Epigenetic roles in the hallmarks of cancer drug tolerance

3.1.1 Heterogeneity

The KDM5 family of H3K4 demethylases, which contains four paralogs (KDM5A-D), is consistently linked to tumor heterogeneity and drug tolerance. For example, KDM5A was implicated in the initial characterization of DTPs (Sharma et al., 2010) and inhibitors that target all four family members have been shown to decrease DTP survival in a multitude of models (Vinogradova et al., 2016). Notably, KDM5B is the best described epigenetic marker of cells poised to become DTPs (Dhanyamraju et al., 2022). Originally associated with slow-cycling, multi-drug-tolerant melanoma subpopulations (Roesch et al., 2010, 2013), the role of KDM5B in heterogeneity and tolerance has since been solidified in melanoma and further demonstrated in breast cancer, glioblastoma, neuroblastoma, and NSCLC (Hinohara & Polyak, 2019; Kuo et al., 2018, 2015; Liu, Zhang, et al., 2019; Patel et al., 2014; Shaffer et al., 2017). Moreover, KDM5B expression is commonly correlated with drug-tolerant cancer stem cells and has been linked to ALDH activity, implicating it as a master regulator in the establishment of DTPs (Kuo et al., 2018, 2015; Liu, Zhang, et al., 2019; Ravindran Menon et al., 2015). However, the prevalence of KDM5 in DTP studies does not preclude other epigenetic drivers of heterogeneity. For example, H3K4me3/H3K27me3 bivalence, TET2-dependent DNA methylation state, and BRD4 activity have been associated with pre-existing DTP precursors (Marsolier et al., 2022; Puig et al., 2018; Risom et al., 2018). Additionally, numerous other epigenetic enzymes have been implicated in general regulation of CSCs, indicating that other pathways may be critical to establishment of heterogeneous, drug-tolerant populations (Wainwright & Scaffidi, 2017).

3.1.2 Plasticity: Signaling

A variety of epigenetic marks and enzymes have been linked to signaling plasticity in DTPs. This includes increased DNA methylation over the EGFR promoter in NSCLC under gefitinib (Li et al., 2013), PI3K/AKT activation in hypermethylated colorectal cancer cells (Mao et al., 2013), mediation of IGF-1R signaling by KDM5A (Sharma et al., 2010), and remodeling of RTK enhancers through BRD4/CBP/P300 (Zawistowski et al., 2017). EZH2, an H3K27 methyltransferase and member of the polycomb repressive complex 2 (PRC2), is noteworthy due to its involvement in various forms of signaling bypass. In NSCLC treated with gefitinib, association of EZH2 with a lncRNA mediates its ability to activate PI3K/AKT signaling and bypass EGFR inhibition (Liu, Lu, et al., 2019). In estrogen receptor (ER) positive breast cancer, EZH2 methylation of H3K27 promotes DNA methylation and silencing of the ERα cofactor GREB1, altering the ERα pathway response to anti-estrogens (Wu et al., 2018). EZH2 also regulates activity of androgen receptor (AR) in castration-resistant prostate cancer, where direct association between AR and PRC2 overrides the AR antagonist enzalutamide and drives survival through effectors such as AKT (Shankar et al., 2020).

3.1.3 Plasticity: Differentiation

With well-defined roles in cancer plasticity, it is unsurprising that epigenetics plays diverse roles in "differentiation"-mediated drug tolerance (Cabanos & Hata, 2021; Flavahan et al., 2017; Ravindran Menon, Hammerlindl, Torrano, Schaider, & Fujita, 2020; Skrypek, Goossens, De Smedt, Vandamme, & Berx, 2017). For example, chromatin remodeling creates permissive states for SOX family transcription factors (Biehs et al., 2018), BRD4 physically mediates YAP/TAZ transcriptional addiction (Zanconato et al., 2018), KDM5B regulates NOTCH signaling to promote stemness in neuroblastoma (Kuo et al., 2015), and a lncRNA mediates activation of Wnt/β-catenin to drive EMT (Lu et al., 2021). Epigenetic alterations also regulate drug tolerance related to neuroendocrine (NE) transdifferentiation, in which EZH2 and the REST transcription factor have been suggested to play opposing roles. REST, which complexes with epigenetic regulators to suppress NE programming, is inactivated upon inhibition of AR, thereby releasing its block (Lapuk et al., 2012; Svensson et al., 2014). Conversely, elevated EZH2 expression and activity are hallmarks of NE prostate cancer, with EZH2 inhibition resulting in dedifferentiation of these cells and sensitization to antiandrogen therapy (Beltran et al., 2016; Clermont et al., 2015; Dardenne et al., 2016; Ku et al., 2017). Furthermore, EZH2 inhibition has been specifically shown to block NE differentiation of castration resistant prostate cancer in response to antiandrogen therapy (Luo et al., 2019), with potential benefit observed in the clinic. In neuroendocrine SCLC, the roles of PRC2 and REST complexes in drug tolerance appear to be reversed: REST activation through NOTCH signaling inhibits NE programming to create a slow-cycling, chemo-resistant subpopulation (Lim et al., 2017). Additionally, EZH2 has been proposed to downregulate TGFβ signaling to stabilize the NE state of the bulk tumor, raising the possibility that EZH2 inhibition may promote drug tolerance in SCLC (Murai et al., 2015).

3.1.4 Proliferation/metabolism

In contrast to the limited number of links between epigenetic regulation and therapy-induced, non-cycling states (senescence, diapause) (Dhimolea et al., 2021; Milanovic et al., 2018), examples of epigenetic involvement in slow-cycling DTPs are relatively abundant. As discussed in relation to heterogeneity, the H3K4 demethylase KDM5B is strongly associated with the slow-cycling phenotype in melanoma (Roesch et al., 2010; Tirosh et al., 2016), where its expression is anti-correlated with cell cycle progression (Chauvistre et al., 2022; Patel et al., 2014). Additionally, the KDM5B-subpopulation in melanoma displays elevated OXPHOS and has been shown to be sensitized to drug treatment by inhibitors of mitochondrial respiration (Roesch et al., 2013). Numerous other factors, including various KDMs, the REST transcription complex, and the DNA-demethylating enzyme TET2, have been correlated to the slow-cycling state, but without a clear relationship to cell cycle or metabolic maintenance (Banelli et al., 2015; Liau et al., 2017; Lim et al., 2017; Puig et al., 2018).

3.1.5 Stress management

Epigenetic regulation is involved in managing all facets of therapy-induced stress. This spans from DNA methylation over genes involved in drug metabolism (Belanger, Tojcic, Harvey, & Guillemette, 2010; Kosuri, Wu, Wang, Villalona-Calero, & Otterson, 2010) to a variety of epigenetic mechanisms to suppress apoptosis, especially through enhancement of anti-apoptotic BCL-2 (Knoechel et al., 2014; Thomas, Thurn, Bicaku, Marchion, & Munster, 2011; Yan et al., 2018). Roles for miRNAs, effectors of RNA silencing, are of particular note. For example, miRNAs have been shown to respond to ROS accumulation by mediating pro-tolerance epigenetic programs (Sun et al., 2019), as well as to directly regulate expression of genes involved in mediation of oxidative stress (Sahu et al., 2016). Additionally, they have been demonstrated to blunt response to chemotherapy by upregulating the expression of ABC drug transporters (Munoz et al., 2015; Wang et al., 2015; Zhu et al., 2011).

3.1.6 Genomic integrity

Epigenetics influences genomic integrity, and thus heterogeneity, to drive drug tolerance in a variety of different ways. This includes protection against DNA damage by the H3K36 demethylase KDM2B (Staberg et al., 2018), suppression of mismatch repair by DNA methylation of the hMLH1 promoter (Arnold, Goel, & Boland, 2003), and regulation of replication stress by KDM5A/B (Gaillard et al., 2021). It may also do so in a less well-characterized manner: balanced expression of repetitive elements. DNA hypomethylation across genomic repetitive elements (REs), including telomeres and transposable viral elements, is observed in many cancers. Overactivation of repetitive elements can undermine tumor fitness by inducing DNA damage, generating immunogenic neoantigens or eliciting "viral mimicry" (Kermi et al., 2022). Chemotherapies can further induce expression of REs, which are epigenetically repressed in DTPs (Baratchian et al., 2022; Deblois et al., 2020; Guler et al., 2017). However, dysregulation of transposable elements may also provide a path to transcriptomic heterogeneity by actively rearranging or activating other parts of the genome, or inducing expression of mutagenic restriction factors (Kermi et al., 2022). Further investigation is required to determine whether these elements remain sufficiently active in DTPs to promote genomic heterogeneity and confer stable resistance.

3.1.7 Tumor microenvironment

While tumors have been shown to influence stromal cells through epigenetic reprogramming (Zhang et al., 2016), it is unclear if there are analogous mechanisms specific to drug-tolerant populations. On the other hand, there is some direct evidence that stromal cells can influence drug-tolerance through epigenetic means. For example, stromal cell adhesion in multiple myeloma can suppress EZH2 activity and promote survival through IGF1, BCL-2, and HIF1α (Kikuchi et al., 2015). In breast cancer, stromal cells have been observed to secrete microvesicles containing miRNA that promote tumor cell stemness and therapeutic resistance (Sansone et al., 2017). As discussed above, a broad range of epigenetic factors influence DTP plasticity by permitting signaling bypass and activating pathways that drive differentiation. Considering how many of these pathways can also be influenced by stromal signaling, it is likely that epigenetics plays a far more pronounced role in how DTPs interact with the tumor microenvironment.

3.1.8 Immune escape

Regarding the immune system, cancer cells have adopted a variety of epigenetic mechanisms to circumvent immune detection, including increased expression of checkpoint blockade genes and downregulation of the antigen-presenting machinery (Cao & Yan, 2020; Kang et al., 2020). However, as with tumor-TME crosstalk, there are few examples that may link drug tolerance, epigenetics, and immunogenicity. As described above in relation to genomic heterogeneity, repetitive element (RE) regulation is one way this can occur. Errant expression of REs can lead to "viral mimicry": production of viral-like nucleic acid species that are sensed by innate immune machinery, resulting in an interferon (IFN) response, reduced cell fitness, and, in some cases, an increase in immunogenicity (Kermi et al., 2022). Enriched epigenetic repression of REs in DTPs has been proposed to occur through repressive heterochromatin formation mediated by factors such as SETDB1, EHMT2/G9a, ATRX, and PRC2, thereby blunting viral mimicry, establishing chemo-tolerance, and providing a path to immune evasion (Baratchian et al., 2022; Colombo et al., 2017; Deblois et al., 2020; Guler et al., 2017). Interestingly, IFN signaling may also promote epigenetically driven immune escape: in cells under immunogenic chemotherapy, Type I IFN can signal through the H3K4 demethylase KDM1B to induce stemness and reduce immunogenicity (Musella et al., 2022). These apparently contradictory roles for IFN signaling highlight how DTPs strike a balance to survive but become vulnerable to therapies that tip this balance toward death.

3.2 Targeting epigenetics in drug tolerance

Epigenetic regulation is deeply intertwined with the many facets of drug tolerance, making this "machinery" an attractive target in combination with targeted, chemo- and immunotherapies. This is underscored by the hypothesis that some epigenetic targets are positioned to have more "universal" success in undermining DTPs. For instance, DTPs sensitive to KDM5 inhibition have been observed in cell lines from diverse cancer types under distinct drug treatments (Vinogradova et al., 2016). However, this does not account for the fact that a single tumor can harbor heterogeneous DTP populations that exist, or arise, with unique markers and provide discrete paths to tolerance (Hata et al., 2016; Kurppa et al., 2020; Ramirez et al., 2016). This may be addressed, in part, by the observation that an epigenetic factor can confer resistance to multiple therapies, targeted and chemotoxic, in a single population. This is broadly demonstrated by dependency on enzymes that modify histones at H3K4, K9 and K27 (Dalvi et al., 2017; Deblois et al., 2020; Gollner et al., 2017; Guler et al., 2017; Ravindran Menon et al., 2015; Roesch et al., 2013).

Notably, crosstalk and redundancies between these epigenetic axes can contribute to a poised or drug-tolerant state (Guler et al., 2017; Marsolier et al., 2022; Pham et al., 2020). This raises questions about which, if any, epigenetic enzyme or pathway is truly a "master" regulator of DTP survival. This is further compounded by a "chicken-or-the-egg" relationship with plasticity. Namely, it is unclear whether stochastic or drug-inducible expression of epigenetic factors in DTP precursors is controlled by transcriptional programs, or vice versa. To deconvolute these relationships, high-priority targets need to be directly compared in diverse cancer backgrounds under a variety of treatment types. This should include both primary and disseminated tumor cells to account for adaptations to changes in the tumor microenvironment. Implementation of functional genomics and small-molecule screens, as well as modern 'omics technologies, will be instrumental in identifying patterns of dependency and establishing whether universal, or master, regulators of drug tolerance exist.

While it may be tempting to think that the epigenetic machinery could provide a cancer therapy panacea, we must heed lessons from past therapeutic advances. As bemoaned at the start of this article, the evolution of cancer drugs has been simultaneously confounded by the evolution of cancer tolerance. In fact, in a single-agent context, tumors have already displayed a diverse array of mechanisms to circumvent epigenetic-targeting therapies. This includes many of the hallmarks discussed here, including induction of stemness, signaling bypass, EMT, lineage plasticity, and, ironically, epigenetics (Drosos et al., 2022; Fong et al., 2015; Ikegaki et al., 2013; Iniguez et al., 2018; Rathert et al., 2015; Shu et al., 2016; Wawruszak et al., 2019; Yamamoto et al., 2014; Yan et al., 2022). Considering the heterogeneity of the bulk tumor population and the inherit plasticity of drug-tolerant cancer cells, even inhibition of a "master" epigenetic regulator is unlikely to eradicate all DTPs. Characterization of cells that tolerate such drug combinations will be essential to our understanding and advancement of cancer therapies. It will be of particular interest to determine whether these survivors represent a distinct, subpopulation of the original tumor, a novel subset of tolerant cells that arise from the targeted subpopulation, or a mix of both types.

Ultimately, these and future advances made in the pre-clinical space will face numerous challenges in their translation to the clinic. Inclusion of treatment-naïve patients in combination trials is perhaps the most significant hurdle to this transition. In the lab, epigenetic factors that mediate drug tolerance are generally targeted prior to, or in combination with, the standard of care. This forces cells that are poised to become DTPs into a less tolerant state, resulting in their elimination upon drug treatment. On the other hand, epigenetic blockade after treatment may be less efficacious due to the stress-activated, drug-tolerant nature of DTPs. As such, patients who have previously undergone rounds of treatment may not see a significant clinical benefit when presented with epigenetic therapies. Epigenetic drugs that have shown individual efficacy in heme indications, such as FDA-approved inhibitors of DNMTs, HDACs and EZH2 (Nepali & Liou, 2021), may have an easier path to combination therapies in solid tumors. Conversely, the clinical path may be impeded for epigenetic targets that lack single-agent activity despite showing efficacy in eradicating drug tolerant cancer cells in a pre-clinical setting. As our knowledge of epigenetics in drug tolerance continues to evolve, so too should the avenues for delivering new therapies to patients.

References

Aldonza, M. B. D., Ku, J., Hong, J. Y., Kim, D., Yu, S. J., Lee, M. S., et al. (2020). Prior acquired resistance to paclitaxel relays diverse EGFR-targeted therapy persistence mechanisms. *Science Advances*, *6*(6), eaav7416. https://doi.org/10.1126/sciadv.aav7416.

Allis, C. D., & Jenuwein, T. (2016). The molecular hallmarks of epigenetic control. *Nature Reviews. Genetics*, *17*(8), 487–500. https://doi.org/10.1038/nrg.2016.59.

Arasada, R. R., Shilo, K., Yamada, T., Zhang, J., Yano, S., Ghanem, R., et al. (2018). Notch3-dependent beta-catenin signaling mediates EGFR TKI drug persistence in EGFR mutant NSCLC. *Nature Communications*, *9*(1), 3198. https://doi.org/10.1038/s41467-018-05626-2.

Arnold, C. N., Goel, A., & Boland, C. R. (2003). Role of hMLH1 promoter hypermethylation in drug resistance to 5-fluorouracil in colorectal cancer cell lines. *International Journal of Cancer*, *106*(1), 66–73. https://doi.org/10.1002/ijc.11176.

Arozarena, I., & Wellbrock, C. (2019). Phenotype plasticity as enabler of melanoma progression and therapy resistance. *Nature Reviews. Cancer*, *19*(7), 377–391. https://doi.org/10.1038/s41568-019-0154-4.

Bai, X., Chen, Y., Hou, X., Huang, M., & Jin, J. (2016). Emerging role of NRF2 in chemoresistance by regulating drug-metabolizing enzymes and efflux transporters. *Drug Metabolism Reviews*, *48*(4), 541–567. https://doi.org/10.1080/03602532.2016.1197239.

Bai, X., Ni, J., Beretov, J., Graham, P., & Li, Y. (2018). Cancer stem cell in breast cancer therapeutic resistance. *Cancer Treatment Reviews*, *69*, 152–163. https://doi.org/10.1016/j.ctrv.2018.07.004.

Banelli, B., Carra, E., Barbieri, F., Wurth, R., Parodi, F., Pattarozzi, A., et al. (2015). The histone demethylase KDM5A is a key factor for the resistance to temozolomide in glioblastoma. *Cell Cycle*, *14*(21), 3418–3429. https://doi.org/10.1080/15384101.2015.1090063.

Baratchian, M., Tiwari, R., Khalighi, S., Chakravarthy, A., Yuan, W., Berk, M., et al. (2022). H3K9 methylation drives resistance to androgen receptor-antagonist therapy in prostate cancer. *Proceedings of the National Academy of Sciences of the United States of America*, *119*(21), e2114324119. https://doi.org/10.1073/pnas.2114324119.

Belanger, A. S., Tojcic, J., Harvey, M., & Guillemette, C. (2010). Regulation of UGT1A1 and HNF1 transcription factor gene expression by DNA methylation in colon cancer cells. *BMC Molecular Biology*, *11*, 9. https://doi.org/10.1186/1471-2199-11-9.

Beltran, H., Hruszkewycz, A., Scher, H. I., Hildesheim, J., Isaacs, J., Yu, E. Y., et al. (2019). The role of lineage plasticity in prostate cancer therapy resistance. *Clinical Cancer Research*, *25*(23), 6916–6924. https://doi.org/10.1158/1078-0432.CCR-19-1423.

Beltran, H., Prandi, D., Mosquera, J. M., Benelli, M., Puca, L., Cyrta, J., et al. (2016). Divergent clonal evolution of castration-resistant neuroendocrine prostate cancer. *Nature Medicine*, *22*(3), 298–305. https://doi.org/10.1038/nm.4045.

Biehs, B., Dijkgraaf, G. J. P., Piskol, R., Alicke, B., Boumahdi, S., Peale, F., et al. (2018). A cell identity switch allows residual BCC to survive Hedgehog pathway inhibition. *Nature*, *562*(7727), 429–433. https://doi.org/10.1038/s41586-018-0596-y.

Bigger, J. (1944). Treatment of staphylococcal infections with penicillin by intermittent sterilisation. *Lancet*, *244*, 497–500.

Bishop, J. L., Thaper, D., Vahid, S., Davies, A., Ketola, K., Kuruma, H., et al. (2017). The master neural transcription factor BRN2 is an androgen receptor-suppressed driver of neuroendocrine differentiation in prostate cancer. *Cancer Discovery*, 7(1), 54–71. https://doi.org/10.1158/2159-8290.CD-15-1263.

Blakely, C. M., Pazarentzos, E., Olivas, V., Asthana, S., Yan, J. J., Tan, I., et al. (2015). NF-kappaB-activating complex engaged in response to EGFR oncogene inhibition drives tumor cell survival and residual disease in lung cancer. *Cell Reports*, *11*(1), 98–110. https://doi.org/10.1016/j.celrep.2015.03.012.

Boshuizen, J., Koopman, L. A., Krijgsman, O., Shahrabi, A., van den Heuvel, E. G., Ligtenberg, M. A., et al. (2018). Cooperative targeting of melanoma heterogeneity with an AXL antibody-drug conjugate and BRAF/MEK inhibitors. *Nature Medicine*, *24*(2), 203–212. https://doi.org/10.1038/nm.4472.

Boumahdi, S., & de Sauvage, F. J. (2020). The great escape: Tumour cell plasticity in resistance to targeted therapy. *Nature Reviews. Drug Discovery*, *19*(1), 39–56. https://doi.org/10.1038/s41573-019-0044-1.

Cabanos, H. F., & Hata, A. N. (2021). Emerging insights into targeted therapy-tolerant persister cells in cancer. *Cancers (Basel)*, *13*(11), 2666. https://doi.org/10.3390/cancers13112666.

Cao, J., & Yan, Q. (2020). Cancer epigenetics, tumor immunity, and immunotherapy. *Trends in Cancer*, *6*(7), 580–592. https://doi.org/10.1016/j.trecan.2020.02.003.

Cara, S., & Tannock, I. F. (2001). Retreatment of patients with the same chemotherapy: Implications for clinical mechanisms of drug resistance. *Annals of Oncology*, *12*(1), 23–27. https://doi.org/10.1023/a:1008389706725.

Chang, P. C., Wang, T. Y., Chang, Y. T., Chu, C. Y., Lee, C. L., Hsu, H. W., et al. (2014). Autophagy pathway is required for IL-6 induced neuroendocrine differentiation and chemoresistance of prostate cancer LNCaP cells. *PLoS One*, *9*(2), e88556. https://doi.org/10.1371/journal.pone.0088556.

Chauvistre, H., Shannan, B., Daignault-Mill, S. M., Ju, R. J., Picard, D., Egetemaier, S., et al. (2022). Persister state-directed transitioning and vulnerability in melanoma. *Nature Communications*, *13*(1), 3055. https://doi.org/10.1038/s41467-022-30641-9.

Chevallier, P., Touzeau, C., Ayari, S., Guillaume, T., Harousseau, J. L., & Delaunay, J. (2008). Re-administration of a combination of chemotherapy + Gemtuzumab at relapse in CD33+ AML patient allows to second remission and is feasible without extra toxicity. *Leukemia Research*, *32*(8), 1321–1322. https://doi.org/10.1016/j.leukres.2007.09.009.

Clermont, P. L., Lin, D., Crea, F., Wu, R., Xue, H., Wang, Y., et al. (2015). Polycomb-mediated silencing in neuroendocrine prostate cancer. *Clinical Epigenetics*, 7, 40. https://doi.org/10.1186/s13148-015-0074-4.

Colombo, A. R., Zubair, A., Thiagarajan, D., Nuzhdin, S., Triche, T. J., & Ramsingh, G. (2017). Suppression of transposable elements in leukemic stem cells. *Scientific Reports*, 7(1), 7029. https://doi.org/10.1038/s41598-017-07356-9.

Corcoran, R. B., Andre, T., Atreya, C. E., Schellens, J. H. M., Yoshino, T., Bendell, J. C., et al. (2018). Combined BRAF, EGFR, and MEK inhibition in patients with BRAF(V600E)-mutant colorectal cancer. *Cancer Discovery*, *8*(4), 428–443. https://doi.org/10.1158/2159-8290.CD-17-1226.

Corcoran, R. B., Ebi, H., Turke, A. B., Coffee, E. M., Nishino, M., Cogdill, A. P., et al. (2012). EGFR-mediated re-activation of MAPK signaling contributes to insensitivity of BRAF mutant colorectal cancers to RAF inhibition with vemurafenib. *Cancer Discovery*, *2*(3), 227–235. https://doi.org/10.1158/2159-8290.CD-11-0341.

Dalvi, M. P., Wang, L., Zhong, R., Kollipara, R. K., Park, H., Bayo, J., et al. (2017). Taxane-platin-resistant lung cancers co-develop hypersensitivity to JumonjiC demethylase inhibitors. *Cell Reports*, *19*(8), 1669–1684. https://doi.org/10.1016/j.celrep.2017.04.077.

Dardenne, E., Beltran, H., Benelli, M., Gayvert, K., Berger, A., Puca, L., et al. (2016). N-Myc induces an EZH2-mediated transcriptional program driving neuroendocrine prostate cancer. *Cancer Cell*, *30*(4), 563–577. https://doi.org/10.1016/j.ccell.2016.09.005.

Dauer, P., Sharma, N. S., Gupta, V. K., Durden, B., Hadad, R., Banerjee, S., et al. (2019). ER stress sensor, glucose regulatory protein 78 (GRP78) regulates redox status in pancreatic cancer thereby maintaining "stemness". *Cell Death & Disease*, *10*(2), 132. https://doi.org/10.1038/s41419-019-1408-5.

De Angelis, M. L., Francescangeli, F., La Torre, F., & Zeuner, A. (2019). Stem cell plasticity and dormancy in the development of cancer therapy resistance. *Frontiers in Oncology*, *9*, 626. https://doi.org/10.3389/fonc.2019.00626.

De Conti, G., Dias, M. H., & Bernards, R. (2021). Fighting drug resistance through the targeting of drug-tolerant persister cells. *Cancers (Basel)*, *13*(5), 1118. https://doi.org/10.3390/cancers13051118.

Deblois, G., Tonekaboni, S. A. M., Grillo, G., Martinez, C., Kao, Y. I., Tai, F., et al. (2020). Epigenetic switch-induced viral mimicry evasion in chemotherapy-resistant breast cancer. *Cancer Discovery*, *10*(9), 1312–1329. https://doi.org/10.1158/2159-8290.CD-19-1493.

Dhanyamraju, P. K., Schell, T. D., Amin, S., & Robertson, G. P. (2022). Drug-tolerant persister cells in cancer therapy resistance. *Cancer Research*, *82*(14), 2503–2514. https://doi.org/10.1158/0008-5472.CAN-21-3844.

Dhimolea, E., de Matos Simoes, R., Kansara, D., Al'Khafaji, A., Bouyssou, J., Weng, X., et al. (2021). An embryonic diapause-like adaptation with suppressed Myc activity enables tumor treatment persistence. *Cancer Cell*, *39*(2). https://doi.org/10.1016/j.ccell.2020.12.002. 240-256.e211.

Dobson, S. M., Garcia-Prat, L., Vanner, R. J., Wintersinger, J., Waanders, E., Gu, Z., et al. (2020). Relapse-fated latent diagnosis subclones in acute B lineage leukemia are drug tolerant and possess distinct metabolic programs. *Cancer Discovery*, *10*(4), 568–587. https://doi.org/10.1158/2159-8290.CD-19-1059.

Domenici, G., Aurrekoetxea-Rodriguez, I., Simoes, B. M., Rabano, M., Lee, S. Y., Millan, J. S., et al. (2019). A Sox2-Sox9 signalling axis maintains human breast luminal progenitor and breast cancer stem cells. *Oncogene*, *38*(17), 3151–3169. https://doi.org/10.1038/s41388-018-0656-7.

Drosos, Y., Myers, J. A., Xu, B., Mathias, K. M., Beane, E. C., Radko-Juettner, S., et al. (2022). NSD1 mediates antagonism between SWI/SNF and polycomb complexes and is required for transcriptional activation upon EZH2 inhibition. *Molecular Cell*, *82*(13), 2472-2489.e2478. https://doi.org/10.1016/j.molcel.2022.04.015.

Elgendy, M., Ciro, M., Hosseini, A., Weiszmann, J., Mazzarella, L., Ferrari, E., et al. (2019). Combination of hypoglycemia and metformin impairs tumor metabolic plasticity and growth by modulating the PP2A-GSK3beta-MCL-1 axis. *Cancer Cell*, *35*(5), 798-815.e795. https://doi.org/10.1016/j.ccell.2019.03.007.

Endo, H., Okami, J., Okuyama, H., Nishizawa, Y., Imamura, F., & Inoue, M. (2017). The induction of MIG6 under hypoxic conditions is critical for dormancy in primary cultured lung cancer cells with activating EGFR mutations. *Oncogene*, *36*(20), 2824–2834. https://doi.org/10.1038/onc.2016.431.

Engelman, J. A., & Janne, P. A. (2008). Mechanisms of acquired resistance to epidermal growth factor receptor tyrosine kinase inhibitors in non-small cell lung cancer. *Clinical Cancer Research*, *14*(10), 2895–2899. https://doi.org/10.1158/1078-0432.CCR-07-2248.

Farge, T., Saland, E., de Toni, F., Aroua, N., Hosseini, M., Perry, R., et al. (2017). Chemotherapy-resistant human acute myeloid leukemia cells are not enriched for leukemic stem cells but require oxidative metabolism. *Cancer Discovery*, 7(7), 716–735. https://doi.org/10.1158/2159-8290.CD-16-0441.

Ferrer, L., Giaj Levra, M., Brevet, M., Antoine, M., Mazieres, J., Rossi, G., et al. (2019). A brief report of transformation from NSCLC to SCLC: Molecular and therapeutic characteristics. *Journal of Thoracic Oncology*, *14*(1), 130–134. https://doi.org/10.1016/j.jtho.2018.08.2028.

Fisher, R. A., Gollan, B., & Helaine, S. (2017). Persistent bacterial infections and persister cells. *Nature Reviews. Microbiology*, *15*(8), 453–464. https://doi.org/10.1038/nrmicro.2017.42.

Flavahan, W. A., Gaskell, E., & Bernstein, B. E. (2017). Epigenetic plasticity and the hallmarks of cancer. *Science*, *357*(6348), eaal2380. https://doi.org/10.1126/science.aal2380.

Fluegen, G., Avivar-Valderas, A., Wang, Y., Padgen, M. R., Williams, J. K., Nobre, A. R., et al. (2017). Phenotypic heterogeneity of disseminated tumour cells is preset by primary tumour hypoxic microenvironments. *Nature Cell Biology*, *19*(2), 120–132. https://doi.org/10.1038/ncb3465.

Fong, C. Y., Gilan, O., Lam, E. Y., Rubin, A. F., Ftouni, S., Tyler, D., et al. (2015). BET inhibitor resistance emerges from leukaemia stem cells. *Nature*, *525*(7570), 538–542. https://doi.org/10.1038/nature14888.

Gaillard, S., Charasson, V., Ribeyre, C., Salifou, K., Pillaire, M. J., Hoffmann, J. S., et al. (2021). KDM5A and KDM5B histone-demethylases contribute to HU-induced replication stress response and tolerance. *Biology Open*, *10*(5), bio057729. https://doi.org/10.1242/bio.057729.

Galluzzi, L., & Kroemer, G. (2022). Immuno-epigenetic escape of cancer stem cells. *Nature Immunology*, *23*(9), 1300–1302. https://doi.org/10.1038/s41590-022-01293-0.

Gejman, R. S., Chang, A. Y., Jones, H. F., DiKun, K., Hakimi, A. A., Schietinger, A., et al. (2018). Rejection of immunogenic tumor clones is limited by clonal fraction. *eLife*, 7, e41090. https://doi.org/10.7554/eLife.41090.

Gerlinger, M., Rowan, A. J., Horswell, S., Math, M., Larkin, J., Endesfelder, D., et al. (2012). Intratumor heterogeneity and branched evolution revealed by multiregion sequencing. *The New England Journal of Medicine*, *366*(10), 883–892. https://doi.org/10.1056/NEJMoa1113205.

Gillies, R. J., Verduzco, D., & Gatenby, R. A. (2012). Evolutionary dynamics of carcinogenesis and why targeted therapy does not work. *Nature Reviews. Cancer*, *12*(7), 487–493. https://doi.org/10.1038/nrc3298.

Glasspool, R. M., Teodoridis, J. M., & Brown, R. (2006). Epigenetics as a mechanism driving polygenic clinical drug resistance. *British Journal of Cancer*, *94*(8), 1087–1092. https://doi.org/10.1038/sj.bjc.6603024.

Gollner, S., Oellerich, T., Agrawal-Singh, S., Schenk, T., Klein, H. U., Rohde, C., et al. (2017). Loss of the histone methyltransferase EZH2 induces resistance to multiple drugs in acute myeloid leukemia. *Nature Medicine*, *23*(1), 69–78. https://doi.org/10.1038/nm.4247.

Guler, G. D., Tindell, C. A., Pitti, R., Wilson, C., Nichols, K., KaiWai Cheung, T., et al. (2017). Repression of stress-induced LINE-1 expression protects cancer cell subpopulations from lethal drug exposure. *Cancer Cell, 32*(2). https://doi.org/10.1016/j.ccell.2017.07.002. 221-237.e213.

Gupta, P. B., Fillmore, C. M., Jiang, G., Shapira, S. D., Tao, K., Kuperwasser, C., et al. (2011). Stochastic state transitions give rise to phenotypic equilibrium in populations of cancer cells. *Cell, 146*(4), 633–644. https://doi.org/10.1016/j.cell.2011.07.026.

Hangauer, M. J., Viswanathan, V. S., Ryan, M. J., Bole, D., Eaton, J. K., Matov, A., et al. (2017). Drug-tolerant persister cancer cells are vulnerable to GPX4 inhibition. *Nature, 551*(7679), 247–250. https://doi.org/10.1038/nature24297.

Haq, R., Yokoyama, S., Hawryluk, E. B., Jonsson, G. B., Frederick, D. T., McHenry, K., et al. (2013). BCL2A1 is a lineage-specific antiapoptotic melanoma oncogene that confers resistance to BRAF inhibition. *Proceedings of the National Academy of Sciences of the United States of America, 110*(11), 4321–4326. https://doi.org/10.1073/pnas.1205575110.

Hashimoto, N., Imaizumi, K., Honda, T., Kawabe, T., Nagasaka, T., Shimokata, K., et al. (2006). Successful re-treatment with gefitinib for carcinomatous meningitis as disease recurrence of non-small-cell lung cancer. *Lung Cancer, 53*(3), 387–390. https://doi.org/10.1016/j.lungcan.2006.05.016.

Hata, A. N., Niederst, M. J., Archibald, H. L., Gomez-Caraballo, M., Siddiqui, F. M., Mulvey, H. E., et al. (2016). Tumor cells can follow distinct evolutionary paths to become resistant to epidermal growth factor receptor inhibition. *Nature Medicine, 22*(3), 262–269. https://doi.org/10.1038/nm.4040.

Hinohara, K., & Polyak, K. (2019). Intratumoral heterogeneity: More than just mutations. *Trends in Cell Biology, 29*(7), 569–579. https://doi.org/10.1016/j.tcb.2019.03.003.

Hirata, E., Girotti, M. R., Viros, A., Hooper, S., Spencer-Dene, B., Matsuda, M., et al. (2015). Intravital imaging reveals how BRAF inhibition generates drug-tolerant microenvironments with high integrin beta1/FAK signaling. *Cancer Cell, 27*(4), 574–588. https://doi.org/10.1016/j.ccell.2015.03.008.

Hobby, G. L., Meyer, K., & Chaffee, E. (1942). Observations on the mechanism of action of penicillin. *Proceedings of the Society for Experimental Biology and Medicine, 50*, 281–285.

Hong, S. P., Chan, T. E., Lombardo, Y., Corleone, G., Rotmensz, N., Bravaccini, S., et al. (2019). Single-cell transcriptomics reveals multi-step adaptations to endocrine therapy. *Nature Communications, 10*(1), 3840. https://doi.org/10.1038/s41467-019-11721-9.

Ikegaki, N., Shimada, H., Fox, A. M., Regan, P. L., Jacobs, J. R., Hicks, S. L., et al. (2013). Transient treatment with epigenetic modifiers yields stable neuroblastoma stem cells resembling aggressive large-cell neuroblastomas. *Proceedings of the National Academy of Sciences of the United States of America, 110*(15), 6097–6102. https://doi.org/10.1073/pnas.1118262110.

Iniguez, A. B., Alexe, G., Wang, E. J., Roti, G., Patel, S., Chen, L., et al. (2018). Resistance to epigenetic-targeted therapy engenders tumor cell vulnerabilities associated with enhancer remodeling. *Cancer Cell, 34*(6). https://doi.org/10.1016/j.ccell.2018.11.005. 922-938.e927.

Iyanagi, T. (2007). Molecular mechanism of phase I and phase II drug-metabolizing enzymes: Implications for detoxification. *International Review of Cytology, 260*, 35–112. https://doi.org/10.1016/S0074-7696(06)60002-8.

Jin, X., Kim, L. J. Y., Wu, Q., Wallace, L. C., Prager, B. C., Sanvoranart, T., et al. (2017). Targeting glioma stem cells through combined BMI1 and EZH2 inhibition. *Nature Medicine, 23*(11), 1352–1361. https://doi.org/10.1038/nm.4415.

Johannessen, C. M., Johnson, L. A., Piccioni, F., Townes, A., Frederick, D. T., Donahue, M. K., et al. (2013). A melanocyte lineage program confers resistance to MAP kinase pathway inhibition. *Nature, 504*(7478), 138–142. https://doi.org/10.1038/nature12688.

Kaminski, M. S., Radford, J. A., Gregory, S. A., Leonard, J. P., Knox, S. J., Kroll, S., et al. (2005). Re-treatment with I-131 tositumomab in patients with non-Hodgkin's lymphoma who had previously responded to I-131 tositumomab. *Journal of Clinical Oncology*, *23*(31), 7985–7993. https://doi.org/10.1200/JCO.2005.01.0892.

Kang, N., Eccleston, M., Clermont, P. L., Latarani, M., Male, D. K., Wang, Y., et al. (2020). EZH2 inhibition: A promising strategy to prevent cancer immune editing. *Epigenomics*, *12*(16), 1457–1476. https://doi.org/10.2217/epi-2020-0186.

Karki, P., Angardi, V., Mier, J. C., & Orman, M. A. (2021). A transient metabolic state in melanoma persister cells mediated by chemotherapeutic treatments. *Frontiers in Molecular Biosciences*, *8*, 780192. https://doi.org/10.3389/fmolb.2021.780192.

Kaur, A., Webster, M. R., Marchbank, K., Behera, R., Ndoye, A., Kugel, C. H., 3rd, et al. (2016). sFRP2 in the aged microenvironment drives melanoma metastasis and therapy resistance. *Nature*, *532*(7598), 250–254. https://doi.org/10.1038/nature17392.

Kermi, C., Lau, L., Asadi Shahmirzadi, A., & Classon, M. (2022). Disrupting mechanisms that regulate genomic repeat elements to combat cancer and drug resistance. *Frontiers in Cell and Development Biology*, *10*, 826461. https://doi.org/10.3389/fcell.2022.826461.

Kikuchi, J., Koyama, D., Wada, T., Izumi, T., Hofgaard, P. O., Bogen, B., et al. (2015). Phosphorylation-mediated EZH2 inactivation promotes drug resistance in multiple myeloma. *The Journal of Clinical Investigation*, *125*(12), 4375–4390. https://doi.org/10.1172/JCI80325.

Kim, J. K., Jung, Y., Wang, J., Joseph, J., Mishra, A., Hill, E. E., et al. (2013). TBK1 regulates prostate cancer dormancy through mTOR inhibition. *Neoplasia*, *15*(9), 1064–1074. https://doi.org/10.1593/neo.13402.

Knoechel, B., Roderick, J. E., Williamson, K. E., Zhu, J., Lohr, J. G., Cotton, M. J., et al. (2014). An epigenetic mechanism of resistance to targeted therapy in T cell acute lymphoblastic leukemia. *Nature Genetics*, *46*(4), 364–370. https://doi.org/10.1038/ng.2913.

Konieczkowski, D. J., Johannessen, C. M., Abudayyeh, O., Kim, J. W., Cooper, Z. A., Piris, A., et al. (2014). A melanoma cell state distinction influences sensitivity to MAPK pathway inhibitors. *Cancer Discovery*, *4*(7), 816–827. https://doi.org/10.1158/2159-8290.CD-13-0424.

Kosuri, K. V., Wu, X., Wang, L., Villalona-Calero, M. A., & Otterson, G. A. (2010). An epigenetic mechanism for capecitabine resistance in mesothelioma. *Biochemical and Biophysical Research Communications*, *391*(3), 1465–1470. https://doi.org/10.1016/j.bbrc.2009.12.095.

Ku, S. Y., Rosario, S., Wang, Y., Mu, P., Seshadri, M., Goodrich, Z. W., et al. (2017). Rb1 and Trp53 cooperate to suppress prostate cancer lineage plasticity, metastasis, and antiandrogen resistance. *Science*, *355*(6320), 78–83. https://doi.org/10.1126/science.aah4199.

Kuo, K. T., Huang, W. C., Bamodu, O. A., Lee, W. H., Wang, C. H., Hsiao, M., et al. (2018). Histone demethylase JARID1B/KDM5B promotes aggressiveness of non-small cell lung cancer and serves as a good prognostic predictor. *Clinical Epigenetics*, *10*(1), 107. https://doi.org/10.1186/s13148-018-0533-9.

Kuo, Y. T., Liu, Y. L., Adebayo, B. O., Shih, P. H., Lee, W. H., Wang, L. S., et al. (2015). JARID1B expression plays a critical role in chemoresistance and stem cell-like phenotype of neuroblastoma cells. *PLoS One*, *10*(5), e0125343. https://doi.org/10.1371/journal.pone.0125343.

Kurppa, K. J., Liu, Y., To, C., Zhang, T., Fan, M., Vajdi, A., et al. (2020). Treatment-induced tumor dormancy through YAP-mediated transcriptional reprogramming of the apoptotic pathway. *Cancer Cell*, *37*(1), 104-122.e112. https://doi.org/10.1016/j.ccell.2019.12.006.

Kwon, Y., Kim, M., Jung, H. S., Kim, Y., & Jeoung, D. (2019). Targeting autophagy for overcoming resistance to anti-EGFR treatments. *Cancers (Basel)*, *11*(9), 1374. https://doi.org/10.3390/cancers11091374.

Landsberg, J., Kohlmeyer, J., Renn, M., Bald, T., Rogava, M., Cron, M., et al. (2012). Melanomas resist T-cell therapy through inflammation-induced reversible dedifferentiation. *Nature, 490*(7420), 412–416. https://doi.org/10.1038/nature11538.
Lapuk, A. V., Wu, C., Wyatt, A. W., McPherson, A., McConeghy, B. J., Brahmbhatt, S., et al. (2012). From sequence to molecular pathology, and a mechanism driving the neuroendocrine phenotype in prostate cancer. *The Journal of Pathology, 227*(3), 286–297. https://doi.org/10.1002/path.4047.
Leccia, F., Del Vecchio, L., Mariotti, E., Di Noto, R., Morel, A. P., Puisieux, A., et al. (2014). ABCG2, a novel antigen to sort luminal progenitors of BRCA1-breast cancer cells. *Molecular Cancer, 13*, 213. https://doi.org/10.1186/1476-4598-13-213.
Lee, A. J., Endesfelder, D., Rowan, A. J., Walther, A., Birkbak, N. J., Futreal, P. A., et al. (2011). Chromosomal instability confers intrinsic multidrug resistance. *Cancer Research, 71*(5), 1858–1870. https://doi.org/10.1158/0008-5472.CAN-10-3604.
Lee, K. M., Giltnane, J. M., Balko, J. M., Schwarz, L. J., Guerrero-Zotano, A. L., Hutchinson, K. E., et al. (2017). MYC and MCL1 cooperatively promote chemotherapy-resistant breast cancer stem cells via regulation of mitochondrial oxidative phosphorylation. *Cell Metabolism, 26*(4), 633-647.e637. https://doi.org/10.1016/j.cmet.2017.09.009.
Lee, G. T., Kwon, S. J., Lee, J. H., Jeon, S. S., Jang, K. T., Choi, H. Y., et al. (2011). Macrophages induce neuroendocrine differentiation of prostate cancer cells via BMP6-IL6 Loop. *Prostate, 71*(14), 1525–1537. https://doi.org/10.1002/pros.21369.
Lee, J. K., Lee, J., Kim, S., Kim, S., Youk, J., Park, S., et al. (2017). Clonal history and genetic predictors of transformation into small-cell carcinomas from lung adenocarcinomas. *Journal of Clinical Oncology, 35*(26), 3065–3074. https://doi.org/10.1200/JCO.2016.71.9096.
Levin-Reisman, I., Ronin, I., Gefen, O., Braniss, I., Shoresh, N., & Balaban, N. Q. (2017). Antibiotic tolerance facilitates the evolution of resistance. *Science, 355*(6327), 826–830. https://doi.org/10.1126/science.aaj2191.
Levy, C., Khaled, M., & Fisher, D. E. (2006). MITF: Master regulator of melanocyte development and melanoma oncogene. *Trends in Molecular Medicine, 12*(9), 406–414. https://doi.org/10.1016/j.molmed.2006.07.008.
Li, S., Song, Y., Quach, C., Guo, H., Jang, G. B., Maazi, H., et al. (2019). Transcriptional regulation of autophagy-lysosomal function in BRAF-driven melanoma progression and chemoresistance. *Nature Communications, 10*(1), 1693. https://doi.org/10.1038/s41467-019-09634-8.
Li, X. Y., Wu, J. Z., Cao, H. X., Ma, R., Wu, J. Q., Zhong, Y. J., et al. (2013). Blockade of DNA methylation enhances the therapeutic effect of gefitinib in non-small cell lung cancer cells. *Oncology Reports, 29*(5), 1975–1982. https://doi.org/10.3892/or.2013.2298.
Liau, B. B., Sievers, C., Donohue, L. K., Gillespie, S. M., Flavahan, W. A., Miller, T. E., et al. (2017). Adaptive chromatin remodeling drives glioblastoma stem cell plasticity and drug tolerance. *Cell Stem Cell, 20*(2), 233-246.e237. https://doi.org/10.1016/j.stem.2016.11.003.
Lim, J. S., Ibaseta, A., Fischer, M. M., Cancilla, B., O'Young, G., Cristea, S., et al. (2017). Intratumoural heterogeneity generated by Notch signalling promotes small-cell lung cancer. *Nature, 545*(7654), 360–364. https://doi.org/10.1038/nature22323.
Liu, S., Kumar, S. M., Martin, J. S., Yang, R., & Xu, X. (2011). Snail1 mediates hypoxia-induced melanoma progression. *The American Journal of Pathology, 179*(6), 3020–3031. https://doi.org/10.1016/j.ajpath.2011.08.038.
Liu, Y., Liang, X., Yin, X., Lv, J., Tang, K., Ma, J., et al. (2017). Blockade of IDO-kynurenine-AhR metabolic circuitry abrogates IFN-gamma-induced immunologic dormancy of tumor-repopulating cells. *Nature Communications, 8*, 15207. https://doi.org/10.1038/ncomms15207.

Liu, X., Lu, X., Zhen, F., Jin, S., Yu, T., Zhu, Q., et al. (2019). LINC00665 induces acquired resistance to gefitinib through recruiting EZH2 and activating PI3K/AKT pathway in NSCLC. *Molecular Therapy—Nucleic Acids*, *16*, 155–161. https://doi.org/10.1016/j.omtn.2019.02.010.

Liu, X., Zhang, S. M., McGeary, M. K., Krykbaeva, I., Lai, L., Jansen, D. J., et al. (2019). KDM5B promotes drug resistance by regulating melanoma-propagating cell subpopulations. *Molecular Cancer Therapeutics*, *18*(3), 706–717. https://doi.org/10.1158/1535-7163.MCT-18-0395.

Liu, L., Zhu, H., Liao, Y., Wu, W., Liu, L., Liu, L., et al. (2020). Inhibition of Wnt/beta-catenin pathway reverses multi-drug resistance and EMT in Oct4(+)/Nanog(+) NSCLC cells. *Biomedicine & Pharmacotherapy*, *127*, 110225. https://doi.org/10.1016/j.biopha.2020.110225.

Lu, M., Qin, X., Zhou, Y., Li, G., Liu, Z., Geng, X., et al. (2021). Long non-coding RNA LINC00665 promotes gemcitabine resistance of Cholangiocarcinoma cells via regulating EMT and stemness properties through miR-424-5p/BCL9L axis. *Cell Death & Disease*, *12*(1), 72. https://doi.org/10.1038/s41419-020-03346-4.

Lue, H. W., Podolak, J., Kolahi, K., Cheng, L., Rao, S., Garg, D., et al. (2017). Metabolic reprogramming ensures cancer cell survival despite oncogenic signaling blockade. *Genes & Development*, *31*(20), 2067–2084. https://doi.org/10.1101/gad.305292.117.

Luo, M., Shang, L., Brooks, M. D., Jiagge, E., Zhu, Y., Buschhaus, J. M., et al. (2018). Targeting breast cancer stem cell state equilibrium through modulation of redox signaling. *Cell Metabolism*, *28*(1). https://doi.org/10.1016/j.cmet.2018.06.006. 69-86.e66.

Luo, J., Wang, K., Yeh, S., Sun, Y., Liang, L., Xiao, Y., et al. (2019). LncRNA-p21 alters the antiandrogen enzalutamide-induced prostate cancer neuroendocrine differentiation via modulating the EZH2/STAT3 signaling. *Nature Communications*, *10*(1), 2571. https://doi.org/10.1038/s41467-019-09784-9.

Luria, S. E., & Delbruck, M. (1943). Mutations of bacteria from virus sensitivity to virus resistance. *Genetics*, *28*(6), 491–511. https://doi.org/10.1093/genetics/28.6.491.

Ma, X. H., Piao, S. F., Dey, S., McAfee, Q., Karakousis, G., Villanueva, J., et al. (2014). Targeting ER stress-induced autophagy overcomes BRAF inhibitor resistance in melanoma. *The Journal of Clinical Investigation*, *124*(3), 1406–1417. https://doi.org/10.1172/JCI70454.

Madden, E., Logue, S. E., Healy, S. J., Manie, S., & Samali, A. (2019). The role of the unfolded protein response in cancer progression: From oncogenesis to chemoresistance. *Biology of the Cell*, *111*(1), 1–17. https://doi.org/10.1111/boc.201800050.

Makena, M. R., Ranjan, A., Thirumala, V., & Reddy, A. P. (2020). Cancer stem cells: Road to therapeutic resistance and strategies to overcome resistance. *Biochimica et Biophysica Acta - Molecular Basis of Disease*, *1866*(4), 165339. https://doi.org/10.1016/j.bbadis.2018.11.015.

Malladi, S., Macalinao, D. G., Jin, X., He, L., Basnet, H., Zou, Y., et al. (2016). Metastatic latency and immune evasion through autocrine inhibition of WNT. *Cell*, *165*(1), 45–60. https://doi.org/10.1016/j.cell.2016.02.025.

Manic, G., Sistigu, A., Corradi, F., Musella, M., De Maria, R., & Vitale, I. (2018). Replication stress response in cancer stem cells as a target for chemotherapy. *Seminars in Cancer Biology*, *53*, 31–41. https://doi.org/10.1016/j.semcancer.2018.08.003.

Mao, M., Tian, F., Mariadason, J. M., Tsao, C. C., Lemos, R., Jr., Dayyani, F., et al. (2013). Resistance to BRAF inhibition in BRAF-mutant colon cancer can be overcome with PI3K inhibition or demethylating agents. *Clinical Cancer Research*, *19*(3), 657–667. https://doi.org/10.1158/1078-0432.CCR-11-1446.

Marsolier, J., Prompsy, P., Durand, A., Lyne, A. M., Landragin, C., Trouchet, A., et al. (2022). H3K27me3 conditions chemotolerance in triple-negative breast cancer. *Nature Genetics*, *54*(4), 459–468. https://doi.org/10.1038/s41588-022-01047-6.

Matassa, D. S., Amoroso, M. R., Lu, H., Avolio, R., Arzeni, D., Procaccini, C., et al. (2016). Oxidative metabolism drives inflammation-induced platinum resistance in human ovarian cancer. *Cell Death and Differentiation*, *23*(9), 1542–1554. https://doi.org/10.1038/cdd.2016.39.

Maynard, A., McCoach, C. E., Rotow, J. K., Harris, L., Haderk, F., Kerr, D. L., et al. (2020). Therapy-induced evolution of human lung cancer revealed by single-cell RNA sequencing. *Cell*, *182*(5), 1232-1251.e1222. https://doi.org/10.1016/j.cell.2020.07.017.

McGranahan, N., Furness, A. J., Rosenthal, R., Ramskov, S., Lyngaa, R., Saini, S. K., et al. (2016). Clonal neoantigens elicit T cell immunoreactivity and sensitivity to immune checkpoint blockade. *Science*, *351*(6280), 1463–1469. https://doi.org/10.1126/science.aaf1490.

Meador, C. B., & Hata, A. N. (2020). Acquired resistance to targeted therapies in NSCLC: Updates and evolving insights. *Pharmacology & Therapeutics*, *210*, 107522. https://doi.org/10.1016/j.pharmthera.2020.107522.

Mehta, A., Kim, Y. J., Robert, L., Tsoi, J., Comin-Anduix, B., Berent-Maoz, B., et al. (2018). Immunotherapy resistance by inflammation-induced dedifferentiation. *Cancer Discovery*, *8*(8), 935–943. https://doi.org/10.1158/2159-8290.CD-17-1178.

Mendez-Blanco, C., Fondevila, F., Garcia-Palomo, A., Gonzalez-Gallego, J., & Mauriz, J. L. (2018). Sorafenib resistance in hepatocarcinoma: Role of hypoxia-inducible factors. *Experimental & Molecular Medicine*, *50*(10), 1–9. https://doi.org/10.1038/s12276-018-0159-1.

Miao, Y., Yang, H., Levorse, J., Yuan, S., Polak, L., Sribour, M., et al. (2019). Adaptive immune resistance emerges from tumor-initiating stem cells. *Cell*, *177*(5), 1172-1186.e1114. https://doi.org/10.1016/j.cell.2019.03.025.

Mikubo, M., Inoue, Y., Liu, G., & Tsao, M. S. (2021). Mechanism of drug tolerant persister cancer cells: The landscape and clinical implication for therapy. *Journal of Thoracic Oncology*, *16*(11), 1798–1809. https://doi.org/10.1016/j.jtho.2021.07.017.

Milanovic, M., Fan, D. N. Y., Belenki, D., Dabritz, J. H. M., Zhao, Z., Yu, Y., et al. (2018). Senescence-associated reprogramming promotes cancer stemness. *Nature*, *553*(7686), 96–100. https://doi.org/10.1038/nature25167.

Moeckel, S., LaFrance, K., Wetsch, J., Seliger, C., Riemenschneider, M. J., Proescholdt, M., et al. (2019). ATF4 contributes to autophagy and survival in sunitinib treated brain tumor initiating cells (BTICs). *Oncotarget*, *10*(3), 368–382. https://doi.org/10.18632/oncotarget.26569.

Mohlin, S., Wigerup, C., Jogi, A., & Pahlman, S. (2017). Hypoxia, pseudohypoxia and cellular differentiation. *Experimental Cell Research*, *356*(2), 192–196. https://doi.org/10.1016/j.yexcr.2017.03.007.

Muller, J., Krijgsman, O., Tsoi, J., Robert, L., Hugo, W., Song, C., et al. (2014). Low MITF/AXL ratio predicts early resistance to multiple targeted drugs in melanoma. *Nature Communications*, *5*, 5712. https://doi.org/10.1038/ncomms6712.

Munoz, J. L., Rodriguez-Cruz, V., Ramkissoon, S. H., Ligon, K. L., Greco, S. J., & Rameshwar, P. (2015). Temozolomide resistance in glioblastoma occurs by miRNA-9-targeted PTCH1, independent of sonic hedgehog level. *Oncotarget*, *6*(2), 1190–1201. https://doi.org/10.18632/oncotarget.2778.

Murai, F., Koinuma, D., Shinozaki-Ushiku, A., Fukayama, M., Miyaozono, K., & Ehata, S. (2015). EZH2 promotes progression of small cell lung cancer by suppressing the TGF-beta-Smad-ASCL1 pathway. *Cell Discovery*, *1*, 15026. https://doi.org/10.1038/celldisc.2015.26.

Musella, M., Guarracino, A., Manduca, N., Galassi, C., Ruggiero, E., Potenza, A., et al. (2022). Type I IFNs promote cancer cell stemness by triggering the epigenetic regulator KDM1B. *Nature Immunology*, *23*(9), 1379–1392. https://doi.org/10.1038/s41590-022-01290-3.

Naing, A., & Kurzrock, R. (2010). Chemotherapy resistance and retreatment: A dogma revisited. *Clinical Colorectal Cancer*, *9*(2), E1–E4. https://doi.org/10.3816/CCC.2010.n.026.

Nepali, K., & Liou, J. P. (2021). Recent developments in epigenetic cancer therapeutics: Clinical advancement and emerging trends. *Journal of Biomedical Science*, *28*(1), 27. https://doi.org/10.1186/s12929-021-00721-x.

Niederst, M. J., & Engelman, J. A. (2013). Bypass mechanisms of resistance to receptor tyrosine kinase inhibition in lung cancer. *Science Signaling*, *6*(294), re6. https://doi.org/10.1126/scisignal.2004652.

Niederst, M. J., Sequist, L. V., Poirier, J. T., Mermel, C. H., Lockerman, E. L., Garcia, A. R., et al. (2015). RB loss in resistant EGFR mutant lung adenocarcinomas that transform to small-cell lung cancer. *Nature Communications*, *6*, 6377. https://doi.org/10.1038/ncomms7377.

Nilsson, M. B., Sun, H., Robichaux, J., Pfeifer, M., McDermott, U., Travers, J., et al. (2020). A YAP/FOXM1 axis mediates EMT-associated EGFR inhibitor resistance and increased expression of spindle assembly checkpoint components. *Science Translational Medicine*, *12*(559), eaaz4589. https://doi.org/10.1126/scitranslmed.aaz4589.

Oh, I. J., Ban, H. J., Kim, K. S., & Kim, Y. C. (2012). Retreatment of gefitinib in patients with non-small-cell lung cancer who previously controlled to gefitinib: A single-arm, open-label, phase II study. *Lung Cancer*, 77(1), 121–127. https://doi.org/10.1016/j.lungcan.2012.01.012.

Ojha, R., Leli, N. M., Onorati, A., Piao, S., Verginadis, I. I., Tameire, F., et al. (2019). ER translocation of the MAPK pathway drives therapy resistance in BRAF-mutant melanoma. *Cancer Discovery*, *9*(3), 396–415. https://doi.org/10.1158/2159-8290.CD-18-0348.

Oser, M. G., Niederst, M. J., Sequist, L. V., & Engelman, J. A. (2015). Transformation from non-small-cell lung cancer to small-cell lung cancer: Molecular drivers and cells of origin. *The Lancet Oncology*, *16*(4), e165–e172. https://doi.org/10.1016/S1470-2045(14)71180-5.

Oshimori, N., Oristian, D., & Fuchs, E. (2015). TGF-beta promotes heterogeneity and drug resistance in squamous cell carcinoma. *Cell*, *160*(5), 963–976. https://doi.org/10.1016/j.cell.2015.01.043.

Patel, A. P., Tirosh, I., Trombetta, J. J., Shalek, A. K., Gillespie, S. M., Wakimoto, H., et al. (2014). Single-cell RNA-seq highlights intratumoral heterogeneity in primary glioblastoma. *Science*, *344*(6190), 1396–1401. https://doi.org/10.1126/science.1254257.

Paz-Ares, L., Luft, A., Vicente, D., Tafreshi, A., Gumus, M., Mazieres, J., et al. (2018). Pembrolizumab plus chemotherapy for squamous non-small-cell lung cancer. *The New England Journal of Medicine*, *379*(21), 2040–2051. https://doi.org/10.1056/NEJMoa1810865.

Pham, V., Pitti, R., Tindell, C. A., Cheung, T. K., Masselot, A., Stephan, J. P., et al. (2020). Proteomic analyses identify a novel role for EZH2 in the initiation of cancer cell drug tolerance. *Journal of Proteome Research*, *19*(4), 1533–1547. https://doi.org/10.1021/acs.jproteome.9b00773.

Pommier, A., Anaparthy, N., Memos, N., Kelley, Z. L., Gouronnec, A., Yan, R., et al. (2018). Unresolved endoplasmic reticulum stress engenders immune-resistant, latent pancreatic cancer metastases. *Science*, *360*(6394), aao4908. https://doi.org/10.1126/science.aao4908.

Prahallad, A., Sun, C., Huang, S., Di Nicolantonio, F., Salazar, R., Zecchin, D., et al. (2012). Unresponsiveness of colon cancer to BRAF(V600E) inhibition through feedback activation of EGFR. *Nature*, *483*(7387), 100–103. https://doi.org/10.1038/nature10868.

Puig, I., Tenbaum, S. P., Chicote, I., Arques, O., Martinez-Quintanilla, J., Cuesta-Borras, E., et al. (2018). TET2 controls chemoresistant slow-cycling cancer cell survival and tumor recurrence. *The Journal of Clinical Investigation*, *128*(9), 3887–3905. https://doi.org/10.1172/JCI96393.

Qin, Y., Roszik, J., Chattopadhyay, C., Hashimoto, Y., Liu, C., Cooper, Z. A., et al. (2016). Hypoxia-driven mechanism of vemurafenib resistance in melanoma. *Molecular Cancer Therapeutics*, *15*(10), 2442–2454. https://doi.org/10.1158/1535-7163.MCT-15-0963.

Raha, D., Wilson, T. R., Peng, J., Peterson, D., Yue, P., Evangelista, M., et al. (2014). The cancer stem cell marker aldehyde dehydrogenase is required to maintain a drug-tolerant tumor cell subpopulation. *Cancer Research*, *74*(13), 3579–3590. https://doi.org/10.1158/0008-5472.CAN-13-3456.

Rajan, P., Sudbery, I. M., Villasevil, M. E., Mui, E., Fleming, J., Davis, M., et al. (2014). Next-generation sequencing of advanced prostate cancer treated with androgen-deprivation therapy. *European Urology*, *66*(1), 32–39. https://doi.org/10.1016/j.eururo.2013.08.011.

Rajbhandari, N., Lin, W. C., Wehde, B. L., Triplett, A. A., & Wagner, K. U. (2017). Autocrine IGF1 signaling mediates pancreatic tumor cell dormancy in the absence of oncogenic drivers. *Cell Reports*, *18*(9), 2243–2255. https://doi.org/10.1016/j.celrep.2017.02.013.

Ramirez, M., Rajaram, S., Steininger, R. J., Osipchuk, D., Roth, M. A., Morinishi, L. S., et al. (2016). Diverse drug-resistance mechanisms can emerge from drug-tolerant cancer persister cells. *Nature Communications*, 7, 10690. https://doi.org/10.1038/ncomms10690.

Ranganathan, A. C., Zhang, L., Adam, A. P., & Aguirre-Ghiso, J. A. (2006). Functional coupling of p38-induced up-regulation of BiP and activation of RNA-dependent protein kinase-like endoplasmic reticulum kinase to drug resistance of dormant carcinoma cells. *Cancer Research*, *66*(3), 1702–1711. https://doi.org/10.1158/0008-5472.CAN-05-3092.

Raoof, S., Mulford, I. J., Frisco-Cabanos, H., Nangia, V., Timonina, D., Labrot, E., et al. (2019). Targeting FGFR overcomes EMT-mediated resistance in EGFR mutant non-small cell lung cancer. *Oncogene*, *38*(37), 6399–6413. https://doi.org/10.1038/s41388-019-0887-2.

Rathert, P., Roth, M., Neumann, T., Muerdter, F., Roe, J. S., Muhar, M., et al. (2015). Transcriptional plasticity promotes primary and acquired resistance to BET inhibition. *Nature*, *525*(7570), 543–547. https://doi.org/10.1038/nature14898.

Ravindran Menon, D., Das, S., Krepler, C., Vultur, A., Rinner, B., Schauer, S., et al. (2015). A stress-induced early innate response causes multidrug tolerance in melanoma. *Oncogene*, *34*(34), 4448–4459. https://doi.org/10.1038/onc.2014.372.

Ravindran Menon, D., Hammerlindl, H., Torrano, J., Schaider, H., & Fujita, M. (2020). Epigenetics and metabolism at the crossroads of stress-induced plasticity, stemness and therapeutic resistance in cancer. *Theranostics*, *10*(14), 6261–6277. https://doi.org/10.7150/thno.42523.

Ribas, A., & Wolchok, J. D. (2018). Cancer immunotherapy using checkpoint blockade. *Science*, *359*(6382), 1350–1355. https://doi.org/10.1126/science.aar4060.

Risom, T., Langer, E. M., Chapman, M. P., Rantala, J., Fields, A. J., Boniface, C., et al. (2018). Differentiation-state plasticity is a targetable resistance mechanism in basal-like breast cancer. *Nature Communications*, *9*(1), 3815. https://doi.org/10.1038/s41467-018-05729-w.

Roesch, A., Fukunaga-Kalabis, M., Schmidt, E. C., Zabierowski, S. E., Brafford, P. A., Vultur, A., et al. (2010). A temporarily distinct subpopulation of slow-cycling melanoma cells is required for continuous tumor growth. *Cell*, *141*(4), 583–594. https://doi.org/10.1016/j.cell.2010.04.020.

Roesch, A., Vultur, A., Bogeski, I., Wang, H., Zimmermann, K. M., Speicher, D., et al. (2013). Overcoming intrinsic multidrug resistance in melanoma by blocking the mitochondrial respiratory chain of slow-cycling JARID1B(high) cells. *Cancer Cell*, *23*(6), 811–825. https://doi.org/10.1016/j.ccr.2013.05.003.

Russo, M., Crisafulli, G., Sogari, A., Reilly, N. M., Arena, S., Lamba, S., et al. (2019). Adaptive mutability of colorectal cancers in response to targeted therapies. *Science*, *366*(6472), 1473–1480. https://doi.org/10.1126/science.aav4474.

Ryan, M. B., Fece de la Cruz, F., Phat, S., Myers, D. T., Wong, E., Shahzade, H. A., et al. (2020). Vertical pathway inhibition overcomes adaptive feedback resistance to KRAS(G12C) inhibition. *Clinical Cancer Research*, *26*(7), 1633–1643. https://doi.org/10.1158/1078-0432.CCR-19-3523.

Saha, S., Mukherjee, S., Khan, P., Kajal, K., Mazumdar, M., Manna, A., et al. (2016). Aspirin suppresses the acquisition of chemoresistance in breast cancer by disrupting an NFkappaB-IL6 signaling axis responsible for the generation of cancer stem cells. *Cancer Research*, *76*(7), 2000–2012. https://doi.org/10.1158/0008-5472.CAN-15-1360.

Sahu, N., Stephan, J. P., Cruz, D. D., Merchant, M., Haley, B., Bourgon, R., et al. (2016). Functional screening implicates miR-371-3p and peroxiredoxin 6 in reversible tolerance to cancer drugs. *Nature Communications*, *7*, 12351. https://doi.org/10.1038/ncomms12351.

Sakai, W., Swisher, E. M., Karlan, B. Y., Agarwal, M. K., Higgins, J., Friedman, C., et al. (2008). Secondary mutations as a mechanism of cisplatin resistance in BRCA2-mutated cancers. *Nature*, *451*(7182), 1116–1120. https://doi.org/10.1038/nature06633.

Saleh, T., Tyutyunyk-Massey, L., & Gewirtz, D. A. (2019). Tumor cell escape from therapy-induced senescence as a model of disease recurrence after dormancy. *Cancer Research*, *79*(6), 1044–1046. https://doi.org/10.1158/0008-5472.CAN-18-3437.

Sanchez-Danes, A., Larsimont, J. C., Liagre, M., Munoz-Couselo, E., Lapouge, G., Brisebarre, A., et al. (2018). A slow-cycling LGR5 tumour population mediates basal cell carcinoma relapse after therapy. *Nature*, *562*(7727), 434–438. https://doi.org/10.1038/s41586-018-0603-3.

Sandoval, J., & Esteller, M. (2012). Cancer epigenomics: Beyond genomics. *Current Opinion in Genetics & Development*, *22*(1), 50–55. https://doi.org/10.1016/j.gde.2012.02.008.

Sansone, P., Berishaj, M., Rajasekhar, V. K., Ceccarelli, C., Chang, Q., Strillacci, A., et al. (2017). Evolution of cancer stem-like cells in endocrine-resistant metastatic breast cancers is mediated by stromal microvesicles. *Cancer Research*, *77*(8), 1927–1941. https://doi.org/10.1158/0008-5472.CAN-16-2129.

Sansone, P., Ceccarelli, C., Berishaj, M., Chang, Q., Rajasekhar, V. K., Perna, F., et al. (2016). Self-renewal of CD133(hi) cells by IL6/Notch3 signalling regulates endocrine resistance in metastatic breast cancer. *Nature Communications*, *7*, 10442. https://doi.org/10.1038/ncomms10442.

Saxena, M., Stephens, M. A., Pathak, H., & Rangarajan, A. (2011). Transcription factors that mediate epithelial-mesenchymal transition lead to multidrug resistance by upregulating ABC transporters. *Cell Death & Disease*, *2*, e179. https://doi.org/10.1038/cddis.2011.61.

Sehgal, K., Portell, A., Ivanova, E. V., Lizotte, P. H., Mahadevan, N. R., Greene, J. R., et al. (2021). Dynamic single-cell RNA sequencing identifies immunotherapy persister cells following PD-1 blockade. *The Journal of Clinical Investigation*, *131*(2), e135038. https://doi.org/10.1172/JCI135038.

Shaffer, S. M., Dunagin, M. C., Torborg, S. R., Torre, E. A., Emert, B., Krepler, C., et al. (2017). Rare cell variability and drug-induced reprogramming as a mode of cancer drug resistance. *Nature*, *546*(7658), 431–435. https://doi.org/10.1038/nature22794.

Shah, K. N., Bhatt, R., Rotow, J., Rohrberg, J., Olivas, V., Wang, V. E., et al. (2019). Aurora kinase A drives the evolution of resistance to third-generation EGFR inhibitors in lung cancer. *Nature Medicine*, *25*(1), 111–118. https://doi.org/10.1038/s41591-018-0264-7.

Shankar, E., Franco, D., Iqbal, O., Moreton, S., Kanwal, R., & Gupta, S. (2020). Dual targeting of EZH2 and androgen receptor as a novel therapy for castration-resistant prostate cancer. *Toxicology and Applied Pharmacology*, *404*, 115200. https://doi.org/10.1016/j.taap.2020.115200.

Sharma, S. V., Lee, D. Y., Li, B., Quinlan, M. P., Takahashi, F., Maheswaran, S., et al. (2010). A chromatin-mediated reversible drug-tolerant state in cancer cell subpopulations. *Cell*, *141*(1), 69–80. https://doi.org/10.1016/j.cell.2010.02.027.

Sharma, S., Xing, F., Liu, Y., Wu, K., Said, N., Pochampally, R., et al. (2016). Secreted protein acidic and rich in cysteine (SPARC) mediates metastatic dormancy of prostate cancer in bone. *The Journal of Biological Chemistry*, *291*(37), 19351–19363. https://doi.org/10.1074/jbc.M116.737379.

Shi, H., Hugo, W., Kong, X., Hong, A., Koya, R. C., Moriceau, G., et al. (2014). Acquired resistance and clonal evolution in melanoma during BRAF inhibitor therapy. *Cancer Discovery*, *4*(1), 80–93. https://doi.org/10.1158/2159-8290.CD-13-0642.

Shu, S., Lin, C. Y., He, H. H., Witwicki, R. M., Tabassum, D. P., Roberts, J. M., et al. (2016). Response and resistance to BET bromodomain inhibitors in triple-negative breast cancer. *Nature*, *529*(7586), 413–417. https://doi.org/10.1038/nature16508.

Singh, A., & Settleman, J. (2010). EMT, cancer stem cells and drug resistance: An emerging axis of evil in the war on cancer. *Oncogene*, *29*(34), 4741–4751. https://doi.org/10.1038/onc.2010.215.

Skrypek, N., Goossens, S., De Smedt, E., Vandamme, N., & Berx, G. (2017). Epithelial-to--mesenchymal transition: Epigenetic reprogramming driving cellular plasticity. *Trends in Genetics*, *33*(12), 943–959. https://doi.org/10.1016/j.tig.2017.08.004.

Skvortsov, S., Debbage, P., Lukas, P., & Skvortsova, I. (2015). Crosstalk between DNA repair and cancer stem cell (CSC) associated intracellular pathways. *Seminars in Cancer Biology*, *31*, 36–42. https://doi.org/10.1016/j.semcancer.2014.06.002.

Smith, M. P., Brunton, H., Rowling, E. J., Ferguson, J., Arozarena, I., Miskolczi, Z., et al. (2016). Inhibiting drivers of non-mutational drug tolerance is a salvage strategy for targeted melanoma therapy. *Cancer Cell*, *29*(3), 270–284. https://doi.org/10.1016/j.ccell.2016.02.003.

Smith, M. P., Sanchez-Laorden, B., O'Brien, K., Brunton, H., Ferguson, J., Young, H., et al. (2014). The immune microenvironment confers resistance to MAPK pathway inhibitors through macrophage-derived TNF alpha. *Cancer Discovery*, *4*(10), 1214–1229. https://doi.org/10.1158/2159-8290.Cd-13-1007.

Song, K. A., Hosono, Y., Turner, C., Jacob, S., Lochmann, T. L., Murakami, Y., et al. (2018). Increased synthesis of MCL-1 protein underlies initial survival of EGFR-mutant lung cancer to EGFR inhibitors and provides a novel drug target. *Clinical Cancer Research*, *24*(22), 5658–5672. https://doi.org/10.1158/1078-0432.CCR-18-0304.

Song, K. A., Niederst, M. J., Lochmann, T. L., Hata, A. N., Kitai, H., Ham, J., et al. (2018). Epithelial-to-mesenchymal transition antagonizes response to targeted therapies in lung cancer by suppressing BIM. *Clinical Cancer Research*, *24*(1), 197–208. https://doi.org/10.1158/1078-0432.CCR-17-1577.

Staberg, M., Rasmussen, R. D., Michaelsen, S. R., Pedersen, H., Jensen, K. E., Villingshoj, M., et al. (2018). Targeting glioma stem-like cell survival and chemoresistance through inhibition of lysine-specific histone demethylase KDM2B. *Molecular Oncology*, *12*(3), 406–420. https://doi.org/10.1002/1878-0261.12174.

Straussman, R., Morikawa, T., Shee, K., Barzily-Rokni, M., Qian, Z. R., Du, J., et al. (2012). Tumour micro-environment elicits innate resistance to RAF inhibitors through HGF secretion. *Nature*, *487*(7408), 500–504. https://doi.org/10.1038/nature11183.

Sun, J., Cai, X., Yung, M. M., Zhou, W., Li, J., Zhang, Y., et al. (2019). miR-137 mediates the functional link between c-Myc and EZH2 that regulates cisplatin resistance in ovarian cancer. *Oncogene*, *38*(4), 564–580. https://doi.org/10.1038/s41388-018-0459-x.

Sun, Y., Campisi, J., Higano, C., Beer, T. M., Porter, P., Coleman, I., et al. (2012). Treatment-induced damage to the tumor microenvironment promotes prostate cancer therapy resistance through WNT16B. *Nature Medicine*, *18*(9), 1359–1368. https://doi.org/10.1038/nm.2890.

Sun, C., Wang, L., Huang, S., Heynen, G. J., Prahallad, A., Robert, C., et al. (2014). Reversible and adaptive resistance to BRAF(V600E) inhibition in melanoma. *Nature*, *508*(7494), 118–122. https://doi.org/10.1038/nature13121.

Svensson, C., Ceder, J., Iglesias-Gato, D., Chuan, Y. C., Pang, S. T., Bjartell, A., et al. (2014). REST mediates androgen receptor actions on gene repression and predicts early recurrence of prostate cancer. *Nucleic Acids Research*, *42*(2), 999–1015. https://doi.org/10.1093/nar/gkt921.

Szakacs, G., Hall, M. D., Gottesman, M. M., Boumendjel, A., Kachadourian, R., Day, B. J., et al. (2014). Targeting the Achilles heel of multidrug-resistant cancer by exploiting the fitness cost of resistance. *Chemical Reviews*, *114*(11), 5753–5774. https://doi.org/10.1021/cr4006236.

Takahashi, K., & Yamanaka, S. (2006). Induction of pluripotent stem cells from mouse embryonic and adult fibroblast cultures by defined factors. *Cell*, *126*(4), 663–676. https://doi.org/10.1016/j.cell.2006.07.024.

Taniguchi, H., Yamada, T., Wang, R., Tanimura, K., Adachi, Y., Nishiyama, A., et al. (2019). AXL confers intrinsic resistance to osimertinib and advances the emergence of tolerant cells. *Nature Communications*, *10*(1), 259. https://doi.org/10.1038/s41467-018-08074-0.

Taverna, C., Voegeli, J., Trojan, A., Olie, R. A., & von Rohr, A. (2012). Effective response with bortezomib retreatment in relapsed multiple myeloma—A multicentre retrospective survey in Switzerland. *Swiss Medical Weekly*, *142*, w13562. https://doi.org/10.4414/smw.2012.13562.

Terai, H., Kitajima, S., Potter, D. S., Matsui, Y., Quiceno, L. G., Chen, T., et al. (2018). ER stress signaling promotes the survival of cancer "persister cells" tolerant to EGFR tyrosine kinase inhibitors. *Cancer Research*, *78*(4), 1044–1057. https://doi.org/10.1158/0008-5472.CAN-17-1904.

Terry, S., Abdou, A., Engelsen, A. S. T., Buart, S., Dessen, P., Corgnac, S., et al. (2019). AXL targeting overcomes human lung cancer cell resistance to NK- and CTL-mediated cytotoxicity. *Cancer Immunology Research*, 7(11), 1789–1802. https://doi.org/10.1158/2326-6066.CIR-18-0903.

Thomas, S., Thurn, K. T., Bicaku, E., Marchion, D. C., & Munster, P. N. (2011). Addition of a histone deacetylase inhibitor redirects tamoxifen-treated breast cancer cells into apoptosis, which is opposed by the induction of autophagy. *Breast Cancer Research and Treatment*, *130*(2), 437–447. https://doi.org/10.1007/s10549-011-1364-y.

Tirosh, I., Izar, B., Prakadan, S. M., Wadsworth, M. H., 2nd, Treacy, D., Trombetta, J. J., et al. (2016). Dissecting the multicellular ecosystem of metastatic melanoma by single-cell RNA-seq. *Science*, *352*(6282), 189–196. https://doi.org/10.1126/science.aad0501.

Todaro, M., Gaggianesi, M., Catalano, V., Benfante, A., Iovino, F., Biffoni, M., et al. (2014). CD44v6 is a marker of constitutive and reprogrammed cancer stem cells driving colon cancer metastasis. *Cell Stem Cell*, *14*(3), 342–356. https://doi.org/10.1016/j.stem.2014.01.009.

Vallette, F. M., Olivier, C., Lezot, F., Oliver, L., Cochonneau, D., Lalier, L., et al. (2019). Dormant, quiescent, tolerant and persister cells: Four synonyms for the same target in cancer. *Biochemical Pharmacology*, *162*, 169–176. https://doi.org/10.1016/j.bcp.2018.11.004.

Varambally, S., Dhanasekaran, S. M., Zhou, M., Barrette, T. R., Kumar-Sinha, C., Sanda, M. G., et al. (2002). The polycomb group protein EZH2 is involved in progression of prostate cancer. *Nature*, *419*(6907), 624–629. https://doi.org/10.1038/nature01075.

Vera-Ramirez, L., Vodnala, S. K., Nini, R., Hunter, K. W., & Green, J. E. (2018). Autophagy promotes the survival of dormant breast cancer cells and metastatic tumour recurrence. *Nature Communications*, *9*(1), 1944. https://doi.org/10.1038/s41467-018-04070-6.

Vermeulen, L., De Sousa, E. M. F., van der Heijden, M., Cameron, K., de Jong, J. H., Borovski, T., et al. (2010). Wnt activity defines colon cancer stem cells and is regulated by the microenvironment. *Nature Cell Biology*, *12*(5), 468–476. https://doi.org/10.1038/ncb2048.

Vinogradova, M., Gehling, V. S., Gustafson, A., Arora, S., Tindell, C. A., Wilson, C., et al. (2016). An inhibitor of KDM5 demethylases reduces survival of drug-tolerant cancer cells. *Nature Chemical Biology*, *12*(7), 531–538. https://doi.org/10.1038/nchembio.2085.
Viswanathan, V. S., Ryan, M. J., Dhruv, H. D., Gill, S., Eichhoff, O. M., Seashore-Ludlow, B., et al. (2017). Dependency of a therapy-resistant state of cancer cells on a lipid peroxidase pathway. *Nature*, *547*(7664), 453–457. https://doi.org/10.1038/nature23007.
Wainwright, E. N., & Scaffidi, P. (2017). Epigenetics and cancer stem cells: Unleashing, hijacking, and restricting cellular plasticity. *Trends in Cancer*, *3*(5), 372–386. https://doi.org/10.1016/j.trecan.2017.04.004.
Wang, W., Li, Q., Yamada, T., Matsumoto, K., Matsumoto, I., Oda, M., et al. (2009). Crosstalk to stromal fibroblasts induces resistance of lung cancer to epidermal growth factor receptor tyrosine kinase inhibitors. *Clinical Cancer Research*, *15*(21), 6630–6638. https://doi.org/10.1158/1078-0432.CCR-09-1001.
Wang, X., Schoenhals, J. E., Li, A., Valdecanas, D. R., Ye, H., Zang, F., et al. (2017). Suppression of type I IFN signaling in tumors mediates resistance to anti-PD-1 treatment that can be overcome by radiotherapy. *Cancer Research*, 77(4), 839–850. https://doi.org/10.1158/0008-5472.CAN-15-3142.
Wang, Z., Yang, J., Xu, G., Wang, W., Liu, C., Yang, H., et al. (2015). Targeting miR-381-NEFL axis sensitizes glioblastoma cells to temozolomide by regulating stemness factors and multidrug resistance factors. *Oncotarget*, *6*(5), 3147–3164. https://doi.org/10.18632/oncotarget.3061.
Wang, L., Zhang, Z., Yu, X., Huang, X., Liu, Z., Chai, Y., et al. (2019). Unbalanced YAP-SOX9 circuit drives stemness and malignant progression in esophageal squamous cell carcinoma. *Oncogene*, *38*(12), 2042–2055. https://doi.org/10.1038/s41388-018-0476-9.
Wawruszak, A., Kalafut, J., Okon, E., Czapinski, J., Halasa, M., Przybyszewska, A., et al. (2019). Histone deacetylase inhibitors and phenotypical transformation of cancer cells. *Cancers (Basel)*, *11*(2), 148. https://doi.org/10.3390/cancers11020148.
Wilson, T. R., Fridlyand, J., Yan, Y., Penuel, E., Burton, L., Chan, E., et al. (2012). Widespread potential for growth-factor-driven resistance to anticancer kinase inhibitors. *Nature*, *487*(7408), 505–509. https://doi.org/10.1038/nature11249.
Wilson, C., Nicholes, K., Bustos, D., Lin, E., Song, Q., Stephan, J. P., et al. (2014). Overcoming EMT-associated resistance to anti-cancer drugs via Src/FAK pathway inhibition. *Oncotarget*, *5*(17), 7328–7341. https://doi.org/10.18632/oncotarget.2397.
Wilson, M. M., Weinberg, R. A., Lees, J. A., & Guen, V. J. (2020). Emerging mechanisms by which EMT programs control stemness. *Trends in Cancer*, *6*(9), 775–780. https://doi.org/10.1016/j.trecan.2020.03.011.
Wolf, Y., Bartok, O., Patkar, S., Eli, G. B., Cohen, S., Litchfield, K., et al. (2019). UVB-induced tumor heterogeneity diminishes immune response in melanoma. *Cell*, *179*(1), 219-235.e221. https://doi.org/10.1016/j.cell.2019.08.032.
Wu, W. S., Heinrichs, S., Xu, D., Garrison, S. P., Zambetti, G. P., Adams, J. M., et al. (2005). Slug antagonizes p53-mediated apoptosis of hematopoietic progenitors by repressing puma. *Cell*, *123*(4), 641–653. https://doi.org/10.1016/j.cell.2005.09.029.
Wu, Y., Zhang, Z., Cenciarini, M. E., Proietti, C. J., Amasino, M., Hong, T., et al. (2018). Tamoxifen resistance in breast cancer is regulated by the EZH2-ERalpha-GREB1 transcriptional axis. *Cancer Research*, *78*(3), 671–684. https://doi.org/10.1158/0008-5472.CAN-17-1327.
Xue, G., Cui, Z. J., Zhou, X. H., Zhu, Y. X., Chen, Y., Liang, F. J., et al. (2019). DNA methylation biomarkers predict objective responses to PD-1/PD-L1 inhibition blockade. *Frontiers in Genetics*, *10*, 724. https://doi.org/10.3389/fgene.2019.00724.
Yamamoto, S., Wu, Z., Russnes, H. G., Takagi, S., Peluffo, G., Vaske, C., et al. (2014). JARID1B is a luminal lineage-driving oncogene in breast cancer. *Cancer Cell*, *25*(6), 762–777. https://doi.org/10.1016/j.ccr.2014.04.024.

Yan, F., Al-Kali, A., Zhang, Z., Liu, J., Pang, J., Zhao, N., et al. (2018). A dynamic N(6)-methyladenosine methylome regulates intrinsic and acquired resistance to tyrosine kinase inhibitors. *Cell Research*, *28*(11), 1062–1076. https://doi.org/10.1038/s41422-018-0097-4.

Yan, W., Chung, C. Y., Xie, T., Ozeck, M., Nichols, T. C., Frey, J., et al. (2022). Intrinsic and acquired drug resistance to LSD1 inhibitors in small cell lung cancer occurs through a TEAD4-driven transcriptional state. *Molecular Oncology*, *16*(6), 1309–1328. https://doi.org/10.1002/1878-0261.13124.

Yano, S., Nakataki, E., Ohtsuka, S., Inayama, M., Tomimoto, H., Edakuni, N., et al. (2005). Retreatment of lung adenocarcinoma patients with gefitinib who had experienced favorable results from their initial treatment with this selective epidermal growth factor receptor inhibitor: A report of three cases. *Oncology Research*, *15*(2), 107–111. https://www.ncbi.nlm.nih.gov/pubmed/16119008.

Yao, Z., Fenoglio, S., Gao, D. C., Camiolo, M., Stiles, B., Lindsted, T., et al. (2010). TGF-beta IL-6 axis mediates selective and adaptive mechanisms of resistance to molecular targeted therapy in lung cancer. *Proceedings of the National Academy of Sciences of the United States of America*, *107*(35), 15535–15540. https://doi.org/10.1073/pnas.1009472107.

Yi, Y., Zeng, S., Wang, Z., Wu, M., Ma, Y., Ye, X., et al. (2018). Cancer-associated fibroblasts promote epithelial-mesenchymal transition and EGFR-TKI resistance of non-small cell lung cancers via HGF/IGF-1/ANXA2 signaling. *Biochimica et Biophysica Acta—Molecular Basis of Disease*, *1864*(3), 793–803. https://doi.org/10.1016/j.bbadis.2017.12.021.

Yu, J., Liao, X., Li, L., Lv, L., Zhi, X., Yu, J., et al. (2017). A preliminary study of the role of extracellular -5'- nucleotidase in breast cancer stem cells and epithelial-mesenchymal transition. *In Vitro Cellular & Developmental Biology. Animal*, *53*(2), 132–140. https://doi.org/10.1007/s11626-016-0089-y.

Yuan, X., Li, J., Coulouarn, C., Lin, T., Sulpice, L., Bergeat, D., et al. (2018). SOX9 expression decreases survival of patients with intrahepatic cholangiocarcinoma by conferring chemoresistance. *British Journal of Cancer*, *119*(11), 1358–1366. https://doi.org/10.1038/s41416-018-0338-9.

Zanconato, F., Battilana, G., Forcato, M., Filippi, L., Azzolin, L., Manfrin, A., et al. (2018). Transcriptional addiction in cancer cells is mediated by YAP/TAZ through BRD4. *Nature Medicine*, *24*(10), 1599–1610. https://doi.org/10.1038/s41591-018-0158-8.

Zawistowski, J. S., Bevill, S. M., Goulet, D. R., Stuhlmiller, T. J., Beltran, A. S., Olivares-Quintero, J. F., et al. (2017). Enhancer remodeling during adaptive bypass to MEK inhibition is attenuated by pharmacologic targeting of the P-TEFb complex. *Cancer Discovery*, 7(3), 302–321. https://doi.org/10.1158/2159-8290.CD-16-0653.

Zhang, G., Frederick, D. T., Wu, L., Wei, Z., Krepler, C., Srinivasan, S., et al. (2016). Targeting mitochondrial biogenesis to overcome drug resistance to MAPK inhibitors. *The Journal of Clinical Investigation*, *126*(5), 1834–1856. https://doi.org/10.1172/JCI82661.

Zhang, Z., Lee, J. C., Lin, L., Olivas, V., Au, V., LaFramboise, T., et al. (2012). Activation of the AXL kinase causes resistance to EGFR-targeted therapy in lung cancer. *Nature Genetics*, *44*(8), 852–860. https://doi.org/10.1038/ng.2330.

Zhu, Y., Yu, F., Jiao, Y., Feng, J., Tang, W., Yao, H., et al. (2011). Reduced miR-128 in breast tumor-initiating cells induces chemotherapeutic resistance via Bmi-1 and ABCC5. *Clinical Cancer Research*, *17*(22), 7105–7115. https://doi.org/10.1158/1078-0432.CCR-11-0071.

CHAPTER TWO

Epigenetically programmed resistance to chemo- and immuno-therapies

Reddick R. Walker[a,b], **Zainab Rentia**[a], **and Katherine B. Chiappinelli**[a,b,*]
[a]The George Washington University Cancer Center (GWCC), Washington, DC, United States
[b]Department of Microbiology, Immunology & Tropical Medicine, The George Washington University, Washington, DC, United States
[*]Corresponding author: e-mail address: kchiapp1@gwu.edu

Contents

Advances in Cancer Research, Volume 158
ISSN 0065-230X
https://doi.org/10.1016/bs.acr.2022.12.001

Abstract

Resistance to cancer treatments remains a major barrier in developing cancer cures. While promising combination chemotherapy treatments and novel immunotherapies have improved patient outcomes, resistance to these treatments remains poorly understood. New insights into the dysregulation of the epigenome show how it promotes tumor growth and resistance to therapy. By altering control of gene expression, tumor cells can evade immune cell recognition, ignore apoptotic cues, and reverse DNA damage induced by chemotherapies. In this chapter, we summarize the data on epigenetic remodeling during cancer progression and treatment that enable cancer cell survival and describe how these epigenetic changes are being targeted clinically to overcome resistance.

1. Epigenetic dysregulation in cancer and therapeutic resistance

1.1 Introduction

A hallmark of cancer is the global dysregulation of the epigenetic state compared to normal cells (Hanahan, 2022). In this chapter, we define epigenetics as the proteins and modifications that govern DNA organization and expression. In eukaryotes, DNA is wrapped around positively charged proteins called histones which package the genetic information inside of the nucleus. How tightly DNA is bound to histones determines whether specific loci are transcribed. This binding affinity is altered through post-translational modifications (PTMs) (i.e., methylation, acetylation) to the histone N-terminal tail (Neganova, Klochkov, Aleksandrova, & Aliev, 2022; Stillman, 2018). The histone-DNA interaction forms repeating structures called nucleosomes which collectively are referred to as chromatin (McGinty & Tan, 2015; Zhou, Gaullier, & Luger, 2019). In addition to histone modifications, chemical modifications to DNA base pairs and the three dimensional structure of chromatin also control transcription.

Dysregulation of gene transcription can alter normal cellular functions to contribute to cellular transformation. These changes include inactivation of tumor suppressor genes and/or aberrant activation of oncogenes. Cancer cells are characterized by mutational instability, dysregulated cell metabolism, uncontrolled cell growth, and evasion of programmed cell death (Hanahan, 2022). Mutations in tumor suppressor genes and oncogenes can arise via damage to DNA from carcinogens, UV radiation, reactive oxygen species, and aging (Gaillard, García-Muse, & Aguilera, 2015; Srinivas, Tan, Vellayappan, & Jeyasekharan, 2019). These mutations cause inactivation of tumor suppressor genes, hyperactivation of oncogenes, and dysfunction of DNA damage repair mechanisms for prolonged cell replication and tumorigenic progression. However, DNA mutations are not the only drivers for neoplastic transformation; epigenetic reprogramming is now recognized as an important hallmark of cancer (Hanahan, 2022). Alterations to epigenetic mechanisms like histone PTMs, DNA hypermethylation, and higher order 3D chromatin structure allow for differential expression of tumor suppressors and oncogenes to promote cancer progression. These epigenetic mechanisms modulate the tumor response to chemo- and immunotherapies (Brown, Curry, Magnani, Wilhelm-Benartzi, & Borley, 2014; Sharma et al., 2010).

1.2 Histone PTMs

There are six types of histone proteins: H1, H2A, H2B, H3, H4, and H5. The nucleosome structure consists of two tetramers of H2A/H2B and H3/H4 (Bowman & Poirier, 2015; Neganova et al., 2022). The N-terminus of each histone and the C-terminus of H2A are frequently targeted for PTMs as they are comprised of positively charged amino acids (Ghoneim, Fuchs, & Musselman, 2021). PTMs on histone tails regulate the charge interaction between histones and DNA and therefore can control access of transcriptional machinery to genomic regions and provide binding sites to guide proteins to specific chromatin loci (Rossetto, Avvakumov, & Côté, 2012). Epigenetically programmed resistance to chemotherapies in cancer is dominated by changes in histone methylation and acetylation, so this chapter will focus on those modifications. Other histone modifications have been reviewed extensively elsewhere (Gräff & Tsai, 2013; Rossetto et al., 2012; Schvartzman, Thompson, & Finley, 2018; Stillman, 2018; Zhao & Shilatifard, 2019).

Histone methylation marks are catalyzed by the lysine methyltransferase (KMT, also known as histone methyltransferase) enzyme family and can add one, two, or three methyl groups to a specific lysine residue (Husmann & Gozani, 2019). These marks are reversed via the lysine specific demethylase (LSD) family of enzymes (Kooistra & Helin, 2012). Activating histone modifications include H3K4me3, H3K4me, and H3K36me3, while repressive histone modifications include H3K27me3 and H3K9me3 (Kimura, 2013; Santos-Rosa, Schneider, Bannister, et al., 2002; Wagner & Carpenter, 2012) (Fig. 1A). Conversely, histone acetylation is generally associated with transcriptional activation. The histone acetyltransferase (HAT) enzymes catalyze the addition of a negatively charge acetyl group onto the positively

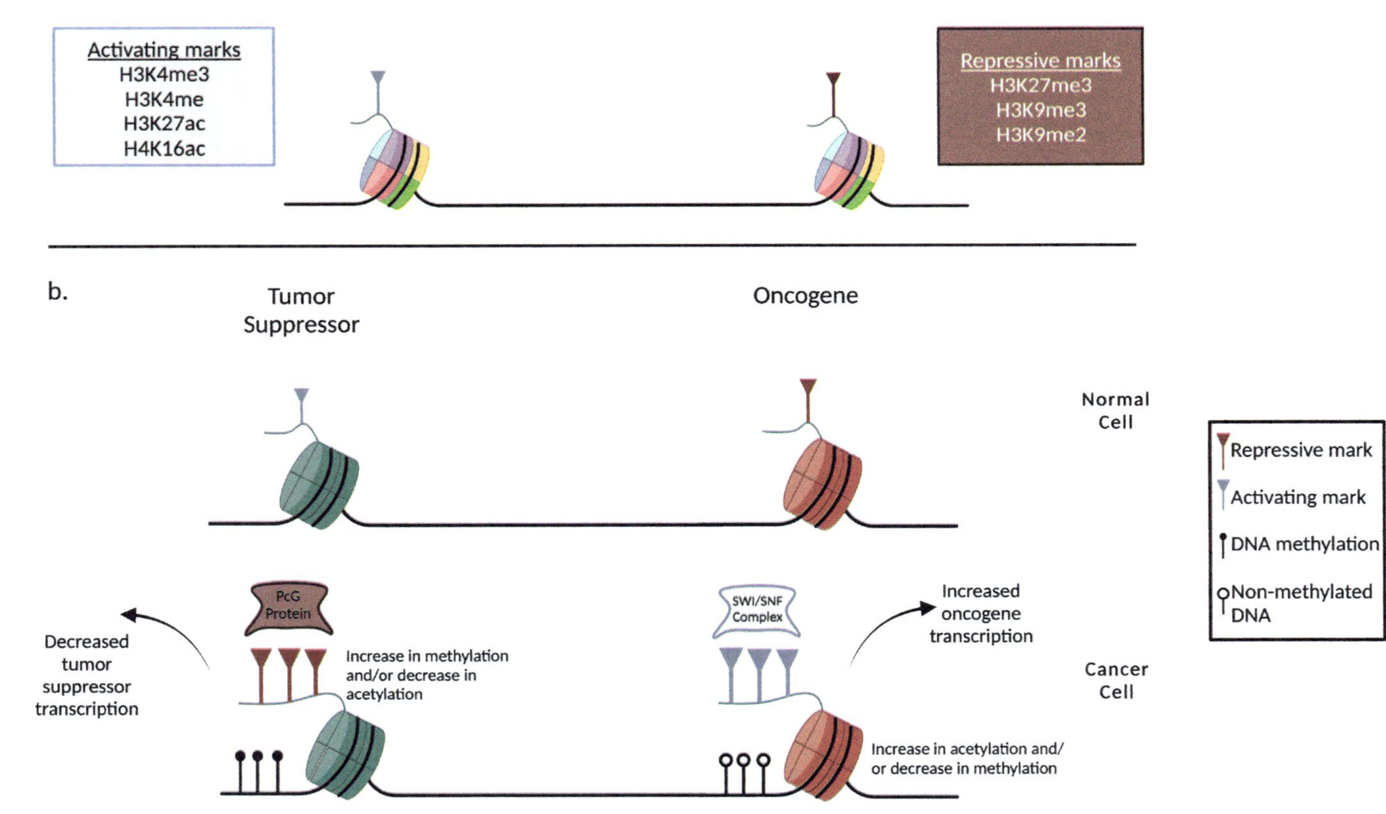

Fig. 1 Histone modifications can alter gene expression. (a) Histone tail modifications can have activating or repressive functions. (b) Increased repressive marks at tumor suppressor loci decreases transcription of tumor suppressors genes. Polycomb group (PcG) proteins recognize H3K27me3 to further repress chromatin accessibility. Increased activating marks at oncogene loci increase oncogene transcription. SWI/SNF protein complex recognizes marks like H3K27ac to further open chromatin accessibility. *Made with BioRender.com.*

charged histone (Carrozza, Utley, Workman, & Côté, 2003). This reduces the positive charge on the histone, thereby reducing its affinity to bind DNA and opening the DNA up to transcriptional machinery. A well-established marker for chromatin opening is H4K16ac (Shogren-Knaak et al., 2006). Acetyl groups are removed by the histone deacetylase (HDAC) superfamily of enzymes (Haberland, Montgomery, & Olson, 2009). Protein complexes can recognize regions of modified histones to further regulate chromatin structure. For example, H3K27me3 can be recognized by Polycomb group (PcG) proteins to repress gene transcription through additional deposition of H3K27me3 (De, Cheng, Sun, Gehred, & Kassis, 2019; Deng, Feng, & Pauklin, 2022) (Fig. 1B). Alternatively, the SWI/SNF complex is recruited to sites of H3K27ac that promote the opening of chromatin (Mittal & Roberts, 2020; Tang et al., 2017). Zhao & Shilatifard (2019) provides a complete review of histone modifying enzyme families (Zhao & Shilatifard, 2019).

1.3 DNA methylation

One repressive epigenetic modification is DNA methylation; addition of methyl groups directly to the cytosine base of DNA (Curradi, Izzo, Badaracco, & Landsberger, 2002; Varley et al., 2013). Gene promoter regions have a high concentration of cytosine/guanine (CpG) dinucleotides that accept methyl groups from the donor S-adenosyl-methionine (Varley et al., 2013). This covalent modification, mediated by the DNA methyltransferase (DNMT) enzymes, represses transcription of the associated gene by recruiting binding proteins that recognize methylated CpGs and hinder transcriptional machinery from binding (Curradi et al., 2002). There are three main DNMT enzymes in mammals: DNMT1, DNMT3A, and DNMT3B (Lyko, 2018). These catalytically active DNMTs transfer a methyl group onto the 5th carbon of the cytosine nucleotide, making a 5-methylcytosine (5-mC). De novo methylation on new CpG regions is carried out by DNMT3A and 3B and occurs most commonly during embryonic development, while maintenance of DNA methylation after cell differentiation is carried out by DNMT1 (Spada et al., 2007). Promoter region methylation generally represses gene expression, while DNA methylation inside exonic regions is associated with increased gene expression (Yang et al., 2014).

Methylation patterns on DNA are dynamically altered. The Ten Eleven Translocation (TET) protein family can remove genomic 5-mC marks (Rasmussen & Helin, 2016; Tahiliani et al., 2009). These proteins can oxidize 5-mC to 5-hydroxymethylcytosine (5-hmC), 5-carboxylcytosine

(5caC), and 5-formylcytosine (5fC), which are actively removed by base excision repair processes, or passively removed during replication due to the inability of DNMT1 to methylate these oxidized cytosine bases (Hill, Amouroux, & Hajkova, 2014). Though these marks are intermediates in the removal of methylation marks, 5-hmC and 5-fC are observed as stable cytosine modifications (Bachman et al., 2015; Bachman, Uribe-Lewis, Yang, et al., 2014). The oxidized cytosine products are actively studied for their unique role in gene expression and DNA methylation dynamics (Sérandour et al., 2012; Wu & Zhang, 2015; Yu et al., 2012).

1.4 Three-dimensional (3D) chromatin

The compaction of DNA inside of the nucleus governs conformational states of chromatin, which regulate cellular processes and gene expression (Deng et al., 2022; Feng & Pauklin, 2020). Three-dimensional (3D) chromatin can be described by loops, domains, and compartments (Dixon et al., 2012; Rao et al., 2014) (Fig. 2A). Loop formation is mediated by the zinc finger phosphoprotein CTCF, which recognizes specific motifs on the two DNA fragments being looped together, and cohesin proteins which pause at CTCF motifs and stabilize the looped DNA interaction (Beagan & Phillips-Cremins, 2020; Gabriele et al., 2022; Ohlsson, Renkawitz, & Lobanenkov, 2001) (Fig. 2B). The co-localization of DNA regions formed from loops create topologically associated domains (TADs) that contribute to de novo enhancer-promoter gene interactions. Enhancers provide a binding site for transcription factors and enzymes that promote the initiation of transcription. When enhancers interact with promoter regions, they promote the binding of RNA polymerase to the promoter region (Ray-Jones & Spivakov, 2021). TADs aggregate together with other domains that have similar repressive or activating modifications, and can inhibit enhancer-promoter interactions when separated by different domains (Dixon et al., 2012; Rao et al., 2014). For example, TADs that occur near the periphery of the nucleus interact with the inner nuclear membrane to form lamina associated domains (LADs) that are enriched in repressive modifications like H3K9me2 and H3K9me3 and are characterized by reduced gene expression (Briand & Collas, 2020; Guelen et al., 2008). Alternatively, active chromatin domains associate preferentially in the interior of the nucleus, and their compartments generally are enriched for open chromatin conformations (Foster & Bridger, 2005; Kosak et al., 2007). The compartments formed in the nucleus that are characterized by euchromatin are called Compartment A, while those of heterochromatin are Compartment B (Nichols & Corces, 2021) (Fig. 2A).

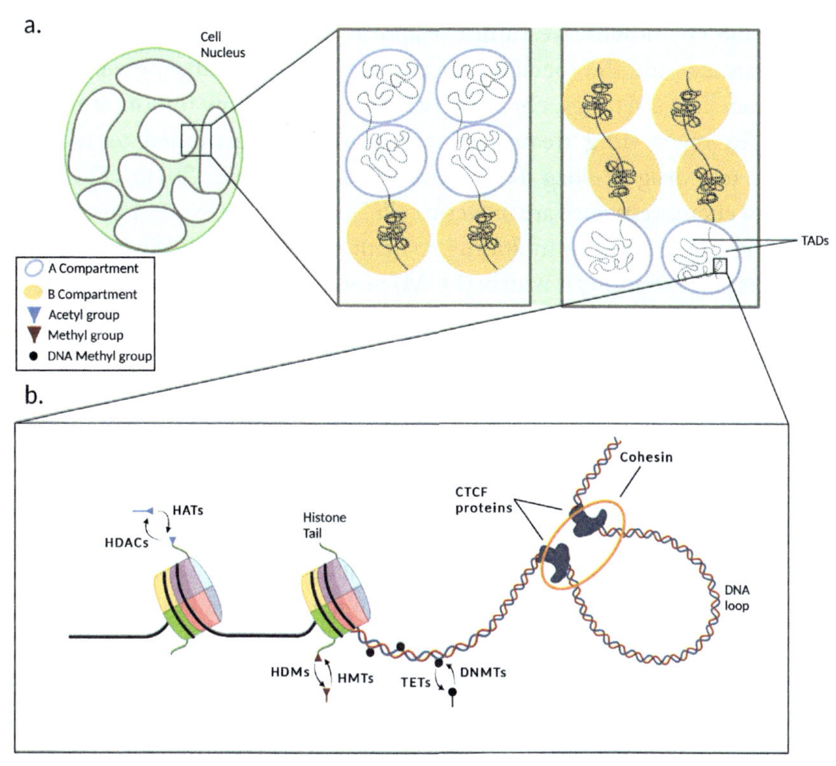

Fig. 2 Chromatin structure and modifying enzymes. (a) Compartmentalization of chromatin inside of the nucleus forms clusters comprised of compartments A (euchromatin) and B (heterochromatin), with peripheral clusters generally being more transcriptionally silenced. A and B compartments form from topologically associated domains (TADs). (b) Within TADs, DNA loop structures can form from CTCF and cohesin interactions, which mediates enhancer-promoter localization. Histone tails can be modified by histone acetyltransferases (HATs) and histone methyltransferases (HMTs), which are removed by deacetylase (HDACs) or demethylase (HDMs) enzymes, respectively. Direct DNA methylation is carried out by DNA methyltransferases (DNMTs) and reversed by the Ten-eleven translocation (TET) proteins. *Made with BioRender.com.*

2. Chemoresistance

The clinical use of the term 'resistance' characterizes disease progression during or after treatment is given. In general, the clinical use of chemorefractory relapse describes a tumor that shows disease progression during treatment, chemoresistant relapse describes tumors that show disease progression after treatment has ended and before a specified time interval has passed, and chemosensitive tumor relapse is defined as disease progression after treatment and the specified interval have passed (Tahara et al., 2020).

In nonclinical settings, chemoresistance describes how tumors are either primed to reject drug treatment or become reprogrammed to block a response to a drug (Yeldag, Rice, & del Río Hernández, 2018). The main drivers of chemoresistance are intrinsic cellular factors like expression of DNA damage response genes or multidrug resistant proteins (MDRPs), and extrinsic factors like extracellular matrix (ECM) protein deposition in the surrounding tumor microenvironment (TME); these mechanisms can be driven by changes to epigenetic states or from direct mutations to the genes that give rise to each factor (Li et al., 2017; Ponnusamy, Mahalingaiah, & Singh, 2020; Vasan, Baselga, & Hyman, 2019; Wu, Yang, Nie, Shi, & Fan, 2014).

A subset of tumor cells, termed persistent cancer cells, are not sensitive to chemotherapy for one of two reasons: (1) they have pre-existing characteristics that allow them to slow proliferation, or (2) they gain characteristics after treatment that leads to a slow proliferative state (Shen, Vagner, & Robert, 2020). Persistor cells that maintain a slow proliferative state evade the damaging effects of chemotherapies that target rapidly dividing cells (Shen et al., 2020). In vitro studies have shown that a population of persistor cells after drug treatment can readily enter a state of proliferation and re-sensitization to chemotherapies, termed tolerant persistor cells (Sharma et al., 2010). Tolerance refers to the ability of cancer cells to switch between drug-resistant and drug-sensitive states through epigenetic gene modulation, while the term resistance is used to describe a more permanent state of insensitivity to chemotherapies (Sharma et al., 2010). While epigenetic reprogramming plays a role to drive both tolerant and resistant cell states, it is unclear how tolerant cells give rise to more permanently resistant cell states (Ramirez et al., 2016).

This chapter will focus on examples of cancer cell resistance rather than tolerant populations. Factors that resistant cells alter to resist chemotherapy include removing the drug from the cell via MDRPs, depositing macromolecular scaffolding in the ECM to decrease drug penetrance, and disabling functional protein targets and DNA damage sensors (Gonzalez-Molina et al., 2022; Zheng, 2017).

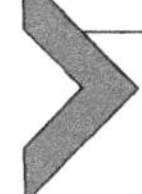

3. Epigenetic modifications drive cancer progression and chemoresistance

Dysregulation in gene expression that stem from changes to epigenetic modifiers can promote cancer development, survival, and metastasis in multiple cancer types (Chatterjee, Rodger, & Eccles, 2018). Aberrant DNA methylation is a hallmark of cancer that can dysregulate gene expression (Cecotka & Polanska, 2018; Klughammer et al., 2018; Mehta, Dobersch, Romero-Olmedo, & Barreto, 2015). Examples of this include DNA demethylation of oncogenic promoter regions or increased methylation of tumor suppressors. For example, demethylation of the Wnt gene promoter increases oncogenic signaling and is associated with poor survival in glioblastoma (Klughammer et al., 2018). In lung cancer, increased promoter region methylation at the CDKN2A (p16) and PTPRN2 genes is an early event in cellular hyperplasia of non-small cell lung cancer patients that inhibits the tumor suppressor functions of these genes (Mehta et al., 2015). The CpG island methylator phenotype (CIMP) occurs in multiple cancer types and appears in roughly 20% of malignant colorectal cancers (Fedorova et al., 2019). The CIMP phenotype is characterized by a subset of tumors with elevated CpG island methylation in genomic regions like tumor suppressors such as p16 and THBS1 (Toyota et al., 1999). Histone modifications can also increase cellular proliferation and carcinogenesis through similar disruptions in gene expression. For example, oral squamous cell carcinomas show global increases in hypoacetylation of H3K9, increasing cancer cell proliferation and lowering cumulative survival (Webber et al., 2017).

Mutations in epigenetic proteins drive dysregulation of chromatin accessibility and gene expression and provide a framework for potential cancer therapeutic targets and diagnostic markers (Baylin & Jones, 2011). An example found in multiple cancers including ovarian, endometrial, colorectal, breast, and gastric, is mutations in the subunits of the SWI/SNFcomplex (Pavlidou & Balis, 2020). Arid1a loss results in improper DNA damage repair via interruption of SWI/SNF interaction with Topoisomerase II, along with global dysregulation of the chromatin remodeler activity (Pavlidou & Balis, 2020; Wiegand et al., 2010). The mutation or deletion of SWI/SNF subunit proteins, Brg1 and Brm, in adrenal and ovarian carcinomas (OCs) has led to increased metastatic potential via hypermethylation of cell adhesion proteins CD44 and E-cadherin (Banine et al., 2005). Arid1a mutations can lead to dependence on its homolog, Arid1b, for SWI/SNF function in cancer, and this subunit interdependency is also seen with SNF5 loss and Brg1 function (Wang et al., 2009). Here, in a malignant rhabdoid tumor model, SNF5 loss is compensated by the activity of Brg1 to continue SWI/SNF driven oncogenesis (Wang et al., 2009). Because there is interdependence between SWI/SNF subunit proteins, therapeutic strategies like bromodomain inhibitors, which can interrupt protein domain interactions with acetyl groups, are being explored in SWI/SNF mutant cancers to stop its compensatory oncogenic progression (Papillon, Nakajima, Adair, et al., 2018; Shorstova et al., 2019).

3.1 Chemotherapies and chemoresistance

One consequence of epigenetic dysregulation in cancers is increased resistance to chemotherapies (Crea et al., 2011; Dhayat, Mardin, Mees, & Haier, 2011; Ponnusamy, Mahalingaiah, Chang, & Singh, 2019). Conventional chemotherapies were developed to target rapidly dividing cells by disrupting processes involved in DNA and protein synthesis. These therapies include platinum agents, topoisomerase II (TOP2) inhibitors, and mitotic poisons (Bruno et al., 2020; Johnstone, Park, & Lippard, 2014; Škubník, Jurášek, Ruml, & Rimpelová, 2020). Platinum agents, like cisplatin, carboplatin, and oxaliplatin, are commonly used to treat OCs, but have been used in other cancers like leukemia and breast cancer (Martín, 2001; Sabatier et al., 2011; Travis et al., 1999). Platinum agents work by reacting with purine bases to cross-link DNA together, causing distortion, and mediating apoptosis (Johnstone et al., 2014). TOP2 inhibitors are effective in cancers with amplification of the TOP2A gene, which normally functions to prevent kinks in DNA during transcription. Gene amplification of TOP2A is common in ERBB2-positive (HER2+) breast cancers, making them good candidates for inhibition (Nitiss & Targeting, 2009). TOP2 inhibitors work by forming TOP2-DNA complexes (doxorubicin, etoposide) or through catalytically inactivating the TOP2 enzyme (Nitiss & Targeting, 2009). These effects prevent DNA replication, ultimately leading to cell death. These examples show that chemotherapeutic agents can have dramatic effects in many tumor types, but a major barrier to successful treatment with these drugs is the development of chemoresistance.

Chemoresistant cancer cells are categorized by primary (de novo) or secondary (acquired or refractory) resistance (Li et al., 2021). This characterization is determined by the response after initial treatment; secondary resistant tumors will show an initial response before rejecting a drug, while a primary resistant tumor will not show any response. Intrinsic and extrinsic factors of the tumor cells cooperate to drive both types of resistance, and specific biomarkers can be used to determine the efficacy of a given drug. Understanding the phenotype of not only the tumor but also its environment is imperative to overcoming chemoresistance. In addition to DNA mutations, epigenetic reprogramming can drive drug resistance.

Here we discuss three types of epigenetic alterations that promote chemoresistance: DNA methylation, posttranslational histone modifications, and 3D epigenome restructuring.

3.2 DNA hypermethylation

Higher expression of DNA damage repair enzymes in cancer cells can reverse chemo-induced damage (Lodovichi, Cervelli, Pellicioli, & Galli, 2020). Since DNA damage repair can contribute to resistance to chemotherapies, silencing these gene loci can help overcome chemoresistance. For example, the O^6-methylguanine methyltransferase (MGMT) enzyme is known to reverse DNA alkylation, and hypermethylation of its gene promoter correlates with higher response to alkylating agents like temozolomide (Weller et al., 2010). However, hypermethylation of the DNA repair protein MLH1 hypermethylation leads to platinum resistance in OC via reduced MLH1/c-Abl apoptotic signaling (Li et al., 2018; Zeller et al., 2012). MLH1 hypermethylation also causes an increase in drug resistance in colorectal cancers, and is attributed to increasing resistance through loss of DNA mismatch repair capability leading to genomic instability (Arnold, Goel, & Boland, 2003).

Multidrug resistant proteins are common targets of DNA methylation, and play a role in chemoresistance (Baker & El-Osta, 2003; Chekhun, Lukyanova, Kovalchuk, Tryndyak, & Pogribny, 2007; Vaclavikova et al., 2019). The MDR1 gene, which encodes for a membrane embedded drug efflux pump protein, has shown to be hypomethylated in AML, and patient studies have shown that lower expression of the MDR1 gene through hypermethylation correspond with increased clinical response (Baker & El-Osta, 2003; Nakayama et al., 1998). Additionally, the hypermethylation of the ABCB1 gene, another transporter protein, has shown to have a positive correlation with clinical outcomes, owing to the lower expression of this transporter and higher exposure to chemotherapies (Vaclavikova et al., 2019).

DNA methylation of promoter regions contributes to chemoresistance in several types of cancer (Gull et al., 2022; Liu et al., 2020; Martinez-Cardús, Vizoso, Moran, & Manzano, 2015; Tao, Huang, Chen, & Chen, 2015). Indeed, markers for drug response have started to include DNA methylation levels. *Luskin* et al. explored the relationship between DNA methylation and complete remission (CR) and overall survival (OS) in 166 patients who received chemotherapy for de novo AML (Luskin et al., n.d.).

They found that higher levels of methylation in patients with de novo AML predicted significantly reduced CR and OS. However, global DNA methylation patterns are cancer specific, and may vary depending on the cancer stage (Liu et al., 2019). While global DNA methylation may provide a generalized prognosis, locus specific methylation provides deeper insight into chemoresistance and patient response (Tost, 2016).

3.3 Histone PTMs

Dysregulation of histone modifying enzymes is associated with drug resistance in multiple cancer types (Emran et al., 2017). In HER2+ breast cancer, resistance to lapatinib, a HER2 receptor inhibitor, is associated with increases in H3K4me3 at the c-Myc promoter, which overcomes HER2 receptor inhibition by constitutively activating c-Myc, releasing the need for upstream receptor signaling (Matkar et al., 2015). In another breast cancer study, overexpression of KDM5, an H3K4 demethylase, disrupts normal tumor apoptotic cues and increases drug resistance to erlotinib by removing the activating methylation patterns on target genes (Hou et al., 2012). The expression of tumor suppressors p21 and BAK1, among other genes, are decreased in response to KDM5 overexpression, and contribute to bypassing drug induced DNA damage (Hou et al., 2012). In glioblastoma, higher KDM5 expression correlated with temozolomide resistance in glioblastoma cells, highlighting the importance that reduced methylation has on increasing cancer resistance (Plch, Hrabeta, & Eckschlager, 2019). Increased methyltransferase activity of SETDB1/2 expression in lung, melanoma, and colon cancers increases the repressive H3K9me3 deposition and decreased the frequency of the H3K4me mark (Emran et al., 2017). The resulting dysregulation in gene expression increases interferon signaling to maintain a slow proliferative state, which reduced survival and increases drug resistance (Emran et al., 2017). Interferon signaling, which is normally activated by pattern recognition receptors, can stimulate both pro- and anti-tumor activity through interferon stimulated gene responses, and has been shown to predict chemotherapy resistance in breast cancer (Abt et al., 2022; Benci et al., 2019; Weichselbaum et al., 2008).

3.4 3D chromatin states

Direct changes to chromatin conformation can result in genetic changes that promote chemoresistance, often by disrupting enhancer-promoter interactions (Shang et al., 2019; Zha, Luo, Qiu, & Huang, 2017). FOXA1, SOX2, and GREB are transcription factors that normally function to control estrogen-receptor signaling pathways and are dysregulated in endocrine (tamoxifen and fulvestrant) resistant breast cancers. The relocation of enhancers away from gene promoters of these transcription factors disrupts normal estrogen signaling, inhibiting the estrogen receptor interaction with its ligand (Achinger-Kawecka et al., 2020). Because these tumors no longer are driven by estrogen-ER interactions, treatment with endocrine therapies which block ER interactions are no longer effective. In an OC model, enhancer regions were reprogrammed with a gain of H3K27ac near cell signaling, DNA damage response, and epithelial-to-mesenchymal transition (EMT) gene super enhancer (SE) regions compared to sensitive cancer cells (Shang et al., 2019). SEs are large groups of enhancers within proximity of one another, and were found to significantly drive SOX9 expression in OC (Shang et al., 2019). SEs in a resistant small cell lung cancer (SCLC) model are redistributed near gene regions associated with DNA binding and RNA pol II activity, which play pivotal roles in dysregulating pathways that lead to chemoresistance (Bao et al., 2019). SE-driven changes to RNA polymerase II activity can increase PI3K/Akt signaling to decrease drug sensitivities in SCLC (Bao et al., 2019).

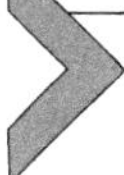

4. Epigenetic modifications drive resistance to immunotherapies

Recognition of tumors by the immune system relies on recruitment of activated immune cells to the TME (Kohli, Pillarisetty, & Kim, 2022). During cancer progression, epigenetic changes result in decreased tumor antigen expression, immunosuppressive ECM remodeling, and increased immune checkpoint protein expression to disrupt normal immune cell interactions with tumor cells. The TME influences immune cell function via cytokine signatures that can control inflammation and effector function of immune cells (Gonzalez, Hagerling, & Werb, 2018). In head and neck cancers, tumors release interleukin-6 (IL-6) and IL-10 to inhibit T-cell activation and expansion (de Vries, 1995; Johnson, De Costa, & Young, 2014). Epigenetic mechanisms can modulate tumor cytokine production. For example, hypomethylation of the IL-10 gene has shown to increase IL-10 production in breast cancers, contributing to breast carcinogenesis (Son et al., 2010). Disruptions like these in the TME can result in prolonged tissue inflammation and immunosuppression (Almand et al., 2001; Jaillon et al., 2020). The hypomethylation of the immune checkpoint programmed death ligand 1 (PDL-1) results in upregulation of expression in multiple cancers and decreases effector T-cell function (Gao et al., 2018; Huang et al., 2020; Iacovelli et al., 2016; Stovgaard, Dyhl-Polk, Roslind, Balslev, & Nielsen, 2019; Wang, Teng, Kong, & Yu, 2016). Additionally, the loss of MHC class I expression in melanoma and lung carcinomas prevents T-cell recognition of tumor neoantigens, which correlates with a stronger metastatic potential (Lou et al., 2005; Seliger et al., 2001). These mechanisms have specific epigenetic drivers which have shown promising reversal when combined with epigenetic therapies in vitro (Gomez, Tabernacki, Kobyra, Roberts, & Chiappinelli, 2020; Topper et al., 2017; Ugurel et al., 2019; Wang, Waschke, Woolaver, et al., 2020). For example, HDAC inhibitors have shown to re-express MHC class I and tumor antigen expression (gp100) in B-cell lymphoma and melanoma, respectively, highlighting the role of HDACs in regulating immune responses (Murakami et al., 2008; Wang et al., 2020).

4.1 Immune surveillance and immunotherapies

Using therapies that stimulate the host immune system (immunotherapies) to increase immune cell activation and antitumor immunity produces dramatic and durable responses in a subset of solid tumors (Cao & Yan, 2020; Esfahani et al., 2020). Immunotherapies can: (1) block the inhibitory signal between host cells and immune T-cells, (2) Stimulate immunity through cancer-associated antigen introduction (also known as cancer vaccines), or (3) promote antitumor activity through immune agonists like anti-CD40 (Djureinovic, Wang, & Kluger, 2021; Esfahani et al., 2020; Melief, van Hall, Arens, Ossendorp, & van der Burg, 2015). This chapter will focus on immune checkpoint blockades therapies. T-cell activation requires both the presentation of antigen peptides on major histocompatibility complexes (MHC) to the T-cell receptor (TCR) and a co-stimulatory signal from the B7 ligand with the CD28 receptor on the T-cell (Greenwald, Freeman, & Sharpe, 2005; Townsend & Allison, 1993). Since B7 expression is limited to hematopoietic cell populations, an inflammatory response allows for cell types such as dendritic cells or macrophages to penetrate the TME, engulf and express tumor peptides on MHC, and effectively activate T-cells through co-stimulation (Sharma & Allison, 2015).

Specific T-cell glycoproteins have been identified to negatively regulate T-cell activation (Freeman et al., 2000; Walunas et al., 1994). CTLA-4 and PD-1 are two common inhibitors termed checkpoint blockades, as they impede T-cell activation. These two negative regulators are now frequently inhibited in cancer treatment to bolster the T-cell recognition of tumors (Pasello et al., 2020). Anti-CTLA4 and anti-PD1 have been used in a variety of cancer types and increase tumor recognition by T-cells and tumor cell death (Korman, Garrett-Thomson, & Lonberg, 2022). The combinatorial use of anti-PD1 (and anti-CTLA4) showed favorable outcomes with 33% of enrolled colorectal cancer patients showing objective response (Ganesh et al., 2019). Nivolumab has also shown tumor remission capabilities in platinum resistant OCs (Hamanishi et al., 2015a). In this study, platinum resistant patients showed an average progression-free survival of 3.5 months with a disease control rate of 45% (Hamanishi et al., 2015b).

4.2 DNA hypermethylation

ECM remodeling via epigenetic changes in tumors can suppress the anti-tumor immune response (Pires, Burnell, & Gallimore, 2022). In breast carcinomas, the de novo methyltransferase DNMT3b hypermethylates the SHP1 phosphatase promoter region to releases interleukin-6, converting fibroblasts to cancer-associated fibroblasts (CAFs) (Albrengues et al., 2015). These CAFs can recruit suppressive immune cells to the TME through CXCL12 secretion (Barrett & Puré, 2020). SHP1 normally can function to disseminate or halt cell signals for pathways controlling cellular proliferation, migration, and invasion (Varone, Spano, & Corda, 2020). This conversion generates a pro-invasive ECM. In addition to promoting tumor invasion, CAF-ECM remodeling has been shown to promote an inhibitory environment for immune cells through T_{reg} activation via hyaluronan enzymes (Pires et al., 2022).

Prognostic indicators of tumor response to immunotherapies have been driven by mutational neoantigen burden and expression of immunomodulatory proteins like PDL-1 (Patel & Kurzrock, 2015; Wang, Chen, & Wang, 2021). As tumors proliferate, replicative stress can introduce high levels of mutations that can signal the immune system to kill the cells presenting defective epitopes. Incorrect protein structures arising from mutations can be presented on the surface of the tumors which is recognized as foreign by immune surveillance. However, even tumors with high mutational burden can evade the immune system with increased promoter region-CpG hypermethylation that silences genes involved in antigen processing and presentation and other immune signaling pathways (Jung et al., 2019). For example, hypermethylation of HLA class I in gastric cancers reduces tumor recognition by cytotoxic CD8+ T-cells (Ye et al., 2010). HLA-I down regulation is common in other cancer types as well (Aptsiauri & Garrido, 2022; Ladányi et al., 2022; Salomé et al., 2022).

Another way that DNA methylation impacts the tumor immune microenvironment is by regulation of transposable elements in the genome. DNMTi treatment increased responsiveness of a preclinical melanoma model of anti-CTLA4 treatment via a viral mimicry response (Chiappinelli et al., 2015). The use of DNMTis can demethylate transposable elements, so that they are transcribed and form dsRNA inside of the cell (Chiappinelli et al., 2015; Grundy, Diab, & Chiappinelli, 2022). The accumulation of dsRNA inside of the cell can trigger an interferon response through dsRNA sensors like TLR3, that aids in antitumor immunity; this response has shown to also be regulated through LSD1 expression (Karakaidos, Verigos, & Magklara, 2019; Sheng et al., 2018). Inhibiting dsRNA sensors MDA5 and TLR3 with DNMTi treatment abrogates activation of the interferon response, implicating expression of dsRNA as the driver for DNMTi driven interferon signaling (Roulois et al., 2015). Others have shown that DNMTi upregulate endogenous retroviral dsRNA and that this can be increased by adding histone deacetylase inhibitors (HDACi), inhibitors of H3K9 methyltransferases, or Vitamin C, which is a cofactor for the TET DNA demethylases (Brocks, Schmidt, Daskalakis, et al., 2017; Liu et al., 2016, 2018).

4.3 Histone PTMs

Posttranslational modifications to histone proteins can control gene expression to enable normal cellular functions (Stillman, 2018). These mechanisms are frequently targeted by transformed cells to evade host antitumor immunity. Changes to histone enzymes that catalyze either the addition or removal of modifications can alter immune cell recognition. For example, disruption of T-cell activation via upregulation of PDL-1 correlates with poorer clinical outcomes (Fan et al., 2019). PDL-1 upregulation is mediated by histone acetyltransferase 1 (HAT1) overexpression in pancreatic cancer. The upregulation of HAT1 results in the activation of Myc and BRD4 which binds the gene encoding PD-L1 (CD274) to promote its transcription. This is not the only way in which T-cell functions can be disrupted by cancer; tumor overconsumption of methionine can also outcompete infiltrating T-cells and reduce T-cell H3K9me2 levels resulting in reduced STAT5 expression and impaired antitumor functionality (Bian et al., 2020). In addition, tumors can actively downregulate major histocompatibility complexes (MHC) to avoid antigen presentation to T-cells (Zika, Greer,

Zhu, & Ting, 2003). Here, histone deacetylase 1 and 2 can actively represses the MHC class II transcriptional regulator, class II trans activator, to reduce the induction of an interferon response (Zika et al., 2003).

5. Targeting epigenetic modifications to overcome chemoresistance and immune evasion

To overcome chemoresistance and immune evasion, novel therapeutic strategies target the epigenome (Dan, Zhang, Zhou, & Guan, 2019; Suraweera, O'Byrne, & Richard, 2018; Zhang, Gong, Li, Yang, & Li, 2021) (Fig. 3). These therapies include hypomethylating agents such as DNMT inhibitors (DNMTi) and HDACi. DNMTis inhibit DNA methylation and HDAC inhibitors inhibit histone deacetylation to ultimately increase chromatin accessibility (Dan et al., 2019; Suraweera et al., 2018). These have been tested individually and in combination and show promising results in numerous trials against a variety of cancers. As of now the only DNMTi inhibitors to be approved by the Food and Drug Administration (FDA) are 5-azacytidine and 2-deoxy-5-azacytidine (also known as decitabine) for treatment of myelodysplastic syndrome (Zhang et al., 2021). These inhibitors are cytidine analogues that incorporate into DNA to form

Epigenetic Therapy	Target	Combined with	Cancer Type	Clinical Outcomes	Citations
Vorinostat	Class I/II HDAC inhibitor	Decitabine, hydroxychloroquine	AML, mCRC	Improved antitumor immunity	Pommert et al. (2022), Richon (2006), Arora et al. (2022)
Belinostat	Non-selective HDAC inhibitor	Decitabine	OC	Increased DNA Repair Expression	Pommert et al. (2022)
Entinostat	Class I/III HDAC inhibitor	5-Azaycytidine	Lung	Improved progression-free survival	Juergens et al. (2011)
Decitabine	DNMT inhibitor	Carboplatin, belinostat, vorinostat	OC, AML	Improved progression free survival	Matei et al. (2012), Steele et al. (2009), Pommert et al. (2022), Richon (2006)
Guadecatibine	DNMT inhibitor	Ipilimumab	Melanoma, OC	Increased HLA expression	Cardenas et al. (2021), Wang et al. (2021), Di Giacomo et al. (2019), Odunsi et al. (2014)
5-Azacytidine	DNMT inhibitor	Entinostat, Panobinostat,	Lung, AML	Improved progression-free survival	Juergens et al. (2011), Babu et al. (2021)
Panobinostat	Non-selective HDAC inhibitor	Bortezomib, dexamethasone, 5-Azacytidine	AML	Increased progression-free surviavl	Babu et al. (2021), Greig (2016)

Fig. 3 Summary table of the described epigenetic therapies used to sensitize tumors to chemotherapies and reverse resistance.

a covalent complex with DNMTs to prevent the enzyme from dissociating (Dan et al., 2019). HDACis work by binding to and inhibiting the catalytic domain of HDACs (Suraweera et al., 2018). FDA approved HDACis include vorinostat, romidepsin, belinostat, and Panobinostat (Emran et al., 2017).

5.1 DNMTi clinical trials

DNMTi agents have shown promise to re-sensitize tumors to chemotherapies in other cancer types than myelodysplastic syndrome. Cardenas et al. found that the DNMTi guadecitabine causes re-expression of 189 silenced genes in platinum resistant high-grade serous ovarian carcinoma (HGSOC) samples to drive drug re-sensitization (Cardenas et al., 2021). Decreased DNMT1 expression from guadecitabine treatment was associated with improved clinical outcomes. DNMTi treatment in OCs can re-sensitize patients to carboplatin. Giving low-dose decitabine before carboplatin in platinum-resistant ovarian cancer patients showed improved sensitivity to the drug via demethylation of the MLH1, RASSF1A, HOXA10, and HOXA11 genes (Matei et al., 2012). Demethylation of these genes induces apoptosis in cells with DNA damage, preventing the growth of cancer and reversing chemo-resistance. This DNMTi treatment resulted in a high response rate and prolonged progression-free survival in the ovarian cancer patients. Notably, a 35% progression-free survival (PFS) of 10.2 months was achieved in the decitabine plus carboplatin treatment patients, and over 50% of these patients have a PFS over 6 months (Matei et al., 2012).

Recent work has focused on non-nucleoside inhibitors which, unlike nucleoside DNMTis, are not incorporated into the DNA, but directly attach to the methylated region of DNMT. The DNMTi drug procainamide causes global DNA hypomethylation, reactivating the expression of the gene GSTP1 in prostate cancer cells, allowing it to defend prostate cells against oxidant and electrophile carcinogens (Lin et al., 2001). In breast cancer, restored expression of the RARβ2 gene is driven by demethylation of the densely hypermethylated CpGs located in its promoter region (Villar-Garea, Fraga, Espada, & Esteller, 2003). RARβ2 promotes apoptosis in breast cancer cells and has anti-activator protein-1 activity (Sirchia et al., 2002). Other studies focused on DNMTi combined with other treatments. Wang et al. found that the long noncoding RNA HOTAIR is over-expressed in HGSOC cell lines, increasing their spheroid formation and colony-forming ability (Wang, Fang, Ozes, & Nephew, 2021). This discovery led them to target HOTAIR through peptide nucleic acid-PNA3 which disrupts the interaction between HOTAIR and EZH2 and combined this with inhibition of DNA methylation through guadecitabine (Wang, Fang, et al., 2021). This combination reduced tumor stem cell frequency and tumor burden.

5.2 Combination DNMTi and HDACi clinical trials

DNMT inhibition is often used in combination with HDAC inhibition to broadly open the chromatin in cancers heavily pre-treated with chemotherapies (Cossío, Esteller, & Berdasco, 2020). Steele et al. found that treatment of cisplatin-resistant cell lines with a DNMTi (decitabine) and HDACi (belinostat) led to a marked increase in expression of epigenetically silenced MLH1 critical to the DNA mismatch repair system (Steele, Finn, Brown, & Plumb, 2009). The combination treatment showed higher sensitivity to cisplatin than cells treated with only decitabine (Steele et al., 2009). In another study, azacytidine and the HDACi entinostat were tested on recurrent lung cancer patients (Juergens et al., 2011). The epigenetically silenced genes CDKN2a, CDH13, APC, and RASSF1a were demethylated and associated with improved progression-free and overall survival (Juergens et al., 2011). Methylation of these genes led to a significantly worse prognosis in patients with stage I lung cancer and were associated with disease recurrence and death. Combination treatment of azacytidine and Panobinostat, a nonselective HDACi, showed a 35% clinical response in cytarabine-resistant AML tumors (Babu et al., 2021). The combination of Panobinostat with bortezomib, a DNMTi, and dexamethasone had a significantly prolonged progression-free survival compared in AML (Greig & Panobinostat, 2016). The combination of the HDACi vorinostat and decitabine treatment in refractory pediatric AML treated with fludarabine, cytarabine, and G-CSF (FLAG) showed an overall response rate of 54% in 35 patients, with a 2-year survival rate of 75.6% (Pommert et al., 2022; Richon, 2006).

5.3 Combination epigenetic and immunotherapy clinical trials

The use of adjuvant epigenetic therapy in late-stage, refractory, or unresectable cancers has shown some clinical responses and increased anti-tumor immune responses (Cruz, Bansal, Tomlinson, Bansal, & Weitman, 2016; Di Giacomo et al., 2019; Gray et al., 2019; Odunsi et al., 2014). A phase Ib dose-escalation trial of the DNMTi guadecitabine with ipilimumab (α-CTLA-4) in 19 patients with unresectable stage III/IV melanoma found the combination treatment upregulated HLA class I expression and increased

CD8+ T-cell and CD20+ B-cell infiltration post-treatment (Di Giacomo et al., 2019). Using the HDACi vorinostat in combination with pembrolizumab (α-PDL-1) had a well-tolerated antitumor activity and increased CD8+ T-cell presence in a Phase I/Ib trial of advanced/metastatic non-small cell lung cancer (Gray et al., 2019). NY-ESO-1 is a germline antigen that can become expressed during cancer progression and is normally silenced by DNA methylation. Its expression has been shown in a broad range of cancers including epithelial ovarian carcinomas, making it a target for cancer vaccine treatment. Guadecitabine treatment increases NY-ESO-1 expression and shows ovarian cancer disease stabilization or partial clinical response when given in combination with an NY-ESO-1 vaccine (Odunsi et al., 2014). Hydroxychloroquine is a toll-like receptor inhibitor that has shown to enhance HDACi treatment (Cruz et al., 2016). The use of combination vorinostat and hydroxychloroquine in refractory metastatic colorectal cancer improved antitumor immunity, and matched median overall survival when compared to standard of care treatment with the tyrosine kinase inhibitor regorafenib (Arora et al., 2022). Many other trials utilizing DNMTi, HDACi, or both in combination with immunotherapies are under way and the use of epigenetic inhibitors in combination with immunotherapies offers exciting potential for improved response in multiple solid tumors (Jones, Ohtani, Chakravarthy, & De Carvalho, 2019).

6. Conclusions

Epigenetic drivers of cancer progression and resistance to immuno- and chemotherapies provide novel therapeutic targets that can increase the efficacy of current drug regimens. The dysregulation of chromatin accessibility, enhancer-promoter interactions, and repressive or activating epigenetic enzymes contribute to many cancer types and offer exciting translational targets. These epigenetic modifications have shown promise as potential biomarkers in cancer diagnosis (Jung, Hernández-Illán, Moreira, Balaguer, & Goel, 2020; Koch et al., 2018; Singh, Gupta, & Sachan, 2019). While the idea of using CpG methylation of key gene regions for cancer diagnosis is an attractive area of research, there remain challenges to this approach. Specific patterns of methylation may be enough to control changes in gene expression; in this case, the level of methylation may stay the same, while the locations of the methyl marks may change. This makes understanding not only locus specific methylation important, but also the sequence of methylation present at a promoter site vital to proper diagnostics (Kristensen, Raynor, Candiloro, & Dobrovic, 2012).

Some clinical trials combining epigenetic inhibitors with chemotherapies have shown that epigenetic drugs can sensitize tumors to chemo- and immunotherapies. *Re*-expressing target proteins for cell receptor signaling and tumor antigen presentation through regulating DNA methylation and histone acetylation improve patient outcomes. This improvement can offer therapeutic benefits to those who currently have little to no treatment options available. Ongoing phase I/II clinical trials are evaluating how epigenetic priming for checkpoint blockade may improve patient outcomes for inoperable, locally advanced non-small cell lung cancer patients (National Cancer Institute (NCI), 2022). High risk advanced myelogenous leukemia (AML) patients or AML patients over the age of 60 who are unfit for aggressive chemotherapy regimens are also in an ongoing phase II single-arm study to assess the benefit from adjuvant DNA demethylating agents with tislelizumab (anti-PD1) (Liu, 2020).

Reversibly inhibiting epigenetic modifiers can return cells to normal epigenetic states that re-sensitize tumors to immuno-and chemotherapies. While the role of the 3D genome becomes better understood, agents that target these structures may prove to show application in reversing chemoresistance (Kantidze, Gurova, Studitsky, & Razin, 2020). Targeting the epigenome shows promising initial clinical results, and ongoing clinical trials combining epigenetic drugs with chemo- and immunotherapies have the potential to significantly improve outcomes in a range of malignancies.

Competing interests

KBC is a consultant for ROME Therapeutics.

Funding

Research was supported by the National Cancer Institute R00CA204592 and R37CA251270 (to KBC), the DOD Ovarian Cancer Research Program (W81XWH2010273 to KBC), and by the Marlene and Michael Berman Endowed Fund for Ovarian Cancer Research. RRW was supported by an NCI NRSA Institutional Research Training Program grant (T32 CA247756).

Acknowledgments

The authors would like to acknowledge the Institute for Biomedical Sciences at the George Washington University for graduate student support and training. Figures were created with BioRender.com. The funding sources above had no involvement or influence on the writing or preparation of this manuscript.

Author Contributions

RRW, ZR, and KBC performed the literature analysis and wrote the manuscript, and RRW and KBC edited the manuscript.

References

Abt, E. R., et al. (2022). Reprogramming of nucleotide metabolism by interferon confers dependence on the replication stress response pathway in pancreatic cancer cells. *Cell Reports*, *38*, 110236.

Achinger-Kawecka, J., et al. (2020). Epigenetic reprogramming at estrogen-receptor binding sites alters 3D chromatin landscape in endocrine-resistant breast cancer. *Nature Communications*, *11*, 320.

Albrengues, J., et al. (2015). Epigenetic switch drives the conversion of fibroblasts into proinvasive cancer-associated fibroblasts. *Nature Communications*, *6*, 10204.

Almand, B., Clark, J. I., Nikitina, E., van Beynen, J., English, N. R., Knight, S. C., et al. (2001). Increased production of immature myeloid cells in cancer patients: A mechanism of immunosuppression in cancer. *The Journal of Immunology*, *166*, 678–689. https://www.jimmunol.org/content/166/1/678.short.

Aptsiauri, N., & Garrido, F. (2022). The challenges of HLA class i loss in cancer immunotherapy: Facts and hopes. *Clinical Cancer Research*, *28*, 5021–5029. OF1–OF9 https://doi.org/10.1158/1078-0432.CCR-21-3501.

Arnold, C. N., Goel, A., & Boland, C. R. (2003). Role of hMLH1 promoter hypermethylation in drug resistance to 5-fluorouracil in colorectal cancer cell lines. *International Journal of Cancer*, *106*, 66–73.

Arora, S. P., et al. (2022). Modulation of autophagy: A phase II study of vorinostat plus hydroxychloroquine versus regorafenib in chemotherapy-refractory metastatic colorectal cancer (mCRC). *British Journal of Cancer*, *127*, 1153–1161.

Babu, G., et al. (2021). JAK-STAT inhibitor as a potential therapeutic opportunity in AML patients resistant to cytarabine and epigenetic therapy. *Cancer Biology & Therapy*, *22*, 66–78.

Bachman, M., Uribe-Lewis, S., Yang, X., et al. (2014). 5-hydroxymethylcytosine is a predominantly stable DNA modification. *Nature Chemistry*, *6*, 1049–1055. https://www.nature.com/articles/nchem.2064.

Bachman, M., et al. (2015). 5-Formylcytosine can be a stable DNA modification in mammals. *Nature Chemical Biology*, *11*, 555–557.

Baker, E. K., & El-Osta, A. (2003). The rise of DNA methylation and the importance of chromatin on multidrug resistance in cancer. *Experimental Cell Research*, *290*, 177–194.

Banine, F., et al. (2005). SWI/SNF chromatin-remodeling factors induce changes in DNA methylation to promote transcriptional activation. *Cancer Research*, *65*, 3542–3547.

Bao, J., et al. (2019). Integrated high-throughput analysis identifies super enhancers associated with chemoresistance in SCLC. *BMC Medical Genomics*, *12*, 67.

Barrett, R. L., & Puré, E. (2020). Cancer-associated fibroblasts and their influence on tumor immunity and immunotherapy. *eLife*, *9*, e57243.

Baylin, S. B., & Jones, P. A. (2011). A decade of exploring the cancer epigenome—Biological and translational implications. *Nature Reviews. Cancer*, *11*, 726–734.

Beagan, J. A., & Phillips-Cremins, J. E. (2020). On the existence and functionality of topologically associating domains. *Nature Genetics*, *52*, 8–16.

Benci, J. L., et al. (2019). Opposing functions of interferon coordinate adaptive and innate immune responses to cancer immune checkpoint blockade. *Cell*, *178*, 933–948.e14.

Bian, Y., et al. (2020). Cancer SLC43A2 alters T cell methionine metabolism and histone methylation. *Nature*, *585*, 277–282.

Bowman, G. D., & Poirier, M. G. (2015). Post-translational modifications of histones that influence nucleosome dynamics. *Chemical Reviews*, *115*, 2274–2295.

Briand, N., & Collas, P. (2020). Lamina-associated domains: Peripheral matters and internal affairs. *Genome Biology*, *21*, 85.

Brocks, D., Schmidt, C., Daskalakis, M., et al. (2017). DNMT and HDAC inhibitors induce cryptic transcription start sites encoded in long terminal repeats. *Nature Genetics*, *49*, 1052–1060. https://www.nature.com/articles/ng.3889.

Brown, R., Curry, E., Magnani, L., Wilhelm-Benartzi, C. S., & Borley, J. (2014). Poised epigenetic states and acquired drug resistance in cancer. *Nature Reviews. Cancer*, *14*, 747–753.

Bruno, P. M., et al. (2020). The primary mechanism of cytotoxicity of the chemotherapeutic agent CX-5461 is topoisomerase II poisoning. *Proceedings of the National Academy of Sciences of the United States of America*, *117*, 4053–4060.

Cao, J., & Yan, Q. (2020). Cancer epigenetics, tumor immunity, and immunotherapy. *Trends Cancer*, *6*, 580–592.

Cardenas, H., et al. (2021). Methylomic signatures of high grade serous ovarian cancer. *Epigenetics*, *16*, 1201–1216.

Carrozza, M. J., Utley, R. T., Workman, J. L., & Côté, J. (2003). The diverse functions of histone acetyltransferase complexes. *Trends in Genetics*, *19*, 321–329.

Cecotka, A., & Polanska, J. (2018). Region-specific methylation profiling in acute myeloid leukemia. *Interdisciplinary Sciences, Computational Life Sciences*, *10*, 33–42.

Chatterjee, A., Rodger, E. J., & Eccles, M. R. (2018). Epigenetic drivers of tumourigenesis and cancer metastasis. *Seminars in Cancer Biology*, *51*, 149–159.

Chekhun, V. F., Lukyanova, N. Y., Kovalchuk, O., Tryndyak, V. P., & Pogribny, I. P. (2007). Epigenetic profiling of multidrug-resistant human MCF-7 breast adenocarcinoma cells reveals novel hyper- and hypomethylated targets. *Molecular Cancer Therapeutics*, *6*, 1089–1098.

Chiappinelli, K. B., et al. (2015). Inhibiting DNA methylation causes an interferon response in Cancer via dsRNA including endogenous retroviruses. *Cell*, *162*, 974–986.

Cossío, F. P., Esteller, M., & Berdasco, M. (2020). Towards a more precise therapy in cancer: Exploring epigenetic complexity. *Current Opinion in Chemical Biology*, *57*, 41–49.

Crea, F., et al. (2011). Epigenetics and chemoresistance in colorectal cancer: An opportunity for treatment tailoring and novel therapeutic strategies. *Drug Resistance Updates*, *14*, 280–296.

Cruz, A. J., Bansal, H., Tomlinson, G., Bansal, S., & Weitman, S. (2016). Abstract B36: Enhancement of histone deacetylase inhibitor's efficacy with hydroxychloroquine via autophagy inhibition in pediatric tumor cell lines. *Cancer Research*, *76*, B36.

Curradi, M., Izzo, A., Badaracco, G., & Landsberger, N. (2002). Molecular mechanisms of gene silencing mediated by DNA methylation. *Molecular and Cellular Biology*, *22*, 3157–3173.

Dan, H., Zhang, S., Zhou, Y., & Guan, Q. (2019). DNA methyltransferase inhibitors: Catalysts for antitumour immune responses. *Oncotargets and Therapy*, *12*, 10903–10916.

De, S., Cheng, Y., Sun, M.-a., Gehred, N. D., & Kassis, J. A. (2019). Structure and function of an ectopic Polycomb chromatin domain. *Science Advances*. https://www.science.org/doi/full/10.1126/sciadv.aau9739.

de Vries, J. E. (1995). Immunosuppressive and anti-inflammatory properties of interleukin 10. *Annals of Medicine*, *27*, 537–541.

Deng, S., Feng, Y., & Pauklin, S. (2022). 3D chromatin architecture and transcription regulation in cancer. *Journal of Hematology & Oncology*, *15*, 49.

Dhayat, S., Mardin, W. A., Mees, S. T., & Haier, J. (2011). Epigenetic markers for chemosensitivity and chemoresistance in pancreatic cancer—A review. *International Journal of Cancer*, *129*, 1031–1041. Wiley Online Library https://onlinelibrary.wiley.com/doi/full/10.1002/ijc.26078.

Di Giacomo, A. M., et al. (2019). Guadecitabine plus Ipilimumab in Unresectable melanoma: The NIBIT-M4 clinical trial. *Clinical Cancer Research*, *25*, 7351–7362.
Dixon, J. R., et al. (2012). Topological domains in mammalian genomes identified by analysis of chromatin interactions. *Nature*, *485*, 376–380.
Djureinovic, D., Wang, M., & Kluger, H. M. (2021). Agonistic CD40 antibodies in cancer treatment. *Cancers*, *13*, 1302.
Emran, A. A., et al. (2017). Distinct histone modifications denote early stress-induced drug tolerance in cancer. *Oncotarget*, *9*, 8206–8222.
Esfahani, K., et al. (2020). A review of Cancer immunotherapy: From the past, to the present, to the future. *Current Oncology*, *27*, 87–97.
Fan, P., et al. (2019). Overexpressed histone acetyltransferase 1 regulates cancer immunity by increasing programmed death-ligand 1 expression in pancreatic cancer. *Journal of Experimental & Clinical Cancer Research*, *38*, 47.
Fedorova, M. S., et al. (2019). The CIMP-high phenotype is associated with energy metabolism alterations in colon adenocarcinoma. *BMC Medical Genetics*, *20*, 52.
Feng, Y., & Pauklin, S. (2020). Revisiting 3D chromatin architecture in cancer development and progression. *Nucleic Acids Research*, *48*, 10632–10647.
Foster, H. A., & Bridger, J. M. (2005). The genome and the nucleus: A marriage made by evolution. *Chromosoma*, *114*, 212–229.
Freeman, G. J., et al. (2000). Engagement of the Pd-1 Immunoinhibitory receptor by a novel B7 family member leads to negative regulation of lymphocyte activation. *The Journal of Experimental Medicine*, *192*, 1027–1034.
Gabriele, M., Brandão, H. B., Grosse-Holz, S., Jha, A., Dailey, G. M., Cattoglio, C., et al. (2022). Dynamics of CTCF- and cohesin-mediated chromatin looping revealed by live-cell imaging. *Science*, *376*, 496–501. https://www.science.org/doi/10.1126/science.abn6583.
Gaillard, H., García-Muse, T., & Aguilera, A. (2015). Replication stress and cancer. *Nature Reviews Cancer*, *15*, 276–289. https://www.nature.com/articles/nrc3916.
Ganesh, K., et al. (2019). Immunotherapy in colorectal cancer: Rationale, challenges and potential. *Nature Reviews. Gastroenterology & Hepatology*, *16*, 361–375.
Gao, H.-L., et al. (2018). The clinicopathological and prognostic significance of PD-L1 expression in pancreatic cancer: A meta-analysis. *Hepatobiliary & Pancreatic Diseases International*, *17*, 95–100.
Ghoneim, M., Fuchs, H. A., & Musselman, C. A. (2021). Histone tail conformations: A fuzzy affair with DNA. *Trends in Biochemical Sciences*, *46*, 564–578.
Gomez, S., Tabernacki, T., Kobyra, J., Roberts, P., & Chiappinelli, K. B. (2020). Combining epigenetic and immune therapy to overcome cancer resistance. *Seminars in Cancer Biology*, *65*, 99–113.
Gonzalez, H., Hagerling, C., & Werb, Z. (2018). Roles of the immune system in cancer: From tumor initiation to metastatic progression. *Genes & Development*, *32*, 1267–1284.
Gonzalez-Molina, J., et al. (2022). Chemotherapy as a regulator of extracellular matrix-cell communication: Implications in therapy resistance. *Seminars in Cancer Biology*, *86*, 224–236.
Gräff, J., & Tsai, L.-H. (2013). Histone acetylation: Molecular mnemonics on the chromatin. *Nature Reviews. Neuroscience*, *14*, 97–111.
Gray, J. E., et al. (2019). Phase I/Ib study of Pembrolizumab plus Vorinostat in advanced/metastatic non–small cell lung Cancer. *Clinical Cancer Research*, *25*, 6623–6632.
Greenwald, R. J., Freeman, G. J., & Sharpe, A. H. (2005). The B7 family revisited. *Annual Review of Immunology*, *23*, 515–548. ProQuest https://www.proquest.com/openview/6897e371f81d878aaa9090028ffced78/1?pq-origsite=gscholar&cbl=47332.
Greig, S., & Panobinostat, L. (2016). A review in relapsed or refractory multiple myeloma. *Targeted Oncology*, *11*, 107–114.

Grundy, E. E., Diab, N., & Chiappinelli, K. B. (2022). Transposable element regulation and expression in cancer. *The FEBS Journal*, *289*, 1160–1179.
Guelen, L., et al. (2008). Domain organization of human chromosomes revealed by mapping of nuclear lamina interactions. *Nature*, *453*, 948–951.
Gull, N., et al. (2022). DNA methylation and transcriptomic features are preserved throughout disease recurrence and chemoresistance in high grade serous ovarian cancers. *Journal of Experimental & Clinical Cancer Research*, *41*, 232.
Haberland, M., Montgomery, R. L., & Olson, E. N. (2009). The many roles of histone deacetylases in development and physiology: Implications for disease and therapy. *Nature Reviews. Genetics*, *10*, 32–42.
Hamanishi, J., et al. (2015a). Durable tumor remission in patients with platinum-resistant ovarian cancer receiving nivolumab. *Journal of Clinical Oncology*, *33*, 5570.
Hamanishi, J., et al. (2015b). Safety and antitumor activity of anti-PD-1 antibody, Nivolumab, in patients with platinum-resistant ovarian Cancer. *Journal of Clinical Oncology: Official Journal of the American Society of Clinical Oncology*, *33*, 4015–4022.
Hanahan, D. (2022). Hallmarks of Cancer: New dimensions. *Cancer Discovery*, *12*, 31–46.
Hill, P. W. S., Amouroux, R., & Hajkova, P. (2014). DNA demethylation, Tet proteins and 5-hydroxymethylcytosine in epigenetic reprogramming: An emerging complex story. *Genomics*, *104*, 324–333.
Hou, J., et al. (2012). Genomic amplification and a role in drug-resistance for the KDM5A histone demethylase in breast cancer. *American Journal of Translational Research*, *4*, 247–256.
Huang, K. C.-Y., et al. (2020). Decitabine augments chemotherapy-induced PD-L1 upregulation for PD-L1 blockade in colorectal Cancer. *Cancers*, *12*, 462.
Husmann, D., & Gozani, O. (2019). Histone lysine methyltransferases in biology and disease. *Nature Structural & Molecular Biology*, *26*, 880–889.
Iacovelli, R., et al. (2016). Prognostic role of PD-L1 expression in renal cell carcinoma. A systematic review and meta-analysis. *Targeted Oncology*, *11*, 143–148.
Jaillon, S., et al. (2020). Neutrophil diversity and plasticity in tumour progression and therapy. *Nature Reviews. Cancer*, *20*, 485–503.
Johnson, S. D., De Costa, A.-M. A., & Young, M. R. I. (2014). Effect of the premalignant and tumor microenvironment on immune cell cytokine production in head and neck Cancer. *Cancers*, *6*, 756–770.
Johnstone, T. C., Park, G. Y., & Lippard, S. J. (2014). Understanding and improving platinum anticancer drugs—Phenanthriplatin. *Anticancer Research*, *34*, 471–476.
Jones, P. A., Ohtani, H., Chakravarthy, A., & De Carvalho, D. D. (2019). Epigenetic therapy in immune-oncology. *Nature Reviews. Cancer*, *19*, 151–161.
Juergens, R. A., et al. (2011). Combination epigenetic therapy has efficacy in patients with refractory advanced non-small cell lung cancer. *Cancer Discovery*, *1*, 598–607.
Jung, G., Hernández-Illán, E., Moreira, L., Balaguer, F., & Goel, A. (2020). Epigenetics of colorectal cancer: Biomarker and therapeutic potential. *Nature Reviews. Gastroenterology & Hepatology*, *17*, 111–130.
Jung, H., et al. (2019). DNA methylation loss promotes immune evasion of tumours with high mutation and copy number load. *Nature Communications*, *10*, 4278.
Kantidze, O. L., Gurova, K. V., Studitsky, V. M., & Razin, S. V. (2020). The 3D genome as a target for anticancer therapy. *Trends in Molecular Medicine*, *26*, 141–149.
Karakaidos, P., Verigos, J., & Magklara, A. (2019). LSD1/KDM1A, a gate-keeper of Cancer Stemness and a promising therapeutic target. *Cancers*, *11*, 1821.
Kimura, H. (2013). Histone modifications for human epigenome analysis. *Journal of Human Genetics*, *58*, 439–445.
Klughammer, J., et al. (2018). The DNA methylation landscape of glioblastoma disease progression shows extensive heterogeneity in time and space. *Nature Medicine*, *24*, 1611–1624.

Koch, A., et al. (2018). Analysis of DNA methylation in cancer: Location revisited. *Nature Reviews. Clinical Oncology, 15*, 459–466.

Kohli, K., Pillarisetty, V. G., & Kim, T. S. (2022). Key chemokines direct migration of immune cells in solid tumors. *Cancer Gene Therapy, 29*, 10–21. https://www.nature.com/articles/s41417-021-00303-x.

Kooistra, S. M., & Helin, K. (2012). Molecular mechanisms and potential functions of histone demethylases. *Nature Reviews. Molecular Cell Biology, 13*, 297–311.

Korman, A. J., Garrett-Thomson, S. C., & Lonberg, N. (2022). The foundations of immune checkpoint blockade and the ipilimumab approval decennial. *Nature Reviews. Drug Discovery, 21*, 509–528.

Kosak, S. T., et al. (2007). Coordinate gene regulation during hematopoiesis is related to genomic organization. *PLoS Biology, 5*, e309.

Kristensen, L. S., Raynor, M., Candiloro, I., & Dobrovic, A. (2012). Methylation profiling of normal individuals reveals mosaic promoter methylation of cancer-associated genes. *Oncotarget, 3*, 450–461.

Ladányi, A., et al. (2022). HLA class I downregulation in progressing metastases of melanoma patients treated with Ipilimumab. *Pathology Oncology Research, 28*, 1610297.

Li, Y.-J., et al. (2017). Autophagy and multidrug resistance in cancer. *Chinese Journal of Cancer, 36*, 52.

Li, Y., et al. (2018). MLH1 enhances the sensitivity of human endometrial carcinoma cells to cisplatin by activating the MLH1/c-Abl apoptosis signaling pathway. *BMC Cancer, 18*, 1294.

Li, G.-H., et al. (2021). Super-enhancers: A new frontier for epigenetic modifiers in cancer chemoresistance. *Journal of Experimental & Clinical Cancer Research, 40*, 174.

Lin, X., et al. (2001). Reversal of GSTP1 CpG island hypermethylation and reactivation of pi-class glutathione S-transferase (GSTP1) expression in human prostate cancer cells by treatment with procainamide. *Cancer Research, 61*, 8611–8616.

Liu, D. (2020). *A Phase II, Single Arm Study of Tislelizumab Combined With DNA Demethylation Agent +/− CAG Regimen in the Treatment of Patients With High-risk AML or AML Patients Older Than 60 Years of Age Who Are Unfit for Intensive Chemotherapy*. https://clinicaltrials.gov/ct2/show/NCT04541277.

Liu, M., et al. (2016). Vitamin C increases viral mimicry induced by 5-aza-2′-deoxycytidine. *Proceedings of the National Academy of Sciences of the United States of America, 113*, 10238–10244.

Liu, M., et al. (2018). Dual inhibition of DNA and histone methyltransferases increases viral mimicry in ovarian Cancer cells. *Cancer Research, 78*, 5754–5766.

Liu, J., et al. (2019). Global DNA 5-Hydroxymethylcytosine and 5-Formylcytosine contents are decreased in the early stage of hepatocellular carcinoma. *Hepatology, 69*, 196–208.

Liu, A., et al. (2020). A novel strategy for the diagnosis, prognosis, treatment, and chemoresistance of hepatocellular carcinoma: DNA methylation. *Medicinal Research Reviews, 40*, 1973–2018.

Lodovichi, S., Cervelli, T., Pellicioli, A., & Galli, A. (2020). Inhibition of DNA repair in Cancer therapy: Toward a multi-target approach. *International Journal of Molecular Sciences, 21*, 6684.

Lou, Y., et al. (2005). Restoration of the expression of transporters associated with antigen processing in lung carcinoma increases tumor-specific immune responses and survival. *Cancer Research, 65*, 7926–7933.

Luskin, M. R., et al. (2016). A clinical measure of DNA methylation predicts outcome in de novo acute myeloid leukemia. *JCI Insight, 1*, e87323.

Lyko, F. (2018). The DNA methyltransferase family: A versatile toolkit for epigenetic regulation. *Nature Reviews Genetics, 19*, 81–92. https://www.nature.com/articles/nrg.2017.80.

Martín, M. (2001). Platinum compounds in the treatment of advanced breast Cancer. *Clinical Breast Cancer, 2*, 190–208.
Martinez-Cardús, A., Vizoso, M., Moran, S., & Manzano, J. L. (2015). Epigenetic mechanisms involved in melanoma pathogenesis and chemoresistance. *Annals of Translational Medicine, 3*, 209.
Matei, D., et al. (2012). Epigenetic Resensitization to platinum in ovarian Cancer. *Cancer Research, 72*, 2197–2205.
Matkar, S., et al. (2015). An epigenetic pathway regulates sensitivity of breast Cancer cells to HER2 inhibition via FOXO/c-Myc Axis. *Cancer Cell, 28*, 472–485.
McGinty, R. K., & Tan, S. (2015). Nucleosome structure and function. *Chemical Reviews, 115*, 2255–2273. https://pubs.acs.org/doi/full/10.1021/cr500373h.
Mehta, A., Dobersch, S., Romero-Olmedo, A. J., & Barreto, G. (2015). Epigenetics in lung cancer diagnosis and therapy. *Cancer Metastasis Reviews, 34*, 229–241.
Melief, C. J. M., van Hall, T., Arens, R., Ossendorp, F., & van der Burg, S. H. (2015). Therapeutic cancer vaccines. *The Journal of Clinical Investigation, 125*, 3401–3412.
Mittal, P., & Roberts, C. W. M. (2020). The SWI/SNF complex in cancer—Biology, biomarkers and therapy. *Nature Reviews. Clinical Oncology, 17*, 435–448.
Murakami, T., et al. (2008). Transcriptional modulation using HDACi Depsipeptide promotes immune cell-mediated tumor destruction of murine B16 melanoma. *The Journal of Investigative Dermatology, 128*, 1506–1516.
Nakayama, M., Wada, M., Harada, T., Nagayama, J., Kusaba, H., Ohshima, K., et al. (1998). Hypomethylation status of CpG sites at the promoter region and overexpression of the human MDR1 gene in acute myeloid leukemias. *Blood, 92*, 4296–4307. American Society of Hematology https://ashpublications.org/blood/article/92/11/4296/260246/Hypomethylation-Status-of-CpG-Sites-at-the.
National Cancer Institute (NCI). (2022). *Phase I/II Evaluation of Oral Decitabine/Tetrahydrouridine as Epigenetic Priming for Pembrolizumab Immune Checkpoint Blockade in Inoperable Locally Advanced or Metastatic Non-Small Cell Lung Cancers, Esophageal Carcinomas, or Pleural Mesotheliomas*. https://clinicaltrials.gov/ct2/show/NCT03233724.
Neganova, M. E., Klochkov, S. G., Aleksandrova, Y. R., & Aliev, G. (2022). Histone modifications in epigenetic regulation of cancer: Perspectives and achieved progress. *Seminars in Cancer Biology, 83*, 452–471.
Nichols, M. H., & Corces, V. G. (2021). Principles of 3D compartmentalization of the human genome. *Cell Reports, 35*, 109330.
Nitiss, J., & Targeting, L. (2009). DNA topoisomerase II in cancer chemotherapy. *Nature Reviews. Cancer, 9*, 338–350.
Odunsi, K., et al. (2014). Epigenetic potentiation of NY-ESO-1 vaccine therapy in human ovarian Cancer. *Cancer Immunology Research, 2*, 37–49.
Ohlsson, R., Renkawitz, R., & Lobanenkov, V. (2001). CTCF is a uniquely versatile transcription regulator linked to epigenetics and disease. *Trends in Genetics, 17*, 520–527.
Papillon, J. P. N., Nakajima, K., Adair, C. D., et al. (2018). Discovery of orally active inhibitors of brahma homolog (BRM)/SMARCA2 ATPase activity for the treatment of brahma related gene 1 (BRG1)/SMARCA4-mutant cancers. *Journal of Medicinal Chemistry, 61*, 10155–10172. https://pubs.acs.org/doi/full/10.1021/acs.jmedchem.8b01318.
Pasello, G., et al. (2020). Real world data in the era of immune checkpoint inhibitors (ICIs): Increasing evidence and future applications in lung cancer. *Cancer Treatment Reviews, 87*, 102031.
Patel, S. P., & Kurzrock, R. (2015). PD-L1 expression as a predictive biomarker in cancer immunotherapy. *Molecular Cancer Therapeutics, 14*, 847–856.
Pavlidou, E. N., & Balis, V. (2020). Diagnostic significance and prognostic role of the ARID1A gene in cancer outcomes (review). *World Academy of Sciences Journal, 2*, 49–64.

Pires, A., Burnell, S., & Gallimore, A. (2022). Exploiting ECM remodelling to promote immune-mediated tumour destruction. *Current Opinion in Immunology, 74*, 32–38.

Plch, J., Hrabeta, J., & Eckschlager, T. (2019). KDM5 demethylases and their role in cancer cell chemoresistance. *International Journal of Cancer, 144*, 221–231.

Pommert, L., et al. (2022). Decitabine and vorinostat with FLAG chemotherapy in pediatric relapsed/refractory AML: Report from the therapeutic advances in childhood leukemia and lymphoma (TACL) consortium. *American Journal of Hematology, 97*, 613–622.

Ponnusamy, L., Mahalingaiah, P. K. S., Chang, Y.-W., & Singh, K. P. (2019). Role of cellular reprogramming and epigenetic dysregulation in acquired chemoresistance in breast cancer. *Cancer Drug Resistance, 2*, 297–312.

Ponnusamy, L., Mahalingaiah, P. K. S., & Singh, K. P. (2020). Epigenetic reprogramming and potential application of epigenetic-modifying drugs in acquired chemotherapeutic resistance. *Advances in Clinical Chemistry, 94*, 219–259.

Ramirez, M., et al. (2016). Diverse drug-resistance mechanisms can emerge from drug-tolerant cancer persister cells. *Nature Communications*, 7, 10690.

Rao, S. S., Huntley, M. H., Durand, N. C., Stamenova, E. K., Bochkov, I. D., Robinson, J. T., et al. (2014). A 3D map of the human genome at kilobase resolution reveals principles of chromatin looping. *Cell, 159*, 1665–1680. https://www.sciencedirect.com/science/article/pii/S0092867414014974.

Rasmussen, K. D., & Helin, K. (2016). Role of TET enzymes in DNA methylation, development, and cancer. *Genes & Development, 30*, 733–750.

Ray-Jones, H., & Spivakov, M. (2021). Transcriptional enhancers and their communication with gene promoters. *Cellular and Molecular Life Sciences, 78*. SpringerLink https://link.springer.com/article/10.1007/s00018-021-03903-w.

Richon, V. M. (2006). Cancer biology: Mechanism of antitumour action of vorinostat (suberoylanilide hydroxamic acid), a novel histone deacetylase inhibitor. *British Journal of Cancer, 95*, S2–S6.

Rossetto, D., Avvakumov, N., & Côté, J. (2012). Histone phosphorylation. *Epigenetics*, 7, 1098–1108.

Roulois, D., et al. (2015). DNA-demethylating agents target colorectal Cancer cells by inducing viral mimicry by endogenous transcripts. *Cell, 162*, 961–973.

Sabatier, R., et al. (2011). A seven-gene prognostic model for platinum-treated ovarian carcinomas. *British Journal of Cancer, 105*, 304–311.

Salomé, B., et al. (2022). NKG2A and HLA-E define an alternative immune checkpoint axis in bladder cancer. *Cancer Cell, 40*, 1027–1043.e9.

Santos-Rosa, H., Schneider, R., Bannister, A., et al. (2002). Active genes are tri-methylated at K4 of histone H3. *Nature, 419*, 407–411. https://www.nature.com/articles/nature01080.

Schvartzman, J. M., Thompson, C. B., & Finley, L. W. S. (2018). Metabolic regulation of chromatin modifications and gene expression. *The Journal of Cell Biology, 217*, 2247–2259.

Seliger, B., et al. (2001). Immune escape of melanoma: First evidence of structural alterations in two distinct components of the MHC class I antigen processing Pathway1. *Cancer Research, 61*, 8647–8650.

Sérandour, A. A., et al. (2012). Dynamic hydroxymethylation of deoxyribonucleic acid marks differentiation-associated enhancers. *Nucleic Acids Research, 40*, 8255–8265.

Shang, S., et al. (2019). Chemotherapy-induced distal enhancers drive transcriptional programs to maintain the chemoresistant state in ovarian cancer. *Cancer Research, 79*, 4599–4611.

Sharma, P., & Allison, J. P. (2015). The future of immune checkpoint therapy. *Science, 348*, 56–61.

Sharma, S. V., et al. (2010). A chromatin-mediated reversible drug-tolerant state in Cancer cell subpopulations. *Cell, 141*, 69–80.

Shen, S., Vagner, S., & Robert, C. (2020). Persistent cancer cells: The deadly survivors. *Cell, 183*, 860–874.

Sheng, W., et al. (2018). LSD1 ablation stimulates anti-tumor immunity and enables checkpoint blockade. *Cell, 174*, 549–563.e19.

Shogren-Knaak, M., et al. (2006). Histone H4-K16 acetylation controls chromatin structure and protein interactions. *Science, 311*, 844–847.

Shorstova, T., et al. (2019). SWI/SNF-compromised cancers are susceptible to Bromodomain inhibitors. *Cancer Research, 79*, 2761–2774.

Singh, A., Gupta, S., & Sachan, M. (2019). Epigenetic biomarkers in the management of ovarian cancer: Current prospectives. *Frontiers in Cell and Development Biology*, 7, 182.

Sirchia, S. M., et al. (2002). Endogenous reactivation of the RARbeta2 tumor suppressor gene epigenetically silenced in breast cancer. *Cancer Research, 62*, 2455–2461.

Škubník, J., Jurášek, M., Ruml, T., & Rimpelová, S. (2020). Mitotic poisons in research and medicine. *Molecules, 25*, 4632. https://www.mdpi.com/1420-3049/25/20/4632.

Son, K. S., et al. (2010). Hypomethylation of the interleukin-10 gene in breast cancer tissues. *The Breast, 19*, 484–488.

Spada, F., et al. (2007). DNMT1 but not its interaction with the replication machinery is required for maintenance of DNA methylation in human cells. *The Journal of Cell Biology, 176*, 565–571.

Srinivas, U. S., Tan, B. W. Q., Vellayappan, B. A., & Jeyasekharan, A. D. (2019). ROS and the DNA damage response in cancer. *Redox Biology, 25*, 101084.

Steele, N., Finn, P., Brown, R., & Plumb, J. A. (2009). Combined inhibition of DNA methylation and histone acetylation enhances gene re-expression and drug sensitivity in vivo. *British Journal of Cancer, 100*, 758–763.

Stillman, B. (2018). Histone modifications: Insights into their influence on gene expression. *Cell, 175*, 6–9.

Stovgaard, E. S., Dyhl-Polk, A., Roslind, A., Balslev, E., & Nielsen, D. (2019). PD-L1 expression in breast cancer: Expression in subtypes and prognostic significance: A systematic review. *Breast Cancer Research and Treatment, 174*, 571–584.

Suraweera, A., O'Byrne, K. J., & Richard, D. J. (2018). Combination therapy with histone deacetylase inhibitors (HDACi) for the treatment of cancer: Achieving the full therapeutic potential of HDACi. *Frontiers in Oncology*. https://www.frontiersin.org/articles/10.3389/fonc.2018.00092/full.

Tahara, M., et al. (2020). Re-challenge of platinum-based chemotherapy for platinum-refractory patients with recurrent or metastatic head and neck Cancer: Claims data analysis in Japan. *Journal of Health Economics and Outcomes Research*, 7, 43–51.

Tahiliani, M., et al. (2009). Conversion of 5-Methylcytosine to 5-Hydroxymethylcytosine in mammalian DNA by MLL partner TET1. *Science, 324*, 930–935.

Tang, Y., et al. (2017). Linking long non-coding RNAs and SWI/SNF complexes to chromatin remodeling in cancer. *Molecular Cancer, 16*, 42.

Tao, L., Huang, G., Chen, Y., & Chen, L. (2015). DNA methylation of DKK3 modulates docetaxel chemoresistance in human nonsmall cell lung cancer cell. *Cancer Biotherapy and Radiopharmaceuticals, 30*, 100–106. https://www.liebertpub.com/doi/abs/10.1089/cbr.2014.1797.

Topper, M. J., et al. (2017). Epigenetic therapy ties MYC depletion to reversing immune evasion and treating lung cancer. *Cell, 171*, 1284–1300.e21.

Tost, J. (2016). Current and emerging technologies for the analysis of the genome-wide and locus-specific DNA methylation patterns. In A. Jeltsch, & R. Z. Jurkowska (Eds.), *DNA Methyltransferases—Role and Function* (pp. 343–430). Springer International Publishing. https://doi.org/10.1007/978-3-319-43624-1_15.

Townsend, S. E., & Allison, J. P. (1993). Tumor rejection after direct Costimulation of CD8+ T cells by B7-transfected melanoma cells. *Science, 259*, 368–370.

Toyota, M., et al. (1999). CpG island methylator phenotype in colorectal cancer. *Proceedings of the National Academy of Sciences of the United States of America*, *96*, 8681–8686.
Travis, L. B., et al. (1999). Risk of leukemia after platinum-based chemotherapy for ovarian Cancer. *The New England Journal of Medicine*, *340*, 351–357.
Ugurel, S., et al. (2019). MHC class-I downregulation in PD-1/PD-L1 inhibitor refractory Merkel cell carcinoma and its potential reversal by histone deacetylase inhibition: A case series. *Cancer Immunology, Immunotherapy*, *68*, 983–990.
Vaclavikova, R., et al. (2019). Development of high-resolution melting analysis for ABCB1 promoter methylation: Clinical consequences in breast and ovarian carcinoma. *Oncology Reports*, *42*, 763–774.
Varley, K. E., et al. (2013). Dynamic DNA methylation across diverse human cell lines and tissues. *Genome Research*, *23*, 555–567.
Varone, A., Spano, D., & Corda, D. (2020). Shp1 in solid cancers and their therapy. *Frontiers in Oncology*, *10*.
Vasan, N., Baselga, J., & Hyman, D. M. (2019). A view on drug resistance in cancer. *Nature*, *575*, 299–309.
Villar-Garea, A., Fraga, M. F., Espada, J., & Esteller, M. (2003). Procaine is a DNA-demethylating agent with growth-inhibitory effects in human cancer cells. *Cancer Research*, *63*, 4984–4989.
Wagner, E. J., & Carpenter, P. B. (2012). Understanding the language of Lys36 methylation at histone H3. *Nature Reviews. Molecular Cell Biology*, *13*, 115–126.
Walunas, T. L., et al. (1994). CTLA-4 can function as a negative regulator of T cell activation. *Immunity*, *1*, 405–413.
Wang, P., Chen, Y., & Wang, C. (2021). Beyond tumor mutation burden: Tumor neoantigen burden as a biomarker for immunotherapy and other types of therapy. *Frontiers in Oncology*, *11*, 672677.
Wang, W., Fang, F., Ozes, A., & Nephew, K. P. (2021). Targeting ovarian Cancer stem cells by dual inhibition of HOTAIR and DNA methylation. *Molecular Cancer Therapeutics*, *20*, 1092–1101.
Wang, X., Teng, F., Kong, L., & Yu, J. (2016). PD-L1 expression in human cancers and its association with clinical outcomes. *OncoTargets Therapy*, *9*, 5023–5039.
Wang, X., Waschke, B. C., Woolaver, R. A., et al. (2020). HDAC inhibitors overcome immunotherapy resistance in B-cell lymphoma. *Protein & Cell*, *11*, 472–482. SpringerLink https://link.springer.com/article/10.1007/s13238-020-00694-x.
Wang, X., et al. (2009). Oncogenesis caused by loss of the SNF5 tumor suppressor is dependent on activity of BRG1, the ATPase of the SWI/SNF chromatin remodeling complex. *Cancer Research*, *69*, 8094–8101.
Webber, L. P., et al. (2017). Hypoacetylation of acetyl-histone H3 (H3K9ac) as marker of poor prognosis in oral cancer. *Histopathology*, *71*, 278–286.
Weichselbaum, R. R., et al. (2008). An interferon-related gene signature for DNA damage resistance is a predictive marker for chemotherapy and radiation for breast cancer. *Proceedings of the National Academy of Sciences of the United States of America*, *105*, 18490–18495.
Weller, M., et al. (2010). MGMT promoter methylation in malignant gliomas: Ready for personalized medicine? *Nature Reviews. Neurology*, *6*, 39–51.
Wiegand, K. C., et al. (2010). ARID1A mutations in endometriosis-associated ovarian carcinomas. *The New England Journal of Medicine*, *363*, 1532–1543.
Wu, Q., Yang, Z., Nie, Y., Shi, Y., & Fan, D. (2014). Multi-drug resistance in cancer chemotherapeutics: Mechanisms and lab approaches. *Cancer Letters*, *347*, 159–166.
Wu, H., & Zhang, Y. (2015). Charting oxidized methylcytosines at base resolution. *Nature Structural & Molecular Biology*, *22*, 656–661.

Yang, X., et al. (2014). Gene body methylation can Alter gene expression and is a therapeutic target in Cancer. *Cancer Cell*, *26*, 577–590.
Ye, Q., et al. (2010). Hypermethylation of HLA class I gene is associated with HLA class I down-regulation in human gastric cancer. *Tissue Antigens*, *75*, 30–39.
Yeldag, G., Rice, A., & del Río Hernández, A. (2018). Chemoresistance and the self-maintaining tumor microenvironment. *Cancers*, *10*, 471.
Yu, M., et al. (2012). Base-resolution analysis of 5-Hydroxymethylcytosine in the mammalian genome. *Cell*, *149*, 1368–1380.
Zeller, C., et al. (2012). Candidate DNA methylation drivers of acquired cisplatin resistance in ovarian cancer identified by methylome and expression profiling. *Oncogene*, *31*, 4567–4576.
Zha, J., Luo, H., Qiu, Y., & Huang, S. (2017). Disregulation of CTCF boundary at the HOX locus enhances sensitity to multiple drug treatment in myeloid malignancies. *Blood*, *130*, 3816.
Zhang, S., Gong, Y., Li, C., Yang, W., & Li, L. (2021). Beyond regulations at DNA levels: A review of epigenetic therapeutics targeting cancer stem cells. *Cell Proliferation*, *54*, e12963.
Zhao, Z., & Shilatifard, A. (2019). Epigenetic modifications of histones in cancer. *Genome Biology*, *20*, 245.
Zheng, H.-C. (2017). The molecular mechanisms of chemoresistance in cancers. *Oncotarget*, *8*, 59950–59964.
Zhou, K., Gaullier, G., & Luger, K. (2019). Nucleosome structure and dynamics are coming of age. *Nature Structural & Molecular Biology*, *26*, 3–13.
Zika, E., Greer, S. F., Zhu, X.-S., & Ting, J. P.-Y. (2003). Histone deacetylase 1/mSin3A disrupts gamma interferon-induced CIITA function and major histocompatibility complex class II enhanceosome formation. *Molecular and Cellular Biology*, *23*, 3091–3102.

CHAPTER THREE

Targeting epigenetic regulation for cancer therapy using small molecule inhibitors

Amit Kumar[a,b], Luni Emdad[a,b,c], Paul B. Fisher[a,b,c,*], and Swadesh K. Das[a,b,c,*]

[a]Department of Human and Molecular Genetics, Virginia Commonwealth University, School of Medicine, Richmond, VA, United States
[b]VCU Institute of Molecular Medicine, Virginia Commonwealth University, School of Medicine, Richmond, VA, United States
[c]VCU Massey Cancer Center, Virginia Commonwealth University, School of Medicine, Richmond, VA, United States
*Corresponding authors: e-mail address: paul.fisher@vcuhealth.org; swadesh.das@vcuhealth.org

Contents

Abstract

Cancer cells display pervasive changes in DNA methylation, disrupted patterns of histone posttranslational modification, chromatin composition or organization and regulatory element activities that alter normal programs of gene expression. It is becoming increasingly clear that disturbances in the epigenome are hallmarks of cancer, which are targetable and represent attractive starting points for drug creation. Remarkable progress has been made in the past decades in discovering and developing

Advances in Cancer Research, Volume 158
ISSN 0065-230X
https://doi.org/10.1016/bs.acr.2023.01.001

epigenetic-based small molecule inhibitors. Recently, epigenetic-targeted agents in hematologic malignancies and solid tumors have been identified and these agents are either in current clinical trials or approved for treatment. However, epigenetic drug applications face many challenges, including low selectivity, poor bioavailability, instability and acquired drug resistance. New multidisciplinary approaches are being designed to overcome these limitations, e.g., applications of machine learning, drug repurposing, high throughput virtual screening technologies, to identify selective compounds with improved stability and better bioavailability. We provide an overview of the key proteins that mediate epigenetic regulation that encompass histone and DNA modifications and discuss effector proteins that affect the organization of chromatin structure and function as well as presently available inhibitors as potential drugs. Current anticancer small-molecule inhibitors targeting epigenetic modified enzymes that have been approved by therapeutic regulatory authorities across the world are highlighted. Many of these are in different stages of clinical evaluation. We also assess emerging strategies for combinatorial approaches of epigenetic drugs with immunotherapy, standard chemotherapy or other classes of agents and advances in the design of novel epigenetic therapies.

Abbreviations

AID auxin-inducible degradation
ALL acute lymphocytic leukemia- Acute myeloid leukemia
ASH2L ASH2 like histone lysine methyltransferase complex subunit
BAH bromo-adjacent homology
BRCA1 the breast cancer 1 protein
BRD bromodomains
BRPF bromodomain and PHD finger-containing protein
BAZ2A bromodomain adjacent to zinc finger domain 2A
CBHA m-carboxycinnamic acid bishydroxamide
CBP cAMP response element-binding protein
CREBBP cyclic adenosine monophosphate response element binding protein binding protein
CTCF CCCTC-binding factor
CTCL cutaneous T-cell lymphoma
DAPK death associated protein kinase 1
DIPGs diffuse intrinsic pontine gliomas
DNA deoxyribonucleic acid
DNMT DNA methyltransferases
DPY30 Dpy-30 histone methyltransferase complex regulatory subunit
EGCG epigallocatechin gallate
ER essential thrombocythemia
EZH enhancer of zeste homolog FdCyd-5-Fluoro-2′-deoxycytidine
FDA Food and Drug Administration
FOXO Forkhead box O
GBM glioblastoma
GLP glucagon-like peptide
GNAT Gcn5 N-acetyltransferases

GSTP1	glutathione S-transferase P1
IDH	isocitrate dehydrogenase
HAT	histone acetyltransferases
HDM	histone demethylases
HMBA	hexamethylene bisacetamide
JAK2	Janus-associated kinase 2
JMJ	Jumonji protein
KAT	lysine acetyltransferases- Histone lysine demethylases
KDM1A	lysine demethylase 1A
LSD	lysine-specific demethylase
MAO	monoaminoxidases
MBT	malignant brain tumor
MDR	multidrug resistance
MGMT	O(6)-methylguanine-DNA methyltransferase
MPNST	malignant peripheral nerve sheath tumor
Morf2	multiple organellar RNA editing factor 2
MOZ	monocytic leukemia zinc finger
MYOD1	myogenic differentiation 1
MYST	MOZ, YBF2/SAS3, SAS2, and Tip60
NMNAt2	nicotinamide nucleotide adenylyl transferase 2
NSCLC	non-small cell lung cancer
PPARy	peroxisome proliferator-activated receptor γ
PCAF	P300/CBP-associating factor
PDHA1	pyruvate dehydrogenase alpha 1
PHD	plant homeodomain
PMF	primary myelofibrosis
PGP9.5	protein gene product 9.5
POMC	pro-opiomelanocortin
PITX2	paired like homeodomain 2
PKM2	pyruvate kinase M2
MOF	metal–organic frameworks
PRC1	polycomb-repressive complex 1
PROTAC	proteolysis targeting chimeric
PTCL	peripheral T-cell lymphoma
RBBP5	RB binding protein 5
RNA	ribonucleic acid
SAGA	Spt–Ada–Gcn5–acetyltransferase
SAM	S-adenosylmethionine
SAHA	suberoylanilide hydroxamic acid
SBHA	HDAC inhibitor suberohydroxamic acid
SCLC	small cell lung cancer
SDC4	Syndecan-4
SGC	structural genomics consortium
SIRTs	sirtuins
TAf1	TATA-box binding protein associated factor 1TdCyd-4′-thio-2′-deoxycytidine
TCP	tranylcypromine
TET	ten eleven translocation

TIMP-3	tissue inhibitor of metalloproteinase 3
TMS1	target of methylation-mediated silencing
TRIM24	tripartite motif containing 24
TSA	trichostatin A
UTX	ubiquitously transcribed tetratricopeptide repeat on chromosome X
VPA	valproic acid
WDR	WD40 repeat

1. Introduction

Conrad Waddington coined the term "epigenetics" to interpret the connection between genotype and phenotype and explain the influence of genetic processes on development (Cavalli & Heard, 2019).

The term "epigenetics" refers to the study of stable phenotypic changes without changing the DNA sequence. Most of these heritable alterations occur during differentiation and are persistently preserved during numerous cycles of cell division, allowing cells to have diverse identities while retaining the same genomic makeup. DNA and histone proteins are physically and functionally entangled in chromatin and serve as the starting point of epigenetics. The basic units of chromatin are the nucleosomes, which are comprised of recurrent 146-bp segments of DNA wrapped around 1.75 turns around a central histone octamer of two molecules each of histone proteins H2A, H2B, H3, and H4 separated by 10–60 bp DNA (Fig. 1) (Kornberg, 1974). DNA and histone proteins are modified to regulate gene accessibility and function, serving as significant drivers of biological complexity with roles in development, differentiation, and proliferation. Additionally, epigenetic dysregulation is linked to neurodegenerative disorders, cardiovascular diseases and is associated with cancer initiation and progression (Ando et al., 2019; Baylin et al., 2001; Dawson & Kouzarides, 2012; Lardenoije et al., 2015; Moore-Morris, van Vliet, Andelfinger, & Puceat, 2018). Aberrant promoter methylation of tumor suppressor genes or activation of oncogenes by DNA methylation or histone modifications lead to dysregulation of transcription and chromatin structure causing changes in gene expression that promote the onset and progression of cancers (Ando et al., 2019; Baylin et al., 2001). Currently, the most advanced therapeutic field of epigenetics is cancer, in which changes in DNA methylation, histone modification, and abnormal expression of microRNAs have all been linked to onset of tumor

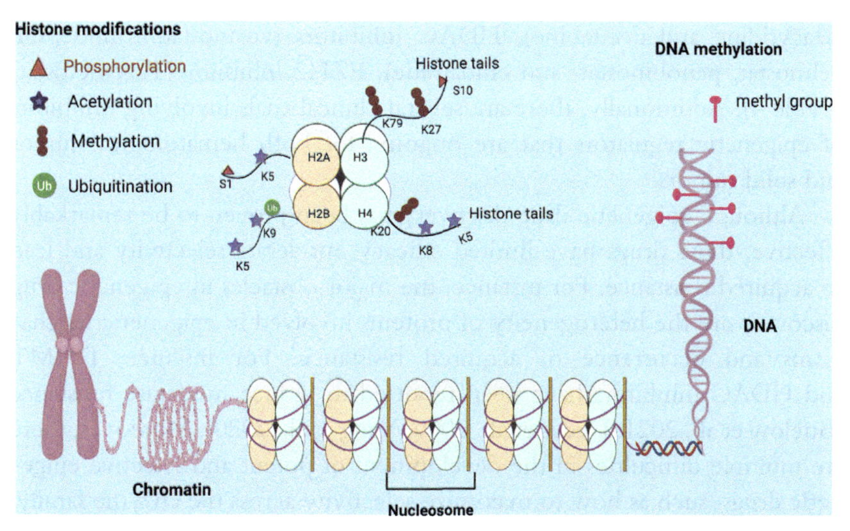

Fig. 1 Comprising 146 bp of DNA wrapped around an octamer of histone proteins, the nucleosome is consisted of four histone proteins (H2A, H2B, H3, and H4) and plays a key part in regulating the replication, transcription, and repair of DNA. Post-translational modifications on histone proteins orchestrate chromatin opening for gene transcription or closing for gene silencing. Some of the key modifications of histones (e.g., acetylation, methylation, phosphorylation, and ubiquitination). DNA methylation originates at specific genomic regions via transfer of a methyl group from S-adenosyl-L-methionine to the 5′-position of a cytosine ring to form 5-methylcytosine (5mCy).

development (Peng & Croce, 2016). Epigenetic modifications are dynamic, reversible and well controlled process. The key players associated with these epigenetic modifications can be considered as writers, readers, and erasers. Epigenetic writers catalyze the addition of epigenetic marks onto DNA and on histones tails (DNA methyl transferases (DNMTs), histone lysine methyltransferases (KMTs), and histone acetyltransferases (HATs)). Readers recognize specific epigenetic marks (bromodomain and chromodomain proteins). Erasers remove these marks and are histone lysine demethylases (KDMs) and histone deacetylases (HDACs).

Over the last two decades, emphasis has been on developing epigenetic therapies as putative anticancer agents based on their direct action on cancer cells. Numerous small-molecule medicines that target the epigenetic regulatory enzymes have been identified recently, some of which show promise as anticancer treatments. Indeed, several epigenetic drugs have been approved by the Food and Drug Administration (FDA) and are commercially available for the treatment of cutaneous T-cell lymphoma, multiple myeloma and for follicular lymphoma including DNMT inhibitors (azacytidine and decitabine), HDAC inhibitors (vorinostat, romidepsin, belinostat, panobinostat, and chidamide), EZH2 inhibitor (Tazemetostat) (Table 4). Additionally, there are several clinical trials involving inhibitors of epigenetic regulators that are ongoing for both hematological tumor and solid tumors.

Although epigenetic drug discovery has been proven to be remarkably effective, these drugs have limited efficacy, moderate selectivity and lead to acquired resistance. For instance, the major obstacles in epigenetic drug discovery are the heterogeneity of proteins involved in epigenetic mechanisms and occurrence of acquired resistance. For instance, DNMT and HDAC inhibitors have been reported to induce moderate resistance (Buelow et al., 2021; Dedes et al., 2009; Yang et al., 2020). Moreover, there are multiple difficulties in the development of potent and selective epigenetic drugs, such as how to maximize selectivity across the enzyme family, how to increase in vivo efficacy by reducing competition with internal cofactors, and how to combat acquired drug resistance.

In this review, we summarize the basic principles manipulating the above-mentioned epigenetic modifications and highlight the comprehensive summary of the evidence from the promising clinical and preclinical outcomes using small-molecule inhibitors against epigenetic modifiers for cancer treatment. Given the importance of epigenetic regulation in cancers, potential efficacious treatments that target epigenetics are becoming attractive strategies of cancer therapy.

2. DNA methylation

DNA methylation is one of the best characterized and most stable epigenetic modifications that play pivotal roles in gene regulation, genomic stability and genomic imprinting during development, cellular differentiation, and tissue homeostasis (Bock et al., 2012; Eden, Gaudet, Waghmare, & Jaenisch, 2003; Karpf & Matsui, 2005; Li, Beard, & Jaenisch, 1993; Ortiz-Barahona, Joshi, & Esteller, 2022; Smith & Meissner, 2013; Stoger et al., 1993; Suzuki et al., 2007). DNA methylation is a covalent chemical modification, that involves the transfer of a methyl (CH3) group from S-adenosylmethionine (SAM) at the carbon 5 position of the cytosine ring by a group of enzymes known as the DNA methyltransferases (DNMTs) (Lyko, 2018). Even though most cytosine methylation occurs in the sequence context 5′CG3′ (also known as the CpG dinucleotide), some are found at non-CpG sites CpA and CpT and CpC dinucleotides (Woodcock, Crowther, & Diver, 1987). Because adequate DNA methylation is necessary for development and healthy cell functions, any anomalies in this process can result in a dysfunction in homeostats leading to cancer progression (Greenberg & Bourc'his, 2019; Sharma, Kelly, & Jones, 2010). Indeed, tumor cells are characterized by global DNA hypomethylation along with hypomethylation of CpG islands at promoters of tumor suppressor genes (Ehrlich, 2002). DNA methylation players can also be categorized as the "writers" such as DNA methyltransferases (DNMTs) which catalyze the addition of a methyl group onto a cytosine residue. The "readers" such as methyl cytosine binding proteins (MBDs) that can bind to DNA methylated sites and give functional context to the DNA modifications. The "eraser" such as (DNA demethylase, 5-Methylcytosine Hydroxylases) can catalyze the removal of epigenetic marks resulting in the modulation of gene expression. The function of abnormal DNA methylation in the development of cancer is the subject of substantial research.

There are considerably more reports of hypermethylation than hypomethylation in cancer. Furthermore, thorough investigation of methylated CpG and unmethylated DNA as ligands it has been shown that DNA methylation facilitates the binding of numerous transcription factors. The genes that are hypermethylated include genes involved in cell cycle regulation (p16INK4a, p15INK4b, Rb, p14ARF), DNA repair (BRCA1, MLH1, MSH2,MGMT), apoptosis (DAPK, APAF-1, Casp-8, Fas, SARP2, TMS1), drug resistance, detoxification, differentiation, angiogenesis, and metastasis (Das et al., 2006; Kusy et al., 2003). Some genes, including Ras association domain family 1 isoform A (RASSF1A) and p16, are frequently methylated in a variety of malignancies, although other genes are methylated in particular tumors. One example is the GSTP1 gene, which is silenced by CpG island DNA hypermethylation in more than 90% of prostate cancers (Lin et al., 2001) but is largely unmethylated in breast cancer and acute myeloid leukemia (Ruzza, Rosato, Rossi, Floreani, & Quintieri, 2009). There are a number of genes in breast cancer which are hypermethylated including p16INK4A, ATP-binding cassette, sub-family B, member 1, B-cell CLL/lymphoma 2, B-cell CLL/lymphoma 2, Cyclin D2 estrogen receptor (ER) alpha, the progesterone receptor (PR), BRCA1, GSTP1, TIMP-3, Cadherin 13, H-cadherin and E-cadherin (Esteller, Corn, Baylin, & Herman, 2001). MYOD1, PITX2 and SDC4 seen to be hypermethylated in 47–64% of acute myeloid leukemia patients (Esteller et al., 2001; Leone, Teofili, Voso, & Lubbert, 2002). In addition, the calcitonin gene has been found to be hypermethylated in 65% of myelodysplastic syndromes and in 95% in acute leukemias (Leone et al., 2002). Claudin-4 is hypermethylated in bladder cancer as well as gastric cancer.

Hypomethylation is a second kind of methylation defect common in solid tumors as well as in hematologic malignancies (Bedford & van Helden, 1987; Duenas-Gonzalez et al., 2005; Lin et al., 2001; Shull et al., 2016). There has been a report that c-MYC and H-RAS is hypomethylated in some primary human cancers (Feinberg & Vogelstein, 1983). Cancer-testis genes such as MAGEA1 LIN28B, CT45 are hypomethylated and associated with disease progression in a variety of cancers, S100A4 is hypomethylated and overexpressed in majority of pancreatic carcinomas, mesothelin is hypomethylated in malignant mesothelioma as well as ovarian cancer, trefoil factor 3, downregulation in prostate cancer is associated with the promoter hypomethylation, maspin and S100P has been found to be hypomethylated in the pancreatic cancer, PGP9.5 gene has been found to be hypomethylated in gall bladder and colon cancer, POMC is hypomethylated and overexpressed in thymic carcinoids and the heparinase gene has also been found to be hypomethylated in some cancers (De Smet, Lurquin, Lethe, Martelange, & Boon, 1999; Mizukami et al., 2008; Rosty et al., 2002; Sato, Fukushima, Matsubayashi, & Goggins, 2004; Ye et al., 2005). DNA hypomethylation typically becomes more prominent as a tumor progresses and shows a progressive increase with the grade of malignancy (Ehrlich et al., 2006; Park, Yoo, Cho, Kim, & Kang, 2009; Qu, Dubeau, Narayan, Yu, & Ehrlich, 1999; Roman-Gomez et al., 2008).

2.1 DNA methyltransferase family (DNMTs) (the writers)

DNA methylation is a major epigenetic modification that cells use to control gene expression and is one of the best-studied epigenetic markers. This process involves enzymes belonging to the DNA methyltransferase family (DNMTs), which catalyze the transfer of a methyl group from S-adenosyl-L-methionine (AdoMet or SAM) to the 5-position of cytosine, most often in CpG dinucleotides (Okano, Xie, & Li, 1998). In general, five members of the DNMT family have been identified, including DNMT1, DNMT2, DNMT3A, DNMT3B, and DNMT3L (Jin & Robertson, 2013). DNMT1 is the most abundant DNA methyltransferase and is responsible for the maintenance of all methylation in the genome and is essential for mammalian development and cancer cell growth (Mortusewicz, Schermelleh,

Walter, Cardoso, & Leonhardt, 2005). DNMT2 is the smallest mammalian DNMT and is not regarded as a catalytically active DNA methyltransferase since this enzyme methylates small transfer RNAs (tRNAs) and functions as an RNA methyltransferase (Tuorto et al., 2015). DNMT3 functions as a de novo methyltransferase and consists of two related proteins DNMT3A and DNMT3B that show an equivalent affinity for hemi-methylated and non-methylated DNAs (Chedin, 2011; Jurkowska et al., 2008). DNMT3L, another member of DNMT3 family, lacks the conserved residues that are known to be important for DNA methyltransferase activity and contains a truncated inactive catalytic domain which acts as an accessory partner to stimulate the de novo methylation activity of DNMT3A and DNMT3B in cell culture system (Jurkowska et al., 2008). In murine germ cells, a new de novo DNA methyltransferase, DNMT3C, was discovered which exhibits high identity with DNMT3B and is specialized at methylating the young retrotransposons in the male germ line (Barau et al., 2016).

Overexpression and somatic mutations in DNMTs have been linked to various tumors and can lead to malignant transformation, overexpression of DNMT1, DNMT3A, and DNMT3B has been reported in a variety of tumors (Baylin & Jones, 2011). Moreover, it has been reported that alterations in the expression of DNMT3B and CCCTC-binding factor (CTCF) are partially responsible in the epigenetic inactivation of BRCA1 in sporadic breast tumors (Butcher & Rodenhiser, 2007). DNMT1 and DNMT3A mutations have been observed in the cancer genome of colon tumor, acute myeloid leukemia (AML), myelodysplastic syndrome (MDS), and adult early T-cell precursor acute lymphoblastic leukemia (Kanai, Ushijima, Nakanishi, Sakamoto, & Hirohashi, 2003; Ley et al., 2010; Neumann et al., 2013; Walter et al., 2011). Increased expression of tumor suppressor genes and decreased tumorigenicity are both associated with DNMT inhibition. Consequently, DNMTs are regarded as worthwhile targets for the creation of anti-cancer treatments.

2.2 DNA methyltransferase inhibitors

Unlike genetic changes, epigenetic modifications are heritable and potentially reversible. Therefore, such alterations have become attractive targets for cancer therapy. Since hypermethylation of tumor suppressor genes and overexpression of DNMTs have been established as fundamental processes in carcinogenesis, demethylating agents should in principle prove promising as anticancer drugs.

DNA demethylating drugs are prodrugs consisting of analogs of deoxycytidine and function by incorporating into replicating DNA and covalently binding to the catalytic sites of active DNMTs sites. These drugs irreversibly inhibit the enzymatic activities of DNMTs and causes its degradation by a proteasome-mediated mechanism (Agrawal, Das, Vyas, & Hajduch, 2018). List of DNMT inhibitors and its target is summarized (Table 1).

Currently, two DNMT inhibitors, azacitidine (5-aza-cytidine) and decitabine, have been approved by the US Food and Drug Administration for the treatment of patients with acute myeloid leukemia (AML) and myelodysplastic syndrome (MDS) (Table 4) (Christman, 2002). Guadecitabine (SGI-110) is a second-generation decitabine consisting of 5-aza-CdR linked to deoxyguanosine, the dinucleotide configuration provides protection from deamination by cytidine deaminase thereby exhibiting a longer half-life than decitabine resulting in higher positive responses in patients with AML and high-risk AML (Yoo et al., 2007). Two other cytosine analogs, 4′-thio-2′-deoxycytidine (TdCyd) and 5-Fluoro-2′-deoxycytidine (FdCyd,) have been used in clinical trials for patients with advanced solid tumors, AML, and multiple sclerosis (Table 1).

An increasing number of non-nucleoside inhibitors have been developed from different sources to target aberrant DNA hypermethylation (Table 1) (Moreira-Silva et al., 2020). Hydralazine, procaine, and procainamide were identified through drug repurposing (Moreira-Silva et al., 2020). Similarly, EGCG, nanaomycin A, genistein and laccaic acid A were obtained from natural products (Fagan et al., 2013; Fang et al., 2003; Kuck et al., 2010). N-Phthaloyl-L-tryptophan (RG108), a DNMT1 inhibitor, blocks the active site of DNA methyltransferase without the formation of covalent reaction intermediates causing demethylation was discovered by virtual screening (Asgatay et al., 2014). The DC_05 series of compounds including DC_517 a selective DNMT1 inhibitor was also identified through virtual screening and was found to induce DNA hypomethylation (Chen et al., 2014; Jones & Taylor, 1980). Interestingly, BIX-01294 a G9a/G9a-like protein (GLP) H3K9 methyltransferase inhibitor showed novel ability to inhibit the DNA methyltransferase DNMT3A at low micromolar levels without inhibiting DNMT1 and G9a (Rotili et al., 2014). Nanaomycin A, a quinone antibiotic, interacts with DNMT3B and reduces global methylation levels in cancer cell lines by reactivating transcription of the RASSF1A tumor suppressor gene (Caulfield & Medina-Franco, 2011; Kuck et al., 2010). SGI-1027 a novel class of relatively stable

Table 1 DNA methylation inhibitors in cancer therapy.

Drug	Target	Condition	Mechanism of action	Clinical trial (https://clinicaltrials.gov) NCT
Azacitidine	Global DNMT	AML, MDS and Solid tumors	Inhibition of DNA methyltransferase at low doses and cytotoxicity through incorporation into RNA and DNA at high doses, slow tumor growth and induce cell differentiation (Hollenbach et al., 2010; Silverman, 2001)	NCT04891068, NCT03873311, NCT04891068, NCT03666559, NCT04187703, NCT05376111, NCT04742101, NCT01305460, NCT02458235, NCT02017457
Decitabine	Global DNMT	AML, MDS and Solid tumors	At high concentrations it can lead to blocked DNA synthesis and cytotoxicity; at low concentrations it leads to changes in gene expression profiles (Jabbour, Issa, Garcia-Manero, & Kantarjian, 2008; Stresemann & Lyko, 2008)	NCT04582604, NCT02264873, NCT02252107, NCT03875287, NCT02316028, NCT01251627, NCT04697940, NCT01893320, NCT00042003, NCT00085293
SGI-1027 (Guadecitabine)	Global DNMT	AML and high-risk AML and Solid tumors	It enhances stability in the aqueous phase, improves resistance to cytidine deaminase (CDA) degradation, and prolongs half-life (Daher-Reyes, Merchan, & Yee, 2019; Jueliger et al., 2016)	NCT02920008, NCT02907359, NCT02429466, NCT02131597, NCT01752933, NCT01261312, NCT03075826, NCT02348489, NCT02293993, NCT02197676
Curcumin	Global DNMT	AML, Solid tumors	Reversing altered patterns of DNA methylation (Boyanapalli & Kong, 2015; Hassan et al., 2019)	NCT05045443, NCT03211104, NCT03980509, NCT01294072, NCT01333917, NCT04208334, NCT02064673, NCT02439385, NCT03769766
Procaine	Global DNMT	Leukemia and Solid tumors	Block the binding of DNMTs to DNA and promotes the proliferation arrest and apoptosis (Castellano et al., 2008; Li, Wang, et al., 2018; Villar-Garea, Fraga, Espada, & Esteller, 2003)	NA

Continued

Table 1 DNA methylation inhibitors in cancer therapy.—cont'd

Drug	Target	Condition	Mechanism of action	Clinical trial (https://clinicaltrials.gov) NCT
S110	Global DNMT	MDS, AML and Solid tumors	The constitutive methylation level of the CTA promoter in cancer cells treated for treatment induction was significantly reduced and retards tumor growth in human xenograft (Chuang et al., 2010; Issa et al., 2015)	NCT01261312
Zebularine	Global DNMT	Solid tumors, AML	It forms a covalent complex with DNMT to inhibit DNA methylation and induces cell growth arrest and apoptosis (Marquez et al., 2005; Nakamura et al., 2015)	NA
EGCG	DNMT1	Solid tumors and Leukemia	It binds non-covalently to the catalytic active site of DNMT to inhibit the methylation catalytic activity of DNMT. Induces apoptosis and inhibit growth in several types of cancers (Hou et al., 2005; Kim, Quon, & Kim, 2014)	NCT02891538, NCT04300855, NCT00666562, NCT00676780
Laccaic Acid A	DNMT1	Solid tumors	DNA-competitive inhibitor of DNA methyltransferase 1 reduces cell viability and induces apoptotic cell death (Fagan, Cryderman, Kopelovich, Wallrath, & Brenner, 2013)	NA
MG98	DNMT1	Renal cell carcinoma	Acts on the mRNA of DNMT1 and radically inhibits the expression and synthesis of DNMT1 (Amato et al., 2012)	NCT00003890
Procainamide	DNMT1	Colon cancer, Breast cancer	Block the enzyme catalytic activity and reactivate tumor suppressor genes (Altundag, Altundag, & Gunduz, 2004; Lee, Yegnasubramanian, Lin, & Nelson, 2005)	NA

RG108	DNMT1	Prostate cancer	Block to DNA methylation. Growth inhibition and apoptosis induction in human prostate cancer cell and endometrial cancer (Graca et al., 2014; Yang, Hou, Cui, Suo, & Lv, 2017)	NA
Theaflavin 3, 3′-digallate N6	DNMT3A	Leukemia and Solid tumors	Inhibit tumor angiogenesis migration and induces cell death (O'Neill, Termini, Albano, & Tsiani, 2021; Pan et al., 2018)	NA
Thearubigin	DNMT3A	Leukemia	Induced cell cycle arrest in human leukemic cells by affecting multiple pathways involved in cell cycle control (Das et al., 2002)	NA
Nanaomycin A	DNMT3B	Lung cancer, Colon cancer, AML	Transcriptional reactivation and expression of the RASSF1A tumor suppressor gene (Kuck, Caulfield, Lyko, & Medina-Franco, 2010)	NA
Trichostatin A	DNMT3B	AML and Solid tumors	Induces differentiation and apoptosis (Chang et al., 2012; Vigushin et al., 2001)	NCT03838926
AG-120	Mutant IDH1	AML, Solid tumors	Profound 2-HG lowering in tumor models and the ability to effect differentiation of primary patient AML samples ex vivo (Popovici-Muller et al., 2018)	NCT03343197, NCT04278781, NCT02074839, NCT03503409, NCT04195555, NCT03245424, NCT02074839
AG-221	Mutant IDH2	AML, MDS	Suppressed 2HG production and induced cellular differentiation in primary human IDH2 mutation–positive acute myeloid leukemia (AML) cells ex vivo and in xenograft mouse models (Yen et al., 2017)	NCT03744390, NCT02273739, NCT03515512

Abbreviations: AML, acute myelogenous leukemia; DNMT, DNA methyltransferase HCC, hepatocellular carcinoma; IDH, isocitrate dehydrogenase; MDS, myelodysplastic syndrome.

quinoline-based compound was synthesized and was described for against DNMT1, DNMT3A and DNMT3B induces the degradation of DNMT1 and reactivates tumor suppressor genes (Gros et al., 2015; Sun, Zhang, Zhang, Zhao, & Jiao, 2018). Later, MC3343 and MC3353, two analogues of SGI-1027, showed cytotoxicity against leukemia KG-1 cells in the micromolar range (Rilova et al., 2014; Valente, Trisciuoglio, et al., 2014). As a novel DNMT inhibitor, MC3343 is more potent and selective than SGI-1027 toward other (SAM-dependent) methyltransferases (Valente, Liu, et al., 2014). MC3353 as a novel DNMT inhibitor displays a stronger demethylating ability than both azacytidine and decitabine in several cancer types (Valente, Liu, et al., 2014). In addition, MG98, an oligonucleotide antisense inhibitor of DNMT1, was used recently in a Phase II study for the treatment of advanced renal cell carcinoma and in a phase I study in patients with high-risk MDS and AML (Klisovic et al., 2008; Winquist et al., 2006). EGCG a polyphenol also acts as an epigenetic modifier by downregulating DNMT1, HDAC1, HDAC2, and G9a which is now being evaluated in a Phase II clinical trial as an anticancer drug (Borutinskaite, Virksaite, Gudelyte, & Navakauskiene, 2018; Shanafelt et al., 2013).

2.3 DNA demethylation and inhibitor (the eraser)

Conversely, changes in DNA methylation can occur with normal differentiation and aging and may aid in the development of tumors. Recently, the mechanisms of DNA methylation and demethylation, as well as the enzymes involved, have been described, but they still require more characterization. Ten-eleven translocation (TET) enzymes are large (~180- to 230-kDa) multidomain enzymes are the 2-oxoglutarate dependent dioxygenase enzymes able to catalyze the oxidation of 5-methylcytosine (5-mC) into 5-hydroxymethylcytosine (5-hmC). They have been identified as key players in DNA demethylation and in the control of cellular differentiation, development, and transformation (Rasmussen & Helin, 2016). Acquired point mutations and deletion events targeting TET genes are frequently observed in diverse lymphoid, myeloid, and solid tumors (Lio, Yuita, & Rao, 2019). These TET proteins are already acknowledged as key participants in molecular medicine, even though their functions are just recently being studied. The race to screen for and discover TET inhibitors to counteract the negative impact of these proteome errors is therefore equally vital. Unfortunately, no class of selective TET inhibitors exists yet that can be used to sufficiently investigate these metabolic processes pharmacologically.

Recently, in silico modeling of the TET enzyme active site was used to rationalize the activity of Bobcat339 and other cytosine-based inhibitors. Bobcat339 inhibited TET1 and TET2 without inhibiting DNMT3a, and reduced 5hmC abundance in the DNA of cultured neurons (Chua et al., 2019). By combining in silico and biochemical screening strategies, a first-in-class, highly potent, and cell-permeable TET inhibitor that recognizes the catalytic core of TET enzymes and interfere with their enzymatic activities compound 35 (C35) was identified and promotes somatic cell reprogramming (Singh et al., 2020).

3. Histone modifications

Posttranslational histone modifications, such as DNA methylation, do not change the sequence of the DNA molecule's nucleotides but can change accessibility to the transcriptional machinery. Several types of histone modifications can occur to the N- and C-terminal tails as well as the core domain of histones, the most frequent ones are acetylation, methylation, ubiquitination, phosphorylation and sumoylation. However other modification also exists including citrullination, ADP-ribosylation, deamination, formylation, O-GlcNAcylation, propionylation, butyrylation, crotonylation, lactylation and proline isomerization. The enzymes responsible for adding or removing these epigenetic marks are often considered as "writers" and "erasers," respectively. These modifications have been extensively studied and represent the most important changes in terms of the regulation of chromatin structure (Fig. 1). Histone acetylation usually leads to higher gene expression, whereas methylation in turn can cause either transcriptionally permissive or repressive properties.

Considering the essential functions of histone modifications it is therefore not unusual that abnormalities in histone modifications are widespread in cancers, and that knowledge of these modification patterns can assist in cancer diagnosis and treatment. Essentially all these abnormalities have been demonstrated to take place at specific gene promoters because of improper targeting histone-modifying enzymes.

3.1 Histone acetylation

Histone acetylation occurs at several N-terminal lysine residues through the catalytic action of histone acetyltransferases (HATs), also called lysine acetyltransferases (KATs). Acetylation of lysine residues of N-terminal tails of H3 and H4 is one of the most studied modifications of this type. Histone acetylation is linked to active transcription, particularly at gene promoters and enhancers of the gene and regulates the compaction state of chromatin through a variety of mechanisms, including neutralizing the positive charge and increasing the size of unmodified lysine residues (Di Cerbo & Schneider, 2013; Kouzarides, 2007). In addition to the structural effects acetylation has on chromatin it can also recruit coregulators, transcription factors and RNA polymerase complexes to the chromatin which can themselves influence transcription and other chromatin-templated processes, e.g., acetylated histones are typically recognized by the bromodomain (Audia & Campbell, 2016). Aberrant levels of histone H4K16 acetylation (H4K16ac) have been a driver of progression of many types of human cancers (Fraga et al., 2005; Sheikh & Akhtar, 2019). Levels of histone acetylation are modulated by two sets of enzymes HDACs which remove acetyl groups and by the HAT which transfer the acetyl groups from acetyl-CoA cofactors to lysine residues at histones tails (Fig. 1). Histone acetylation levels are essential for chromatin remodeling and the control of gene transcription. Histone tails containing acetylated lysines are linked to a more flexible chromatin state and gene-transcription activation, whereas histone tails containing deacetylated lysine are linked to a more condensed chromatin state and transcriptional gene suppression (Wang et al., 2009). Small molecule inhibitor targeting these epigenetic modifications have emerged as powerful and effective treatments for cancers and are being intensely evaluated.

3.1.1 Histone acetyltransferases (the writers)

The discovery that histones can be modified by the addition or removal of acetyl or methyl group by Allfrey and colleagues over 40 years ago laid the groundwork for our current understanding of HATs. Since that time, multiple studies have linked the acetylation of histones with gene activity. Allis and colleagues identified and cloned the first two HATs from *Saccharomyces cerevisiae*, Hat1a and Gcn5 (Kuo, Zhou, Jambeck, Churchill, & Allis, 1998). Later, the first chromatin-modifying and remodeling HAT complex the Spt–Ada–Gcn5–acetyltransferase (SAGA) was identified (Martinez, Kundu, Fu, & Roeder, 1998).

Histone acetyltransferases are epigenetic enzymes that can be divided into three groups based on their catalytic domains (Hodawadekar & Marmorstein, 2007; Sterner & Berger, 2000). General control non-depressible 5 (GCN5) or lysine acetyltransferase 2A (KAT2A) is the original member of the Gcn5 N-acetyltransferases (GNATs), which catalyzes the transfer of an acetyl group from acetyl coenzyme A (acetyl-CoA) to the primary amine, and this family includes Gcn5/PCAF, Nut1a, Elp3, Hat1a, MCM3AP Hpa2/Hpa3 and ARD1 (Yang, 2004). The MYST family of histone acetyltransferases (HATs) is the largest and most divergent group named after the founding members in yeasts and mammals and this family currently comprises five human HATs, i.e., Morf, Ybf2 (Sas3), Sas2 and Tip60 (Wapenaar & Dekker, 2016). The other HATs families that have been recently characterized include p300/CBP (CREB-binding protein) (Liu et al., 2008), Taf1 and several nuclear receptor co-activators including steroid-receptor coactivator SRC-1, ACTR, have also been shown to possess intrinsic HAT activity (Tavassoli et al., 2010). These are classified as an "orphan class" of HAT enzymes due to absence of true HAT domain (Montgomery, Sorum, & Meier, 2015). Most recently the protein CLOCK a central component of the circadian rhythms has also been shown to have a MYST-like HAT domain, which is important for its function and shares sequence similarity to ACTR, a member of the SRC family of HATs (Doi, Hirayama, & Sassone-Corsi, 2006).

It is evident that appropriate acetylation is necessary for normal cellular behavior and abnormalities in HATs function are associated with tumorigenesis and tumor progression. Enhanced HAT activity, can occur as the result of chromosomal translocations with diverse fusion partners, which leads to cancer progression, including recurrent translocation of CBP gene that leads to the MLL-CBP fusion [t(11;16)(q23;p13)] or translocation of p300 leads to MLL-p300[t(11;22)(q23;q13)] fusions, which account for 1% of the total MLL-fusions and the MOZ-CBP [t(8;16)(p11;p13)] or fusion of the MOZ gene to the p300 gene [t(8;22)(p11;q13)] which are even rarer with about 0.4% of cases among AMLs (Chaffanet et al., 2000; Krivtsov & Armstrong, 2007). EP300/CBP is a unique family of KATs serving as transcriptional coactivators when directly interacting with AML1-ETO [t(8;21) (q22;q22)] (Wang et al., 2011), where it has also been associated with the fusion partners of MOZ-MLL and NUP98-HOXA9. Somatic alteration in p300/CBP genes are associated with the progression of different types of cancers, including breast, colorectal, gastric cancers, nasopharyngeal, ovarian, lung, colon, breast, and cutaneous squamous cell carcinomas. Recently higher frequency of mutation in p300/CBP genes has been reported in small-cell lung cancers and non-Hodgkin B-cell lymphomas (Pasqualucci et al., 2011; Peifer et al., 2012). Both GCN5 and PCAF have been closely linked to the progression of cancer. GCN5 expression is significantly upregulated in human glioma, HCC, NSCLC, colon, and lung cancer, whereas the PCAF gene is frequently deleted in solid tumors such as ovarian gastric, HCC and esophageal carcinomas (Di Martile, Del Bufalo, & Trisciuoglio, 2016). Also, MYST family members have often been implicated in several forms of cancer and have been reported to drive leukemogenesis (Avvakumov & Cote, 2007; Thomas & Voss, 2007). Non-histone proteins can also be modified by HATs, and it has been demonstrated that the function of these substrates (p53 and pRb, YY1, HMG, STAT-3, c-MYC) is regulated by the acetylation of certain lysines (Tang, Zhao, Chen, Zhao, & Gu, 2008). In conclusion, the oncogenic or tumor-suppressive effects of a HAT in cancer depend on its expression level where overexpression potentiates, whereas reduction results in a loss of acetylation capacity. This suggests that HATs may be a promising pharmacological target, even though, at this point, the discovery of effective HAT inhibitors has trailed behind that of HDAC inhibitors, the enzyme's counterpart.

3.1.2 Histone acetyltransferase inhibitors

Compared with DNA methylation, histone modifications have been investigated as potential targets for developing novel therapeutic strategies in solid tumors, hematological malignancies, neurodegenerative disorders, and many inflammatory diseases (viral infections, diabetes, inflammatory lung disease, etc.). Currently, only a few HATs modulators have been described. Recently, various approaches have been explored to develop small molecule inhibitors of HATs (e.g., screening of natural products, rational drug design, and high-throughput methods). The available HAT inhibitors described so far can be classified into several groups: natural products, acetyl-CoA derived bi-substrate inhibitors, and synthetic compounds (Table 2).

As an alternate strategy, anacardic acid, which is considered as a small molecule inhibitor of the HATs p300 and PCAF was extracted from the cashew nut, which also lead to the development of the alkylidene malonate that inhibits HATs p300/CBP and activates PCAF (Balasubramanyam, Swaminathan, Ranganathan, & Kundu, 2003). MG149 is another derivative of anacardic acid that is selective toward the MYST type of HATs Tip60 and MOF and inhibits the NF-κB and the p53 pathways (Sung et al., 2008). Garcinol is a naturally occurring compound extracted from *Garcinia indica* fruit peel is a novel p300 HAT inhibitor displaying micromolar potency that inhibits histone acetylation and alters global gene expression in cells (Balasubramanyam, Altaf, et al., 2004).Curcumin is a natural phenolic compound that can suppress p53 and histone acetylation at the micromolar level

Table 2 Histone methylation and demethylation inhibitors in cancer therapy.

Drug	Target	Role in epigenetic regulation	Condition	Mechanism of action	Clinical trial (https://clinicaltrials.gov) NCT
KMT inhibitors					
EPZ00477	DOTIL	H3K79 methyltransferase	MLL	Selectively target MLL-translocated cells while not affecting non-MLL-translocated cells and better survival in mice with MLL-rearranged leukemia (Buechele et al., 2015)	NA
SGC0946	DOTIL	H3K79 methyltransferase	MLL-rearranged leukemias	Growth inhibitory synergy against MLL-r cell lines, primary human leukemia cells, and mouse leukemia models (Gilan et al., 2016; Klaus et al., 2014)	NA
EPZ-5676 (Pinometostat)	DOTIL	H3K79 methyltransferase	Hematologic malignancies	Causes tumor regressions in a rat xenograft model of MLL-rearranged leukemia. Clinical development for both adult and pediatric relapsed/refractory acute leukemia patients harboring MLL-r (Campbell et al., 2017)	NCT03701295 NCT03724084
DZNep	EZH2	H3K27 methyltransferase	Breast cancer, Colon cancer, Prostate cancer	Decreases the expression of PRC2 proteins EZH2, SUZ12, and EED, as well as global H3K27me3 levels, leading to the reactivation of the PRC2-repressed genes and apoptosis in breast cancer cell line MCF-7 and colorectal cancer cell line HCT116 (Tan et al., 2007; Yao et al., 2016)	NA
GSK126	EZH2	H3K27 methyltransferase	DLBCL	Decreases global H3K27me3, inhibits the proliferation of EZH2 mutant DLBCL cell lines and inhibits the growth of EZH2 mutant DLBCL xenografts in mice (Park et al., 2020)	NCT02082977
GSK926	EZH2	H3K27 methyltransferase	Breast cancer, Prostate cancer	Reduces H3K27me3 levels in a breast cancer cell line and inhibits cell proliferation in breast and prostate cancer cell-based models (Verma et al., 2012)	NA
GSK343	EZH2	H3K27 methyltransferase	Ovarian cancer	Exhibits anticancer activity, inhibiting cell invasion and cell growth in epithelial ovarian cancer cells (Amatangelo et al., 2013)	NA

Continued

Table 2 Histone methylation and demethylation inhibitors in cancer therapy.—cont'd

Drug	Target	Role in epigenetic regulation	Condition	Mechanism of action	Clinical trial (https://clinicaltrials.gov) NCT
EPZ005687	EZH2	H3K27 methyltransferase	EZH2 mutant non-HL	Reduces H3K27me3 levels, Treatment of lymphoma cells bearing a mutant EZH2, leads to antiproliferative effects (Knutson et al., 2012)	NA
EPZ011989	EZH2	H3K27 methyltransferase	DLBCL	Inhibits EZH2 in a mouse xenograft model of DLBCL (Campbell et al., 2015)	NA
EI1	EZH2	H3K27 methyltransferase	DLBCL	Exhibit genome-wide loss of H3K27 methylation and activation of PRC2 target genes, leading to decreased proliferation, increased cell cycle arrest, and apoptosis (Qi et al., 2012)	NA
Tazemetostat (EPZ-6438)	EZH2	H3K27 methyltransferase	DLBCL, HL, NON-HL	Treatment of mice bearing a lymphoma xenograft with mutant EZH2 reduces cell growth, FDA-approved for follicular lymphoma and epithelioid sarcoma with SNF5 deletions (Bhat, Umit Kaniskan, Jin, & Gozani, 2021)	NCT02875548 NCT03009344 NCT05228158 NCT05023655
UNC1999	EZH2	H3K27 methyltransferase	DLBCL	Reduced H3K27me3 levels in cells and selectively killed diffused large B cell lymphoma cell lines harboring the EZH2 (Y641N) mutant (Bisserier & Wajapeyee, 2018)	NA
CPI-1205	EZH2	H3K27 methyltransferase	Lymphoma	Well-tolerated and highly bioavailable in a lymphoma xenograft model. Currently under clinical trials (Vaswani et al., 2016)	NCT02395601 NCT03525795 NCT03480646
PF06821497	EZH2	H3K27 methyltransferase	DLBCL	EZH2 catalytic inhibitor effective in mouse xenograft model of DLBCL (Yamazaki, Gukasyan, Wang, Uryu, & Sharma, 2020)	NCT03460977
Valemetostat	EZH1/2	H3K27 methyltransferase	DLBCL, AML, TALL, Urogenital cancers	EZH1/EZH2 dual inhibitor with activity in DLBCL, AML, TALL and urogenital cancers (Yamagishi et al., 2019)	NCT04388852 NCT04102150 NCT04842877 NCT04703192

Ebastine	EZH2	H3K27 methyltransferase	Breast cancer, Triple-negative breast cancer, Prostate cancer	Reduced H3K27me3 levels in breast cancer and prostate cancer cells. Also active in a triple-negative breast cancer murine model (Li, Liu, et al., 2020)	NA
BIX-01294	G9A	H3K9me2 methyltransferase	Leukemia, Bladder cancer	Lowers bulk H3K9me2 levels, induced autophagy or autophagy-associated cell death (Cui et al., 2015; Xu et al., 2021)	NA
BRD4770	G9A	H3K9me2 methyltransferase	Pancreatic cancer	Induces senescence in the pancreatic cancer cell line PANC-1 (Yuan, Wang, Li, et al., 2012; Yuan, Wang, Paulk, et al., 2012)	NA
UNC0638	G9A	H3K9me2 methyltransferase	AML and Breast cancer	Lower global H3K9me2 levels reduced the clonogenicity of MCF7 cells, inhibited human AML cell proliferation (Lehnertz et al., 2014; Liu et al., 2018)	NA
Chaetocin	SUV39H1	SUV39 methyltransferase	Lymphomas	Reduced H3K9me3 levels, antimyeloma activity (Lai, Chen, Tsai, Chen, & Hung, 2015)	NA
GSK3326595	PRMT5	Protein arginine methyltransferase 5	Solid tumors and non-HL	Inhibits proliferation and induces cell death in a broad range of solid and hematologic tumor cell lines (Kim, Kim, et al., 2020; Tang et al., 2022)	NCT04676516 NCT03614728
AMI-408	PRMT1	H4R3 methyltransferase	AML	Depletion of H4R3me2a, leading to repression of MLL-GAS7 fusion or MOZ-TIF2 (Cheung et al., 2016)	NA
MS023	PRMT1, 3, 4, 6 and 8	H4R3 methyltransferase	Breast cancer	Inhibits PRMT1 methyltransferase activity in MCF7 cells (Wu et al., 2022)	NA
KDM inhibitors					
ORY-1001	LSD1	Demethylates histone 3 at lysine 4 (H3K4) and H3K9	Refractory acute leukemia, TNBC	Reduces growth of an AML xenograft model, and extends survival in a mouse PDX (patient-derived xenograft) model of T cell acute leukemia, inhibits proliferation, and promotes apoptosis of triple negative breast cancer (Maes et al., 2018; Wang, Zhang, & Sun, 2022)	NCT05546580

Continued

Table 2 Histone methylation and demethylation inhibitors in cancer therapy.—cont'd

Drug	Target	Role in epigenetic regulation	Condition	Mechanism of action	Clinical trial (https://clinicaltrials.gov) NCT
GSK2879552	LSD1	H3K4me1 and H3K4me2 demethylase	Refractory acute leukemia	Enhance H3K4 methylation and to increase the expression of tumor-suppressor genes. Anti-proliferative growth effects in AML cell lines and is currently under clinical evaluation for cancer treatment (Smitheman et al., 2019)	NCT02929498
NCD25, NCD38	LSD1	H3K4me1 and H3K4me2 demethylase	MDS	In vivo eradication of primary MDS-related leukemia cells (Sugino et al., 2017)	NA
Tranylcypromine	LSD1	H3K4me1 and H3K4me2 demethylase	AML, MDS	Markedly diminished the engraftment of primary human AML cells in murine models (Dai et al., 2020)	NCT02261779 NCT02717884 NCT02273102
GSK-j1, GSK-J4	JmjC-domain protein	H3K27me3 demethylase	AML	Could affect the proliferation and apoptosis of a variety of cancer cells, attenuated the disease progression in a human AML xenograft mouse model in vivo (Heinemann et al., 2014; Li, Zhang, et al., 2018).	NA
EPT-103182	KDM5B	H3K4me2 and H3K4me3 demethylase	Hematologic and Solid tumors	Anti-proliferative effect in hematological and solid cancer cell lines as well as inhibiting tumor growth in a dose-dependent manner in xenograft models (Morera, Lubbert, & Jung, 2016)	NA
PBIT	KDM5	H3K4me2 and H3K4me3 demethylase	Breast cancer	Inhibited proliferation of breast cancer by derepressing and upregulation of the tumor suppressor HEXIMI in vitro (Sayegh et al., 2013)	NA
KDOAM-25	KDM5A-C	H3K4me2 and H3K4me3 demethylase	Breast cancer, Multiple myeloma	Reduce proliferation and growth in breast cancer and multiple myeloma which highly expressed KDM5B (Montano et al., 2019; Tumber et al., 2017)	NA

ML324	KDM4B and KDM4E	H3K4me2 and H3K4me3 demethylase	Triple negative breast cancer, Prostate cancer	Reduced tumor volume and growth in a triple negative breast cancer mouse model, inhibits proliferation in prostate cancer in vitro and in vivo (Carter et al., 2021; Wang, Oguz, et al., 2018)	NA
INCB059872	KDM1A	H3K4me2 and H3K4me3 demethylase	AML, Prostate cancer, and Ewing sarcoma	Induce cell differentiation in progenitor cells in AML, prostate cancer, and Ewing sarcoma (Fang, Liao, & Yu, 2019)	NCT02959437 NCT02712905 NCT03514407
CC-90011	KDM1A	H3K4me2 and H3K4me3 demethylase	AML and SCLC	Induce cellular differentiation exhibits anti-tumor efficacy in vitro and in vivo in AML and SCLC (Kanouni et al., 2020)	NCT04628988 NCT04350463 NCT03850067 NCT04748848 NCT02875223
SP-2577	KDM1A	H3K4me2 and H3K4me3 demethylase	Ovarian cancer, Ewing Sarcoma	Anti-tumor immunity in mutated ovarian cancer cells in vitro and has also been found to inhibit growth in Ewing Sarcoma xenografts (Kurmasheva et al., 2021; Soldi et al., 2020)	NCT03895684 NCT04611139 NCT03600649 NCT04734990

Abbreviations: AML, acute myelogenous leukemia; DLBCL, diffuse large B cell lymphoma; DOTIL, DOT1 Like; EZH2, enhancer of zeste 2; HDAC, histone deacetylase; HL, Hodgkin's lymphoma; JmjC, Jumonji C; KDM, lysine demethylase; KMT, lysine methyltransferase; MDS, myelodysplastic syndrome; MLL, mixed-lineage leukemia; MOZ, monocytic leukemia zinc finger; PRMT, protein arginine N-methyltransferases; SCLC, small cell lung cancer; SUV39H1, suppressor of variegation 3-9 Homolog 1; TALL, T-cell acute lymphocytic leukemia; TBD, to be determined.

through inhibition of p300/CREB binding to HAT (Balasubramanyam, Varier, et al., 2004). At present, curcumin is being evaluated in several clinical trials for several major human diseases (Table 2).

In addition to the above-mentioned inhibitors, high-throughput screening identified isothiazolones as HAT inhibitors targeting p300 and PCAF (Stimson et al., 2005). However, attempts to optimize this class of inhibitors failed due to the exceptionally high reactivity of isothiazolones with thiolates (Ghizzoni, Haisma, & Dekker, 2009). Other inhibitors of HATs, such as thiazide sulfonamide and C646, that are potent and cell-permeable HAT inhibitors have been identified by virtual screening and shown to have promising effects toward multiple cancers (Bowers et al., 2010).

CTPB, another derivative of anacardic acid, selectively activates p300 resulting in increased gene transcription (Chen et al., 2022). Pentadecylidenemalonate, an analog of anacardic acid, modulates acetyltransferase activity depending on the specific HAT target (Suryanarayanan & Singh, 2019). It exhibits inhibitory properties against p300/CBP and activates PCAF and induces apoptosis in leukemia cancer cells (Sbardella et al., 2008).

Despite significant efforts, the potency HAT inhibitors are limited to the micromolar range, thus there is an urgent need to develop potential inhibitors with improved potency.

3.1.3 Acetyl-lysine recognition proteins or BRD proteins (the readers)

Bromodomain (BRD)-containing proteins recognize and bind to acetyl-lysine marks on histone thus acting as readers of the lysine acetylation state. The human proteome encodes 61 unique bromodomains modules that are found in 42 diverse proteins. These proteins primarily recognize acetylated histones and regulate gene expression through a wide range of activities (Barbieri, Cannizzaro, & Dawson, 2013). From a functional perspective, most of the BRD containing proteins act as scaffolds that facilitate the assembly of larger multi-subunit chromatin complexes. These can be loosely divided into three categories: chromatin modifiers, chromatin readers, and chromatin remodelers. Histone methylation (MLL) and histone acetylation (CREBBP/p300) are two examples of chromatin modifications that can be introduced by the enzymes in the first group. The second category consists of proteins that can attract chromatin modifying complexes or general transcription factors but do not have enzymatic action on chromatin (BET). The third group consists of ATPase-active proteins that can modify chromatin (SMARCA4) (Sarnik, Poplawski, & Tokarz, 2021). BRD-containing proteins are dysregulated in cancer and it has been demonstrated that their abnormal expression both stimulates and suppresses malignant behavior (Barbieri et al., 2013).

3.1.4 BET protein inhibitors

Binding to BRDs and blocking acetylated lysine recognition is another inhibitory mechanism of acetylation (Filippakopoulos & Knapp, 2012) (Table 5). Bromodomain and extra-terminal (BET), which taxonomically belongs to human BRD proteins family and highly specific small molecule inhibitors of this family have recently emerged as promising therapeutic agents in cancer (Table 5). JQ1 and I-BET762 (GSK525762A, molibresib, I-BET) are the two principal inhibitors of the BET family that have been well studied. JQ1 is a cell-permeable small molecule target of BRD4-NUT fusion oncoproteins, resulting in cancer cell death (Zengerle, Chan, & Ciulli, 2015). However, despite robust anti-tumor activity of JQ1, it exhibits a shorter half-life and poor oral bioavailability (Boi et al., 2015). Optimizing chemical and biological properties have resulted in the synthesis of a JQ1 analogue named TEN-010 (JQ2) (Finley & Copeland, 2014) that is undergoing clinical trials in patients with AML, myelodysplastic syndrome (MDS) and solid tumors (Table 5). Similarly, I-BET762 a synthetic mimic is one of the most advanced BET BRD inhibitors that binds the acetyl-recognizing BET pocket with nanomolar affinity and suppresses Myc transcription. I-BET762 has been tested against several preclinical cancer models including neuroblastoma, multiple myeloma and pancreatic cancer (Chaidos et al., 2014; Leal et al., 2017). Other BET inhibitors such as MS417 (GTPL7512), OTX-015 (birabresib), RVX-208 (apabetalone), OXFBD02/4, I-BET151 (GSK1210151A), PFI-1, MS436, and XD14 have been generated with improved selectivity and enhanced drug potency to make these compounds suitable for clinical investigation (Brand et al., 2015). Recently, NEO2734 a novel dual inhibitor of BET and CBP/p300 has been added to the list of inhibitors which potently inhibits growth and induces differentiation of NMC cells in vitro and produces tumor regression and results in a significant survival benefit in preclinical animal models (Morrison-Smith et al., 2020).

BAZ2A and BAZ2B are members of the BAZ protein family, BAZ2A is overexpressed in prostate cancer but very little is known about BAZ2A (Pena-Hernandez et al., 2021). To date, only two BAZ2-specific BRD inhibitors, BAZ2-ICR and GSK2801, have been reported for use in vitro and in vivo (Bevill et al., 2019).

BRD9 is a part of the chromatin remodeling complex SNF/SWI BAF and several BRD9 chemical inhibitors have been reported so far, I-BRD9 is a potent and selective BRD9 inhibitor identified by GlaxoSmithKline (GSK) that shows >200-fold selectivity toward BRD7 relative to BRD9 and >700-fold selectivity toward BRD9 over BET family members (Theodoulou et al., 2016). BI-7271, BI-7273, BI-7189 and BI-9564 are the other potent BRD7/9 inhibitors exerting antitumor activity in an AML cell and in a xenograft model (Martin et al., 2016; Hohmann et al., 2016).

The quinolone-fused lactam LP99 is the first selective BRD7/9 BRD inhibitor, however, its role in cancer has not been studied. The structural genomics consortium (SGC) added TP-472 as a new inhibitor of the BRD7/9 bromodomain, which exhibits >30-fold selectivity for BRD9 over other BRDs and is also suitable for in vivo experiments (Mason, Chava, Reddi, & Gupta, 2021).

PFI-3 potently targets the BRDs of SMARCA2A/B, SMARCA4, as well as the fifth BRD of PB1, LM146, a Potent Inhibitor of PB1 with an improved selectivity profile over SMARCA2 has also been reported (Melin et al., 2021). Within the past 2 years, the development of potent inhibitors against the family of BRD and PHD finger-containing proteins (BRPF) and the tripartite motif containing protein 24 (TRIM24) has been reported. 1,3-Dimethyl benzimidazolones was the first potent and selective inhibitors of BRPF1, leading subsequently to the development of the highly selective BRPF1 inhibitor OF-1, PFI-4, and the dimethylquinolinone NI-57 (Igoe, Bayle, Fedorov, et al., 2017; Igoe, Bayle, Tallant, et al., 2017; Meier et al., 2017). PFI 4 is a potent and selective BRPF1B BRM inhibitor, whereas NI-57 and OF-1 are thought to act as pan-BRPF BRM inhibitors (Meier et al., 2017). IACS-9571 and 1-(indolin-1-yl) ethan-1-ones were developed as a novel selective TRIM24/BRPF1 BRM inhibitor with 9- and 21-fold selectivity against BRPF2 and BRPF3, respectively (Palmer et al., 2016; Xiang et al., 2022). Recently, N-benzyl-3,6-dimethylbenzo[d] isoxazol-5-amines as TRIM24 BRM inhibitors were obtained from an in-house library screen that showed inhibitory effects on the growth of prostate and non-small lung cancer cell lines (Hu et al., 2020). Bromosporine is a pan BRM inhibitor with good cellular activity (D'Ascenzio et al., 2019; Picaud et al., 2016), These BRM inhibitors are specific for the BET family rather than other BRDs (Picaud et al., 2016).

A recent innovation in the design of BET inhibitors is the conjugation of JQ1 with chemical moieties that facilitate the recruitment of E3 ubiquitin ligases, generating a heterobifunctional molecule, known as "proteolysis targeting chimeras" (PROTACs) leading to polyubiquitination and proteasome-dependent degradation of BET proteins. (Gadd et al., 2017; Miller et al., 2022). This new generation of inhibitors (known as dBET1, ARV-825, or MZ1) cause more potent suppression of BET proteins in cells and demonstrate efficacy in solid-tumor malignancies as well as in hematological malignancies (Raina et al., 2016; Saenz et al., 2017; Sun, Fiskus, et al., 2018; Winter et al., 2015).

The second most studies class of drugs is CREBBP BRM inhibitors, after the BET family of inhibitors (Table 5). The first compounds MS2126, MS7972 and Ischemin inhibit the interaction of CREBBP-p53 in ovarian cancer cell lines and consequently alter the expression of p53 target genes involved in DNA damage repair and apoptosis (Borah et al., 2011; Sachchidanand et al., 2006). A collaboration between the GSK and SGC reported I-CBP112 being a novel inhibitor of CREBBP and P300 BRMs in nanomolar range in leukemic cells (Picaud et al., 2015). Another inhibitor SGC-CBP300 is selective for CREBBP/P300 BRMs but its use in vivo has not been recommended (Taylor et al., 2016). CN470 a BET/CBP/p300 multi-BRM inhibitor has shown anti-tumor activity against MLL-r ALL in vitro and in vivo (Imayoshi et al., 2022).

3.2 Histone deacetylase (the eraser)

Based on phylogenetic analysis and sequence similarity to yeast factors, the human genome encodes 18 HDACs grouped into four classes. Class I (HDAC1, HDAC2, HDAC3, HDAC8) with a single deacetylase domain at the N-terminus and diversified C-terminal regions. Class IIa HDACs (HDAC4, HDAC5, HDAC7, HDAC9), with deacetylase domain at a C-terminal position and can shuttle between the nucleus and cytoplasm. Class IIb HDACs (HDAC6 and HDAC10). Class III HDACs Sir2-like enzymes (SIRT1, SIRT2, SIRT3, SIRT4, SIRT5, SIRT6, and SIRT7) which are NAD-dependent protein deacetylases and/or ADP ribosylases (de Ruijter, van Gennip, Caron, Kemp, & van Kuilenburg, 2003; Haberland, Montgomery, & Olson, 2009; Yang & Gregoire, 2005; Yang & Seto, 2008). The recently discovered HDAC11 is the only member of the class IV HDACs, which shares sequences similarity to both class I and II proteins. Aberrant expression and mutation of HDACs has been linked to a variety of malignancies, including solid and hematological tumors (Ropero & Esteller, 2007) (Table 3).

Table 3 Histone deacetylase inhibitors in cancer therapy.

Drug	Target	Condition	Mechanism of action	Clinical trial (https:/clinicaltrials.gov) NCT
Abexinostat	Class I, II	HL, non-HL, CLL	Target HDAC enzymes and inhibit the proliferation of cancer cells and induce cancer cell death, or apoptosis (Lopez et al., 2009; Yang et al., 2011)	NCT03934567, NCT03936153, NCT03592472, NCT04024696
Belinostat	Class I, II, IV	PTCL	Caused the accumulation of acetylated histones and other proteins, inducing cell cycle arrest and/or apoptosis of some transformed cells (McDermott & Jimeno, 2014)	NCT02680795, NCT00865969, NCT01839097, NCT00413075
CG200745	Class I, II, IV	Solid tumors	Induced the acetylation of the tumor suppressor, p53, and cancer cell death (Chun et al., 2015; Oh et al., 2012)	NCT02737462, NCT02737228, NCT01226407
CUDC-101	Class I, II	Squamous cell carcinoma	Displays potent antiproliferative and proapoptotic activities against cultured and implanted tumor cells that are sensitive or resistant to several approved single-targeted drugs (Galloway et al., 2015)	NCT00728793, NCT01171924, NCT01384799
CHR-3996	Class I	Solid tumors	Potent and promising class I HDAC inhibitor with good oral bioavailability and the ability to completely inhibit human tumor xenografts (Banerji et al., 2012)	NCT00697879
CUDC-907	Class I, II	MM, Lymphoma, Solid tumors	Down-regulates DNA damage response proteins and induces DNA damage in cancer cells (Sun et al., 2017)	CT02307240, NCT02909777, NCT02674750, NCT03002623

Givinostat	Class I, II	CLL, MM, HL	Inhibition of proliferation and apoptosis induction, induces cell cycle and differentiation block (Amaru Calzada et al., 2012)	NCT01761968, NCT00792831, NCT00792467
MPT0E028	HDAC1, 2, 6	Solid tumors	Inhibit proliferation and induce apoptosis (Huang et al., 2012; Huang et al., 2015)	NCT02350868
Panobinostat	Class I, II, IV	MM	Inhibits tumor cell growth, proliferation, and differentiation, ultimately leading to cell-cycle arrest (Singh, Patel, Jain, Patel, & Rajak, 2016)	NCT04326764, NCT02717455, NCT01242774, NCT01802879, NCT02722941
Pracinostat	Class I, II, IV	AML, Solid tumor	Accumulates in tumor tissue and exerts a sustained histone hyperacetylation (Novotny-Diermayr et al., 2010; Novotny-Diermayr et al., 2011)	NCT01912274, NCT01993641, NCT01112384, NCT00741234
Quisinostat	Class I, II	Solid tumors, Lymphoma, CTCL	Inhibits cancer cell self-renewal, effectively halting disease maintenance and relapse, increases p53 acetylation (Bao et al., 2016; Child et al., 2016; Morales Torres et al., 2020)	NCT01486277, NCT00677105, NCT01464112, NCT02948075
Resminostat	Class I, II	Colorectal cancer, HCC, HL	Causes the accumulation of highly acetylated histones leading to the inhibition of tumor suppressor genes transcription and the cell division is also blocked leading to the apoptosis of the tumor cells (Karagianni et al., 2021; Mandl-Weber et al., 2010; Zhao & Lawless, 2016)	NCT04955340, NCT02953301, NCT01037478, NCT02400788, NCT01277406
Vorinostat (SAHA)	Class I, II, IVA	CTCL, Solid tumors	Interrupting various transcription factors, thereby producing cell-cycle arrest in some cancer lines (Bolden, Peart, & Johnstone, 2006; Wozniak et al., 2010)	NCT02042989, NCT01059552, NCT01023737, NCT02042989, NCT01720875

Continued

Table 3 Histone deacetylase inhibitors in cancer therapy.—cont'd

Drug	Target	Condition	Mechanism of action	Clinical trial (https:/clinicaltrials.gov) NCT
Chidamide	HDAC1, 2, 3, 10	Breast cancer, NSCLC, PTCL	Suppress cell and xenograft growth and induces apoptosis, Cell Cycle Arrest and Cell Apoptosis (Cao et al., 2021; Huang et al., 2021)	NCT02944812, NCT05330364, NCT05270200, NCT05572983
Entinostat	HDAC1, 2, 3	NSCLC, Solid tumors	Reprograms the tumor innate and adaptive immune milieu (Hicks et al., 2021; Trapani et al., 2017)	NCT02915523, NCT02833155, NCT03552380, NCT01594398,
Mocetinostat	Class I, IV	Hematologic and Solid tumors	Induces apoptosis, inhibited growth of human tumor xenografts (Fournel et al., 2008; Zhang et al., 2011)	NCT02429375, NCT02282358, NCT02954991, NCT00324194
Ricolinostat	HDAC6	MM, Lymphoma, Solid tumors	Increases CD38 RNA levels and CD38 surface expression, enhanced lysis of MM cells (Garcia-Guerrero et al., 2021)	NCT01997840, NCT02091063, NCT02632071, NCT01323751
Tacedinaline	Class I	MM, Lung cancer, Pancreatic cancer	Reduced tumor cell viability and induced apoptosis, enhanced radiosensitivity (Jin et al., 2019)	NCT00005624, NCT00005093, NCT00004861
Romidepsin	HDAC1, 2	CTCL, PTCL	Cell cycle arrest, apoptosis, and inhibition of angiogenesis (Smolewski & Robak, 2017)	NCT03547700, NCT03770000, NCT02296398, NCT01537744
Largazole	Class I	Lung cancer, Solid tumors	E2F1-targeting cell cycle inhibitor, sufficiently crosses the blood-brain barrier (Al-Awadhi et al., 2020; Wu et al., 2013)	NA

AR-42	Class I, II	AML, MM	Prolonged survival of ATL engrafted mice, induces apoptosis and cell cycle arrest (Liva et al., 2020; Zhang et al., 2011; Zimmerman et al., 2011)	NCT02569320, NCT01798901, NCT01129193
Phenylbutyrate	Class I, II	Solid tumors, Hematologic malignancies	Inhibit endoplasmic reticulum stress, can induce growth arrest, differentiation, and apoptosis in cancer cells (Gore & Carducci, 2000; Mostoufi, Baghgoli, & Fereidoonnezhad, 2019)	NCT00005639, NCT00002909, NCT00006450, NCT00006239
Pivanex	Class I, II	NSCLC, Myeloma, CLL	It is rapidly and extensively transported intracellularly because of its high lipophilicity, causes apoptosis of cancer cells through signaling cellular differentiation (Rabizadeh, Merkin, Belyaeva, Shaklai, & Zimra, 2007; Tarasenko et al., 2008)	NCT05147311, NCT00073385
Valproic acid	Class I, II	Solid tumors, Hematologic malignancies	Inhibits proliferation, and induces apoptosis and histone H4 hyperacetylation, in vitro, reduces the tumor growth significantly in two NOD/SCID mouse models (Duenas-Gonzalez et al., 2008; Hudak et al., 2012; Liu et al., 2016)	NCT02124174, NCT02608736, NCT00530907, NCT00996060

Abbreviations: ALL, acute lymphocytic leukemia; AML, acute myelogenous leukemia; CLL, chronic lymphocytic leukemia; CTCL, cutaneous T cell lymphoma; HCC, hepatocellular carcinoma; HDAC, histone deacetylases; HL, Hodgkin's lymphoma; MM, multiple myeloma; NSCLC, non-small-cell lung cancer; PTCL, peripheral T cell lymphoma.

In most cases, high HDAC levels are associated with advanced disease and poor patient outcomes. There are several different ways that individual HDACs control carcinogenesis. HDACs could either decrease the expression of tumor suppressor genes or control oncogenic cell-signaling pathways by altering critical regulatory molecules because they cause a variety of cellular and molecular effects through post translational modification of histone and nonhistone substrates. However, because aberrant HDAC activity is allied to important carcinogenic events, the contribution of HDACs to cancer may not always be correlated with the degree of HDAC expression. Certain HDACs function as the catalytic subunits of large corepressor complexes and can be abnormally recruited to target genes by oncogenic proteins to promote cancer. For example, the aberrant recruitment of HDAC1, 2, or 3 by oncogenic fusion proteins AML1-ETO (acute myeloid leukemia 1 and eight twenty-one) and PML-RAR (promyelocytic leukemia and retinoic acid receptor α) contributes to the pathogenesis of acute myeloid leukemia (AML) (Elagib & Goldfarb, 2007). Knockdown of HDAC1, 2, 3, and 6 induces apoptosis and cell cycle arrest in a variety of cancers such as colon, breast, lung and acute promyelocytic leukemia (APL) implicating HDAC activity as a key mediator of survival and tumorigenic capacity in these cancers (Li & Seto, 2016; Li, Tian, & Zhu, 2020).

Altered expression and mutations of genes that encode HDACs have been linked to tumor progression, HDAC4 is routinely decreased in gastric tumors, HDAC1 somatic mutations has been detected in approximately 8.3% of dedifferentiated human liposarcomas and HDAC4 homozygous deletions occurred in 4% of melanomas (Taylor et al., 2011). Frame-shift mutation causing dysfunctional HDAC2 expression have been observed in human epithelial cancers, which confers enhanced resistance to HDAC inhibitors (Ropero et al., 2006). In individuals with lung and stomach cancer, low expression of HDAC10 is linked to poor prognosis (Jin et al., 2014; Li, Zhang, et al., 2020). HDAC6 is downregulated in a large cohort of human hepatocellular carcinoma (HCC) patients, and low expression of HDAC6 associates with angiogenesis and poor prognosis in liver transplantation patients (Lv et al., 2016) (Table 4).

Recently, the sirtuins or class III HDACs have received increased attention. These HDACs are key regulators of normal cell growth and proliferation (Voelter-Mahlknecht & Mahlknecht, 2010). In particular, SIRT1 not only deacetylates histones H1, H3 and H4 but also regulates histone acetylation of many non-histone proteins p53, Ku70, PPARγ, PGC1α, Beclin 1, β-catenin, p300 histone acetyltransferase, E2F1, NF-KB and the androgen

Table 4 FDA approved drug.

Category	Compound	Clinical name	Disease	Case number and approved year	FDA approved number	Company
DNMTi	Azacitidine (5-Azacitidine, 5-Aza-CR)	Vidaza	MDS	320-67-2(2004)	50794	Pharmion Corporation
	5-Aza-2′-deoxycytidine (5-Aza-CdR, decitabine)	Dacogen	MDS	2353-33-5(2006)	21790	Janssen Pharmaceuticals
	Hydralazine		Hypertension	86-54-4(1997)	8303	
HDACi	Suberoylanilide hydroxamic acid (SAHA) (Vorinostat)	Zolinza	CTCL	149647-78-9(2006)	21991	Merck
	Romidepsin (Depsipeptide, FK-229, FR901228)	Istodax	CTCL	128517-07-7	22393	Celgene
	Belinostat (PXD101)	Beleodaq	CTCL	414864-00-9(2014)	206256	TopoTarget
	Panobinostat (LBH589)	Farydak	Multiple myeloma	404950-80-7	205353	Novartis
EZH2i	Tazemetostat	TAZVERIK®	Follicular lymphoma	E7438-G000-101 (2020)	211723	Epizyme, Inc.
IDH1 inhibitor	Ivosidenib	Tibsovo	AML	AG120-C-005 (2022)	211192	Agios Pharmaceuticals, Inc

Abbreviations: CTCL, cutaneous T-cell lymphoma; DNMTI, DNA methyltransferase inhibitor; EZH2, enhancer of zeste homolog 2; HDAC, histone deacetylase inhibitor; IDH, isocitrate dehydrogenase; MDS, myelodysplastic syndrome.

receptor (Lin & Fang, 2013). SIRT1 also targets the Forkhead box O (FOXO) family of transcription factors, FOXO3a hyperacetylation mediated by SIRT1 accompanies apoptosis while FOXO4 deacetylation activates this target enhancing its transcriptional and biological activity in the nucleus (Brooks & Gu, 2009). SIRTs genes in cancers are either upregulated or downregulated. For instance, SIRT1 is upregulated in lung cancer, prostate cancer (Huffman et al., 2007; Yeung et al., 2004) and leukemia (Dickson, Riddle, Brooks, Pasha, & Zhang, 2013) and is downregulated in gastric cancers (Yang et al., 2013). SIRT2 acts as a tumor suppressor that deacetylates substrates involved in the modulation of the cell cycle H4K16, P53, P65, FOXO1, FOXO3, and CDK4 (Du, Wu, Zhang, Li, & Sun, 2017; Wang et al., 2012). SIRT3 is a mitochondria-localized tumor suppressor that regulates mitochondrial function mediating deacetylation of a variety of substrates, such as P53, GSK-3β, PDHA1, IDH2, and NMNAT2 (Li et al., 2010, 2013; Torrens-Mas, Oliver, Roca, & Sastre-Serra, 2017). SIRT4 is also a mitochondrial sirtuin that can induce G1 cell cycle arrest in gastric cancer and can also inhibit glutamine metabolism in the mitochondria of cancer cells (Hu et al., 2019). SIRT5 mediates lysine deglutarylation, desuccinylation, and demalonylation in colorectal cancer and gastric cancer (Lu, Yang, et al., 2019; Wang, Wang, et al., 2018). Overexpression of SIRT5 significantly reduces the levels of ROS and protects neuroblastoma cells from staurosporine-induced apoptosis (Liang, Wang, Ow, Chen, & Ong, 2017). SIRT5 can inhibit cell apoptosis by deacetylating cytochrome c in hepatocellular carcinoma (Zhang et al., 2019). Deacetylation of several genes such as PKM2, NF-κB, HIF1α, CtBP, and JUN by SIRT-6 results in a reduction in cell proliferation in GBM, colorectal and hepatocellular carcinomas (Feng et al., 2016; Marquardt et al., 2013; Tian & Yuan, 2018; Xiangyun et al., 2017; Zhang, Yu, Huang, & Tang, 2019). SIRT7 promotes proliferation, migration, and invasion of osteosarcoma cells via downregulating CDC4 expression (Wei, Jing, Ke, & Yi, 2017).

3.2.1 Histone deacetylase inhibitors

HDAC inhibitors modify both histone and non-histone proteins and can sensitize cancer cells to chemotherapy and immunotherapy. In these contexts, targeting these enzymes could facilitate the identification of promising anti-cancer agents. Butyrate a product of bacterial anaerobic fermentation was found to inhibit histone deacetylation in cancer cells, β-hydroxybutyrate derivative of butyrate was shown to have HDAC inhibitory activity (Candido, Reeves, & Davie, 1978; Shimazu et al., 2013).

Later, a natural extract trichostatin A (TSA), originally reported as a fungistatic antibiotic, selectively inhibited the activity of class I and II HDACs and induced cancer cell differentiation and apoptosis that lead to identification of more natural and synthetic compounds (Kim et al., 2006) (Table 2).

The short-chain fatty acid group containing sodium butyrate, valproic acid (VPA), sodium phenylbutyrate, and pivanex (pivaloyloxymethyl butyrate; AN-9) are selective inhibitors of class I HDACs. Butyrate typically has an effective concentration at the micromolar range. The class that has been investigated the most is the group of hydroxamic acids, which has more than 10 members. Since TSA and suberoylanilide hydroxamic acid (SAHA) share significant homology with class I and class II HDACs indicating that they are mimics of lysine substrates and noncompetitive inhibitors of HDACs. They can chelate zinc ions in a bidentate fashion, which is essential for enzymatic activity (Furumai et al., 2001). Hexamethylene bisacetamide (HMBA) belongs to a class of hybrid bipolar compounds, which induces terminal differentiation in a variety of leukemic cell lines and was studied in several Phase I and Phase II clinical trials for the treatment of myelodysplastic syndrome (MDS), acute myelogenous leukemia (AML), general advanced cancer, and solid tumors (Smith, Ketchart, Zhou, Montano, & Xu, 2011). Whereas second-generation hybrid polar compounds (HPCs), such as oxamflatin, SAHA, suberic bishydroxamic acid (SBHA), and m-carboxycinnamic acid bishydroxamide (CBHA), have demonstrated more potent and specific anticancer effects than first-generation agents (Richon et al., 1998). TSA analogs such as oxyamflatin, scriptaid, and amide have anticancer properties.(Kim, Lee, Sugita, Yoshida, & Horinouchi, 1999). Benzamides are another class of the bidentate zinc binder HDAC inhibitors that includes Entinostat (MS-275, SNDX-275), Mocetinostat (MGCD0103), and CI-994 (N-acetyldinaline) that are well-studied and undergoing clinical trials for treatment of various cancers including follicular lymphoma, Hodgkin's lymphoma, and acute myelogenous leukemia. (Suzuki et al., 1999). The inhibitor coordinates to the catalytic Zn^{2+} ion through carbonyl and amino groups and inhibits histone deacetylation. HDACs inhibition by benzamide inhibitors is thought to be selective and reversible, but the bond may become firm and pseudo-irreversible in a time-dependent manner (Bondarev et al., 2021).

Of note, second-generation HDACs inhibitor including, Quisinostat (JNJ-26481585), benzamides, entinostat (MS-275), tacedinaline (CI-994), and mocetinostat (MGCD010), are currently in clinical trials, whereas panobinostat (LBH589) received FDA approval in 2015 for the treatment of patients with refractory multiple myeloma, belinostat (PXD101) received FDA approval in 2014 for the treatment of PTCL (Tables 3 and 5). SAHA (vorinostat), which is only prescribed to people with cutaneous T-cell lymphoma (CTCL) for the treatment of cancer, was initially approved by the US Food and Drug Administration (FDA) in 2006 (Mann, Johnson, Cohen, Justice, & Pazdur, 2007) (Table 5).

Approval of Romidepsin's (FK-228) for the treatment of cutaneous T-cell lymphoma in November 2009 by the US Food and Drug hastened the development of HDAC inhibitors as anti-cancer medications. (Grant et al., 2010). Romidepsin (Istodax) was the second approved HDAC inhibitor is a potent and selective inhibitor of HDAC1 and HDAC2 at low nanomolar levels and HDAC4 and HDAC6 at micromolar levels (Gryder, Sodji, & Oyelere, 2012). BRD8430 and compound 60 are selective HDAC1 and HDAC2 inhibitors that induce differentiation and cell death in neuroblastoma cells whereas MRLB-223 induce apoptosis in brachial lymph of mice (Frumm et al., 2013). However, MRLB-223 was less effective in inducing apoptosis and promoting therapeutic efficacy compared to the HDAC inhibitor vorinostat (Newbold et al., 2013). RGFP966 an N-(o-aminophenyl) carboxamide HDAC3-selective inhibitor has been found to induce growth arrest and increase apoptosis in refractory CTCL cells by disrupting their DNA replication activity (Wells et al., 2013). BG45, a HDAC3 selective inhibitor induces multiple myeloma cell toxicity, without affecting normal donor PBMCs by hyperacetylation and hypophosphorylation of STAT3 (Minami et al., 2014; Tang et al., 2018). T247 and T326, identified as potent and selective HDAC3 inhibitors by screening 504-member triazole library using click chemistry, inhibited the growth of colon and prostate cancer cells by selectively enhancing the acetylation of NF-κB (Suzuki et al., 2013). The target that has shown to be most promising for achieving selectivity is HDAC8. Two structurally divergent HDAC8 inhibitors, Compound 2 (Cpd2; 1-naphthohydroxamic acid) (Krennhrubec, Marshall, Hedglin, Verdin, & Ulrich, 2007) and PCI-34051 (Balasubramanian et al., 2008), induced cell differentiation, and apoptosis in neuroblastoma cells, while untransformed cells were unaffected (Rettig et al., 2015). PCI-34051 (>200-fold selectivity over the other HDAC isoforms) induces caspase-dependent apoptosis in cell lines derived from T-cell lymphomas or leukemias, but not in other hematopoietic or solid tumor cell lines (Balasubramanian et al., 2008). A recent study indicates that PCI-34051 enhances the anticancer effects of ACY-241 in ovarian cancer

Table 5 BET inhibitors.

Condition	Drug	Target	Mechanism of action	Clinical trial (https://clinicaltrials.gov) NCT
Prostate cancer, AML with MLL translocations, MM, NUT Midline carcinoma	JQ1	BET proteins	Induces G1 cell cycle arrest and apoptosis and produces expression changes in key genes such as c-MYC, p21, hTERT, BCL-2, and BCL-XL (Wang, Xu, Kao, Tsai, & Tsai, 2020; Zhang et al., 2021)	NA
Hematologic malignancies, NUT Midline carcinoma, Solid tumors	I-BET762 (GSK525762)	BET proteins	Exert antiproliferative effects and induce apoptosis, inhibit MYC expression (Wyce et al., 2018)	NCT01943851, NCT02964507, NCT01587703
AML, DLBCL, ALL, MM, CRPC, TNBC, PDAC, NSCLC	OTX015(MK-8628)	BRD2, 3, 4	Cell growth inhibition, cell cycle arrest and apoptosis (Djamai et al., 2021; Han et al., 2019)	NCT01713582, NCT02259114
MM	I-BET151 (GSK1210151A)	BET proteins	Induces apoptosis and cell cycle arrest, decrease in the BCL-2, MYC, and CDK6 genes expression (Shorstova, Foulkes, & Witcher, 2021)	NA
MM	CPI203	BRD4	Increase the apoptosis level, downregulate MYC expression, induces G1 cell cycle arrest (Diaz et al., 2017)	NA
MM	RVX2135	Brd2, 3, 4, and BRDT	Inhibits proliferation and induces apoptosis (Bhadury et al., 2014)	NA
Prostate cancer, Leukemia	PFI-1	BRD4/BRD2	Cell cycle arrest in G1, downregulation of MYC expression and induction of apoptosis (Hupe et al., 2019; Picaud et al., 2013)	NA

Continued

Table 5 BET inhibitors.—cont'd

Condition	Drug	Target	Mechanism of action	Clinical trial (https://clinicaltrials.gov) NCT
Melanoma	MS436	BRD4	Inhibit NF-Kβ-directed production of nitric oxide and pro-inflammatory cytokine interleukin-6 (Segura et al., 2013)	NA
AML MDS NHL	FT-1101	BRD2, 3, 4, and BRDT	Potent anti-proliferative activity, suppression of MYC (Sun, Han, et al., 2020)	NCT02543879
MM, Solid Tumor	CPI-0610	BRD2/4	G1 cell cycle arrest, potent cytotoxicity (Siu et al., 2017)	NCT02157636, NCT01949883, NCT05391022
Neoplasms	BAY1238097	BET inhibitor	Suppress MYC expression (Bernasconi et al., 2017)	NCT02369029
Solid Tumors and Hematologic Malignancy	INCB054329	BRD4	DNA damage induction and apoptosis (Stubbs et al., 2019; Wilson, Stubbs, Liu, Ruggeri, & Khabele, 2018)	NCT02431260
Solid Tumors, Advanced Solid Tumors, AML MDS	TEN-010 (RO6870810)	BET inhibitor	Deregulated MYC expression (Shapiro et al., 2021)	NCT02308761, NCT01987362
Cancer, Neoplasms	GSK2820151	BET inhibitor	Disrupt interaction between the BET proteins and acetylated histones (Alqahtani et al., 2019)	NCT02630251
Metastatic Castration-Resistant Prostate Cancer, Ovarian Cancer, TNBC Metastatic Castration-Resistant Prostate Cancer	ZEN003694	BET inhibitor	Inhibits the production of certain growth-promoting proteins and may prevent proliferation of tumor cells (Aggarwal et al., 2020)	NCT04986423, NCT05071937, NCT03901469, NCT02705469

AML	BAY-299	BRD1 and TAF1	Increase expression of cell cycle inhibit genes and multiple pyroptosis-promoting genes (Zhou, Yao, Li, & Chen, 2021; Zhou, Yao, Ma, Li, & Chen, 2021)	NA
Advanced tumors, MM, Solid Tumor, Childhood Lymphoma	BMS-986158	BET inhibitor	Suppresses tumor growth (Hilton et al., 2022; Yin et al., 2020)	NCT03936465, NCT02419417, NCT05372354
Breast Cancer, NSLC, AML, MM, Prostate Cancer, SCLC, NHL	ABBV-075	BRD2/4/T	Display potent antiproliferative activity (Faivre et al., 2020; Piha-Paul et al., 2019)	NCT02391480
Solid tumors and Lymphomas	GS-5829	BET inhibitor	Inhibited signaling pathways within nurse like cells and their growth (Aggarwal et al., 2022; Kim, Ten Hacken, et al., 2020)	NCT02392611
AML, MDS	PLX51107	BRD2/3/4/T	Decreases cell proliferation, induces apoptosis, and cell cycle arrest (Alcitepe, Salcin, Karatekin, & Kaymaz, 2022)	NCT04022785
Prostate cancer	BAZ2-ICR	BAZ2A/B	PCa stem cells and PTEN-loss mediated oncogenic transformation (Pena-Hernandez et al., 2021)	NA
TNBC	GSK2801		Induce apoptosis (Bevill et al., 2019)	NA
Leukemia	I-BRD9	BRD9	Induces apoptosis and cell cycle inhibition (Zhou, Yao, Li, & Chen, 2021; Zhou, Yao, Ma, et al., 2021)	NA

Continued

Table 5 BET inhibitors.—cont'd

Condition	Drug	Target	Mechanism of action	Clinical trial (https://clinicaltrials.gov) NCT
AML	BI-7273/BI-9564	BRD9	Suppressed the proliferation (Martin et al., 2016)	
AML	BI-7271/BI-7273/BI-7189	BRD9	Suppressed the proliferation (Martin et al., 2016)	
NT	LP99	BRD9/BRD7	Regulating pro-inflammatory cytokine secretion (Clark et al., 2015)	NA
Melanoma	TP-472	BRD9/BRD7	Downregulates cancer-promoting signaling pathways and induces cell death (Mason et al., 2021)	NA
NT	OF-1	PAN-BRPF inhibitors	Impaired RANKL-induced differentiation of primary murine bone marrow cells and human primary monocytes into bone resorbing osteoclasts (Meier et al., 2017)	NA
NT	PFI-4	BRPF1B inhibitor	Impaired RANKL-induced differentiation of primary murine bone marrow cells and human primary monocytes into bone resorbing osteoclasts (Meier et al., 2017)	NA
NT	NI-57	PAN-BRPF inhibitors	Inhibits RANKL/MCSF induced differentiation of primary human monocytes into osteoclasts (Meier et al., 2017)	NA
Lung cancer, Synovial sarcoma, Rhabdoid cancer	PFI-3	SMARCA2/4 Inhibitor	PFI-3 synergistically sensitizes several human cancer cell lines to DNA damage (Lee et al., 2021)	NA

Osteosarcoma	MS2126/MS7972	Blocks human p53 and CREB binding	Stimulate DNA damage resulted in a dramatic decrease in p53 levels (Sachchidanand et al., 2006)	NA
NT	Ischemin	Blocks human p53 and CREB binding	Stimulate DNA damage resulted in a dramatic decrease in p53 levels (Sachchidanand et al., 2006)	NA
Leukemia and Prostate cancer	I-CBP112	CBP/p300 inhibitor	Suppress the lymphocyte-specific transcription factor IRF4 (Picaud et al., 2015)	NA
Multiple myeloma	SGC-CBP30	CBP/p300 inhibitor	Suppress the lymphocyte-specific transcription factor IRF4 (Picaud et al., 2015)	NA

Abbreviations: ALL, acute lymphocytic leukemia; AML, acute myelogenous leukemia; BRD, bromodomain; BET, bromodomain and extraterminal domain; BRPF, bromodomain and PHD finger-containing; DLBCL, diffuse large B-cell lymphoma; CBP, CREB-binding protein; CLL, chronic lymphocytic leukemia; CREB, cAMP response element-binding protein; CRPC, Castration-resistant prostate cancer; CTCL, cutaneous T cell lymphoma; HCC, hepatocellular carcinoma; HL, Hodgkin's lymphoma; M-CSF, macrophage-colony stimulating factor; MM, multiple myeloma; NSCLC, non-small-cell lung cancer; PDAC, pancreatic ductal adenocarcinoma; PTCL, peripheral-T cell lymphoma; RANKL, receptor activator of nuclear factor kappa-B ligand; TNBC, triple-negative breast cancer.

cells (Kim et al., 2022). PCI-34051 and its variant PCI-48012 induce S phase cell-cycle arrest and apoptosis in human and murine-derived malignant peripheral nerve sheath tumors (MPNST), however, the underlying mechanism remains unclear (Lopez et al., 2015). PCI-48012, with improved pharmacokinetic properties and an in vivo stable HDAC8 inhibitor has significant antitumor activity without toxicity in xenograft mouse models (Rettig et al., 2015). PCI-48012 in combination with retinoic acid inhibit cell growth and induce differentiation in neuroblastoma cells and delays tumor growth in vivo (Rettig et al., 2015). YK-4-272, an unconventional HDAC inhibitor which binds to HDAC4 synergistically inhibited growth and induced apoptosis in human prostate cancer cells. YK-4-272 binding to HDAC4 results in accumulation of HDAC4 in the cytoplasm and increased acetylation of tubulin and nuclear histones (Kong et al., 2012). Tasquinimod, a small molecule that allosterically binds to the regulatory Zn^{2+} binding domain of HDAC4 and has a good safety profile was used to treat prostate cancer in phase II and III clinical trials (Isaacs et al., 2013; Pili et al., 2011; Sternberg et al., 2016).

Several HDAC6 inhibitors have been developed as potential cancer therapeutics. Tubacin was the first synthesized HDAC6-specific inhibitor discovered in 2003 but was not favorable as a drug-like compound because of its high lipophilicity, fast metabolization in vivo, and complex synthesis (Haggarty, Koeller, Wong, Grozinger, & Schreiber, 2003). Later, another HDAC6 inhibitor tubastatin A, a hydroxamic acid, was developed based on structural homology modeling and showed inhibitory properties both in vitro and in vivo (Butler et al., 2010). However, low bioavailability, short half-life, and rapid metabolism, prevented further clinical development. Ricolinostat (ACY-1215) was the first potent and selective HDAC6 inhibitor to enter in clinical studies of patients with relapsed/refractory multiple myeloma or lymphoma (Amengual et al., 2021). ACY-1215 has shown satisfactory efficacy in various tumors (Corno et al., 2020; Huang et al., 2020; Kim et al., 2022; Ruan, Wang, & Lu, 2021). C1A is a selective inhibitor of HDAC6, which targets a-tubulin and HSP90 induces growth inhibition in a panel of cancer cell lines at the low micromolar range and exhibits good in vivo potency (Kaliszczak et al., 2013). Dihydropyrazole-Carbohydrazide derivatives have been evaluated as promising drugs for the treatment of breast cancer by targeting HDAC6 (Balbuena-Rebolledo et al., 2022).

In the development of anticancer drugs, three other benzamide family members have demonstrated therapeutic importance. Belinostat (Beleodaq, previously known as PXD101) is a histone deacetylase inhibitor developed by Topo Target that was approved in 2014 by the US FDA and European Medicines Agency to treat peripheral T-cell lymphoma (Lee, Kwitkowski, et al., 2015). A phase 1 study of belinostat (PXD-101) and bortezomib (Velcade, PS-341) in patients with relapsed or refractory acute leukemia and myelodysplastic syndrome yielded stable disease and this treatment strategy appears feasible for further study (Holkova et al., 2021). Panobinostat is a potent orally active non-selective HDAC Inhibitor. It has been approved by the U.S. Food and Drug Administration (FDA) under the brand name Farydak proven to be effective in patients with advanced CTCL and PTCL (Duvic et al., 2013; Liu, Li, Wu, & Liu, 2022; Tan et al., 2015).

Since up-regulated SIRT1 has been identified in cancer cell lines, this raises the prospect that sirtuin inhibition may decrease the growth of cancer cells and sirtuin inhibitors may also be beneficial as therapeutic drugs. Sirtinol 2-[(2-Hydroxynaphthalen-1-ylmethylene)amino]-N-(1-phenethyl)benzamide a sirtuin inhibitor which does not affect HDAC1 induced apoptosis and autophagic ell death in human breast cancer MCF-7 and lung cancer H1299 cells (Ota et al., 2006). Moreover, it has been reported that administration of SIRT3 inhibitors sensitize oral squamous cell carcinoma cells to radiation and cisplatin inhibited cell proliferation and enhanced apoptosis in oral squamous cell carcinoma (OSCC) cell lines in vitro and in vivo (Alhazzazi et al., 2011). Cambinol is a cell-permeable beta-naphthol derivative that inhibits both SIRT1 and SIRT2 in vitro and was well tolerated in mice and inhibited growth of BCL6-expressing Burkitt lymphoma cell lines, inducing apoptosis via hyperacetylation of p53 and BCL6 (Heltweg et al., 2006). Tenovin-1 and its more water soluble analog tenovin-6 compounds delay growth of tumors in preclinical model of highly aggressive melanoma and are active in vivo without significant general toxicity (Lain et al., 2008). Treatment of mice with the tenovin-6 reduces tumor growth in vivo (Yuan, Wang, Li, et al., 2012; Yuan, Wang, Paulk, et al., 2012). Other inhibitors of SIRT families have been described, and they have been previously reviewed (Dai, Sinclair, Ellis, & Steegborn, 2018).

3.3 Histone methylation (lysine and arginine)

Histone methylation is a dynamic process, and methyl groups can be added or removed by specific enzymes. Like histone acetylation, histone methylation also consists of three important components: "writers," histone methyltransferases (HMTs), "readers," histone methylation-recognizing proteins,

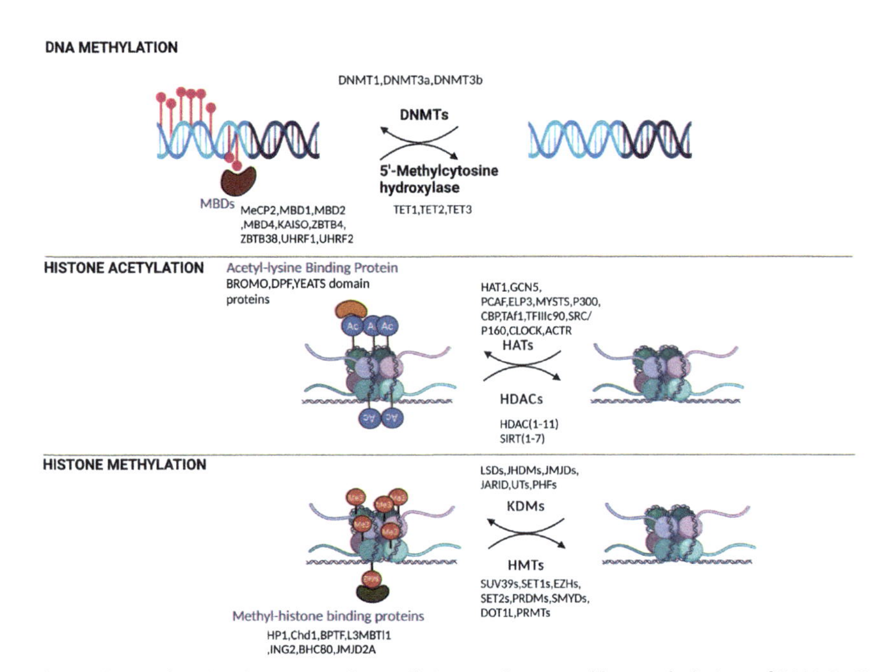

Fig. 2 Gene silencing in mammalian cells is usually caused by methylation of DNA CpG islands together with hypoacetylated and hypermethylated histones. The "writers" (DNMTs, HATs, and HMTs) and "erasers" (DNA-demethylating enzymes, HDACs, and KDMs) are enzymes catalyze the removal of epigenetic marks; MBDs and other binding proteins are "readers" that recognize methyl-CpGs bind to the epigenetic marks and recruit other regulators. DNMTs, DNA methyltransferases; MBDs, methyl-CpG binding domain proteins; HATs, histone acetylases; HDACs, histone deacetylases; HMTs, histone methyltransferases; KDMs, histone-demethylating enzymes. Several drugs targeting writers and erasers are already in the clinic and many more, including reader inhibitors, are progressing through clinical trials.

and "erasers," histone demethylases (HDMs) (Fig. 2). Methylation of histones can occur at arginine and lysine residues. Lysine can be mono-, di-, or trimethylated, whereas arginine is capable of being mono-methylated or di-methylated. Histone methylation can either promote or inhibit gene expression, depending on which amino acids in the histones are methylated and how many methyl groups are attached (Heintzman et al., 2009). For example, lysine methylation at H3K9, H3K27, and H4K20 is marks for heterochromatin and is associated with repressed gene state, whereas methylation of H3K4, H3K36, and H3K79 typically found at active transcription sites and induces gene expression (Barski et al., 2007). Mutation of H3K27M (lysine 27 to methionine) and H3K36M are two important oncogenic events of pediatric gliomas and sarcomas (Lu, Ramirez, et al., 2019;

Mosaab et al., 2020; Pengelly, Copur, Jackle, Herzig, & Muller, 2013; Phillips, Soshnev, & Allis, 2020; Schwartzentruber et al., 2012).

3.3.1 Histone methyltransferase

One of the most significant and extensively researched post-translational modifications is histone methylation, which is catalyzed by protein methyl transferases (PMTs) and is involved in a variety of biological processes, including the development and maintenance of heterochromatin, transcriptional regulation, DNA repair, X-chromosome inactivation, and RNA maturation (Martin & Zhang, 2005). Based on the type of residues they modify; they are separated into two groups: protein lysine methyltransferases (PKMTs) and protein arginine methyltransferases (PRMTs). PKMTs catalyze mono-, di- or tri-methylation marks on specific lysine residues of histone while PRMTs catalyze mono- and/or asymmetric or symmetric dimethylation of arginine guanidinium groups (Fuhrmann, Clancy, & Thompson, 2015; Martin & Zhang, 2005). PKMTs are broadly defined as SET [Su(var)3–9, Enhancer of Zeste, Trithorax] domain-containing proteins and non-SET domain-containing proteins. The only non-SET domain-containing KMTs identified so far are DOT1 (disrupter of telomeric silencing 1) and its mammalian homolog, DOT1L (DOT1-Like), family members, which methylate K79 of histone H3 in the core domain and which also do not share structural similarities with SET-containing proteins but structurally more akin to PRMT (Nguyen & Zhang, 2011). DOT1L activity has been demonstrated to be a required for the development and maintenance of MLL-rearranged leukemia that affects patients of all ages (Daigle et al., 2013). The Pre-SET domain as found in the SUV39 SET family contains nine invariant cysteine residues which coordinate with three zinc ions for its function. The SET1 family members are large (approximately 1707–5537 amino acid) proteins with a 130-amino-acid long C-terminal SET domain and share a similar post-SET motif of three conserved cysteine residues (Zhang et al., 2015). The SET2 family share the conserved domains such an AWS motif that contains 7–9 cysteines are overexpressed in multiple myeloma, neuroblastoma, bladder, and breast cancers, and correlate with poor prognosis (Bilokapic & Halic, 2019). The members of the enhancer of zeste homolog (EZH) family are the catalytic components of polycomb repressive complexes (PRCs), have been found to be responsible for gene silencing through its methyltransferase activity. EZH2 catalyzes di- and tri- methylation of histone H1 at lysine 27 and is frequently overexpressed in many types of tumors, and is associated with aggressiveness of tumors (Kleer et al., 2003;

Varambally et al., 2002). The SMYD family members, consisting of an N-terminal SET domain split by a MYND (myeloid-nervy-DEAF1) domain, with the exception of Smyd5, a zinc-finger motif responsible for protein–protein interaction (Du, Tan, & Zhang, 2014) and plays a major role in the development of the heart and muscle, and its potential role in the development of cancer (Doughan, Spellmon, Li, & Yang, 2016). A PR domain homolog of the SET domain is found in the RIZ (PRDM) family, which is a sizable family. Both the PR and SET domains have between 20% and 30% sequence similarity and can cause methylation of histone H3. RIZ gene normally produces two different products, RIZ1 and RIZ2, which are both widely expressed and have been proven to induce the methylation of histones. Epigenetic silencing of RIZ1 and/or an increase of RIZ2 expression levels has been frequently revealed in many human cancer tissues and cell lines (Kim, Geng, & Huang, 2003). PRDM1/BLIMP1 a master transcriptional regulator with a PR/SET domain has been identified to interact with EHMT2, acts as a tumor suppressor gene in different types of lymphomas derived from B, T, and NK cells (Casamassimi et al., 2020). PRDM6 gene encodes for a transcriptional repressor by interacting with class I HDACs and EHMT2 and involved in the regulation of endothelial cell proliferation, survival, and differentiation (Armengol et al., 2007). Other SET-containing histone lysine methyltransferases, including SET7/9, SET8, SUV4-20H1, and SUV4-20H2, cannot be categorized into these families because they lack a distinguishing sequence or structure flanking the SET domain. Notably, certain KMTs have several domains, which enables them to interact with other proteins, particularly those that operate as epigenetic modifiers (Qian & Zhou, 2006). G9a also belong to the Su(var) 3–9 family of proteins and have pre-SET and post-SET domains which contributes to its methyltransferase activity in mammalian euchromatin (Tachibana et al., 2002). G9a mediates the epigenetic regulation of genes related to epithelial–mesenchymal transition (EMT), leading to the suppression of breast cancer growth and metastasis (Jin et al., 2022). SUV39H1 which functions as a histone methyltransferase that mainly catalyzes the trimethylation of histone 3 lysine 9 (H3K9) in vivo and is thought to be essential for heterochromatin stability and integrity (Wang et al., 2013). MLL1 contains both the CXXC domain, which recognizes unmethylated DNA, and the histone methyltransferase SET domain (Risner et al., 2013). SETDB1 contains a tandem Tudor domains (TTD), a methyl-DNA binding domain (MBD), and a classical catalytic SET domain that recognizes histone H3 sequences containing both methylated and acetylated lysine's (Guo et al., 2021).

In contrast to KMTs, there is another important family of protein methyltransferase, namely, protein arginine methyltransferases (PRMTs) that transfer the methyl group from AdoMet to the guanidino group of arginine in protein substrates (Lee, Kim, & Paik, 1977) and total of nine PRMTs have been identified and characterized in mammals to date (Bedford & Clarke, 2009; Yang & Bedford, 2013). PRMTs are ubiquitously expressed in mammalian cells and regulate primary cellular processes, including cell proliferation, protein-protein interactions, DNA repair, mRNA splicing, metabolism, survival, and differentiation. Deregulation of these enzymes has been implicated in the tumorigenesis and malignancy. The abnormal expression of PRMT1, 4, 5 and 6 frequently leads to tumorigenesis and malignancy (Hwang, Cho, Bae, Kim, & Kim, 2021). Taken together, many members in the PKMT and PRMT family have shown essential role in tumorigenesis. Thus, it is not surprising that PKMTs and PRMTs have become significant targets in cancer therapeutic development in recent years and there has been a considerable advance in the identification of clinically relevant PKMT and PRMT inhibitors (Copeland, 2018; Smith et al., 2018).

3.3.2 Histone demethylase (the eraser)

Histone demethylases catalyze the removal of methyl groups from histones as well as non-histone proteins and help in resetting the chromatin states (Chen, Hu, & Zhou, 2011; Schotta et al., 2008). Two classes of histone demethylases (KDMs) govern demethylation of histones: Lysine-specific demethylases 1 and 2 (LSD1 and 2) or KDM1A and B, respectively, and the JumonjiC (JMJC) domain-containing histone lysine demethylases (Deleris et al., 2010). The LSD-family of KDMs are flavin adenine dinucleotide (FAD)-dependent amine oxidases recognizing only mono and di-methylated lysine, whereas the domain-containing demethylases are iron and α-ketoglutarate (2-oxoglutarate (2-OG))-dependent enzymes that remove methyl from all three states of lysine methylation (Hou & Yu, 2010).

The histone demethylase LSD1 (KDM1A) is highly expressed in several cancers such as breast, prostate, esophageal, bladder and lung cancer, neuroblastoma, and acute myeloid leukemia (AML) and is specifically required for terminal differentiation of hematopoietic cells (Maiques-Diaz & Somervaille, 2016). The normal function of LSD1 is to demethylate H3K4me1/2, which inhibits transcription. LSD1 has a C-terminal AOL (amine oxidase-like) domain that is structurally related to monoaminoxidases (MAO) (Kong et al., 2011; Perillo, Tramontano, Pezone, & Migliaccio, 2020). Members of the KDM2 family can either promote or suppress tumorigenesis in different human malignancies. KDM2B inhibits cell senescence by regulating the expression of INK4/Arf locus genes both indirectly through EZH2 modulation and by direct epigenetic modifications (Tzatsos et al., 2011). Moreover, it promotes cell proliferation and migration as well as self-renewal and stem-like features. KDM2B was found to be overexpressed in HPV+ head and neck carcinomas, and its overexpression was associated with c-MYC copy number gains (Peta et al., 2018). KDM3A functioning as an oxygen sensor also enhances HIF-1 transcriptional activity within the tumor microenvironment and promotes tumor cell migration and invasion. Inhibition of KDM3A decreases the capacity for phagocytosis and migration promoting M1 and restraining M2 macrophage phenotype polarization (Liuet al., 2019). The survival of MLL-AF9 leukemia cells depends on the catalytic JmjC-domain and the zinc finger domain of KDM3C and is mutated in patients with intracranial germline tumors (Sroczynska et al., 2014; Wang et al., 2014). Overexpression of KDM4A, KDM4B, and KDM4C have been found in breast, esophageal, lung, prostate, non-small cell lung and lymphatic cancer with decreased levels of di- and tri-methylated lysine at H3K9 and increased levels of mono-methylation at lysine H3K9 (Lee et al., 2020). Additionally, the KDM4 family of demethylases was the first one to be discovered to target tri-methylated lysine's. The KDM4 family's abnormal expression may cause genomic instability and contribute to the development of tumors (Young, McDonald, & Hendzel, 2013). Typically, members of the KDM6 family operate as tumor suppressors and loss of KDM6A results in decreased expression of the tumor suppressor genes CDKN1A, LOXL1, SASH1, TXNIP and IGFBP2 (Watanabe et al., 2019). In contrast, overexpression of KDM6B inhibits proliferation and induces differentiation genes in neuroblastoma differentiation (Yang et al., 2019). KDM7C has been found to be downregulated in colon and stomach cancers and is required for the efficacy of oxaliplatin and doxorubicin for the activation of p53 (Lee, Park, et al., 2015).

3.3.3 Histone methyltransferase and histone demethyltransferase inhibitors

Due to their combined efficacy and potency in treating various cancers, small molecule inhibitors of histone methyltransferases and histone demethyltransferases have attracted a lot of attention in recent years. This section will cover a few HMTs and PRMTs' small molecule inhibitors that have either had positive preclinical or clinical trial outcomes or that have received

FDA approval. DOT1L is one of the most studied H3K79 methyltransferase and has an important role in MLL (KMT2A) leukemias. (Bernt et al., 2011). Therefore, DOT1L inhibitors represent prospective therapy options for this subtype of cancer. PZ004777 was the first potent and selective DOT1L inhibitor (Table 2). A bromine substitution at the 7 position of the deazaadenosine ring of EPZ004777 led to the discovery of SGC0946, which was about 10-fold more potent in reducing H3K79 levels as well as being more cell permeable than EPZ004777 (Yu et al., 2012). The DOT1L inhibitor pinometostat (EPZ-5676), which is currently in clinical trials, was generated by replacing the ribose moiety with a cyclobutyl ring in EPZ004777 resulting in improved pharmacokinetic properties and capable of inhibiting growth of cell lines with MLL rearrangements (Campbell et al., 2017).

Several EZH2 inhibitors have been described by Epizyme, GSK and Novartis. In 2014, EPZ-6438 (Tazemetostat) entered clinical trials in patients with advanced solid tumors or relapsed or refractory B cell lymphomas (Italiano et al., 2018). In 2017, the FDA granted fast-track designation to Tazemetostat for the treatment of patients with follicular lymphoma. CPI-1205 another potent, selective, and cofactor-competitive inhibitor of EZH2 catalytic activity has entered clinical trials in patients with progressive B cell lymphomas and mCRPC.

3-DeazaneplanocinA (DZNep), an inhibitor of S-adenosylmethionine-dependent methyltransferase, is an adenosine analog that binds to the enzyme in place of the S-adenosyl methionine (SAM) leading to the indirect inhibition of various S-adenosylmethionine-dependent methylation reactions and degradation of target protein (Fujiwara et al., 2014). DZNep treatment induces cell cycle arrest and apoptosis of primary AML and MCL cells and other cancer cells (Fiskus et al., 2012; Kikuchi et al., 2012). The more potent and selective SAM-competitive inhibitors EZH2i EI1 and GSK126 have been tested in multiple cell lines and xenograft models (Huang et al., 2018; Qi et al., 2012; Stazi et al., 2019).

Other small-molecule KMT inhibitors, such as those that target H3K9 KMTs, have been created as possible anticancer treatments. In lung cancer cell lines, there is an increase in the expression of the H3K9-specific KMT G9a, and administration of the inhibitor BIX- 01294 lowers H3K9 methylation (Chen et al., 2018). Similarly, SETDB1 is typically overexpressed in lung and melanoma tumors, and mithramycin therapy reduces SETDB1 levels to stop cell growth (Lazaro-Camp, Salari, Meng, & Yang, 2021). Additionally, SUV39H, a KMT that controls erythroid and B cell differentiation, is inhibited by the natural substance chaetocin (Lai et al., 2015). These agents are currently being evaluated in vitro and in preclinical studies.

The only PRMT inhibitor in clinical trials is GSK3326595 (formerly EPZ015938), which is a potent, selective, reversible inhibitor of protein arginine methyltransferase 5 and is being evaluated in patients with solid tumors and non-Hodgkin's lymphoma (Vinet et al., 2019). Several inhibitors have shown great potential as cancer therapies in preclinical models and clinical trials. AMI-408, a PRMT1 inhibitor can decrease the transformation of MLL-GAS7 or MOZ-TIF2 fusions in AML models, and the PRMT5 inhibitor EPZ015666, induces cell death in MCL cell lines and also displays antitumor activity in xenografts of mantle cell lymphoma (Chan-Penebre et al., 2015; Cheung et al., 2016). In 2016, MS023 a potent, selective, and cell-active type I PRMT selective inhibitor was discovered, which potently inhibits asymmetric arginine di-methylation while increasing mono-methylation and symmetric di-methylation with IC_{50} values in the low nanomolar range (Eram et al., 2016). The first selective, cell-active allosteric inhibitor of PRMT3 (Compound 4) that targets the allosteric binding site of PRMT3 was identified in 2012 (Kaniskan et al., 2018). SGC707 is the most potent cell-active allosteric inhibitor of PRMT3 and has been selective against >30 other methyl-transferases and many other targets. EPZ020411, the first PRMT6 selective inhibitor was identified and has been shown to be well tolerated in vivo (Mitchell et al., 2015). A collaboration between Eli Lilly and the SGC resulted in the identification of LLY-283 a SAM competitive nucleoside PRMT5 inhibitor with low nanomolar cellular activity and a 100-fold selectivity over other histone methyltransferases and non-epigenetic targets (Bondarev et al., 2021). Of note, more recently,TP-064 the first potent, selective and cell active CARM1 inhibitor has been developed and found to be active against a subset of multiple myeloma cell lines (Drew et al., 2017; Nakayama et al., 2018).

3.3.4 Histone demethylation inhibitors

Demethylase inhibitor development is being explored by numerous research organizations due to the therapeutic promise of targeting histone methylation in cancer and other disorders. Lysine-specific histone demethylase 1 (LSD1) represents the first identified histone demethylase activity that shares homology to the MAOs. The first drugs developed against LSD1 were based on MAO inhibitors (MAOIs) derived from tranylcypromine (TCP) which irreversibly bind to LSD1, but its use is limited by indiscriminate anti-MAO-A and MAO-B activity (Tayari et al., 2021). Several TCP derivatives have been developed (Table 2), and some have entered clinical trials (Liang et al., 2020). ORY-1001 is an inhibitor of lysine-specific demethylase 1 that induces di-methylated H3K4 accumulation in cells and reduces growth of AML xenograft models (Maes et al., 2018). GSK2879552 promotes differentiation and inhibits proliferation of AML cells (Pedicona et al., 2022). In addition, reversible KDM1A inhibitors GSK354 or GSK690 have recently been reported to inhibit AML cells with low cytotoxicity (Dhanak & Jackson, 2014). Interestingly, the small-molecule epigenetic inhibitor (4SC-202), which inhibits both class I HDAC's and Lysine Demethylase (LSD1) is in clinical trials for patients with advanced hematologic malignancies (Wobser et al., 2019).

Although four compounds have already entered clinical trials for LSD1, the development of inhibitors of JMJC-domain-containing demethylases has made significant progress since first reported in 2008. Most identified inhibitors are pan-specific iron chelators that are also competitive for cofactor 2-oxoglutarate binding and are only active in the low nanomolar range. Two such compounds are GSK-J1 and its prodrug ethyl ester analogue, GSK-J4 which rapidly inhibits KDM6A (UTX) and KDM6B (JMJD3) with marginal activity against KDM5A and 5B (D'Oto et al., 2021). These proteins are all implicated in cancer. For instance, inactivating mutations of KDM6A have been reported in a broad range of cancers such as AML, MM, bladder and renal cell carcinoma, whereas KDM6B is highly expressed in T-ALL, pancreatic precancerous lesions, invasive breast cancer and metastatic prostate cancer (Hua et al., 2021). The fusion gene NUP98/KDM5A is found in 10% of pediatric acute megakaryoblastic leukemias (de Rooij et al., 2013). The small molecule inhibitor ryuvidine that inhibits KDM5A represses the growth of drug-resistant cells derived from the PC9 non-small-cell lung cancer (NSCLC) cell line (Mitsui et al., 2019). Recently, a potent and selective inhibitor of KDM5B, EPT1013182 discovered by high throughput screening has shown antiproliferative effects in cell lines and inhibit tumor growth in a preclinical model of MM (Hancock, Dunne, Walport, Flashman, & Kawamura, 2015).

3.4 Methyl-histone recognition proteins

To date, dozens of methyl-lysine readers have been discovered and have been shown to recognize various lysine methylation sites with selective methylation state preferences (Fig. 2). These proteins contain methyl-lysine-binding motifs, including PHD (plant homeodomain), Chromodomain

(CD), Tudor, WD40, Bromo Adjacent Homology (BAH), ATRX–DNMT3A–DNMT3L (ADD), Ankyrin repeat, MBT (malignant brain tumor), Proline–Tryptophan–Tryptophan–Proline (PWWP) and Zn-CW domains, and can distinguish target methyl-lysines based on their methylation state and surrounding amino-acid sequences. The CDs of HP1 and polycomb recognize tri-methylated lysine at H3K9 and H3K27, respectively (Jacobs & Khorasanizadeh, 2002; Min, Zhang, & Xu, 2003). CHD1 and Eaf3 preferentially bind tri-methylated lysine at H3K4 and H3K36, respectively (Flanagan et al., 2005; Joshi & Struhl, 2005). The double Tudor domain of JMJD2A selectively binds to di- and tri-methylated lysine at H3K4 and H4K20 (Huang, Fang, Bedford, Zhang, & Xu, 2006), and that of 53BP1 binds mono- or di-methylated lysine at H4K20 (Botuyan et al., 2006). The ankyrin repeat domains of G9a and GLP preferentially bind mono- or di-methylated lysine at K9 (Collins et al., 2008; Tachibana et al., 2005). The PHD finger families comprise Zn-coordinating domain proteins that recognize histones trimethylated at lysine at H3K4. Other subsets bind the unmodified histone H3 (H3K4) tail, a few show preferences to H3 tails trimethylated at lysine H3K9, at lysine H3K36, or acetylated at lysine H3K9 and lysine H3K14 (Shi et al., 2007).

WDR5 is a highly conserved WD40-repeat containing protein involved in chromatin remodeling and gene transcription (Suganuma, Pattenden, & Workman, 2008). WDR5 Forms a protein complex with MLL, RBBP5, ASH2L, and DPY30 to facilitate histone H3K4 di- and tri-methylation at transcriptionally active sites in the genome (Aho, Weissmiller, Fesik, & Tansey, 2019; Han et al., 2006; Schuetz et al., 2006; Wysocka et al., 2005).

3.5 Histone phosphorylation and ubiquitylation

Phosphorylation and ubiquitylation also affect the functions of histones besides methylation and acetylation. Aurora B kinase plays a major role in generating the double histone H3 modification tri-methylated K9/phosphorylated S10 (H3K9me3/S10ph), which has been implicated in chromosome condensation during mitosis (Monier, Mouradian, & Sullivan, 2007; Sabbattini et al., 2007). Recently it has been reported that Janus-associated kinase 2 (JAK2) which is a member of a family of cytoplasmic tyrosine kinases phosphorylate tyrosine 41 of the histone H3 tail (H3Y41p) and can modify epigenome, where a mutation known as JAK2V617F is associated with the condition myeloproliferative neoplasms (MPNs), polycythemia vera (PV), essential thrombocythemia (ET), and primary myelofibrosis

(PMF) (Dawson et al., 2009). Histone can be modified through ubiquitination and is mediated by polycomb-repressive complex 1 (PRC1) which catalyzes ubiquitination of histone H2A at lysine 119 and is highly correlated with transcriptional repression and cancer development (Barbour, Daou, Hendzel, & Affar, 2020).

3.5.1 Histone phosphorylation and ubiquitylation inhibitors

Three Aurora-kinase inhibitors have been described recently and most of these inhibitors are ATP-competitive inhibitors. ZM447439, Hesperidin is selective for Aurora B whereas VX-680 is a pan Aurora A/B inhibitor. VX-680 shows antitumor activity in rodent xenograft models (Keen & Taylor, 2004). A phase I–II clinical trial showed that INCB018424 selective inhibitor of Janus kinase 1 and JAK2 is associated with marked and durable clinical benefits in patients with JAK2 V617F–positive or JAK2 V617F–negative primary myelofibrosis (Verstovsek et al., 2010). Other JAK2 inhibitors Lestaurtinib and Pacritinib (SB1518) are undergoing clinical evaluation for the therapy of myelogenous leukemia (AML), relapsed lymphoma, and advanced myeloid malignancies, (Hart et al., 2011; Shabbir & Stuart, 2010). PRT4165 (2-pyridine-3-yl-methylene-indan-1,3-dione) is a novel chromatin-remodeling compound which inhibits PRC1-mediated H2A ubiquitylation in vitro and in vivo (Ismail, McDonald, Strickfaden, Xu, & Hendzel, 2013).

4. Combination epigenetic therapy

Epigenetic drugs combined with conventional or novel anticancer therapies may provide an alternative to traditional chemotherapy and may enhance therapeutic impact. Preclinical and clinical studies demonstrate significant benefits of combining various DNMT inhibitors and HDAC inhibitors with diverse chemotherapeutic medicines, especially in hematological cancers but also in solid malignancies. Recent clinical trials using epigenetic modifier combinations as well as combinations with cytotoxic chemotherapies, hormonal therapies, and immune checkpoint inhibitors are presented in Table 6.

Combining epigenetic drugs with chemotherapy can re-sensitize resistant tumor cells to radiotherapy and chemotherapy. Patients with resectable gastric and esophageal adenocarcinomas are treated with epigenetic priming with azacitidine before receiving the usual neoadjuvant chemotherapy regimen of epirubicin, oxaliplatin, and capecitabine (Schneider et al., 2017).

Table 6 Ongoing clinical trials with combination therapies.

Condition or disease	Combination	Sample size	Phase	Status	NCT
DNA methylation inhibitor + Histone acetylation inhibitor					
Myelodysplastic Syndrome	Pracinostat + Placebo + Azacitidine	102	II	Completed	NCT01873703
Solid Tumors Hematologic Malignancies Myelodysplastic Syndrome	SB939 + Azacitidine	85	I	Completed	NCT00741234
Acute Myeloid Leukemia	Pracinostat + Azacitidine	50	II	Completed	NCT01912274
Myelodysplastic Syndrome	Pracinostat + Azacitidine + Decitabine	45	II	Completed	NCT01993641
Myelodysplastic Syndrome	Mocetinostat + Azacitidine	18	I and II	Completed	NCT02018926
Myelodysplastic Syndrome Acute Myelogenous Leukemia	MGCD0103 (MG-0103) + Azacitidine	66	I and II	Completed	NCT00324220
Advanced Cancers	Valproic Acid + Azacitidine	69	I	Completed	NCT00496444
Solid Tumors	Valproic Acid + Azacitidine + Carboplatin	36	I	Completed	NCT00529022
Acute Myelogenous Leukemia Myelodysplastic Syndrome Leukemia	Valproic Acid + Azacitidine + Ara-C	11	II	Completed	NCT00382590
Acute Myelogenous Leukemia, Myelodysplastic Syndrome	Valproic Acid + Azacitidine + Retinoic acid	25	II	Completed	NCT00339196
Myelodysplastic Syndromes	Valproic Acid + Azacitidine + ATRA	62	II	Completed	NCT00439673
Azacitidine	Valproic acid + Lenalidomide + Idarubicine + Azacitidine	320	II	Active, not recruiting	NCT01342692
Advanced Cancers Lymphoma	Azacitidine + Vorinostat + Gemcitabine + Busulfan + Melphalan + Dexamethasone + Caphosol + Glutamine + Pyridoxine + Rituximab	61	I and II	Completed	NCT01983969

Leukemia, Myeloid, Acute	Vorinostat + Azacitidine	260	II	Completed	NCT01617226
Leukemia	Vorinostat + Azacitidine	110	II	Completed	NCT00948064
Acute Erythroid Leukemia, Acute Megakaryoblastic Leukemia, Chronic Myelomonocytic Leukemia	Vorinostat + Azacitidine	135	I and II	Active, not recruiting	NCT00392353
Acute Myeloid Leukemia, Myelodysplastic Syndromes, Mixed Phenotype Acute Leukemia, Juvenile Myelomonocytic Leukemia	Vorinostat + Azacitidine	15	I	Recruiting	NCT03843528
Chronic Myelomonocytic Leukemia, Myelodysplastic Syndrome	Vorinostat + Azacitidine + Lenalidomide	282	II	Active, not recruiting	NCT01522976
Solid Tumors, Virally Mediated Cancers and Liposarcoma	Romidepsin + Azacitidine	18	I	Completed	NCT01537744
Pancreas Cancer	Romidepsin + Azacitidine + nab-Paclitaxel + Gemcitabine + Durvalumab + Lenalidomide capsule	75	I and II	Recruiting	NCT04257448
Lymphoma, T-Cell	Romidepsin + Azacitidine + Durvalumab + Pralatrexate	148	I and II	Recruiting	NCT03161223
Peripheral T-cell lymphoma	Romidepsin + Azacitidine + Belinostat + Pralatrexate + Gemcitabine	74	II	Recruiting	NCT04747236
Refractory Angioimmunoblastic T-cell Lymphoma	Romidepsin + Azacitidine + Bendamustine + Gemcitabine	86		Active, not recruiting	NCT03593018
Recurrent Non-Small Cell Lung Carcinoma	Azacitidine + Entinosta	94	I and II	Completed	NCT00387465

Continued

Table 6 Ongoing clinical trials with combination therapies.—cont'd

Condition or disease	Combination	Sample size	Phase	Status	NCT
Chronic Myelomonocytic Leukemia	Azacitidine + Entinostat	197	II	Completed	NCT00313586
Myelodysplastic Syndrome Acute Myeloid Leukemia Chronic Myelomonocytic Leukemia	Azacitidine + Entinostat	63	I	Completed	NCT00101179
Recurrent Colon Cancer	Azacitidine + Entinostat	47	II	Completed	NCT01105377
Non-Small Cell Lung Cancer	Azacitidine + Entinostat	101	II	Active, not recruiting	NCT01928576
Leukemia, Lung Cancer, Prostate Cancer, Myelodysplastic Syndromes	Azacitidine + sodium phenylbutyrate	None	II	Completed	NCT00006019
BET inhibitor + DNA methylation inhibitor					
Relapsed or refractory hematologic malignancies	FT-1101 + azacitidine	160	I	Recruiting	NCT02543879
Myelodysplastic Syndrome Acute Myeloid Leukemia	GSK3326595 + azacitidine	302	I/II	Recruiting	NCT03614728
Epigenetic inhibitor + Immune checkpoint inhibitor					
Peripheral T-cell Lymphoma	PD-1 antibody + HDAC inhibitor	51	II	Recruiting	NCT04512534
Hodgkin Lymphoma	Decitabine + Camrelizumab	200	II	Recruiting	NCT04514081
Melanoma	Pembrolizumab + Entinostat	14	II	Recruiting	NCT03765229

The results of a phase I study showed that this method is safe and may increase the effectiveness of the chemotherapy regimen. In another study, the authors showed that decitabine-induced epigenetic priming could improve the sensitivity of gastric cancer to SN38 (an active metabolite of irinotecan) and cisplatin by inducing apoptosis-related genes (Moro et al., 2020).

Del Bufalo evaluated the molecular and functional effects of a pan-HDAC inhibitor givinostat (ITF2357) and pemetrexed (multi-target folate antagonist) in non-small-cell lung cancer and patient-derived lung-cancer stem cells. ITF2357 potentiated pemetrexed cytotoxic activity in a sequence-dependent manner due to activation of both apoptosis and autophagy and showed effective synergy in vitro in both NSCLC and LCSC cells (Del Bufalo et al., 2014). A phase I/II clinical trial of panobinostat with carfilzomib is an effective steroid-sparing regimen with a reasonable safety profile for relapsed/refractory multiple myeloma (Berdeja et al., 2015). A randomized phase III placebo-controlled study revealed that hydralazine and valproate added to cisplatin and topotecan demonstrate a significant advantage in promoting progression-free survival over one of the current standard combination chemotherapies in cervical cancer patients (Coronel et al., 2011; Long et al., 2005).

Combination epigenetic therapy is also a potential viable therapeutic strategy and involves the use of two or more different types of epigenetic drugs, such as DNMT and HDAC inhibitors. Combining epigenetic therapies has the important advantage of enabling the use of lower drug doses. This has potential to reduce side effects and increase effectiveness. Combined decitabine-vorinostat treatment results in decreased cell proliferation and induction of cancer cell death in acute myeloid leukemia (Burke et al., 2020). When HDAC and DNMT inhibitors are used together, histone acetylation is elevated relative to when HDAC inhibitors are used alone (Young, Clarke, Kettyle, Thompson, & Mills, 2017). The combination of the DNA methyltransferase inhibitor hydralazine and the histone deacetylase inhibitor valproate displays preliminary efficacy in MDS (Candelaria, Burgos, Ponce, Espinoza, & Duenas-Gonzalez, 2017). Moreover, the most recent report provides evidence that a combination of low-dose DNMT inhibitor and HDAC inhibitor may permit an adjuvant approach to cancer therapy, inhibiting metastases of solid tumors (Lu et al., 2020).

The behavior of cells involved in the immunological response that may also be regulated by post-translational modifications of histones can induce long-lasting immunological memory in immune cells, activate antigen-presenting dendritic cells, re-program regulatory T cells, enhance effector T cells and myeloid-derived suppressor cells, and other immune cells (Alvarez-Errico, Vento-Tormo, Sieweke, & Ballestar, 2015; Beier et al., 2012; Yan et al., 2011). Thus, HDAC and DNMT inhibitors can be used as priming modulators of immunotherapy to modulate the tumor microenvironment to activate the immune system via several mechanisms such as enhancing expression of tumor-associated antigens, antigen processing and presentation, immune checkpoint inhibitors, chemokines, and other immune-related genes (Cao & Yan, 2020; Hogg, Beavis, Dawson, & Johnstone, 2020).

Recently, a combination treatment using an HDAC inhibitor with immune checkpoint inhibitors is being actively investigated and has shown promising results in several cancer types. Knox et al. observed a significant improvement in anti-PD-1 immune checkpoint blockade therapy responses in preclinical models of melanoma when treated with anti-PD-1 and the HDAC6 inhibitor Nexturastat A (Knox et al., 2019). DNMT inhibitors (GSK126 and 5-AZA dC) induced chemokine expression and Th1 tumor infiltration in ovarian cancer (Peng et al., 2015). Moreover, HDAC inhibitors combined with anti PD-1 also enhanced T cell chemokine expression and primed the anti-tumor response in lung adenocarcinoma (Zheng et al., 2016). Many other epigenetic drugs (e.g., JQ1, LSD1 inhibitors, and EZH2 inhibitor) in combination with anti-PD1 therapy, can increases immune cell infiltration in the TME and increase anti-tumor T cell response in cancers, such as lung cancer, TNBC, and lymphoma (Sheng et al., 2018; Zhu et al., 2016).

5. Conclusions

Epigenetic therapy represents a promising therapeutic strategy. However, to be effective some common limitations and several impediments must be overcome to enhance effectiveness. Although epigenetic drugs have been most successful against hematological malignancies a major limitation has been their relative ineffectiveness against solid tumors. Contributing factors relate to the components of the core TME that promote hypoxic stress and provide an unusual epigenetic profile characterized by DNA hypermethylation and hypoxia decreased acetyl-CoA levels leading to a global reduction in histone acetylation thereby promoting heterochromatin formation. This is important since it appears that current HDAC and DNMT inhibitors are ineffective against tumor cells developing in hypoxic environments. To promote innovative therapeutics a better understanding of the epigenetic patterns of cancer cells living in hypoxic environments is required. Epigenetic drugs targeting DNA methylation or DNMTs show considerable cytotoxicity because they cause changes in patterns of gene regulation and expression, due to the passive demethylation or trapping of DNMTs. Development of resistance during therapy due to cellular mechanisms affecting drug absorption and metabolism are additional significant factors that can restrict the effectiveness of epigenetic inhibitors. Recently, several alternative approaches have been proposed to circumvent these current therapeutic limitations.

Additionally, combination therapy can be used to enhance and expand the efficacy of epigenetic therapies. This therapeutic strategy depends on several cancer medicines cooperating; however, in-depth studies are required to identify the most effective drug combinations, suitable doses, and timing of administration. New epigenetic medicines that are potent, non-toxic and target specific are essential for enhancing cancer treatment. Increasing the specificity of pharmacological designs through medicinal chemistry is one of the key strategies that may allow accomplishment of these objectives. The study of epigenetic biomarkers connected to cancer (early detection and response to therapy) and efforts to comprehend the epigenetic mechanisms underlying drug resistance in cancer cells are important areas requiring increased research. Achieving the ultimate goal of creating safe, effective treatments as well as developing agents to both prevent and permit early detection of cancer will benefit significantly from an enhanced understanding of cancer epigenetics.

Acknowledgments

We appreciate the support for our research efforts from: NIH/NCI through 1R01 CA244993 (Sarkar and Fisher), 1R01 CA259599 (Fisher and Wang); National Foundation for Cancer Research (Fisher); Commonwealth Health Research Board (Fisher); SRA (sponsored research agreement) from InterLeukin Combinatorial Therapies, Inc. (ILCT) (Emdad); SRA from InVaMet Therapeutics, Inc. (IVMT) (Das); Thelma Newmeyer Corman Chair in Cancer Research (Fisher); Department of Human and Molecular Genetics developmental funds (Das, Emdad); VCU Institute of Molecular Medicine (VIMM) developmental funds (Das, Emdad, Fisher); and VCU Massey Cancer Center developmental funds (Fisher).

Conflict of interest

P.B. Fisher is co-founder and has an ownership interest in InterLeukin Combinatorial Therapies (ILCT). VCU also has ownership interest in ILCT. P.B. Fisher is co-founder and has ownership interest in InVaMet Therapeutics (IVMT). VCU and the Sanford-Burnham-Prebys Medical Discovery Institute in La Jolla, CA have ownership interest in IVMT. None of the other investigators have conflicts of interest to report.

References

Aggarwal, R. R., Schweizer, M. T., Nanus, D. M., Pantuck, A. J., Heath, E. I., Campeau, E., et al. (2020). A phase Ib/IIa study of the Pan-BET inhibitor ZEN-3694 in combination with enzalutamide in patients with metastatic castration-resistant prostate cancer. *Clinical Cancer Research*, *26*(20), 5338–5347.

Aggarwal, R., Starodub, A. N., Koh, B. D., Xing, G., Armstrong, A. J., & Carducci, M. A. (2022). Phase Ib study of the BET inhibitor GS-5829 as monotherapy and combined with enzalutamide in patients with metastatic castration-resistant prostate cancer. *Clinical Cancer Research*, *28*(18), 3979–3989.

Agrawal, K., Das, V., Vyas, P., & Hajduch, M. (2018). Nucleosidic DNA demethylating epigenetic drugs—A comprehensive review from discovery to clinic. *Pharmacology & Therapeutics*, *188*, 45–79.

Aho, E. R., Weissmiller, A. M., Fesik, S. W., & Tansey, W. P. (2019). Targeting WDR5: A WINning anti-cancer strategy? *Epigenet Insights*, *12*, 2516865719865282.

Al-Awadhi, F. H., Salvador-Reyes, L. A., Elsadek, L. A., Ratnayake, R., Chen, Q. Y., & Luesch, H. (2020). Largazole is a brain-penetrant class I HDAC inhibitor with extended applicability to glioblastoma and CNS diseases. *ACS Chemical Neuroscience*, *11*(13), 1937–1943.

Alcitepe, I., Salcin, H., Karatekin, I., & Kaymaz, B. T. (2022). HDAC inhibitor Vorinostat and BET inhibitor Plx51107 epigenetic agents' combined treatments exert a therapeutic approach upon acute myeloid leukemia cell model. *Medical Oncology*, *39*(12), 257.

Alhazzazi, T. Y., Kamarajan, P., Joo, N., Huang, J. Y., Verdin, E., D'Silva, N. J., et al. (2011). Sirtuin-3 (SIRT3), a novel potential therapeutic target for oral cancer. *Cancer*, *117*(8), 1670–1678.

Alqahtani, A., Choucair, K., Ashraf, M., Hammouda, D. M., Alloghbi, A., Khan, T., et al. (2019). Bromodomain and extra-terminal motif inhibitors: A review of preclinical and clinical advances in cancer therapy. *Future Science OA*, *5*(3), FSO372.

Altundag, O., Altundag, K., & Gunduz, M. (2004). DNA methylation inhibitor, procainamide, may decrease the tamoxifen resistance by inducing overexpression of the estrogen receptor beta in breast cancer patients. *Medical Hypotheses*, *63*(4), 684–687.

Alvarez-Errico, D., Vento-Tormo, R., Sieweke, M., & Ballestar, E. (2015). Epigenetic control of myeloid cell differentiation, identity and function. *Nature Reviews. Immunology*, *15*(1), 7–17.

Amaru Calzada, A., Todoerti, K., Donadoni, L., Pellicioli, A., Tuana, G., Gatta, R., et al. (2012). The HDAC inhibitor Givinostat modulates the hematopoietic transcription factors NFE2 and C-MYB in JAK2(V617F) myeloproliferative neoplasm cells. *Experimental Hematology*, *40*(8), 634–645 e610.

Amatangelo, M. D., Garipov, A., Li, H., Conejo-Garcia, J. R., Speicher, D. W., & Zhang, R. (2013). Three-dimensional culture sensitizes epithelial ovarian cancer cells to EZH2 methyltransferase inhibition. *Cell Cycle*, *12*(13), 2113–2119.

Amato, R. J., Stephenson, J., Hotte, S., Nemunaitis, J., Belanger, K., Reid, G., et al. (2012). MG98, a second-generation DNMT1 inhibitor, in the treatment of advanced renal cell carcinoma. *Cancer Investigation*, *30*(5), 415–421.

Amengual, J. E., Lue, J. K., Ma, H., Lichtenstein, R., Shah, B., Cremers, S., et al. (2021). First-in-class selective HDAC6 inhibitor (ACY-1215) has a highly favorable safety profile in patients with relapsed and refractory lymphoma. *The Oncologist*, *26*(3), 184–e366.

Ando, M., Saito, Y., Xu, G., Bui, N. Q., Medetgul-Ernar, K., Pu, M., et al. (2019). Chromatin dysregulation and DNA methylation at transcription start sites associated with transcriptional repression in cancers. *Nature Communications*, *10*(1), 2188.

Armengol, G., Eissa, S., Lozano, J. J., Shoman, S., Sumoy, L., Caballin, M. R., et al. (2007). Genomic imbalances in Schistosoma-associated and non-Schistosoma-associated bladder carcinoma. An array comparative genomic hybridization analysis. *Cancer Genetics and Cytogenetics*, *177*(1), 16–19.

Asgatay, S., Champion, C., Marloie, G., Drujon, T., Senamaud-Beaufort, C., Ceccaldi, A., et al. (2014). Synthesis and evaluation of analogues of N-phthaloyl-l-tryptophan (RG108) as inhibitors of DNA methyltransferase 1. *Journal of Medicinal Chemistry*, *57*(2), 421–434.

Audia, J. E., & Campbell, R. M. (2016). Histone modifications and cancer. *Cold Spring Harbor Perspectives in Biology*, *8*(4), a019521.

Avvakumov, N., & Cote, J. (2007). The MYST family of histone acetyltransferases and their intimate links to cancer. *Oncogene*, *26*(37), 5395–5407.

Balasubramanian, S., Ramos, J., Luo, W., Sirisawad, M., Verner, E., & Buggy, J. J. (2008). A novel histone deacetylase 8 (HDAC8)-specific inhibitor PCI-34051 induces apoptosis in T-cell lymphomas. *Leukemia*, *22*(5), 1026–1034.

Balasubramanyam, K., Altaf, M., Varier, R. A., Swaminathan, V., Ravindran, A., Sadhale, P. P., et al. (2004). Polyisoprenylated benzophenone, garcinol, a natural histone acetyltransferase inhibitor, represses chromatin transcription and alters global gene expression. *The Journal of Biological Chemistry*, *279*(32), 33716–33726.

Balasubramanyam, K., Swaminathan, V., Ranganathan, A., & Kundu, T. K. (2003). Small molecule modulators of histone acetyltransferase p300. *The Journal of Biological Chemistry*, *278*(21), 19134–19140.

Balasubramanyam, K., Varier, R. A., Altaf, M., Swaminathan, V., Siddappa, N. B., Ranga, U., et al. (2004). Curcumin, a novel p300/CREB-binding protein-specific inhibitor of acetyltransferase, represses the acetylation of histone/nonhistone proteins and histone acetyltransferase-dependent chromatin transcription. *The Journal of Biological Chemistry*, *279*(49), 51163–51171.

Balbuena-Rebolledo, I., Rivera-Antonio, A. M., Sixto-Lopez, Y., Correa-Basurto, J., Rosales-Hernandez, M. C., Mendieta-Wejebe, J. E., et al. (2022). Dihydropyrazole-carbohydrazide derivatives with dual activity as antioxidant and anti-proliferative drugs on breast cancer targeting the HDAC6. *Pharmaceuticals (Basel)*, *15*(6), 690.

Banerji, U., van Doorn, L., Papadatos-Pastos, D., Kristeleit, R., Debnam, P., Tall, M., et al. (2012). A phase I pharmacokinetic and pharmacodynamic study of CHR-3996, an oral class I selective histone deacetylase inhibitor in refractory solid tumors. *Clinical Cancer Research*, *18*(9), 2687–2694.

Bao, L., Diao, H., Dong, N., Su, X., Wang, B., Mo, Q., et al. (2016). Histone deacetylase inhibitor induces cell apoptosis and cycle arrest in lung cancer cells via mitochondrial injury and p53 up-acetylation. *Cell Biology and Toxicology*, *32*(6), 469–482.

Barau, J., Teissandier, A., Zamudio, N., Roy, S., Nalesso, V., Herault, Y., et al. (2016). The DNA methyltransferase DNMT3C protects male germ cells from transposon activity. *Science*, *354*(6314), 909–912.

Barbieri, I., Cannizzaro, E., & Dawson, M. A. (2013). Bromodomains as therapeutic targets in cancer. *Briefings in Functional Genomics*, *12*(3), 219–230.

Barbour, H., Daou, S., Hendzel, M., & Affar, E. B. (2020). Polycomb group-mediated histone H2A monoubiquitination in epigenome regulation and nuclear processes. *Nature Communications*, *11*(1), 5947.

Barski, A., Cuddapah, S., Cui, K., Roh, T. Y., Schones, D. E., Wang, Z., et al. (2007). High-resolution profiling of histone methylations in the human genome. *Cell*, *129*(4), 823–837.

Baylin, S. B., Esteller, M., Rountree, M. R., Bachman, K. E., Schuebel, K., & Herman, J. G. (2001). Aberrant patterns of DNA methylation, chromatin formation and gene expression in cancer. *Human Molecular Genetics*, *10*(7), 687–692.

Baylin, S. B., & Jones, P. A. (2011). A decade of exploring the cancer epigenome—Biological and translational implications. *Nature Reviews. Cancer*, *11*(10), 726–734.

Bedford, M. T., & Clarke, S. G. (2009). Protein arginine methylation in mammals: Who, what, and why. *Molecular Cell*, *33*(1), 1–13.

Bedford, M. T., & van Helden, P. D. (1987). Hypomethylation of DNA in pathological conditions of the human prostate. *Cancer Research*, *47*(20), 5274–5276.
Beier, U. H., Wang, L., Han, R., Akimova, T., Liu, Y., & Hancock, W. W. (2012). Histone deacetylases 6 and 9 and sirtuin-1 control Foxp3+ regulatory T cell function through shared and isoform-specific mechanisms. *Science Signaling*, *5*(229), ra45.
Berdeja, J. G., Hart, L. L., Mace, J. R., Arrowsmith, E. R., Essell, J. H., Owera, R. S., et al. (2015). Phase I/II study of the combination of panobinostat and carfilzomib in patients with relapsed/refractory multiple myeloma. *Haematologica*, *100*(5), 670–676.
Bernasconi, E., Gaudio, E., Lejeune, P., Tarantelli, C., Cascione, L., Kwee, I., et al. (2017). Preclinical evaluation of the BET bromodomain inhibitor BAY 1238097 for the treatment of lymphoma. *British Journal of Haematology*, *178*(6), 936–948.
Bernt, K. M., Zhu, N., Sinha, A. U., Vempati, S., Faber, J., Krivtsov, A. V., et al. (2011). MLL-rearranged leukemia is dependent on aberrant H3K79 methylation by DOT1L. *Cancer Cell*, *20*(1), 66–78.
Bevill, S. M., Olivares-Quintero, J. F., Sciaky, N., Golitz, B. T., Singh, D., Beltran, A. S., et al. (2019). GSK2801, a BAZ2/BRD9 bromodomain inhibitor, synergizes with BET inhibitors to induce apoptosis in triple-negative breast cancer. *Molecular Cancer Research*, *17*(7), 1503–1518.
Bhadury, J., Nilsson, L. M., Muralidharan, S. V., Green, L. C., Li, Z., Gesner, E. M., et al. (2014). BET and HDAC inhibitors induce similar genes and biological effects and synergize to kill in Myc-induced murine lymphoma. *Proceedings of the National Academy of Sciences of the United States of America*, *111*(26), E2721–E2730.
Bhat, K. P., Umit Kaniskan, H., Jin, J., & Gozani, O. (2021). Epigenetics and beyond: Targeting writers of protein lysine methylation to treat disease. *Nature Reviews. Drug Discovery*, *20*(4), 265–286.
Bilokapic, S., & Halic, M. (2019). Nucleosome and ubiquitin position Set2 to methylate H3K36. *Nature Communications*, *10*(1), 3795.
Bisserier, M., & Wajapeyee, N. (2018). Mechanisms of resistance to EZH2 inhibitors in diffuse large B-cell lymphomas. *Blood*, *131*(19), 2125–2137.
Bock, C., Beerman, I., Lien, W. H., Smith, Z. D., Gu, H., Boyle, P., et al. (2012). DNA methylation dynamics during in vivo differentiation of blood and skin stem cells. *Molecular Cell*, *47*(4), 633–647.
Boi, M., Gaudio, E., Bonetti, P., Kwee, I., Bernasconi, E., Tarantelli, C., et al. (2015). The BET bromodomain inhibitor OTX015 affects pathogenetic pathways in preclinical B-cell tumor models and synergizes with targeted drugs. *Clinical Cancer Research*, *21*(7), 1628–1638.
Bolden, J. E., Peart, M. J., & Johnstone, R. W. (2006). Anticancer activities of histone deacetylase inhibitors. *Nature Reviews. Drug Discovery*, *5*(9), 769–784.
Bondarev, A. D., Attwood, M. M., Jonsson, J., Chubarev, V. N., Tarasov, V. V., & Schioth, H. B. (2021). Recent developments of HDAC inhibitors: Emerging indications and novel molecules. *British Journal of Clinical Pharmacology*, *87*(12), 4577–4597.
Borah, J. C., Mujtaba, S., Karakikes, I., Zeng, L., Muller, M., Patel, J., et al. (2011). A small molecule binding to the coactivator CREB-binding protein blocks apoptosis in cardiomyocytes. *Chemistry & Biology*, *18*(4), 531–541.
Borutinskaite, V., Virksaite, A., Gudelyte, G., & Navakauskiene, R. (2018). Green tea polyphenol EGCG causes anti-cancerous epigenetic modulations in acute promyelocytic leukemia cells. *Leukemia & Lymphoma*, *59*(2), 469–478.
Botuyan, M. V., Lee, J., Ward, I. M., Kim, J. E., Thompson, J. R., Chen, J., et al. (2006). Structural basis for the methylation state-specific recognition of histone H4-K20 by 53BP1 and Crb2 in DNA repair. *Cell*, *127*(7), 1361–1373.
Bowers, E. M., Yan, G., Mukherjee, C., Orry, A., Wang, L., Holbert, M. A., et al. (2010). Virtual ligand screening of the p300/CBP histone acetyltransferase: Identification of a selective small molecule inhibitor. *Chemistry & Biology*, *17*(5), 471–482.

Boyanapalli, S. S., & Kong, A. T. (2015). "Curcumin, the king of spices": Epigenetic regulatory mechanisms in the prevention of cancer, neurological, and inflammatory diseases. *Current Pharmacology Reports*, *1*(2), 129–139.

Brand, M., Measures, A. R., Wilson, B. G., Cortopassi, W. A., Alexander, R., Hoss, M., et al. (2015). Small molecule inhibitors of bromodomain-acetyl-lysine interactions. *ACS Chemical Biology*, *10*(1), 22–39.

Brooks, C. L., & Gu, W. (2009). How does SIRT1 affect metabolism, senescence and cancer? *Nature Reviews. Cancer*, *9*(2), 123–128.

Buechele, C., Breese, E. H., Schneidawind, D., Lin, C. H., Jeong, J., Duque-Afonso, J., et al. (2015). MLL leukemia induction by genome editing of human CD34+ hematopoietic cells. *Blood*, *126*(14), 1683–1694.

Buelow, D. R., Anderson, J. T., Pounds, S. B., Shi, L., Lamba, J. K., Hu, S., et al. (2021). DNA methylation-based epigenetic repression of SLC22A4 promotes resistance to cytarabine in acute myeloid leukemia. *Clinical and Translational Science*, *14*(1), 137–142.

Burke, M. J., Kostadinov, R., Sposto, R., Gore, L., Kelley, S. M., Rabik, C., et al. (2020). Decitabine and vorinostat with chemotherapy in relapsed pediatric acute lymphoblastic leukemia: A TACL pilot study. *Clinical Cancer Research*, *26*(10), 2297–2307.

Butcher, D. T., & Rodenhiser, D. I. (2007). Epigenetic inactivation of BRCA1 is associated with aberrant expression of CTCF and DNA methyltransferase (DNMT3B) in some sporadic breast tumours. *European Journal of Cancer*, *43*(1), 210–219.

Butler, K. V., Kalin, J., Brochier, C., Vistoli, G., Langley, B., & Kozikowski, A. P. (2010). Rational design and simple chemistry yield a superior, neuroprotective HDAC6 inhibitor, tubastatin A. *Journal of the American Chemical Society*, *132*(31), 10842–10846.

Campbell, C. T., Haladyna, J. N., Drubin, D. A., Thomson, T. M., Maria, M. J., Yamauchi, T., et al. (2017). Mechanisms of pinometostat (EPZ-5676) treatment-emergent resistance in MLL-rearranged leukemia. *Molecular Cancer Therapeutics*, *16*(8), 1669–1679.

Campbell, J. E., Kuntz, K. W., Knutson, S. K., Warholic, N. M., Keilhack, H., Wigle, T. J., et al. (2015). EPZ011989, A potent, orally-available EZH2 inhibitor with robust in vivo activity. *ACS Medicinal Chemistry Letters*, *6*(5), 491–495.

Candelaria, M., Burgos, S., Ponce, M., Espinoza, R., & Duenas-Gonzalez, A. (2017). Encouraging results with the compassionate use of hydralazine/valproate (TRANSKRIP) as epigenetic treatment for myelodysplastic syndrome (MDS). *Annals of Hematology*, *96*(11), 1825–1832.

Candido, E. P., Reeves, R., & Davie, J. R. (1978). Sodium butyrate inhibits histone deacetylation in cultured cells. *Cell*, *14*(1), 105–113.

Cao, J., & Yan, Q. (2020). Cancer epigenetics, tumor immunity, and immunotherapy. *Trends in Cancer*, *6*(7), 580–592.

Cao, L., Zhao, S., Yang, Q., Shi, Z., Liu, J., Pan, T., et al. (2021). Chidamide combined with doxorubicin induced p53-driven cell cycle arrest and cell apoptosis reverse multidrug resistance of breast cancer. *Frontiers in Oncology*, *11*, 614458.

Carter, D. M., Specker, E., Malecki, P. H., Przygodda, J., Dudaniec, K., Weiss, M. S., et al. (2021). Enhanced properties of a benzimidazole benzylpyrazole lysine demethylase inhibitor: Mechanism-of-action, binding site analysis, and activity in cellular models of prostate cancer. *Journal of Medicinal Chemistry*, *64*(19), 14266–14282.

Casamassimi, A., Rienzo, M., Di Zazzo, E., Sorrentino, A., Fiore, D., Proto, M. C., et al. (2020). Multifaceted role of PRDM proteins in human cancer. *International Journal of Molecular Sciences*, *21*(7), 2648.

Castellano, S., Kuck, D., Sala, M., Novellino, E., Lyko, F., & Sbardella, G. (2008). Constrained analogues of procaine as novel small molecule inhibitors of DNA methyltransferase-1. *Journal of Medicinal Chemistry*, *51*(7), 2321–2325.

Caulfield, T., & Medina-Franco, J. L. (2011). Molecular dynamics simulations of human DNA methyltransferase 3B with selective inhibitor nanaomycin A. *Journal of Structural Biology*, *176*(2), 185–191.

Cavalli, G., & Heard, E. (2019). Advances in epigenetics link genetics to the environment and disease. *Nature*, *571*(7766), 489–499.

Chaffanet, M., Gressin, L., Preudhomme, C., Soenen-Cornu, V., Birnbaum, D., & Pebusque, M. J. (2000). MOZ is fused to p300 in an acute monocytic leukemia with t(8;22). *Genes, Chromosomes & Cancer*, *28*(2), 138–144.

Chaidos, A., Caputo, V., Gouvedenou, K., Liu, B., Marigo, I., Chaudhry, M. S., et al. (2014). Potent antimyeloma activity of the novel bromodomain inhibitors I-BET151 and I-BET762. *Blood*, *123*(5), 697–705.

Chang, J., Varghese, D. S., Gillam, M. C., Peyton, M., Modi, B., Schiltz, R. L., et al. (2012). Differential response of cancer cells to HDAC inhibitors trichostatin A and depsipeptide. *British Journal of Cancer*, *106*(1), 116–125.

Chan-Penebre, E., Kuplast, K. G., Majer, C. R., Boriack-Sjodin, P. A., Wigle, T. J., Johnston, L. D., et al. (2015). A selective inhibitor of PRMT5 with in vivo and in vitro potency in MCL models. *Nature Chemical Biology*, *11*(6), 432–437.

Chedin, F. (2011). The DNMT3 family of mammalian de novo DNA methyltransferases. *Progress in Molecular Biology and Translational Science*, *101*, 255–285.

Chen, X., Hu, Y., & Zhou, D. X. (2011). Epigenetic gene regulation by plant Jumonji group of histone demethylase. *Biochimica et Biophysica Acta*, *1809*(8), 421–426.

Chen, Y., Liu, X., Li, Y., Quan, C., Zheng, L., & Huang, K. (2018). Lung cancer therapy targeting histone methylation: Opportunities and challenges. *Computational and Structural Biotechnology Journal*, *16*, 211–223.

Chen, S., Wang, Y., Zhou, W., Li, S., Peng, J., Shi, Z., et al. (2014). Identifying novel selective non-nucleoside DNA methyltransferase 1 inhibitors through docking-based virtual screening. *Journal of Medicinal Chemistry*, *57*(21), 9028–9041.

Chen, Q., Yang, B., Liu, X., Zhang, X. D., Zhang, L., & Liu, T. (2022). Histone acetyltransferases CBP/p300 in tumorigenesis and CBP/p300 inhibitors as promising novel anticancer agents. *Theranostics*, *12*(11), 4935–4948.

Cheung, N., Fung, T. K., Zeisig, B. B., Holmes, K., Rane, J. K., Mowen, K. A., et al. (2016). Targeting aberrant epigenetic networks mediated by PRMT1 and KDM4C in acute myeloid leukemia. *Cancer Cell*, *29*(1), 32–48.

Child, F., Ortiz-Romero, P. L., Alvarez, R., Bagot, M., Stadler, R., Weichenthal, M., et al. (2016). Phase II multicentre trial of oral quisinostat, a histone deacetylase inhibitor, in patients with previously treated stage IB-IVA mycosis fungoides/Sezary syndrome. *The British Journal of Dermatology*, *175*(1), 80–88.

Christman, J. K. (2002). 5-Azacytidine and 5-aza-2′-deoxycytidine as inhibitors of DNA methylation: Mechanistic studies and their implications for cancer therapy. *Oncogene*, *21*(35), 5483–5495.

Chua, G. N. L., Wassarman, K. L., Sun, H., Alp, J. A., Jarczyk, E. I., Kuzio, N. J., et al. (2019). Cytosine-based TET enzyme inhibitors. *ACS Medicinal Chemistry Letters*, *10*(2), 180–185.

Chuang, J. C., Warner, S. L., Vollmer, D., Vankayalapati, H., Redkar, S., Bearss, D. J., et al. (2010). S110, a 5-Aza-2′-deoxycytidine-containing dinucleotide, is an effective DNA methylation inhibitor in vivo and can reduce tumor growth. *Molecular Cancer Therapeutics*, *9*(5), 1443–1450.

Chun, S. M., Lee, J. Y., Choi, J., Lee, J. H., Hwang, J. J., Kim, C. S., et al. (2015). Epigenetic modulation with HDAC inhibitor CG200745 induces anti-proliferation in non-small cell lung cancer cells. *PLoS One*, *10*(3), e0119379.

Clark, P. G., Vieira, L. C., Tallant, C., Fedorov, O., Singleton, D. C., Rogers, C. M., et al. (2015). LP99: Discovery and synthesis of the first selective BRD7/9 bromodomain inhibitor. *Angewandte Chemie (International Ed. in English)*, *54*(21), 6217–6221.

Collins, R. E., Northrop, J. P., Horton, J. R., Lee, D. Y., Zhang, X., Stallcup, M. R., et al. (2008). The ankyrin repeats of G9a and GLP histone methyltransferases are mono- and dimethyllysine binding modules. *Nature Structural & Molecular Biology, 15*(3), 245–250.

Copeland, R. A. (2018). Protein methyltransferase inhibitors as precision cancer therapeutics: A decade of discovery. *Philosophical Transactions of the Royal Society of London. Series B, Biological Sciences, 373*(1748), 20170080.

Corno, C., Arrighetti, N., Ciusani, E., Corna, E., Carenini, N., Zaffaroni, N., et al. (2020). Synergistic interaction of histone deacetylase 6- and MEK-inhibitors in castration-resistant prostate cancer cells. *Frontiers in Cell and Development Biology, 8*, 610.

Coronel, J., Cetina, L., Pacheco, I., Trejo-Becerril, C., Gonzalez-Fierro, A., de la Cruz-Hernandez, E., et al. (2011). A double-blind, placebo-controlled, randomized phase III trial of chemotherapy plus epigenetic therapy with hydralazine valproate for advanced cervical cancer. Preliminary results. *Medical Oncology, 28*(Suppl. 1), S540–S546.

Cui, J., Sun, W., Hao, X., Wei, M., Su, X., Zhang, Y., et al. (2015). EHMT2 inhibitor BIX-01294 induces apoptosis through PMAIP1-USP9X-MCL1 axis in human bladder cancer cells. *Cancer Cell International, 15*(1), 4.

Daher-Reyes, G. S., Merchan, B. M., & Yee, K. W. L. (2019). Guadecitabine (SGI-110): An investigational drug for the treatment of myelodysplastic syndrome and acute myeloid leukemia. *Expert Opinion on Investigational Drugs, 28*(10), 835–849.

Dai, X. J., Liu, Y., Xiong, X. P., Xue, L. P., Zheng, Y. C., & Liu, H. M. (2020). Tranylcypromine based lysine-specific demethylase 1 inhibitor: Summary and perspective. *Journal of Medicinal Chemistry, 63*(23), 14197–14215.

Dai, H., Sinclair, D. A., Ellis, J. L., & Steegborn, C. (2018). Sirtuin activators and inhibitors: Promises, achievements, and challenges. *Pharmacology & Therapeutics, 188*, 140–154.

Daigle, S. R., Olhava, E. J., Therkelsen, C. A., Basavapathruni, A., Jin, L., Boriack-Sjodin, P. A., et al. (2013). Potent inhibition of DOT1L as treatment of MLL-fusion leukemia. *Blood, 122*(6), 1017–1025.

Das, M., Chaudhuri, T., Goswami, S. K., Murmu, N., Gomes, A., Mitra, S., et al. (2002). Studies with black tea and its constituents on leukemic cells and cell lines. *Journal of Experimental & Clinical Cancer Research, 21*(4), 563–568.

Das, P. M., Ramachandran, K., Vanwert, J., Ferdinand, L., Gopisetty, G., Reis, I. M., et al. (2006). Methylation mediated silencing of TMS1/ASC gene in prostate cancer. *Molecular Cancer, 5*, 28.

D'Ascenzio, M., Pugh, K. M., Konietzny, R., Berridge, G., Tallant, C., Hashem, S., et al. (2019). An activity-based probe targeting non-catalytic, highly conserved amino acid residues within bromodomains. *Angewandte Chemie (International Ed. in English), 58*(4), 1007–1012.

Dawson, M. A., Bannister, A. J., Gottgens, B., Foster, S. D., Bartke, T., Green, A. R., et al. (2009). JAK2 phosphorylates histone H3Y41 and excludes HP1alpha from chromatin. *Nature, 461*(7265), 819–822.

Dawson, M. A., & Kouzarides, T. (2012). Cancer epigenetics: From mechanism to therapy. *Cell, 150*(1), 12–27.

de Rooij, J. D., Hollink, I. H., Arentsen-Peters, S. T., van Galen, J. F., Berna Beverloo, H., Baruchel, A., et al. (2013). NUP98/JARID1A is a novel recurrent abnormality in pediatric acute megakaryoblastic leukemia with a distinct HOX gene expression pattern. *Leukemia, 27*(12), 2280–2288.

de Ruijter, A. J., van Gennip, A. H., Caron, H. N., Kemp, S., & van Kuilenburg, A. B. (2003). Histone deacetylases (HDACs): Characterization of the classical HDAC family. *The Biochemical Journal, 370*(Pt. 3), 737–749.

De Smet, C., Lurquin, C., Lethe, B., Martelange, V., & Boon, T. (1999). DNA methylation is the primary silencing mechanism for a set of germ line- and tumor-specific genes with a CpG-rich promoter. *Molecular and Cellular Biology, 19*(11), 7327–7335.

Dedes, K. J., Dedes, I., Imesch, P., von Bueren, A. O., Fink, D., & Fedier, A. (2009). Acquired vorinostat resistance shows partial cross-resistance to 'second-generation' HDAC inhibitors and correlates with loss of histone acetylation and apoptosis but not with altered HDAC and HAT activities. *Anti-Cancer Drugs*, *20*(5), 321–333.

Del Bufalo, D., Desideri, M., De Luca, T., Di Martile, M., Gabellini, C., Monica, V., et al. (2014). Histone deacetylase inhibition synergistically enhances pemetrexed cytotoxicity through induction of apoptosis and autophagy in non-small cell lung cancer. *Molecular Cancer*, *13*, 230.

Deleris, A., Greenberg, M. V., Ausin, I., Law, R. W., Moissiard, G., Schubert, D., et al. (2010). Involvement of a Jumonji-C domain-containing histone demethylase in DRM2-mediated maintenance of DNA methylation. *EMBO Reports*, *11*(12), 950–955.

Dhanak, D., & Jackson, P. (2014). Development and classes of epigenetic drugs for cancer. *Biochemical and Biophysical Research Communications*, *455*(1–2), 58–69.

Di Cerbo, V., & Schneider, R. (2013). Cancers with wrong HATs: The impact of acetylation. *Briefings in Functional Genomics*, *12*(3), 231–243.

Di Martile, M., Del Bufalo, D., & Trisciuoglio, D. (2016). The multifaceted role of lysine acetylation in cancer: Prognostic biomarker and therapeutic target. *Oncotarget*, 7(34), 55789–55810.

Diaz, T., Rodriguez, V., Lozano, E., Mena, M. P., Calderon, M., Rosinol, L., et al. (2017). The BET bromodomain inhibitor CPI203 improves lenalidomide and dexamethasone activity in in vitro and in vivo models of multiple myeloma by blockade of Ikaros and MYC signaling. *Haematologica*, *102*(10), 1776–1784.

Dickson, B. C., Riddle, N. D., Brooks, J. S., Pasha, T. L., & Zhang, P. J. (2013). Sirtuin 1 (SIRT1): A potential immunohistochemical marker and therapeutic target in soft tissue neoplasms with myoid differentiation. *Human Pathology*, *44*(6), 1125–1130.

Djamai, H., Berrou, J., Dupont, M., Coude, M. M., Delord, M., Clappier, E., et al. (2021). Biological effects of BET inhibition by OTX015 (MK-8628) and JQ1 in NPM1-mutated (NPM1c) acute myeloid leukemia (AML). *Biomedicine*, *9*(11), 1704.

Doi, M., Hirayama, J., & Sassone-Corsi, P. (2006). Circadian regulator CLOCK is a histone acetyltransferase. *Cell*, *125*(3), 497–508.

D'Oto, A., Fang, J., Jin, H., Xu, B., Singh, S., Mullasseril, A., et al. (2021). KDM6B promotes activation of the oncogenic CDK4/6-pRB-E2F pathway by maintaining enhancer activity in MYCN-amplified neuroblastoma. *Nature Communications*, *12*(1), 7204.

Doughan, M., Spellmon, N., Li, C., & Yang, Z. (2016). SMYD proteins in immunity: Dawning of a new era. *AIMS Biophysics*, *3*(4), 450–455.

Drew, A. E., Moradei, O., Jacques, S. L., Rioux, N., Boriack-Sjodin, A. P., Allain, C., et al. (2017). Identification of a CARM1 inhibitor with potent in vitro and in vivo activity in preclinical models of multiple myeloma. *Scientific Reports*, 7(1), 17993.

Du, S. J., Tan, X., & Zhang, J. (2014). SMYD proteins: Key regulators in skeletal and cardiac muscle development and function. *The Anatomical Record (Hoboken)*, *297*(9), 1650–1662.

Du, Y., Wu, J., Zhang, H., Li, S., & Sun, H. (2017). Reduced expression of SIRT2 in serous ovarian carcinoma promotes cell proliferation through disinhibition of CDK4 expression. *Molecular Medicine Reports*, *15*(4), 1638–1646.

Duenas-Gonzalez, A., Candelaria, M., Perez-Plascencia, C., Perez-Cardenas, E., de la Cruz-Hernandez, E., & Herrera, L. A. (2008). Valproic acid as epigenetic cancer drug: Preclinical, clinical and transcriptional effects on solid tumors. *Cancer Treatment Reviews*, *34*(3), 206–222.

Duenas-Gonzalez, A., Lizano, M., Candelaria, M., Cetina, L., Arce, C., & Cervera, E. (2005). Epigenetics of cervical cancer. An overview and therapeutic perspectives. *Molecular Cancer*, *4*, 38.

Duvic, M., Dummer, R., Becker, J. C., Poulalhon, N., Ortiz Romero, P., Grazia Bernengo, M., et al. (2013). Panobinostat activity in both bexarotene-exposed and -naive patients with refractory cutaneous T-cell lymphoma: Results of a phase II trial. *European Journal of Cancer*, *49*(2), 386–394.

Eden, A., Gaudet, F., Waghmare, A., & Jaenisch, R. (2003). Chromosomal instability and tumors promoted by DNA hypomethylation. *Science*, *300*(5618), 455.

Ehrlich, M. (2002). DNA methylation in cancer: Too much, but also too little. *Oncogene*, *21*(35), 5400–5413.

Ehrlich, M., Woods, C. B., Yu, M. C., Dubeau, L., Yang, F., Campan, M., et al. (2006). Quantitative analysis of associations between DNA hypermethylation, hypomethylation, and DNMT RNA levels in ovarian tumors. *Oncogene*, *25*(18), 2636–2645.

Elagib, K. E., & Goldfarb, A. N. (2007). Oncogenic pathways of AML1-ETO in acute myeloid leukemia: Multifaceted manipulation of marrow maturation. *Cancer Letters*, *251*(2), 179–186.

Eram, M. S., Shen, Y., Szewczyk, M., Wu, H., Senisterra, G., Li, F., et al. (2016). A potent, selective, and cell-active inhibitor of human type I protein arginine methyltransferases. *ACS Chemical Biology*, *11*(3), 772–781.

Esteller, M., Corn, P. G., Baylin, S. B., & Herman, J. G. (2001). A gene hypermethylation profile of human cancer. *Cancer Research*, *61*(8), 3225–3229.

Fagan, R. L., Cryderman, D. E., Kopelovich, L., Wallrath, L. L., & Brenner, C. (2013). Laccaic acid A is a direct, DNA-competitive inhibitor of DNA methyltransferase 1. *The Journal of Biological Chemistry*, *288*(33), 23858–23867.

Faivre, E. J., McDaniel, K. F., Albert, D. H., Mantena, S. R., Plotnik, J. P., Wilcox, D., et al. (2020). Selective inhibition of the BD2 bromodomain of BET proteins in prostate cancer. *Nature*, *578*(7794), 306–310.

Fang, Y., Liao, G., & Yu, B. (2019). LSD1/KDM1A inhibitors in clinical trials: Advances and prospects. *Journal of Hematology & Oncology*, *12*(1), 129.

Fang, M. Z., Wang, Y., Ai, N., Hou, Z., Sun, Y., Lu, H., et al. (2003). Tea polyphenol (−)-epigallocatechin-3-gallate inhibits DNA methyltransferase and reactivates methylation-silenced genes in cancer cell lines. *Cancer Research*, *63*(22), 7563–7570.

Feinberg, A. P., & Vogelstein, B. (1983). Hypomethylation of ras oncogenes in primary human cancers. *Biochemical and Biophysical Research Communications*, *111*(1), 47–54.

Feng, J., Yan, P. F., Zhao, H. Y., Zhang, F. C., Zhao, W. H., & Feng, M. (2016). SIRT6 suppresses glioma cell growth via induction of apoptosis, inhibition of oxidative stress and suppression of JAK2/STAT3 signaling pathway activation. *Oncology Reports*, *35*(3), 1395–1402.

Filippakopoulos, P., & Knapp, S. (2012). The bromodomain interaction module. *FEBS Letters*, *586*(17), 2692–2704.

Finley, A., & Copeland, R. A. (2014). Small molecule control of chromatin remodeling. *Chemistry & Biology*, *21*(9), 1196–1210.

Fiskus, W., Rao, R., Balusu, R., Ganguly, S., Tao, J., Sotomayor, E., et al. (2012). Superior efficacy of a combined epigenetic therapy against human mantle cell lymphoma cells. *Clinical Cancer Research*, *18*(22), 6227–6238.

Flanagan, J. F., Mi, L. Z., Chruszcz, M., Cymborowski, M., Clines, K. L., Kim, Y., et al. (2005). Double chromodomains cooperate to recognize the methylated histone H3 tail. *Nature*, *438*(7071), 1181–1185.

Fournel, M., Bonfils, C., Hou, Y., Yan, P. T., Trachy-Bourget, M. C., Kalita, A., et al. (2008). MGCD0103, a novel isotype-selective histone deacetylase inhibitor, has broad spectrum antitumor activity in vitro and in vivo. *Molecular Cancer Therapeutics*, *7*(4), 759–768.

Fraga, M. F., Ballestar, E., Villar-Garea, A., Boix-Chornet, M., Espada, J., Schotta, G., et al. (2005). Loss of acetylation at Lys16 and trimethylation at Lys20 of histone H4 is a common hallmark of human cancer. *Nature Genetics*, *37*(4), 391–400.

Frumm, S. M., Fan, Z. P., Ross, K. N., Duvall, J. R., Gupta, S., VerPlank, L., et al. (2013). Selective HDAC1/HDAC2 inhibitors induce neuroblastoma differentiation. *Chemistry & Biology*, *20*(5), 713–725.

Fuhrmann, J., Clancy, K. W., & Thompson, P. R. (2015). Chemical biology of protein arginine modifications in epigenetic regulation. *Chemical Reviews*, *115*(11), 5413–5461.

Fujiwara, T., Saitoh, H., Inoue, A., Kobayashi, M., Okitsu, Y., Katsuoka, Y., et al. (2014). 3-Deazaneplanocin A (DZNep), an inhibitor of S-adenosylmethionine-dependent methyltransferase, promotes erythroid differentiation. *The Journal of Biological Chemistry*, *289*(12), 8121–8134.

Furumai, R., Komatsu, Y., Nishino, N., Khochbin, S., Yoshida, M., & Horinouchi, S. (2001). Potent histone deacetylase inhibitors built from trichostatin A and cyclic tetrapeptide antibiotics including trapoxin. *Proceedings of the National Academy of Sciences of the United States of America*, *98*(1), 87–92.

Gadd, M. S., Testa, A., Lucas, X., Chan, K. H., Chen, W., Lamont, D. J., et al. (2017). Structural basis of PROTAC cooperative recognition for selective protein degradation. *Nature Chemical Biology*, *13*(5), 514–521.

Galloway, T. J., Wirth, L. J., Colevas, A. D., Gilbert, J., Bauman, J. E., Saba, N. F., et al. (2015). A phase I study of CUDC-101, a multitarget inhibitor of HDACs, EGFR, and HER2, in combination with chemoradiation in patients with head and neck squamous cell carcinoma. *Clinical Cancer Research*, *21*(7), 1566–1573.

Garcia-Guerrero, E., Gotz, R., Doose, S., Sauer, M., Rodriguez-Gil, A., Nerreter, T., et al. (2021). Upregulation of CD38 expression on multiple myeloma cells by novel HDAC6 inhibitors is a class effect and augments the efficacy of daratumumab. *Leukemia*, *35*(1), 201–214.

Ghizzoni, M., Haisma, H. J., & Dekker, F. J. (2009). Reactivity of isothiazolones and isothiazolone-1-oxides in the inhibition of the PCAF histone acetyltransferase. *European Journal of Medicinal Chemistry*, *44*(12), 4855–4861.

Gilan, O., Lam, E. Y., Becher, I., Lugo, D., Cannizzaro, E., Joberty, G., et al. (2016). Functional interdependence of BRD4 and DOT1L in MLL leukemia. *Nature Structural & Molecular Biology*, *23*(7), 673–681.

Gore, S. D., & Carducci, M. A. (2000). Modifying histones to tame cancer: Clinical development of sodium phenylbutyrate and other histone deacetylase inhibitors. *Expert Opinion on Investigational Drugs*, *9*(12), 2923–2934.

Graca, I., Sousa, E. J., Baptista, T., Almeida, M., Ramalho-Carvalho, J., Palmeira, C., et al. (2014). Anti-tumoral effect of the non-nucleoside DNMT inhibitor RG108 in human prostate cancer cells. *Current Pharmaceutical Design*, *20*(11), 1803–1811.

Grant, C., Rahman, F., Piekarz, R., Peer, C., Frye, R., Robey, R. W., et al. (2010). Romidepsin: A new therapy for cutaneous T-cell lymphoma and a potential therapy for solid tumors. *Expert Review of Anticancer Therapy*, *10*(7), 997–1008.

Greenberg, M. V. C., & Bourc'his, D. (2019). The diverse roles of DNA methylation in mammalian development and disease. *Nature Reviews. Molecular Cell Biology*, *20*(10), 590–607.

Gros, C., Fleury, L., Nahoum, V., Faux, C., Valente, S., Labella, D., et al. (2015). New insights on the mechanism of quinoline-based DNA methyltransferase inhibitors. *The Journal of Biological Chemistry*, *290*(10), 6293–6302.

Gryder, B. E., Sodji, Q. H., & Oyelere, A. K. (2012). Targeted cancer therapy: Giving histone deacetylase inhibitors all they need to succeed. *Future Medicinal Chemistry*, *4*(4), 505–524.

Guo, Y., Mao, X., Xiong, L., Xia, A., You, J., Lin, G., et al. (2021). Structure-guided discovery of a potent and selective cell-active inhibitor of SETDB1 tudor domain. *Angewandte Chemie (International Ed. in English)*, *60*(16), 8760–8765.

Haberland, M., Montgomery, R. L., & Olson, E. N. (2009). The many roles of histone deacetylases in development and physiology: Implications for disease and therapy. *Nature Reviews. Genetics*, *10*(1), 32–42.

Haggarty, S. J., Koeller, K. M., Wong, J. C., Grozinger, C. M., & Schreiber, S. L. (2003). Domain-selective small-molecule inhibitor of histone deacetylase 6 (HDAC6)-mediated tubulin deacetylation. *Proceedings of the National Academy of Sciences of the United States of America*, *100*(8), 4389–4394.

Han, Z., Guo, L., Wang, H., Shen, Y., Deng, X. W., & Chai, J. (2006). Structural basis for the specific recognition of methylated histone H3 lysine 4 by the WD-40 protein WDR5. *Molecular Cell*, *22*(1), 137–144.

Han, Y., Lindner, S., Bei, Y., Garcia, H. D., Timme, N., Althoff, K., et al. (2019). Synergistic activity of BET inhibitor MK-8628 and PLK inhibitor Volasertib in preclinical models of medulloblastoma. *Cancer Letters*, *445*, 24–33.

Hancock, R. L., Dunne, K., Walport, L. J., Flashman, E., & Kawamura, A. (2015). Epigenetic regulation by histone demethylases in hypoxia. *Epigenomics*, 7(5), 791–811.

Hart, S., Goh, K. C., Novotny-Diermayr, V., Hu, C. Y., Hentze, H., Tan, Y. C., et al. (2011). SB1518, a novel macrocyclic pyrimidine-based JAK2 inhibitor for the treatment of myeloid and lymphoid malignancies. *Leukemia*, *25*(11), 1751–1759.

Hassan, F. U., Rehman, M. S., Khan, M. S., Ali, M. A., Javed, A., Nawaz, A., et al. (2019). Curcumin as an alternative epigenetic modulator: Mechanism of action and potential effects. *Frontiers in Genetics*, *10*, 514.

Heinemann, B., Nielsen, J. M., Hudlebusch, H. R., Lees, M. J., Larsen, D. V., Boesen, T., et al. (2014). Inhibition of demethylases by GSK-J1/J4. *Nature*, *514*(7520), E1–E2.

Heintzman, N. D., Hon, G. C., Hawkins, R. D., Kheradpour, P., Stark, A., Harp, L. F., et al. (2009). Histone modifications at human enhancers reflect global cell-type-specific gene expression. *Nature*, *459*(7243), 108–112.

Heltweg, B., Gatbonton, T., Schuler, A. D., Posakony, J., Li, H., Goehle, S., et al. (2006). Antitumor activity of a small-molecule inhibitor of human silent information regulator 2 enzymes. *Cancer Research*, *66*(8), 4368–4377.

Hicks, K. C., Chariou, P. L., Ozawa, Y., Minnar, C. M., Knudson, K. M., Meyer, T. J., et al. (2021). Tumour-targeted interleukin-12 and entinostat combination therapy improves cancer survival by reprogramming the tumour immune cell landscape. *Nature Communications*, *12*(1), 5151.

Hilton, J., Cristea, M., Postel-Vinay, S., Baldini, C., Voskoboynik, M., Edenfield, W., et al. (2022). BMS-986158, a small molecule inhibitor of the bromodomain and extraterminal domain proteins, in patients with selected advanced solid tumors: Results from a phase 1/2a trial. *Cancers (Basel)*, *14*(17), 4079.

Hodawadekar, S. C., & Marmorstein, R. (2007). Chemistry of acetyl transfer by histone modifying enzymes: Structure, mechanism and implications for effector design. *Oncogene*, *26*(37), 5528–5540.

Hogg, S. J., Beavis, P. A., Dawson, M. A., & Johnstone, R. W. (2020). Targeting the epigenetic regulation of antitumour immunity. *Nature Reviews. Drug Discovery*, *19*(11), 776–800.

Hohmann, A. F., Martin, L. J., Minder, J. L., Roe, J. S., Shi, J., Steurer, S., et al. (2016). Sensitivity and engineered resistance of myeloid leukemia cells to BRD9 inhibition. *Nature Chemical Biology*, *12*(9), 672–679.

Holkova, B., Shafer, D., Yazbeck, V., Dave, S., Bose, P., Tombes, M. B., et al. (2021). Phase 1 study of belinostat (PXD-101) and bortezomib (Velcade, PS-341) in patients with relapsed or refractory acute leukemia and myelodysplastic syndrome. *Leukemia & Lymphoma*, *62*(5), 1187–1194.
Hollenbach, P. W., Nguyen, A. N., Brady, H., Williams, M., Ning, Y., Richard, N., et al. (2010). A comparison of azacitidine and decitabine activities in acute myeloid leukemia cell lines. *PLoS One*, *5*(2), e9001.
Hou, Z., Sang, S., You, H., Lee, M. J., Hong, J., Chin, K. V., et al. (2005). Mechanism of action of (−)-epigallocatechin-3-gallate: Auto-oxidation-dependent inactivation of epidermal growth factor receptor and direct effects on growth inhibition in human esophageal cancer KYSE 150 cells. *Cancer Research*, *65*(17), 8049–8056.
Hou, H., & Yu, H. (2010). Structural insights into histone lysine demethylation. *Current Opinion in Structural Biology*, *20*(6), 739–748.
Hu, Y., Lin, J., Lin, Y., Chen, X., Zhu, G., & Huang, G. (2019). Overexpression of SIRT4 inhibits the proliferation of gastric cancer cells through cell cycle arrest. *Oncology Letters*, *17*(2), 2171–2176.
Hu, Q., Wang, C., Xiang, Q., Wang, R., Zhang, C., Zhang, M., et al. (2020). Discovery and optimization of novel N-benzyl-3,6-dimethylbenzo[d]isoxazol-5-amine derivatives as potent and selective TRIM24 bromodomain inhibitors with potential anti-cancer activities. *Bioorganic Chemistry*, *94*, 103424.
Hua, C., Chen, J., Li, S., Zhou, J., Fu, J., Sun, W., et al. (2021). KDM6 demethylases and their roles in human cancers. *Frontiers in Oncology*, *11*, 779918.
Huang, X., Bi, N., Wang, J., Ren, H., Pan, D., Lu, X., et al. (2021). Chidamide and radiotherapy synergistically induce cell apoptosis and suppress tumor growth and cancer stemness by regulating the MiR-375-EIF4G3 axis in lung squamous cell carcinomas. *Journal of Oncology*, *2021*, 4936207.
Huang, Y., Fang, J., Bedford, M. T., Zhang, Y., & Xu, R. M. (2006). Recognition of histone H3 lysine-4 methylation by the double tudor domain of JMJD2A. *Science*, *312*(5774), 748–751.
Huang, H. L., Lee, H. Y., Tsai, A. C., Peng, C. Y., Lai, M. J., Wang, J. C., et al. (2012). Anticancer activity of MPT0E028, a novel potent histone deacetylase inhibitor, in human colorectal cancer HCT116 cells in vitro and in vivo. *PLoS One*, 7(8), e43645.
Huang, H. L., Peng, C. Y., Lai, M. J., Chen, C. H., Lee, H. Y., Wang, J. C., et al. (2015). Novel oral histone deacetylase inhibitor, MPT0E028, displays potent growth-inhibitory activity against human B-cell lymphoma in vitro and in vivo. *Oncotarget*, *6*(7), 4976–4991.
Huang, Z., Xia, Y., Hu, K., Zeng, S., Wu, L., Liu, S., et al. (2020). Histone deacetylase 6 promotes growth of glioblastoma through the MKK7/JNK/c-Jun signaling pathway. *Journal of Neurochemistry*, *152*(2), 221–234.
Huang, X., Yan, J., Zhang, M., Wang, Y., Chen, Y., Fu, X., et al. (2018). Targeting epigenetic crosstalk as a therapeutic strategy for EZH2-aberrant solid tumors. *Cell*, *175*(1), 186–199.e119.
Hudak, L., Tezeeh, P., Wedel, S., Makarevic, J., Juengel, E., Tsaur, I., et al. (2012). Low dosed interferon alpha augments the anti-tumor potential of histone deacetylase inhibition on prostate cancer cell growth and invasion. *Prostate*, *72*(16), 1719–1735.
Huffman, D. M., Grizzle, W. E., Bamman, M. M., Kim, J. S., Eltoum, I. A., Elgavish, A., et al. (2007). SIRT1 is significantly elevated in mouse and human prostate cancer. *Cancer Research*, *67*(14), 6612–6618.
Hupe, M. C., Hoda, M. R., Zengerling, F., Perner, S., Merseburger, A. S., & Cronauer, M. V. (2019). The BET-inhibitor PFI-1 diminishes AR/AR-V7 signaling in prostate cancer cells. *World Journal of Urology*, *37*(2), 343–349.

Hwang, J. W., Cho, Y., Bae, G. U., Kim, S. N., & Kim, Y. K. (2021). Protein arginine methyltransferases: Promising targets for cancer therapy. *Experimental & Molecular Medicine, 53*(5), 788–808.

Igoe, N., Bayle, E. D., Fedorov, O., Tallant, C., Savitsky, P., Rogers, C., et al. (2017). Design of a biased potent small molecule inhibitor of the bromodomain and PHD finger-containing (BRPF) proteins suitable for cellular and in vivo studies. *Journal of Medicinal Chemistry, 60*(2), 668–680.

Igoe, N., Bayle, E. D., Tallant, C., Fedorov, O., Meier, J. C., Savitsky, P., et al. (2017). Design of a chemical probe for the bromodomain and plant homeodomain finger-containing (BRPF) family of proteins. *Journal of Medicinal Chemistry, 60*(16), 6998–7011.

Imayoshi, N., Yoshioka, M., Tanaka, K., Yang, S. M., Akahane, K., Toda, Y., et al. (2022). CN470 is a BET/CBP/p300 multi-bromodomain inhibitor and has an anti-tumor activity against MLL-rearranged acute lymphoblastic leukemia. *Biochemical and Biophysical Research Communications, 590*, 49–54.

Isaacs, J. T., Antony, L., Dalrymple, S. L., Brennen, W. N., Gerber, S., Hammers, H., et al. (2013). Tasquinimod is an allosteric modulator of HDAC4 survival signaling within the compromised cancer microenvironment. *Cancer Research, 73*(4), 1386–1399.

Ismail, I. H., McDonald, D., Strickfaden, H., Xu, Z., & Hendzel, M. J. (2013). A small molecule inhibitor of polycomb repressive complex 1 inhibits ubiquitin signaling at DNA double-strand breaks. *The Journal of Biological Chemistry, 288*(37), 26944–26954.

Issa, J. J., Roboz, G., Rizzieri, D., Jabbour, E., Stock, W., O'Connell, C., et al. (2015). Safety and tolerability of guadecitabine (SGI-110) in patients with myelodysplastic syndrome and acute myeloid leukaemia: A multicentre, randomised, dose-escalation phase 1 study. *The Lancet Oncology, 16*(9), 1099–1110.

Italiano, A., Soria, J. C., Toulmonde, M., Michot, J. M., Lucchesi, C., Varga, A., et al. (2018). Tazemetostat, an EZH2 inhibitor, in relapsed or refractory B-cell non-Hodgkin lymphoma and advanced solid tumours: A first-in-human, open-label, phase 1 study. *The Lancet Oncology, 19*(5), 649–659.

Jabbour, E., Issa, J. P., Garcia-Manero, G., & Kantarjian, H. (2008). Evolution of decitabine development: Accomplishments, ongoing investigations, and future strategies. *Cancer, 112*(11), 2341–2351.

Jacobs, S. A., & Khorasanizadeh, S. (2002). Structure of HP1 chromodomain bound to a lysine 9-methylated histone H3 tail. *Science, 295*(5562), 2080–2083.

Jin, X. F., Auernhammer, C. J., Ilhan, H., Lindner, S., Nolting, S., Maurer, J., et al. (2019). Combination of 5-fluorouracil with epigenetic modifiers induces radiosensitization, somatostatin receptor 2 expression, and radioligand binding in neuroendocrine tumor cells in vitro. *Journal of Nuclear Medicine, 60*(9), 1240–1246.

Jin, Z., Jiang, W., Jiao, F., Guo, Z., Hu, H., Wang, L., et al. (2014). Decreased expression of histone deacetylase 10 predicts poor prognosis of gastric cancer patients. *International Journal of Clinical and Experimental Pathology*, 7(9), 5872–5879.

Jin, Y., Park, S., Park, S. Y., Lee, C. Y., Eum, D. Y., Shim, J. W., et al. (2022). G9a knockdown suppresses cancer aggressiveness by facilitating Smad protein phosphorylation through increasing BMP5 expression in luminal A type breast cancer. *International Journal of Molecular Sciences, 23*(2), 589.

Jin, B., & Robertson, K. D. (2013). DNA methyltransferases, DNA damage repair, and cancer. *Advances in Experimental Medicine and Biology, 754*, 3–29.

Jones, P. A., & Taylor, S. M. (1980). Cellular differentiation, cytidine analogs and DNA methylation. *Cell, 20*(1), 85–93.

Joshi, A. A., & Struhl, K. (2005). Eaf3 chromodomain interaction with methylated H3-K36 links histone deacetylation to Pol II elongation. *Molecular Cell, 20*(6), 971–978.

Jueliger, S., Lyons, J., Cannito, S., Pata, I., Pata, P., Shkolnaya, M., et al. (2016). Efficacy and epigenetic interactions of novel DNA hypomethylating agent guadecitabine (SGI-110) in preclinical models of hepatocellular carcinoma. *Epigenetics*, *11*(10), 709–720.

Jurkowska, R. Z., Anspach, N., Urbanke, C., Jia, D., Reinhardt, R., Nellen, W., et al. (2008). Formation of nucleoprotein filaments by mammalian DNA methyltransferase Dnmt3a in complex with regulator Dnmt3L. *Nucleic Acids Research*, *36*(21), 6656–6663.

Kaliszczak, M., Trousil, S., Aberg, O., Perumal, M., Nguyen, Q. D., & Aboagye, E. O. (2013). A novel small molecule hydroxamate preferentially inhibits HDAC6 activity and tumour growth. *British Journal of Cancer*, *108*(2), 342–350.

Kanai, Y., Ushijima, S., Nakanishi, Y., Sakamoto, M., & Hirohashi, S. (2003). Mutation of the DNA methyltransferase (DNMT) 1 gene in human colorectal cancers. *Cancer Letters*, *192*(1), 75–82.

Kaniskan, H. U., Eram, M. S., Zhao, K., Szewczyk, M. M., Yang, X., Schmidt, K., et al. (2018). Discovery of potent and selective allosteric inhibitors of protein arginine methyltransferase 3 (PRMT3). *Journal of Medicinal Chemistry*, *61*(3), 1204–1217.

Kanouni, T., Severin, C., Cho, R. W., Yuen, N. Y., Xu, J., Shi, L., et al. (2020). Discovery of CC-90011: A potent and selective reversible inhibitor of lysine specific demethylase 1 (LSD1). *Journal of Medicinal Chemistry*, *63*(23), 14522–14529.

Karagianni, F., Piperi, C., Mpakou, V., Spathis, A., Foukas, P. G., Dalamaga, M., et al. (2021). Ruxolitinib with resminostat exert synergistic antitumor effects in cutaneous T-cell lymphoma. *PLoS One*, *16*(3), e0248298.

Karpf, A. R., & Matsui, S. (2005). Genetic disruption of cytosine DNA methyltransferase enzymes induces chromosomal instability in human cancer cells. *Cancer Research*, *65*(19), 8635–8639.

Keen, N., & Taylor, S. (2004). Aurora-kinase inhibitors as anticancer agents. *Nature Reviews. Cancer*, *4*(12), 927–936.

Kikuchi, J., Takashina, T., Kinoshita, I., Kikuchi, E., Shimizu, Y., Sakakibara-Konishi, J., et al. (2012). Epigenetic therapy with 3-deazaneplanocin A, an inhibitor of the histone methyltransferase EZH2, inhibits growth of non-small cell lung cancer cells. *Lung Cancer*, *78*(2), 138–143.

Kim, K. C., Geng, L., & Huang, S. (2003). Inactivation of a histone methyltransferase by mutations in human cancers. *Cancer Research*, *63*(22), 7619–7623.

Kim, H., Kim, H., Feng, Y., Li, Y., Tamiya, H., Tocci, S., et al. (2020). PRMT5 control of cGAS/STING and NLRC5 pathways defines melanoma response to antitumor immunity. *Science Translational Medicine*, *12*(551), eaaz5683.

Kim, H. R., Kim, E. J., Yang, S. H., Jeong, E. T., Park, C., Lee, J. H., et al. (2006). Trichostatin A induces apoptosis in lung cancer cells via simultaneous activation of the death receptor-mediated and mitochondrial pathway? *Experimental & Molecular Medicine*, *38*(6), 616–624.

Kim, C., Lee, S., Kim, D., Lee, D. S., Lee, E., Yoo, C., et al. (2022). Blockade of GRP78 translocation to the cell surface by HDAC6 inhibition suppresses proliferation of cholangiocarcinoma cells. *Anticancer Research*, *42*(1), 471–482.

Kim, Y. B., Lee, K. H., Sugita, K., Yoshida, M., & Horinouchi, S. (1999). Oxamflatin is a novel antitumor compound that inhibits mammalian histone deacetylase. *Oncogene*, *18*(15), 2461–2470.

Kim, H. S., Quon, M. J., & Kim, J. A. (2014). New insights into the mechanisms of polyphenols beyond antioxidant properties; lessons from the green tea polyphenol, epigallocatechin 3-gallate. *Redox Biology*, *2*, 187–195.

Kim, E., Ten Hacken, E., Sivina, M., Clarke, A., Thompson, P. A., Jain, N., et al. (2020). The BET inhibitor GS-5829 targets chronic lymphocytic leukemia cells and their supportive microenvironment. *Leukemia*, *34*(6), 1588–1598.

Klaus, C. R., Iwanowicz, D., Johnston, D., Campbell, C. A., Smith, J. J., Moyer, M. P., et al. (2014). DOT1L inhibitor EPZ-5676 displays synergistic antiproliferative activity in combination with standard of care drugs and hypomethylating agents in MLL-rearranged leukemia cells. *The Journal of Pharmacology and Experimental Therapeutics*, *350*(3), 646–656.

Kleer, C. G., Cao, Q., Varambally, S., Shen, R., Ota, I., Tomlins, S. A., et al. (2003). EZH2 is a marker of aggressive breast cancer and promotes neoplastic transformation of breast epithelial cells. *Proceedings of the National Academy of Sciences of the United States of America*, *100*(20), 11606–11611.

Klisovic, R. B., Stock, W., Cataland, S., Klisovic, M. I., Liu, S., Blum, W., et al. (2008). A phase I biological study of MG98, an oligodeoxynucleotide antisense to DNA methyltransferase 1, in patients with high-risk myelodysplasia and acute myeloid leukemia. *Clinical Cancer Research*, *14*(8), 2444–2449.

Knox, T., Sahakian, E., Banik, D., Hadley, M., Palmer, E., Noonepalle, S., et al. (2019). Selective HDAC6 inhibitors improve anti-PD-1 immune checkpoint blockade therapy by decreasing the anti-inflammatory phenotype of macrophages and down-regulation of immunosuppressive proteins in tumor cells. *Scientific Reports*, *9*(1), 6136.

Knutson, S. K., Wigle, T. J., Warholic, N. M., Sneeringer, C. J., Allain, C. J., Klaus, C. R., et al. (2012). A selective inhibitor of EZH2 blocks H3K27 methylation and kills mutant lymphoma cells. *Nature Chemical Biology*, *8*(11), 890–896.

Kong, X., Ouyang, S., Liang, Z., Lu, J., Chen, L., Shen, B., et al. (2011). Catalytic mechanism investigation of lysine-specific demethylase 1 (LSD1): A computational study. *PLoS One*, *6*(9), e25444.

Kong, H. S., Tian, S., Kong, Y., Du, G., Zhang, L., Jung, M., et al. (2012). Preclinical studies of YK-4-272, an inhibitor of class II histone deacetylases by disruption of nucleocytoplasmic shuttling. *Pharmaceutical Research*, *29*(12), 3373–3383.

Kornberg, R. D. (1974). Chromatin structure: A repeating unit of histones and DNA. *Science*, *184*(4139), 868–871.

Kouzarides, T. (2007). Chromatin modifications and their function. *Cell*, *128*(4), 693–705.

Krennhrubec, K., Marshall, B. L., Hedglin, M., Verdin, E., & Ulrich, S. M. (2007). Design and evaluation of 'Linkerless' hydroxamic acids as selective HDAC8 inhibitors. *Bioorganic & Medicinal Chemistry Letters*, *17*(10), 2874–2878.

Krivtsov, A. V., & Armstrong, S. A. (2007). MLL translocations, histone modifications and leukaemia stem-cell development. *Nature Reviews. Cancer*, 7(11), 823–833.

Kuck, D., Caulfield, T., Lyko, F., & Medina-Franco, J. L. (2010). Nanaomycin A selectively inhibits DNMT3B and reactivates silenced tumor suppressor genes in human cancer cells. *Molecular Cancer Therapeutics*, *9*(11), 3015–3023.

Kuo, M. H., Zhou, J., Jambeck, P., Churchill, M. E., & Allis, C. D. (1998). Histone acetyltransferase activity of yeast Gcn5p is required for the activation of target genes in vivo. *Genes & Development*, *12*(5), 627–639.

Kurmasheva, R. T., Erickson, S. W., Han, R., Teicher, B. A., Smith, M. A., Roth, M., et al. (2021). In vivo evaluation of the lysine-specific demethylase (KDM1A/LSD1) inhibitor SP-2577 (Seclidemstat) against pediatric sarcoma preclinical models: A report from the Pediatric Preclinical Testing Consortium (PPTC). *Pediatric Blood & Cancer*, *68*(11), e29304.

Kusy, S., Cividin, M., Sorel, N., Brizard, F., Guilhot, F., Brizard, A., et al. (2003). p14ARF, p15INK4b, and p16INK4a methylation status in chronic myelogenous leukemia. *Blood*, *101*(1), 374–375.

Lai, Y. S., Chen, J. Y., Tsai, H. J., Chen, T. Y., & Hung, W. C. (2015). The SUV39H1 inhibitor chaetocin induces differentiation and shows synergistic cytotoxicity with other epigenetic drugs in acute myeloid leukemia cells. *Blood Cancer Journal*, *5*, e313.

Lain, S., Hollick, J. J., Campbell, J., Staples, O. D., Higgins, M., Aoubala, M., et al. (2008). Discovery, in vivo activity, and mechanism of action of a small-molecule p53 activator. *Cancer Cell, 13*(5), 454–463.

Lardenoije, R., Iatrou, A., Kenis, G., Kompotis, K., Steinbusch, H. W., Mastroeni, D., et al. (2015). The epigenetics of aging and neurodegeneration. *Progress in Neurobiology, 131*, 21–64.

Lazaro-Camp, V. J., Salari, K., Meng, X., & Yang, S. (2021). SETDB1 in cancer: Overexpression and its therapeutic implications. *American Journal of Cancer Research, 11*(5), 1803–1827.

Leal, A. S., Williams, C. R., Royce, D. B., Pioli, P. A., Sporn, M. B., & Liby, K. T. (2017). Bromodomain inhibitors, JQ1 and I-BET 762, as potential therapies for pancreatic cancer. *Cancer Letters, 394*, 76–87.

Lee, D. H., Kim, G. W., Jeon, Y. H., Yoo, J., Lee, S. W., & Kwon, S. H. (2020). Advances in histone demethylase KDM4 as cancer therapeutic targets. *The FASEB Journal, 34*(3), 3461–3484.

Lee, H. W., Kim, S., & Paik, W. K. (1977). S-adenosylmethionine: Protein-arginine methyltransferase. Purification and mechanism of the enzyme. *Biochemistry, 16*(1), 78–85.

Lee, H. Z., Kwitkowski, V. E., Del Valle, P. L., Ricci, M. S., Saber, H., Habtemariam, B. A., et al. (2015). FDA approval: Belinostat for the treatment of patients with relapsed or refractory peripheral t-cell lymphoma. *Clinical Cancer Research, 21*(12), 2666–2670.

Lee, D., Lee, D. Y., Hwang, Y. S., Seo, H. R., Lee, S. A., & Kwon, J. (2021). The bromodomain inhibitor PFI-3 sensitizes cancer cells to DNA damage by targeting SWI/SNF. *Molecular Cancer Research, 19*(5), 900–912.

Lee, K. H., Park, J. W., Sung, H. S., Choi, Y. J., Kim, W. H., Lee, H. S., et al. (2015). PHF2 histone demethylase acts as a tumor suppressor in association with p53 in cancer. *Oncogene, 34*(22), 2897–2909.

Lee, B. H., Yegnasubramanian, S., Lin, X., & Nelson, W. G. (2005). Procainamide is a specific inhibitor of DNA methyltransferase 1. *The Journal of Biological Chemistry, 280*(49), 40749–40756.

Lehnertz, B., Pabst, C., Su, L., Miller, M., Liu, F., Yi, L., et al. (2014). The methyltransferase G9a regulates HoxA9-dependent transcription in AML. *Genes & Development, 28*(4), 317–327.

Leone, G., Teofili, L., Voso, M. T., & Lubbert, M. (2002). DNA methylation and demethylating drugs in myelodysplastic syndromes and secondary leukemias. *Haematologica, 87*(12), 1324–1341.

Ley, T. J., Ding, L., Walter, M. J., McLellan, M. D., Lamprecht, T., Larson, D. E., et al. (2010). DNMT3A mutations in acute myeloid leukemia. *The New England Journal of Medicine, 363*(25), 2424–2433.

Li, S., Banck, M., Mujtaba, S., Zhou, M. M., Sugrue, M. M., & Walsh, M. J. (2010). p53-induced growth arrest is regulated by the mitochondrial SirT3 deacetylase. *PLoS One, 5*(5), e10486.

Li, E., Beard, C., & Jaenisch, R. (1993). Role for DNA methylation in genomic imprinting. *Nature, 366*(6453), 362–365.

Li, H., Feng, Z., Wu, W., Li, J., Zhang, J., & Xia, T. (2013). SIRT3 regulates cell proliferation and apoptosis related to energy metabolism in non-small cell lung cancer cells through deacetylation of NMNAT2. *International Journal of Oncology, 43*(5), 1420–1430.

Li, Q., Liu, K. Y., Liu, Q., Wang, G., Jiang, W., Meng, Q., et al. (2020). Antihistamine drug ebastine inhibits cancer growth by targeting polycomb group protein EZH2. *Molecular Cancer Therapeutics, 19*(10), 2023–2033.

Li, G., Tian, Y., & Zhu, W. G. (2020). The roles of histone deacetylases and their inhibitors in cancer therapy. *Frontiers in Cell and Development Biology, 8*, 576946.

Li, Y. C., Wang, Y., Li, D. D., Zhang, Y., Zhao, T. C., & Li, C. F. (2018). Procaine is a specific DNA methylation inhibitor with anti-tumor effect for human gastric cancer. *Journal of Cellular Biochemistry*, *119*(2), 2440–2449.

Li, Y., & Seto, E. (2016). HDACs and HDAC inhibitors in cancer development and therapy. *Cold Spring Harbor Perspectives in Medicine*, *6*(10), a026831.

Li, Y., Zhang, M., Sheng, M., Zhang, P., Chen, Z., Xing, W., et al. (2018). Therapeutic potential of GSK-J4, a histone demethylase KDM6B/JMJD3 inhibitor, for acute myeloid leukemia. *Journal of Cancer Research and Clinical Oncology*, *144*(6), 1065–1077.

Li, Y., Zhang, X., Zhu, S., Dejene, E. A., Peng, W., Sepulveda, A., et al. (2020). HDAC10 regulates cancer stem-like cell properties in KRAS-driven lung adenocarcinoma. *Cancer Research*, *80*(16), 3265–3278.

Liang, L., Wang, H., Du, Y., Luo, B., Meng, N., Cen, M., et al. (2020). New tranylcypromine derivatives containing sulfonamide motif as potent LSD1 inhibitors to target acute myeloid leukemia: Design, synthesis and biological evaluation. *Bioorganic Chemistry*, *99*, 103808.

Liang, F., Wang, X., Ow, S. H., Chen, W., & Ong, W. C. (2017). Sirtuin 5 is anti-apoptotic and anti-oxidative in cultured SH-EP neuroblastoma cells. *Neurotoxicity Research*, *31*(1), 63–76.

Lin, Z., & Fang, D. (2013). The roles of SIRT1 in cancer. *Genes & Cancer*, *4*(3–4), 97–104.

Lin, C. H., Hsieh, S. Y., Sheen, I. S., Lee, W. C., Chen, T. C., Shyu, W. C., et al. (2001). Genome-wide hypomethylation in hepatocellular carcinogenesis. *Cancer Research*, *61*(10), 4238–4243.

Lio, C. J., Yuita, H., & Rao, A. (2019). Dysregulation of the TET family of epigenetic regulators in lymphoid and myeloid malignancies. *Blood*, *134*(18), 1487–1497.

Liu, X., Chen, J., Zhang, B., Liu, G., Zhao, H., & Hu, Q. (2019). KDM3A inhibition modulates macrophage polarization to aggravate post-MI injuries and accelerates adverse ventricular remodeling via an IRF4 signaling pathway. *Cellular Signalling*, *64*, 109415.

Liu, J., Li, J. N., Wu, H., & Liu, P. (2022). The status and prospects of epigenetics in the treatment of lymphoma. *Frontiers in Oncology*, *12*, 874645.

Liu, N., Wang, C., Wang, L., Gao, L., Cheng, H., Tang, G., et al. (2016). Valproic acid enhances the antileukemic effect of cytarabine by triggering cell apoptosis. *International Journal of Molecular Medicine*, *37*(6), 1686–1696.

Liu, X., Wang, L., Zhao, K., Thompson, P. R., Hwang, Y., Marmorstein, R., et al. (2008). The structural basis of protein acetylation by the p300/CBP transcriptional coactivator. *Nature*, *451*(7180), 846–850.

Liu, X. R., Zhou, L. H., Hu, J. X., Liu, L. M., Wan, H. P., & Zhang, X. Q. (2018). UNC0638, a G9a inhibitor, suppresses epithelialmesenchymal transitionmediated cellular migration and invasion in triple negative breast cancer. *Molecular Medicine Reports*, *17*(2), 2239–2244.

Liva, S. G., Coss, C. C., Wang, J., Blum, W., Klisovic, R., Bhatnagar, B., et al. (2020). Phase I study of AR-42 and decitabine in acute myeloid leukemia. *Leukemia & Lymphoma*, *61*(6), 1484–1492.

Long, H. J., 3rd, Bundy, B. N., Grendys, E. C., Jr., Benda, J. A., McMeekin, D. S., Sorosky, J., et al. (2005). Randomized phase III trial of cisplatin with or without topotecan in carcinoma of the uterine cervix: A Gynecologic Oncology Group Study. *Journal of Clinical Oncology*, *23*(21), 4626–4633.

Lopez, G., Bill, K. L., Bid, H. K., Braggio, D., Constantino, D., Prudner, B., et al. (2015). HDAC8, a potential therapeutic target for the treatment of malignant peripheral nerve sheath tumors (MPNST). *PLoS One*, *10*(7), e0133302.

Lopez, G., Liu, J., Ren, W., Wei, W., Wang, S., Lahat, G., et al. (2009). Combining PCI-24781, a novel histone deacetylase inhibitor, with chemotherapy for the treatment of soft tissue sarcoma. *Clinical Cancer Research*, *15*(10), 3472–3483.

Lu, C., Ramirez, D., Hwang, S., Jungbluth, A., Frosina, D., Ntiamoah, P., et al. (2019). Histone H3K36M mutation and trimethylation patterns in chondroblastoma. *Histopathology*, *74*(2), 291–299.
Lu, X., Yang, P., Zhao, X., Jiang, M., Hu, S., Ouyang, Y., et al. (2019). OGDH mediates the inhibition of SIRT5 on cell proliferation and migration of gastric cancer. *Experimental Cell Research*, *382*(2), 111483.
Lu, Z., Zou, J., Li, S., Topper, M. J., Tao, Y., Zhang, H., et al. (2020). Epigenetic therapy inhibits metastases by disrupting premetastatic niches. *Nature*, *579*(7798), 284–290.
Lv, Z., Weng, X., Du, C., Zhang, C., Xiao, H., Cai, X., et al. (2016). Downregulation of HDAC6 promotes angiogenesis in hepatocellular carcinoma cells and predicts poor prognosis in liver transplantation patients. *Molecular Carcinogenesis*, *55*(5), 1024–1033.
Lyko, F. (2018). The DNA methyltransferase family: A versatile toolkit for epigenetic regulation. *Nature Reviews. Genetics*, *19*(2), 81–92.
Maes, T., Mascaro, C., Tirapu, I., Estiarte, A., Ciceri, F., Lunardi, S., et al. (2018). ORY-1001, a potent and selective covalent KDM1A inhibitor, for the treatment of acute leukemia. *Cancer Cell*, *33*(3), 495–511 e412.
Maiques-Diaz, A., & Somervaille, T. C. (2016). LSD1: Biologic roles and therapeutic targeting. *Epigenomics*, *8*(8), 1103–1116.
Mandl-Weber, S., Meinel, F. G., Jankowsky, R., Oduncu, F., Schmidmaier, R., & Baumann, P. (2010). The novel inhibitor of histone deacetylase resminostat (RAS2410) inhibits proliferation and induces apoptosis in multiple myeloma (MM) cells. *British Journal of Haematology*, *149*(4), 518–528.
Mann, B. S., Johnson, J. R., Cohen, M. H., Justice, R., & Pazdur, R. (2007). FDA approval summary: Vorinostat for treatment of advanced primary cutaneous T-cell lymphoma. *The Oncologist*, *12*(10), 1247–1252.
Marquardt, J. U., Fischer, K., Baus, K., Kashyap, A., Ma, S., Krupp, M., et al. (2013). Sirtuin-6-dependent genetic and epigenetic alterations are associated with poor clinical outcome in hepatocellular carcinoma patients. *Hepatology*, *58*(3), 1054–1064.
Marquez, V. E., Barchi, J. J., Jr., Kelley, J. A., Rao, K. V., Agbaria, R., Ben-Kasus, T., et al. (2005). Zebularine: A unique molecule for an epigenetically based strategy in cancer chemotherapy. The magic of its chemistry and biology. *Nucleosides, Nucleotides & Nucleic Acids*, *24*(5–7), 305–318.
Martin, L. J., Koegl, M., Bader, G., Cockcroft, X. L., Fedorov, O., Fiegen, D., et al. (2016). Structure-based design of an in vivo active selective BRD9 inhibitor. *Journal of Medicinal Chemistry*, *59*(10), 4462–4475.
Martin, C., & Zhang, Y. (2005). The diverse functions of histone lysine methylation. *Nature Reviews. Molecular Cell Biology*, *6*(11), 838–849.
Martinez, E., Kundu, T. K., Fu, J., & Roeder, R. G. (1998). A human SPT3-TAFII31-GCN5-L acetylase complex distinct from transcription factor IID. *The Journal of Biological Chemistry*, *273*(37), 23781–23785.
Mason, L. D., Chava, S., Reddi, K. K., & Gupta, R. (2021). The BRD9/7 inhibitor TP-472 blocks melanoma tumor growth by suppressing ECM-mediated oncogenic signaling and inducing apoptosis. *Cancers (Basel)*, *13*(21), 5516.
McDermott, J., & Jimeno, A. (2014). Belinostat for the treatment of peripheral T-cell lymphomas. *Drugs Today (Barc)*, *50*(5), 337–345.
Meier, J. C., Tallant, C., Fedorov, O., Witwicka, H., Hwang, S. Y., van Stiphout, R. G., et al. (2017). Selective targeting of bromodomains of the bromodomain-PHD fingers family impairs osteoclast differentiation. *ACS Chemical Biology*, *12*(10), 2619–2630.
Melin, L., Gesner, E., Attwell, S., Kharenko, O. A., van der Horst, E. H., Hansen, H. C., et al. (2021). Design and synthesis of LM146, a potent inhibitor of PB1 with an improved selectivity profile over SMARCA2. *ACS Omega*, *6*(33), 21327–21338.

Miller, D. S. J., Voell, S. A., Sosic, I., Proj, M., Rossanese, O. W., Schnakenburg, G., et al. (2022). Encoding BRAF inhibitor functions in protein degraders. *RSC Medicinal Chemistry*, *13*(6), 731–736.

Min, J., Zhang, Y., & Xu, R. M. (2003). Structural basis for specific binding of polycomb chromodomain to histone H3 methylated at Lys 27. *Genes & Development*, *17*(15), 1823–1828.

Minami, J., Suzuki, R., Mazitschek, R., Gorgun, G., Ghosh, B., Cirstea, D., et al. (2014). Histone deacetylase 3 as a novel therapeutic target in multiple myeloma. *Leukemia*, *28*(3), 680–689.

Mitchell, L. H., Drew, A. E., Ribich, S. A., Rioux, N., Swinger, K. K., Jacques, S. L., et al. (2015). Aryl pyrazoles as potent inhibitors of arginine methyltransferases: Identification of the first PRMT6 tool compound. *ACS Medicinal Chemistry Letters*, *6*(6), 655–659.

Mitsui, E., Yoshida, S., Shinoda, Y., Matsumori, Y., Tsujii, H., Tsuchida, M., et al. (2019). Identification of ryuvidine as a KDM5A inhibitor. *Scientific Reports*, *9*(1), 9952.

Mizukami, H., Shirahata, A., Goto, T., Sakata, M., Saito, M., Ishibashi, K., et al. (2008). PGP9.5 methylation as a marker for metastatic colorectal cancer. *Anticancer Research*, *28*(5A), 2697–2700.

Monier, K., Mouradian, S., & Sullivan, K. F. (2007). DNA methylation promotes Aurora-B-driven phosphorylation of histone H3 in chromosomal subdomains. *Journal of Cell Science*, *120*(Pt. 1), 101–114.

Montano, M. M., Yeh, I. J., Chen, Y., Hernandez, C., Kiselar, J. G., de la Fuente, M., et al. (2019). Inhibition of the histone demethylase, KDM5B, directly induces re-expression of tumor suppressor protein HEXIM1 in cancer cells. *Breast Cancer Research*, *21*(1), 138.

Montgomery, D. C., Sorum, A. W., & Meier, J. L. (2015). Defining the orphan functions of lysine acetyltransferases. *ACS Chemical Biology*, *10*(1), 85–94.

Moore-Morris, T., van Vliet, P. P., Andelfinger, G., & Puceat, M. (2018). Role of epigenetics in cardiac development and congenital diseases. *Physiological Reviews*, *98*(4), 2453–2475.

Morales Torres, C., Wu, M. Y., Hobor, S., Wainwright, E. N., Martin, M. J., Patel, H., et al. (2020). Selective inhibition of cancer cell self-renewal through a Quisinostat-histone H1.0 axis. *Nature Communications*, *11*(1), 1792.

Moreira-Silva, F., Camilo, V., Gaspar, V., Mano, J. F., Henrique, R., & Jeronimo, C. (2020). Repurposing old drugs into new epigenetic inhibitors: Promising candidates for cancer treatment? *Pharmaceutics*, *12*(5), 410.

Morera, L., Lubbert, M., & Jung, M. (2016). Targeting histone methyltransferases and demethylases in clinical trials for cancer therapy. *Clinical Epigenetics*, *8*, 57.

Moro, H., Hattori, N., Nakamura, Y., Kimura, K., Imai, T., Maeda, M., et al. (2020). Epigenetic priming sensitizes gastric cancer cells to irinotecan and cisplatin by restoring multiple pathways. *Gastric Cancer*, *23*(1), 105–115.

Morrison-Smith, C. D., Knox, T. M., Filic, I., Soroko, K. M., Eschle, B. K., Wilkens, M. K., et al. (2020). Combined targeting of the BRD4-NUT-p300 Axis in NUT midline carcinoma by dual selective bromodomain inhibitor, NEO2734. *Molecular Cancer Therapeutics*, *19*(7), 1406–1414.

Mortusewicz, O., Schermelleh, L., Walter, J., Cardoso, M. C., & Leonhardt, H. (2005). Recruitment of DNA methyltransferase I to DNA repair sites. *Proceedings of the National Academy of Sciences of the United States of America*, *102*(25), 8905–8909.

Mosaab, A., El-Ayadi, M., Khorshed, E. N., Amer, N., Refaat, A., El-Beltagy, M., et al. (2020). Histone H3K27M mutation overrides histological grading in pediatric gliomas. *Scientific Reports*, *10*(1), 8368.

Mostoufi, A., Baghgoli, R., & Fereidoonnezhad, M. (2019). Synthesis, cytotoxicity, apoptosis and molecular docking studies of novel phenylbutyrate derivatives as potential anticancer agents. *Computational Biology and Chemistry*, *80*, 128–137.

Nakamura, K., Nakabayashi, K., Htet Aung, K., Aizawa, K., Hori, N., Yamauchi, J., et al. (2015). DNA methyltransferase inhibitor zebularine induces human cholangiocarcinoma cell death through alteration of DNA methylation status. *PLoS One*, *10*(3), e0120545.

Nakayama, K., Szewczyk, M. M., Dela Sena, C., Wu, H., Dong, A., Zeng, H., et al. (2018). TP-064, a potent and selective small molecule inhibitor of PRMT4 for multiple myeloma. *Oncotarget*, *9*(26), 18480–18493.

Neumann, M., Heesch, S., Schlee, C., Schwartz, S., Gokbuget, N., Hoelzer, D., et al. (2013). Whole-exome sequencing in adult ETP-ALL reveals a high rate of DNMT3A mutations. *Blood*, *121*(23), 4749–4752.

Newbold, A., Matthews, G. M., Bots, M., Cluse, L. A., Clarke, C. J., Banks, K. M., et al. (2013). Molecular and biologic analysis of histone deacetylase inhibitors with diverse specificities. *Molecular Cancer Therapeutics*, *12*(12), 2709–2721.

Nguyen, A. T., & Zhang, Y. (2011). The diverse functions of Dot1 and H3K79 methylation. *Genes & Development*, *25*(13), 1345–1358.

Novotny-Diermayr, V., Sangthongpitag, K., Hu, C. Y., Wu, X., Sausgruber, N., Yeo, P., et al. (2010). SB939, a novel potent and orally active histone deacetylase inhibitor with high tumor exposure and efficacy in mouse models of colorectal cancer. *Molecular Cancer Therapeutics*, *9*(3), 642–652.

Novotny-Diermayr, V., Sausgruber, N., Loh, Y. K., Pasha, M. K., Jayaraman, R., Hentze, H., et al. (2011). Pharmacodynamic evaluation of the target efficacy of SB939, an oral HDAC inhibitor with selectivity for tumor tissue. *Molecular Cancer Therapeutics*, *10*(7), 1207–1217.

Oh, E. T., Park, M. T., Choi, B. H., Ro, S., Choi, E. K., Jeong, S. Y., et al. (2012). Novel histone deacetylase inhibitor CG200745 induces clonogenic cell death by modulating acetylation of p53 in cancer cells. *Investigational New Drugs*, *30*(2), 435–442.

Okano, M., Xie, S., & Li, E. (1998). Cloning and characterization of a family of novel mammalian DNA (cytosine-5) methyltransferases. *Nature Genetics*, *19*(3), 219–220.

O'Neill, E. J., Termini, D., Albano, A., & Tsiani, E. (2021). Anti-cancer properties of theaflavins. *Molecules*, *26*(4), 987.

Ortiz-Barahona, V., Joshi, R. S., & Esteller, M. (2022). Use of DNA methylation profiling in translational oncology. *Seminars in Cancer Biology*, *83*, 523–535.

Ota, H., Tokunaga, E., Chang, K., Hikasa, M., Iijima, K., Eto, M., et al. (2006). Sirt1 inhibitor, Sirtinol, induces senescence-like growth arrest with attenuated Ras-MAPK signaling in human cancer cells. *Oncogene*, *25*(2), 176–185.

Palmer, W. S., Poncet-Montange, G., Liu, G., Petrocchi, A., Reyna, N., Subramanian, G., et al. (2016). Structure-guided design of IACS-9571, a selective high-affinity dual TRIM24-BRPF1 bromodomain inhibitor. *Journal of Medicinal Chemistry*, *59*(4), 1440–1454.

Pan, H., Kim, E., Rankin, G. O., Rojanasakul, Y., Tu, Y., & Chen, Y. C. (2018). Theaflavin-3,3'-digallate enhances the inhibitory effect of cisplatin by regulating the copper transporter 1 and glutathione in human ovarian cancer cells. *International Journal of Molecular Sciences*, *19*(1).

Park, S., Jo, S. H., Kim, J. H., Kim, S. Y., Ha, J. D., Hwang, J. Y., et al. (2020). Combination treatment with GSK126 and pomalidomide induces B-cell differentiation in EZH2 gain-of-function mutant diffuse large B-cell lymphoma. *Cancers (Basel)*, *12*(9), 2541.

Park, S. Y., Yoo, E. J., Cho, N. Y., Kim, N., & Kang, G. H. (2009). Comparison of CpG island hypermethylation and repetitive DNA hypomethylation in premalignant stages of gastric cancer, stratified for Helicobacter pylori infection. *The Journal of Pathology*, *219*(4), 410–416.

Pasqualucci, L., Dominguez-Sola, D., Chiarenza, A., Fabbri, G., Grunn, A., Trifonov, V., et al. (2011). Inactivating mutations of acetyltransferase genes in B-cell lymphoma. *Nature*, *471*(7337), 189–195.

Pedicona, F., Casado, P., Hijazi, M., Gribben, J. G., Rouault-Pierre, K., & Cutillas, P. R. (2022). Targeting the lysine-specific demethylase 1 rewires kinase networks and primes leukemia cells for kinase inhibitor treatment. *Science Signaling*, *15*(730), eabl7989.

Peifer, M., Fernandez-Cuesta, L., Sos, M. L., George, J., Seidel, D., Kasper, L. H., et al. (2012). Integrative genome analyses identify key somatic driver mutations of small-cell lung cancer. *Nature Genetics*, *44*(10), 1104–1110.

Pena-Hernandez, R., Aprigliano, R., Carina Frommel, S., Pietrzak, K., Steiger, S., Roganowicz, M., et al. (2021). BAZ2A-mediated repression via H3K14ac-marked enhancers promotes prostate cancer stem cells. *EMBO Reports*, *22*(11), e53014.

Peng, Y., & Croce, C. M. (2016). The role of MicroRNAs in human cancer. *Signal Transduction and Targeted Therapy*, *1*, 15004.

Peng, D., Kryczek, I., Nagarsheth, N., Zhao, L., Wei, S., Wang, W., et al. (2015). Epigenetic silencing of TH1-type chemokines shapes tumour immunity and immunotherapy. *Nature*, *527*(7577), 249–253.

Pengelly, A. R., Copur, O., Jackle, H., Herzig, A., & Muller, J. (2013). A histone mutant reproduces the phenotype caused by loss of histone-modifying factor Polycomb. *Science*, *339*(6120), 698–699.

Perillo, B., Tramontano, A., Pezone, A., & Migliaccio, A. (2020). LSD1: More than demethylation of histone lysine residues. *Experimental & Molecular Medicine*, *52*(12), 1936–1947.

Peta, E., Sinigaglia, A., Masi, G., Di Camillo, B., Grassi, A., Trevisan, M., et al. (2018). HPV16 E6 and E7 upregulate the histone lysine demethylase KDM2B through the c-MYC/miR-146a-5p axys. *Oncogene*, *37*(12), 1654–1668.

Phillips, R. E., Soshnev, A. A., & Allis, C. D. (2020). Epigenomic reprogramming as a driver of malignant glioma. *Cancer Cell*, *38*(5), 647–660.

Picaud, S., Da Costa, D., Thanasopoulou, A., Filippakopoulos, P., Fish, P. V., Philpott, M., et al. (2013). PFI-1, a highly selective protein interaction inhibitor, targeting BET Bromodomains. *Cancer Research*, *73*(11), 3336–3346.

Picaud, S., Fedorov, O., Thanasopoulou, A., Leonards, K., Jones, K., Meier, J., et al. (2015). Generation of a selective small molecule inhibitor of the CBP/p300 bromodomain for leukemia therapy. *Cancer Research*, *75*(23), 5106–5119.

Picaud, S., Leonards, K., Lambert, J. P., Dovey, O., Wells, C., Fedorov, O., et al. (2016). Promiscuous targeting of bromodomains by bromosporine identifies BET proteins as master regulators of primary transcription response in leukemia. *Science Advances*, *2*(10), e1600760.

Piha-Paul, S. A., Sachdev, J. C., Barve, M., LoRusso, P., Szmulewitz, R., Patel, S. P., et al. (2019). First-in-human study of mivebresib (ABBV-075), an oral pan-inhibitor of bromodomain and extra terminal proteins, in patients with relapsed/refractory solid tumors. *Clinical Cancer Research*, *25*(21), 6309–6319.

Pili, R., Haggman, M., Stadler, W. M., Gingrich, J. R., Assikis, V. J., Bjork, A., et al. (2011). Phase II randomized, double-blind, placebo-controlled study of tasquinimod in men with minimally symptomatic metastatic castrate-resistant prostate cancer. *Journal of Clinical Oncology*, *29*(30), 4022–4028.

Popovici-Muller, J., Lemieux, R. M., Artin, E., Saunders, J. O., Salituro, F. G., Travins, J., et al. (2018). Discovery of AG-120 (ivosidenib): A first-in-class mutant IDH1 inhibitor for the treatment of IDH1 mutant cancers. *ACS Medicinal Chemistry Letters*, *9*(4), 300–305.

Qi, W., Chan, H., Teng, L., Li, L., Chuai, S., Zhang, R., et al. (2012). Selective inhibition of Ezh2 by a small molecule inhibitor blocks tumor cells proliferation. *Proceedings of the National Academy of Sciences of the United States of America*, *109*(52), 21360–21365.

Qian, C., & Zhou, M. M. (2006). SET domain protein lysine methyltransferases: Structure, specificity and catalysis. *Cellular and Molecular Life Sciences*, *63*(23), 2755–2763.

Qu, G., Dubeau, L., Narayan, A., Yu, M. C., & Ehrlich, M. (1999). Satellite DNA hypomethylation vs. overall genomic hypomethylation in ovarian epithelial tumors of different malignant potential. *Mutation Research, 423*(1–2), 91–101.

Rabizadeh, E., Merkin, V., Belyaeva, I., Shaklai, M., & Zimra, Y. (2007). Pivanex, a histone deacetylase inhibitor, induces changes in BCR-ABL expression and when combined with STI571, acts synergistically in a chronic myelocytic leukemia cell line. *Leukemia Research, 31*(8), 1115–1123.

Raina, K., Lu, J., Qian, Y., Altieri, M., Gordon, D., Rossi, A. M., et al. (2016). PROTAC-induced BET protein degradation as a therapy for castration-resistant prostate cancer. *Proceedings of the National Academy of Sciences of the United States of America, 113*(26), 7124–7129.

Rasmussen, K. D., & Helin, K. (2016). Role of TET enzymes in DNA methylation, development, and cancer. *Genes & Development, 30*(7), 733–750.

Rettig, I., Koeneke, E., Trippel, F., Mueller, W. C., Burhenne, J., Kopp-Schneider, A., et al. (2015). Selective inhibition of HDAC8 decreases neuroblastoma growth in vitro and in vivo and enhances retinoic acid-mediated differentiation. *Cell Death & Disease, 6*, e1657.

Richon, V. M., Emiliani, S., Verdin, E., Webb, Y., Breslow, R., Rifkind, R. A., et al. (1998). A class of hybrid polar inducers of transformed cell differentiation inhibits histone deacetylases. *Proceedings of the National Academy of Sciences of the United States of America, 95*(6), 3003–3007.

Rilova, E., Erdmann, A., Gros, C., Masson, V., Aussagues, Y., Poughon-Cassabois, V., et al. (2014). Design, synthesis and biological evaluation of 4-amino-N- (4-aminophenyl) benzamide analogues of quinoline-based SGI-1027 as inhibitors of DNA methylation. *ChemMedChem, 9*(3), 590–601.

Risner, L. E., Kuntimaddi, A., Lokken, A. A., Achille, N. J., Birch, N. W., Schoenfelt, K., et al. (2013). Functional specificity of CpG DNA-binding CXXC domains in mixed lineage leukemia. *The Journal of Biological Chemistry, 288*(41), 29901–29910.

Roman-Gomez, J., Jimenez-Velasco, A., Agirre, X., Castillejo, J. A., Navarro, G., San Jose-Eneriz, E., et al. (2008). Repetitive DNA hypomethylation in the advanced phase of chronic myeloid leukemia. *Leukemia Research, 32*(3), 487–490.

Ropero, S., & Esteller, M. (2007). The role of histone deacetylases (HDACs) in human cancer. *Molecular Oncology, 1*(1), 19–25.

Ropero, S., Fraga, M. F., Ballestar, E., Hamelin, R., Yamamoto, H., Boix-Chornet, M., et al. (2006). A truncating mutation of HDAC2 in human cancers confers resistance to histone deacetylase inhibition. *Nature Genetics, 38*(5), 566–569.

Rosty, C., Ueki, T., Argani, P., Jansen, M., Yeo, C. J., Cameron, J. L., et al. (2002). Overexpression of S100A4 in pancreatic ductal adenocarcinomas is associated with poor differentiation and DNA hypomethylation. *The American Journal of Pathology, 160*(1), 45–50.

Rotili, D., Tarantino, D., Marrocco, B., Gros, C., Masson, V., Poughon, V., et al. (2014). Properly substituted analogues of BIX-01294 lose inhibition of G9a histone methyltransferase and gain selective anti-DNA methyltransferase 3A activity. *PLoS One, 9*(5), e96941.

Ruan, Y., Wang, L., & Lu, Y. (2021). HDAC6 inhibitor, ACY1215 suppress the proliferation and induce apoptosis of gallbladder cancer cells and increased the chemotherapy effect of gemcitabine and oxaliplatin. *Drug Development Research, 82*(4), 598–604.

Ruzza, P., Rosato, A., Rossi, C. R., Floreani, M., & Quintieri, L. (2009). Glutathione transferases as targets for cancer therapy. *Anti-Cancer Agents in Medicinal Chemistry, 9*(7), 763–777.

Sabbattini, P., Canzonetta, C., Sjoberg, M., Nikic, S., Georgiou, A., Kemball-Cook, G., et al. (2007). A novel role for the Aurora B kinase in epigenetic marking of silent chromatin in differentiated postmitotic cells. *The EMBO Journal, 26*(22), 4657–4669.

Sachchidanand, Resnick-Silverman, L., Yan, S., Mutjaba, S., Liu, W. J., Zeng, L., et al. (2006). Target structure-based discovery of small molecules that block human p53 and CREB binding protein association. *Chemistry & Biology*, *13*(1), 81–90.

Saenz, D. T., Fiskus, W., Qian, Y., Manshouri, T., Rajapakshe, K., Raina, K., et al. (2017). Novel BET protein proteolysis-targeting chimera exerts superior lethal activity than bromodomain inhibitor (BETi) against post-myeloproliferative neoplasm secondary (s) AML cells. *Leukemia*, *31*(9), 1951–1961.

Sato, N., Fukushima, N., Matsubayashi, H., & Goggins, M. (2004). Identification of maspin and S100P as novel hypomethylation targets in pancreatic cancer using global gene expression profiling. *Oncogene*, *23*(8), 1531–1538.

Sayegh, J., Cao, J., Zou, M. R., Morales, A., Blair, L. P., Norcia, M., et al. (2013). Identification of small molecule inhibitors of Jumonji AT-rich interactive domain 1B (JARID1B) histone demethylase by a sensitive high throughput screen. *The Journal of Biological Chemistry*, *288*(13), 9408–9417.

Sbardella, G., Castellano, S., Vicidomini, C., Rotili, D., Nebbioso, A., Miceli, M., et al. (2008). Identification of long chain alkylidenemalonates as novel small molecule modulators of histone acetyltransferases. *Bioorganic & Medicinal Chemistry Letters*, *18*(9), 2788–2792.

Schneider, B. J., Shah, M. A., Klute, K., Ocean, A., Popa, E., Altorki, N., et al. (2017). Phase I study of epigenetic priming with azacitidine prior to standard neoadjuvant chemotherapy for patients with resectable gastric and esophageal adenocarcinoma: Evidence of tumor hypomethylation as an indicator of major histopathologic response. *Clinical Cancer Research*, *23*(11), 2673–2680.

Schotta, G., Sengupta, R., Kubicek, S., Malin, S., Kauer, M., Callen, E., et al. (2008). A chromatin-wide transition to H4K20 monomethylation impairs genome integrity and programmed DNA rearrangements in the mouse. *Genes & Development*, *22*(15), 2048–2061.

Schuetz, A., Allali-Hassani, A., Martin, F., Loppnau, P., Vedadi, M., Bochkarev, A., et al. (2006). Structural basis for molecular recognition and presentation of histone H3 by WDR5. *The EMBO Journal*, *25*(18), 4245–4252.

Schwartzentruber, J., Korshunov, A., Liu, X. Y., Jones, D. T., Pfaff, E., Jacob, K., et al. (2012). Driver mutations in histone H3.3 and chromatin remodelling genes in paediatric glioblastoma. *Nature*, *482*(7384), 226–231.

Segura, M. F., Fontanals-Cirera, B., Gaziel-Sovran, A., Guijarro, M. V., Hanniford, D., Zhang, G., et al. (2013). BRD4 sustains melanoma proliferation and represents a new target for epigenetic therapy. *Cancer Research*, *73*(20), 6264–6276.

Shabbir, M., & Stuart, R. (2010). Lestaurtinib, a multitargeted tyrosine kinase inhibitor: From bench to bedside. *Expert Opinion on Investigational Drugs*, *19*(3), 427–436.

Shanafelt, T. D., Call, T. G., Zent, C. S., Leis, J. F., LaPlant, B., Bowen, D. A., et al. (2013). Phase 2 trial of daily, oral Polyphenon E in patients with asymptomatic, Rai stage 0 to II chronic lymphocytic leukemia. *Cancer*, *119*(2), 363–370.

Shapiro, G. I., LoRusso, P., Dowlati, A., Khanh, T. D., Jacobson, C. A., Vaishampayan, U., et al. (2021). A Phase 1 study of RO6870810, a novel bromodomain and extra-terminal protein inhibitor, in patients with NUT carcinoma, other solid tumours, or diffuse large B-cell lymphoma. *British Journal of Cancer*, *124*(4), 744–753.

Sharma, S., Kelly, T. K., & Jones, P. A. (2010). Epigenetics in cancer. *Carcinogenesis*, *31*(1), 27–36.

Sheikh, B. N., & Akhtar, A. (2019). The many lives of KATs—Detectors, integrators and modulators of the cellular environment. *Nature Reviews. Genetics*, *20*(1), 7–23.

Sheng, W., LaFleur, M. W., Nguyen, T. H., Chen, S., Chakravarthy, A., Conway, J. R., et al. (2018). LSD1 ablation stimulates anti-tumor immunity and enables checkpoint blockade. *Cell*, *174*(3), 549–563 e519.

Shi, X., Kachirskaia, I., Walter, K. L., Kuo, J. H., Lake, A., Davrazou, F., et al. (2007). Proteome-wide analysis in Saccharomyces cerevisiae identifies several PHD fingers as novel direct and selective binding modules of histone H3 methylated at either lysine 4 or lysine 36. *The Journal of Biological Chemistry*, *282*(4), 2450–2455.
Shimazu, T., Hirschey, M. D., Newman, J., He, W., Shirakawa, K., Le Moan, N., et al. (2013). Suppression of oxidative stress by beta-hydroxybutyrate, an endogenous histone deacetylase inhibitor. *Science*, *339*(6116), 211–214.
Shorstova, T., Foulkes, W. D., & Witcher, M. (2021). Achieving clinical success with BET inhibitors as anti-cancer agents. *British Journal of Cancer*, *124*(9), 1478–1490.
Sarnik, J., Poplawski, T., & Tokarz, P. (2021). BET proteins as attractive targets for cancer therapeutics. *International Journal of Molecular Sciences*, *22*(20), 11102.
Shull, A. Y., Luo, J. F., Pei, L. R., Lee, E. J., Liu, J. M., Choi, J., et al. (2016). DNA hypomethylation within B-cell enhancers and super enhancers reveal a dependency on immune and metabolic mechanisms in chronic lymphocytic leukemia. *Blood*, *128*(22), 1049.
Silverman, L. R. (2001). Targeting hypomethylation of DNA to achieve cellular differentiation in myelodysplastic syndromes (MDS). *The Oncologist*, *6*(Suppl. 5), 8–14.
Singh, A., Patel, V. K., Jain, D. K., Patel, P., & Rajak, H. (2016). Panobinostat as pan-deacetylase inhibitor for the treatment of pancreatic cancer: Recent progress and future prospects. *Oncology and Therapy*, *4*(1), 73–89.
Singh, A. K., Zhao, B., Liu, X., Wang, X., Li, H., Qin, H., et al. (2020). Selective targeting of TET catalytic domain promotes somatic cell reprogramming. *Proceedings of the National Academy of Sciences of the United States of America*, *117*(7), 3621–3626.
Siu, K. T., Ramachandran, J., Yee, A. J., Eda, H., Santo, L., Panaroni, C., et al. (2017). Preclinical activity of CPI-0610, a novel small-molecule bromodomain and extra-terminal protein inhibitor in the therapy of multiple myeloma. *Leukemia*, *31*(8), 1760–1769.
Smith, K. M., Ketchart, W., Zhou, X., Montano, M. M., & Xu, Y. (2011). Determination of hexamethylene bisacetamide, an antineoplastic compound, in mouse and human plasma by LC-MS/MS. *Journal of Chromatography. B, Analytical Technologies in the Biomedical and Life Sciences*, *879*(23), 2206–2212.
Smith, Z. D., & Meissner, A. (2013). DNA methylation: Roles in mammalian development. *Nature Reviews. Genetics*, *14*(3), 204–220.
Smith, E., Zhou, W., Shindiapina, P., Sif, S., Li, C., & Baiocchi, R. A. (2018). Recent advances in targeting protein arginine methyltransferase enzymes in cancer therapy. *Expert Opinion on Therapeutic Targets*, *22*(6), 527–545.
Smitheman, K. N., Severson, T. M., Rajapurkar, S. R., McCabe, M. T., Karpinich, N., Foley, J., et al. (2019). Lysine specific demethylase 1 inactivation enhances differentiation and promotes cytotoxic response when combined with all-trans retinoic acid in acute myeloid leukemia across subtypes. *Haematologica*, *104*(6), 1156–1167.
Smolewski, P., & Robak, T. (2017). The discovery and development of romidepsin for the treatment of T-cell lymphoma. *Expert Opinion on Drug Discovery*, *12*(8), 859–873.
Soldi, R., Ghosh Halder, T., Weston, A., Thode, T., Drenner, K., Lewis, R., et al. (2020). The novel reversible LSD1 inhibitor SP-2577 promotes anti-tumor immunity in SWItch/Sucrose-NonFermentable (SWI/SNF) complex mutated ovarian cancer. *PLoS One*, *15*(7), e0235705.
Sroczynska, P., Cruickshank, V. A., Bukowski, J. P., Miyagi, S., Bagger, F. O., Walfridsson, J., et al. (2014). shRNA screening identifies JMJD1C as being required for leukemia maintenance. *Blood*, *123*(12), 1870–1882.
Stazi, G., Taglieri, L., Nicolai, A., Romanelli, A., Fioravanti, R., Morrone, S., et al. (2019). Dissecting the role of novel EZH2 inhibitors in primary glioblastoma cell cultures: Effects on proliferation, epithelial-mesenchymal transition, migration, and on the pro-inflammatory phenotype. *Clinical Epigenetics*, *11*(1), 173.

Sternberg, C., Armstrong, A., Pili, R., Ng, S., Huddart, R., Agarwal, N., et al. (2016). Randomized, double-blind, placebo-controlled phase III study of tasquinimod in men with metastatic castration-resistant prostate cancer. *Journal of Clinical Oncology, 34*(22), 2636–2643.

Sterner, D. E., & Berger, S. L. (2000). Acetylation of histones and transcription-related factors. *Microbiology and Molecular Biology Reviews, 64*(2), 435–459.

Stimson, L., Rowlands, M. G., Newbatt, Y. M., Smith, N. F., Raynaud, F. I., Rogers, P., et al. (2005). Isothiazolones as inhibitors of PCAF and p300 histone acetyltransferase activity. *Molecular Cancer Therapeutics, 4*(10), 1521–1532.

Stoger, R., Kubicka, P., Liu, C. G., Kafri, T., Razin, A., Cedar, H., et al. (1993). Maternal-specific methylation of the imprinted mouse Igf2r locus identifies the expressed locus as carrying the imprinting signal. *Cell, 73*(1), 61–71.

Stresemann, C., & Lyko, F. (2008). Modes of action of the DNA methyltransferase inhibitors azacytidine and decitabine. *International Journal of Cancer, 123*(1), 8–13.

Stubbs, M. C., Burn, T. C., Sparks, R., Maduskuie, T., Diamond, S., Rupar, M., et al. (2019). The novel bromodomain and extraterminal domain inhibitor INCB054329 induces vulnerabilities in myeloma cells that inform rational combination strategies. *Clinical Cancer Research, 25*(1), 300–311.

Suganuma, T., Pattenden, S. G., & Workman, J. L. (2008). Diverse functions of WD40 repeat proteins in histone recognition. *Genes & Development, 22*(10), 1265–1268.

Sugino, N., Kawahara, M., Tatsumi, G., Kanai, A., Matsui, H., Yamamoto, R., et al. (2017). A novel LSD1 inhibitor NCD38 ameliorates MDS-related leukemia with complex karyotype by attenuating leukemia programs via activating super-enhancers. *Leukemia, 31*(11), 2303–2314.

Sun, K., Atoyan, R., Borek, M. A., Dellarocca, S., Samson, M. E., Ma, A. W., et al. (2017). Dual HDAC and PI3K inhibitor CUDC-907 downregulates MYC and suppresses growth of MYC-dependent cancers. *Molecular Cancer Therapeutics, 16*(2), 285–299.

Sun, B., Fiskus, W., Qian, Y., Rajapakshe, K., Raina, K., Coleman, K. G., et al. (2018). BET protein proteolysis targeting chimera (PROTAC) exerts potent lethal activity against mantle cell lymphoma cells. *Leukemia, 32*(2), 343–352.

Sun, Y., Han, J., Wang, Z., Li, X., Sun, Y., & Hu, Z. (2020). Safety and efficacy of bromodomain and extra-terminal inhibitors for the treatment of hematological malignancies and solid tumors: A systematic study of clinical trials. *Frontiers in Pharmacology, 11*, 621093.

Sun, N., Zhang, J., Zhang, C., Zhao, B., & Jiao, A. (2018). DNMTs inhibitor SGI-1027 induces apoptosis in Huh7 human hepatocellular carcinoma cells. *Oncology Letters, 16*(5), 5799–5806.

Sung, B., Pandey, M. K., Ahn, K. S., Yi, T., Chaturvedi, M. M., Liu, M., et al. (2008). Anacardic acid (6-nonadecyl salicylic acid), an inhibitor of histone acetyltransferase, suppresses expression of nuclear factor-kappaB-regulated gene products involved in cell survival, proliferation, invasion, and inflammation through inhibition of the inhibitory subunit of nuclear factor-kappaBalpha kinase, leading to potentiation of apoptosis. *Blood, 111*(10), 4880–4891.

Suryanarayanan, V., & Singh, S. K. (2019). Deciphering the binding mode and mechanistic insights of pentadecylidenemalonate (1b) as activator of histone acetyltransferase PCAF. *Journal of Biomolecular Structure & Dynamics, 37*(9), 2296–2309.

Suzuki, T., Ando, T., Tsuchiya, K., Fukazawa, N., Saito, A., Mariko, Y., et al. (1999). Synthesis and histone deacetylase inhibitory activity of new benzamide derivatives. *Journal of Medicinal Chemistry, 42*(15), 3001–3003.

Suzuki, T., Kasuya, Y., Itoh, Y., Ota, Y., Zhan, P., Asamitsu, K., et al. (2013). Identification of highly selective and potent histone deacetylase 3 inhibitors using click chemistry-based combinatorial fragment assembly. *PLoS One, 8*(7), e68669.

Suzuki, S., Ono, R., Narita, T., Pask, A. J., Shaw, G., Wang, C., et al. (2007). Retrotransposon silencing by DNA methylation can drive mammalian genomic imprinting. *PLoS Genetics*, *3*(4), e55.

Tachibana, M., Sugimoto, K., Nozaki, M., Ueda, J., Ohta, T., Ohki, M., et al. (2002). G9a histone methyltransferase plays a dominant role in euchromatic histone H3 lysine 9 methylation and is essential for early embryogenesis. *Genes & Development*, *16*(14), 1779–1791.

Tachibana, M., Ueda, J., Fukuda, M., Takeda, N., Ohta, T., Iwanari, H., et al. (2005). Histone methyltransferases G9a and GLP form heteromeric complexes and are both crucial for methylation of euchromatin at H3-K9. *Genes & Development*, *19*(7), 815–826.

Tan, D., Phipps, C., Hwang, W. Y., Tan, S. Y., Yeap, C. H., Chan, Y. H., et al. (2015). Panobinostat in combination with bortezomib in patients with relapsed or refractory peripheral T-cell lymphoma: An open-label, multicentre phase 2 trial. *The Lancet Haematology*, *2*(8), e326–e333.

Tan, J., Yang, X., Zhuang, L., Jiang, X., Chen, W., Lee, P. L., et al. (2007). Pharmacologic disruption of Polycomb-repressive complex 2-mediated gene repression selectively induces apoptosis in cancer cells. *Genes & Development*, *21*(9), 1050–1063.

Tang, S., Cheng, B., Zhe, N., Ma, D., Xu, J., Li, X., et al. (2018). Histone deacetylase inhibitor BG45-mediated HO-1 expression induces apoptosis of multiple myeloma cells by the JAK2/STAT3 pathway. *Anti-Cancer Drugs*, *29*(1), 61–74.

Tang, Y., Huang, S., Chen, X., Huang, J., Lin, Q., Huang, L., et al. (2022). Design, synthesis and biological evaluation of novel and potent protein arginine methyltransferases 5 inhibitors for cancer therapy. *Molecules*, *27*(19), 6637.

Tang, Y., Zhao, W., Chen, Y., Zhao, Y., & Gu, W. (2008). Acetylation is indispensable for p53 activation. *Cell*, *133*(4), 612–626.

Tarasenko, N., Nudelman, A., Tarasenko, I., Entin-Meer, M., Hass-Kogan, D., Inbal, A., et al. (2008). Histone deacetylase inhibitors: The anticancer, antimetastatic and antiangiogenic activities of AN-7 are superior to those of the clinically tested AN-9 (Pivanex). *Clinical & Experimental Metastasis*, *25*(7), 703–716.

Tavassoli, P., Wafa, L. A., Cheng, H., Zoubeidi, A., Fazli, L., Gleave, M., et al. (2010). TAF1 differentially enhances androgen receptor transcriptional activity via its N-terminal kinase and ubiquitin-activating and -conjugating domains. *Molecular Endocrinology*, *24*(4), 696–708.

Tayari, M. M., Santos, H. G. D., Kwon, D., Bradley, T. J., Thomassen, A., Chen, C., et al. (2021). Clinical responsiveness to all-trans retinoic acid is potentiated by LSD1 inhibition and associated with a quiescent transcriptome in myeloid malignancies. *Clinical Cancer Research*, *27*(7), 1893–1903.

Taylor, A. M., Cote, A., Hewitt, M. C., Pastor, R., Leblanc, Y., Nasveschuk, C. G., et al. (2016). Fragment-based discovery of a selective and cell-active benzodiazepinone CBP/EP300 bromodomain inhibitor (CPI-637). *ACS Medicinal Chemistry Letters*, *7*(5), 531–536.

Taylor, B. S., DeCarolis, P. L., Angeles, C. V., Brenet, F., Schultz, N., Antonescu, C. R., et al. (2011). Frequent alterations and epigenetic silencing of differentiation pathway genes in structurally rearranged liposarcomas. *Cancer Discovery*, *1*(7), 587–597.

Theodoulou, N. H., Bamborough, P., Bannister, A. J., Becher, I., Bit, R. A., Che, K. H., et al. (2016). Discovery of I-BRD9, a selective cell active chemical probe for bromodomain containing protein 9 inhibition. *Journal of Medicinal Chemistry*, *59*(4), 1425–1439.

Thomas, T., & Voss, A. K. (2007). The diverse biological roles of MYST histone acetyltransferase family proteins. *Cell Cycle*, *6*(6), 696–704.

Tian, J., & Yuan, L. (2018). Sirtuin 6 inhibits colon cancer progression by modulating PTEN/AKT signaling. *Biomedicine & Pharmacotherapy*, *106*, 109–116.

Torrens-Mas, M., Oliver, J., Roca, P., & Sastre-Serra, J. (2017). SIRT3: Oncogene and tumor suppressor in cancer. *Cancers (Basel)*, *9*(7), 90.

Trapani, D., Esposito, A., Criscitiello, C., Mazzarella, L., Locatelli, M., Minchella, I., et al. (2017). Entinostat for the treatment of breast cancer. *Expert Opinion on Investigational Drugs*, *26*(8), 965–971.

Tumber, A., Nuzzi, A., Hookway, E. S., Hatch, S. B., Velupillai, S., Johansson, C., et al. (2017). Potent and selective KDM5 inhibitor stops cellular demethylation of H3K4me3 at transcription start sites and proliferation of MM1S myeloma cells. *Cell Chemical Biology*, *24*(3), 371–380.

Tuorto, F., Herbst, F., Alerasool, N., Bender, S., Popp, O., Federico, G., et al. (2015). The tRNA methyltransferase Dnmt2 is required for accurate polypeptide synthesis during haematopoiesis. *The EMBO Journal*, *34*(18), 2350–2362.

Tzatsos, A., Paskaleva, P., Lymperi, S., Contino, G., Stoykova, S., Chen, Z., et al. (2011). Lysine-specific demethylase 2B (KDM2B)-let-7-enhancer of zester homolog 2 (EZH2) pathway regulates cell cycle progression and senescence in primary cells. *The Journal of Biological Chemistry*, *286*(38), 33061–33069.

Valente, S., Liu, Y., Schnekenburger, M., Zwergel, C., Cosconati, S., Gros, C., et al. (2014). Selective non-nucleoside inhibitors of human DNA methyltransferases active in cancer including in cancer stem cells. *Journal of Medicinal Chemistry*, *57*(3), 701–713.

Valente, S., Trisciuoglio, D., De Luca, T., Nebbioso, A., Labella, D., Lenoci, A., et al. (2014). 1,3,4-Oxadiazole-containing histone deacetylase inhibitors: Anticancer activities in cancer cells. *Journal of Medicinal Chemistry*, *57*(14), 6259–6265.

Varambally, S., Dhanasekaran, S. M., Zhou, M., Barrette, T. R., Kumar-Sinha, C., Sanda, M. G., et al. (2002). The polycomb group protein EZH2 is involved in progression of prostate cancer. *Nature*, *419*(6907), 624–629.

Vaswani, R. G., Gehling, V. S., Dakin, L. A., Cook, A. S., Nasveschuk, C. G., Duplessis, M., et al. (2016). Identification of (R)-N-((4-Methoxy-6-methyl-2-oxo-1, 2-dihydropyridin-3-yl)methyl)-2-methyl-1-(1-(1 -(2,2,2-trifluoroethyl)piperidin-4-yl) ethyl)-1H-indole-3-carboxamide (CPI-1205), a Potent and Selective Inhibitor of Histone Methyltransferase EZH2, Suitable for Phase I Clinical Trials for B-Cell Lymphomas. *Journal of Medicinal Chemistry*, *59*(21), 9928–9941.

Verma, S. K., Tian, X., LaFrance, L. V., Duquenne, C., Suarez, D. P., Newlander, K. A., et al. (2012). Identification of potent, selective, cell-active inhibitors of the histone lysine methyltransferase EZH2. *ACS Medicinal Chemistry Letters*, *3*(12), 1091–1096.

Verstovsek, S., Kantarjian, H., Mesa, R. A., Pardanani, A. D., Cortes-Franco, J., Thomas, D. A., et al. (2010). Safety and efficacy of INCB018424, a JAK1 and JAK2 inhibitor, in myelofibrosis. *The New England Journal of Medicine*, *363*(12), 1117–1127.

Vigushin, D. M., Ali, S., Pace, P. E., Mirsaidi, N., Ito, K., Adcock, I., et al. (2001). Trichostatin A is a histone deacetylase inhibitor with potent antitumor activity against breast cancer in vivo. *Clinical Cancer Research*, 7(4), 971–976.

Villar-Garea, A., Fraga, M. F., Espada, J., & Esteller, M. (2003). Procaine is a DNA-demethylating agent with growth-inhibitory effects in human cancer cells. *Cancer Research*, *63*(16), 4984–4989.

Vinet, M., Suresh, S., Maire, V., Monchecourt, C., Nemati, F., Lesage, L., et al. (2019). Protein arginine methyltransferase 5: A novel therapeutic target for triple-negative breast cancers. *Cancer Medicine*, *8*(5), 2414–2428.

Voelter-Mahlknecht, S., & Mahlknecht, U. (2010). The sirtuins in the pathogenesis of cancer. *Clinical Epigenetics*, *1*(3–4), 71–83.

Walter, M. J., Ding, L., Shen, D., Shao, J., Grillot, M., McLellan, M., et al. (2011). Recurrent DNMT3A mutations in patients with myelodysplastic syndromes. *Leukemia*, *25*(7), 1153–1158.

Wang, F., Chan, C. H., Chen, K., Guan, X., Lin, H. K., & Tong, Q. (2012). Deacetylation of FOXO3 by SIRT1 or SIRT2 leads to Skp2-mediated FOXO3 ubiquitination and degradation. *Oncogene, 31*(12), 1546–1557.

Wang, L., Gural, A., Sun, X. J., Zhao, X., Perna, F., Huang, G., et al. (2011). The leukemogenicity of AML1-ETO is dependent on site-specific lysine acetylation. *Science, 333*(6043), 765–769.

Wang, W., Oguz, G., Lee, P. L., Bao, Y., Wang, P., Terp, M. G., et al. (2018). KDM4B-regulated unfolded protein response as a therapeutic vulnerability in PTEN-deficient breast cancer. *The Journal of Experimental Medicine, 215*(11), 2833–2849.

Wang, Y. Q., Wang, H. L., Xu, J., Tan, J., Fu, L. N., Wang, J. L., et al. (2018). Sirtuin5 contributes to colorectal carcinogenesis by enhancing glutaminolysis in a deglutarylation-dependent manner. *Nature Communications, 9*(1), 545.

Wang, L., Xu, M., Kao, C. Y., Tsai, S. Y., & Tsai, M. J. (2020). Small molecule JQ1 promotes prostate cancer invasion via BET-independent inactivation of FOXA1. *The Journal of Clinical Investigation, 130*(4), 1782–1792.

Wang, L., Yamaguchi, S., Burstein, M. D., Terashima, K., Chang, K., Ng, H. K., et al. (2014). Novel somatic and germline mutations in intracranial germ cell tumours. *Nature, 511*(7508), 241–245.

Wang, Z., Zang, C., Cui, K., Schones, D. E., Barski, A., Peng, W., et al. (2009). Genome-wide mapping of HATs and HDACs reveals distinct functions in active and inactive genes. *Cell, 138*(5), 1019–1031.

Wang, T., Zhang, F., & Sun, F. (2022). ORY-1001, a KDM1A inhibitor, inhibits proliferation, and promotes apoptosis of triple negative breast cancer cells by inactivating androgen receptor. *Drug Development Research, 83*(1), 208–216.

Wang, D., Zhou, J., Liu, X., Lu, D., Shen, C., Du, Y., et al. (2013). Methylation of SUV39H1 by SET7/9 results in heterochromatin relaxation and genome instability. *Proceedings of the National Academy of Sciences of the United States of America, 110*(14), 5516–5521.

Wapenaar, H., & Dekker, F. J. (2016). Histone acetyltransferases: Challenges in targeting bi-substrate enzymes. *Clinical Epigenetics, 8*, 59.

Watanabe, S., Shimada, S., Akiyama, Y., Ishikawa, Y., Ogura, T., Ogawa, K., et al. (2019). Loss of KDM6A characterizes a poor prognostic subtype of human pancreatic cancer and potentiates HDAC inhibitor lethality. *International Journal of Cancer, 145*(1), 192–205.

Wei, W., Jing, Z. X., Ke, Z., & Yi, P. (2017). Sirtuin 7 plays an oncogenic role in human osteosarcoma via downregulating CDC4 expression. *American Journal of Cancer Research, 7*(9), 1788–1803.

Wells, C. E., Bhaskara, S., Stengel, K. R., Zhao, Y., Sirbu, B., Chagot, B., et al. (2013). Inhibition of histone deacetylase 3 causes replication stress in cutaneous T cell lymphoma. *PLoS One, 8*(7), e68915.

Wilson, A. J., Stubbs, M., Liu, P., Ruggeri, B., & Khabele, D. (2018). The BET inhibitor INCB054329 reduces homologous recombination efficiency and augments PARP inhibitor activity in ovarian cancer. *Gynecologic Oncology, 149*(3), 575–584.

Winquist, E., Knox, J., Ayoub, J. P., Wood, L., Wainman, N., Reid, G. K., et al. (2006). Phase II trial of DNA methyltransferase 1 inhibition with the antisense oligonucleotide MG98 in patients with metastatic renal carcinoma: A National Cancer Institute of Canada Clinical Trials Group investigational new drug study. *Investigational New Drugs, 24*(2), 159–167.

Winter, G. E., Buckley, D. L., Paulk, J., Roberts, J. M., Souza, A., Dhe-Paganon, S., et al. (2015). DRUG DEVELOPMENT. Phthalimide conjugation as a strategy for in vivo target protein degradation. *Science, 348*(6241), 1376–1381.

Wobser, M., Weber, A., Glunz, A., Tauch, S., Seitz, K., Butelmann, T., et al. (2019). Elucidating the mechanism of action of domatinostat (4SC-202) in cutaneous T cell lymphoma cells. *Journal of Hematology & Oncology, 12*(1), 30.

Woodcock, D. M., Crowther, P. J., & Diver, W. P. (1987). The majority of methylated deoxycytidines in human DNA are not in the CpG dinucleotide. *Biochemical and Biophysical Research Communications*, *145*(2), 888–894.
Wozniak, M. B., Villuendas, R., Bischoff, J. R., Aparicio, C. B., Martinez Leal, J. F., de La Cueva, P., et al. (2010). Vorinostat interferes with the signaling transduction pathway of T-cell receptor and synergizes with phosphoinositide-3 kinase inhibitors in cutaneous T-cell lymphoma. *Haematologica*, *95*(4), 613–621.
Wu, Q., Nie, D. Y., Ba-Alawi, W., Ji, Y., Zhang, Z., Cruickshank, J., et al. (2022). PRMT inhibition induces a viral mimicry response in triple-negative breast cancer. *Nature Chemical Biology*, *18*(8), 821–830.
Wu, L. C., Wen, Z. S., Qiu, Y. T., Chen, X. Q., Chen, H. B., Wei, M. M., et al. (2013). Largazole arrests cell cycle at G1 phase and triggers proteasomal degradation of E2F1 in lung cancer cells. *ACS Medicinal Chemistry Letters*, *4*(10), 921–926.
Wyce, A., Matteo, J. J., Foley, S. W., Felitsky, D. J., Rajapurkar, S. R., Zhang, X. P., et al. (2018). MEK inhibitors overcome resistance to BET inhibition across a number of solid and hematologic cancers. *Oncogene*, 7(4), 35.
Wysocka, J., Swigut, T., Milne, T. A., Dou, Y., Zhang, X., Burlingame, A. L., et al. (2005). WDR5 associates with histone H3 methylated at K4 and is essential for H3 K4 methylation and vertebrate development. *Cell*, *121*(6), 859–872.
Xiang, Q., Luo, G., Zhang, C., Hu, Q., Wang, C., Wu, T., et al. (2022). Discovery, optimization and evaluation of 1-(indolin-1-yl)ethan-1-ones as novel selective TRIM24/BRPF1 bromodomain inhibitors. *European Journal of Medicinal Chemistry*, *236*, 114311.
Xiangyun, Y., Xiaomin, N., Linping, G., Yunhua, X., Ziming, L., Yongfeng, Y., et al. (2017). Desuccinylation of pyruvate kinase M2 by SIRT5 contributes to antioxidant response and tumor growth. *Oncotarget*, *8*(4), 6984–6993.
Xu, L., Gao, X., Yang, P., Sang, W., Jiao, J., Niu, M., et al. (2021). EHMT2 inhibitor BIX-01294 induces endoplasmic reticulum stress mediated apoptosis and autophagy in diffuse large B-cell lymphoma cells. *Journal of Cancer*, *12*(4), 1011–1022.
Yamagishi, M., Hori, M., Fujikawa, D., Ohsugi, T., Honma, D., Adachi, N., et al. (2019). Targeting excessive EZH1 and EZH2 activities for abnormal histone methylation and transcription network in malignant lymphomas. *Cell Reports*, *29*(8). 2321–2337.e2327.
Yamazaki, S., Gukasyan, H. J., Wang, H., Uryu, S., & Sharma, S. (2020). Translational pharmacokinetic-pharmacodynamic modeling for an orally available novel inhibitor of epigenetic regulator enhancer of zeste homolog 2. *The Journal of Pharmacology and Experimental Therapeutics*, *373*(2), 220–229.
Yan, K., Cao, Q., Reilly, C. M., Young, N. L., Garcia, B. A., & Mishra, N. (2011). Histone deacetylase 9 deficiency protects against effector T cell-mediated systemic autoimmunity. *The Journal of Biological Chemistry*, *286*(33), 28833–28843.
Yang, X. J. (2004). The diverse superfamily of lysine acetyltransferases and their roles in leukemia and other diseases. *Nucleic Acids Research*, *32*(3), 959–976.
Yang, Y., & Bedford, M. T. (2013). Protein arginine methyltransferases and cancer. *Nature Reviews. Cancer*, *13*(1), 37–50.
Yang, C., Choy, E., Hornicek, F. J., Wood, K. B., Schwab, J. H., Liu, X., et al. (2011). Histone deacetylase inhibitor (HDACI) PCI-24781 potentiates cytotoxic effects of doxorubicin in bone sarcoma cells. *Cancer Chemotherapy and Pharmacology*, *67*(2), 439–446.
Yang, X. J., & Gregoire, S. (2005). Class II histone deacetylases: From sequence to function, regulation, and clinical implication. *Molecular and Cellular Biology*, *25*(8), 2873–2884.
Yang, L., Hou, J., Cui, X. H., Suo, L. N., & Lv, Y. W. (2017). RG108 induces the apoptosis of endometrial cancer Ishikawa cell lines by inhibiting the expression of DNMT3B and demethylation of HMLH1. *European Review for Medical and Pharmacological Sciences*, *21*(22), 5056–5064.

Yang, X. J., & Seto, E. (2008). The Rpd3/Hda1 family of lysine deacetylases: From bacteria and yeast to mice and men. *Nature Reviews. Molecular Cell Biology*, *9*(3), 206–218.
Yang, Q., Wang, B., Gao, W., Huang, S., Liu, Z., Li, W., et al. (2013). SIRT1 is downregulated in gastric cancer and leads to G1-phase arrest via NF-kappaB/Cyclin D1 signaling. *Molecular Cancer Research*, *11*(12), 1497–1507.
Yang, L., Zha, Y., Ding, J., Ye, B., Liu, M., Yan, C., et al. (2019). Histone demethylase KDM6B has an anti-tumorigenic function in neuroblastoma by promoting differentiation. *Oncogene*, *8*(1), 3.
Yang, C., Zhang, J., Ma, Y., Wu, C., Cui, W., & Wang, L. (2020). Histone methyltransferase and drug resistance in cancers. *Journal of Experimental & Clinical Cancer Research*, *39*(1), 173.
Yao, Y., Hu, H., Yang, Y., Zhou, G., Shang, Z., Yang, X., et al. (2016). Downregulation of enhancer of zeste homolog 2 (EZH2) is essential for the induction of autophagy and apoptosis in colorectal cancer cells. *Genes (Basel)*, 7(10), 83.
Ye, L., Li, X., Kong, X., Wang, W., Bi, Y., Hu, L., et al. (2005). Hypomethylation in the promoter region of POMC gene correlates with ectopic overexpression in thymic carcinoids. *The Journal of Endocrinology*, *185*(2), 337–343.
Yen, K., Travins, J., Wang, F., David, M. D., Artin, E., Straley, K., et al. (2017). AG-221, a first-in-class therapy targeting acute myeloid leukemia harboring oncogenic IDH2 mutations. *Cancer Discovery*, 7(5), 478–493.
Yeung, F., Hoberg, J. E., Ramsey, C. S., Keller, M. D., Jones, D. R., Frye, R. A., et al. (2004). Modulation of NF-kappaB-dependent transcription and cell survival by the SIRT1 deacetylase. *The EMBO Journal*, *23*(12), 2369–2380.
Yin, M., Guo, Y., Hu, R., Cai, W. L., Li, Y., Pei, S., et al. (2020). Potent BRD4 inhibitor suppresses cancer cell-macrophage interaction. *Nature Communications*, *11*(1), 1833.
Yoo, C. B., Jeong, S., Egger, G., Liang, G., Phiasivongsa, P., Tang, C., et al. (2007). Delivery of 5-aza-2′-deoxycytidine to cells using oligodeoxynucleotides. *Cancer Research*, *67*(13), 6400–6408.
Young, C. S., Clarke, K. M., Kettyle, L. M., Thompson, A., & Mills, K. I. (2017). Decitabine-Vorinostat combination treatment in acute myeloid leukemia activates pathways with potential for novel triple therapy. *Oncotarget*, *8*(31), 51429–51446.
Young, L. C., McDonald, D. W., & Hendzel, M. J. (2013). Kdm4b histone demethylase is a DNA damage response protein and confers a survival advantage following gamma-irradiation. *The Journal of Biological Chemistry*, *288*(29), 21376–21388.
Yu, W., Chory, E. J., Wernimont, A. K., Tempel, W., Scopton, A., Federation, A., et al. (2012). Catalytic site remodelling of the DOT1L methyltransferase by selective inhibitors. *Nature Communications*, *3*, 1288.
Yuan, H., Wang, Z., Li, L., Zhang, H., Modi, H., Horne, D., et al. (2012). Activation of stress response gene SIRT1 by BCR-ABL promotes leukemogenesis. *Blood*, *119*(8), 1904–1914.
Yuan, Y., Wang, Q., Paulk, J., Kubicek, S., Kemp, M. M., Adams, D. J., et al. (2012). A small-molecule probe of the histone methyltransferase G9a induces cellular senescence in pancreatic adenocarcinoma. *ACS Chemical Biology*, 7(7), 1152–1157.
Zengerle, M., Chan, K. H., & Ciulli, A. (2015). Selective small molecule induced degradation of the BET bromodomain protein BRD4. *ACS Chemical Biology*, *10*(8), 1770–1777.
Zhang, Y., Duan, S., Jang, A., Mao, L., Liu, X., & Huang, G. (2021). JQ1, a selective inhibitor of BRD4, suppresses retinoblastoma cell growth by inducing cell cycle arrest and apoptosis. *Experimental Eye Research*, *202*, 108304.
Zhang, Y., Mittal, A., Reid, J., Reich, S., Gamblin, S. J., & Wilson, J. R. (2015). Evolving catalytic properties of the MLL family SET domain. *Structure*, *23*(10), 1921–1933.

Zhang, S., Suvannasankha, A., Crean, C. D., White, V. L., Chen, C. S., & Farag, S. S. (2011). The novel histone deacetylase inhibitor, AR-42, inhibits gp130/Stat3 pathway and induces apoptosis and cell cycle arrest in multiple myeloma cells. *International Journal of Cancer, 129*(1), 204–213.

Zhang, R., Wang, C., Tian, Y., Yao, Y., Mao, J., Wang, H., et al. (2019). SIRT5 promotes hepatocellular carcinoma progression by regulating mitochondrial apoptosis. *Journal of Cancer, 10*(16), 3871–3882.

Zhang, C., Yu, Y., Huang, Q., & Tang, K. (2019). SIRT6 regulates the proliferation and apoptosis of hepatocellular carcinoma via the ERK1/2 signaling pathway. *Molecular Medicine Reports, 20*(2), 1575–1582.

Zhao, J., & Lawless, M. W. (2016). Resminostat: Opening the door to epigenetic treatments for liver cancer. *Hepatology, 63*(2), 668–669.

Zheng, H., Zhao, W., Yan, C., Watson, C. C., Massengill, M., Xie, M., et al. (2016). HDAC inhibitors enhance T-cell chemokine expression and augment response to PD-1 immunotherapy in lung adenocarcinoma. *Clinical Cancer Research, 22*(16), 4119–4132.

Zhou, L., Yao, Q., Li, H., & Chen, J. (2021). Targeting BRD9 by I-BRD9 efficiently inhibits growth of acute myeloid leukemia cells. *Translational Cancer Research, 10*(7), 3364–3372.

Zhou, L., Yao, Q., Ma, L., Li, H., & Chen, J. (2021). TAF1 inhibitor Bay-299 induces cell death in acute myeloid leukemia. *Translational Cancer Research, 10*(12), 5307–5318.

Zhu, H., Bengsch, F., Svoronos, N., Rutkowski, M. R., Bitler, B. G., Allegrezza, M. J., et al. (2016). BET bromodomain inhibition promotes anti-tumor immunity by suppressing PD-L1 expression. *Cell Reports, 16*(11), 2829–2837.

Zimmerman, B., Sargeant, A., Landes, K., Fernandez, S. A., Chen, C. S., & Lairmore, M. D. (2011). Efficacy of novel histone deacetylase inhibitor, AR42, in a mouse model of, human T-lymphotropic virus type 1 adult T cell lymphoma. *Leukemia Research, 35*(11), 1491–1497.

Further reading

Healy, J. R., Hart, L. S., Shazad, A. L., Gagliardi, M. E., Tsang, M., Elias, J., et al. (2020). Limited antitumor activity of combined BET and MEK inhibition in neuroblastoma. *Pediatric Blood & Cancer, 67*(6), e28267.

CHAPTER FOUR

Histone deacetylase inhibitors as sanguine epitherapeutics against the deadliest lung cancer

Shabir Ahmad Ganai[a,*], Basit Amin Shah[b], and Manzoor Ahmad Yatoo[a]
[a]Division of Basic Sciences and Humanities, FoA, SKUAST-Kashmir, Sopore-193201, India
[b]Directorate of Forensic Science Laboratory, Bemina, Srinagar, Jammu & Kashmir, India
*Corresponding author: e-mail address: shabir.muntazir82@gmail.com

Contents

Abstract

The back-breaking resistance mechanisms generated by lung cancer cells against epidermal growth factor receptor (EGFR), KRAS and Janus kinase 2 (JAK2) directed therapies strongly prioritizes the requirement of novel therapies which are perfectly tolerated, potentially cytotoxic and can reinstate the drug-sensitivity in lung cancer cells. Enzymatic proteins modifying the post-translational modifications of nucleosome-integrated histone substrates are appearing as current targets for defeating various malignancies. Histone deacetylases (HDACs) are hyperexpressed in diverse lung cancer types. Blocking the active pocket of these acetylation erasers through HDAC inhibitors (HDACi) has come out as an optimistic therapeutic recourse for annihilating lung cancer. This article in the beginning gives an overview about lung cancer statistics and predominant lung cancer types. Succeeding this, compendium about conventional therapies and their serious drawbacks has been provided. Then, connection of uncommon

Advances in Cancer Research, Volume 158
ISSN 0065-230X
https://doi.org/10.1016/bs.acr.2022.12.003

expression of classical HDACs in lung cancer onset and expansion has been detailed. Moreover, keeping the main theme in view this article deeply discusses HDACi in the context of aggressive lung cancer as single agents and spotlights various molecular targets suppressed or induced by these inhibitors for engendering cytotoxic effect. Most particularly, the raised pharmacological effects achieved on using these inhibitors in concerted form with other therapeutic molecules and the cancer-linked pathways altered by this procedure are described. The positive direction towards further heightening of efficacy and the pressing requirement of exhaustive clinical assessment has been proposed as a new focus point.

1. Lung cancer statistics

As per National Centre for Health Statistics, 609,360 cancer deaths and 1,918,030 fresh cancer cases are expected to happen in America in 2022. Further, it has been predicted that nearly 350 deaths in a single day will result entirely from lung cancer (Siegel, Miller, Fuchs, & Jemal, 2022). Another comparative study involving data of open source projected that 2,370,000 and 4,820,000 novel cancer cases will occur in America and China respectively in the year 2022. This study also predicted that 3,210,000 cancer related deaths will occur in China and 640,000 such deaths will take place in America in the aforesaid year. While the lung cancer is most frequent type occurring in China, breast cancer holds the defined status in America. Despite this alteration, in both countries the foremost cause of cancer attributed deaths is lung cancer (Xia et al., 2022).

2. Overview of major types of lung cancer

Lung cancers are majorly divided into two types based on the appearance of lung cancer cells under microscope; non-small cell lung cancer (NSCLC) and small cell lung cancer (SCLC). NSCLC is the most frequently occurring and this form of lung cancer accounts for around 85% of the whole lung cancer cases. NSCLC encompasses adenocarcinomas contributing 40% of the cases of lung cancer. Although adenocarcinomas like other lung cancers are found in smokers they are observed in non-smokers as well (Zappa & Mousa, 2016). Majority of adenocarcinomas originate either in outer or peripheral regions of the lungs. Adenocarcinomas exhibit lymph node migration and even afar. On the other hand squamous cell carcinomas although being previously more frequent over adenocarcinomas contribute only 25–30% of total lung cancer cases. The third type of NSCLC is large cell carcinomas which are occasionally termed as undifferentiated carcinomas. This form of NSCLC shows least frequency of occurrence and thus forms only 10–15% of the entire lung cancers (Chan & Hughes, 2015). Another infrequent subtype of NSCLC is adenosquamous carcinoma (ASC). As the name mentions this carcinoma has components of both lung squamous cell carcinoma (SCC) and lung adenocarcinoma (ADC). Despite this, adenosquamous carcinoma of lungs has biological properties resembling to SCC and ADC but it should not be regarded just a mere hybrid of the two. ASC forms only 0.4–4% of total lung carcinomas (Li & Lu, 2018; Travis, Brambilla, Burke, Marx, & Nicholson, 2015). Another subtype of NSCLC (poorly differentiated) having rare occurrence is pulmonary sarcomatoid carcinoma (PSC). It is less frequent than ASC as only 0.1–0.4% of entire lung cancers are of PSCs (Huang, Shen, & Li, 2013; Karim et al., 2018; Ung et al., 2016). Only 10–15% lung cancers have SCLC appearance. Among all the lung cancer types SCLC is highly bellicose, neuroendocrine and grows very fast. Additionally this lung cancer type is potentially linked to smoking of cigarettes (Rudin, Brambilla, Faivre-Finn, & Sage, 2021). Due to their high metastatic rate they are confirmed only when they have metastasized to a greater extent (Fig. 1). In 2004, World Health Organization (WHO) classified SCLC into pure-SCLC and combined-SCLC (Qin & Lu, 2018). While pure-SCLC accounts for nearly 80% of the SCLC, about 20% of the SCLC cases are of combined-SCLC (Rudin et al., 2021). Combined-SCLC as the name signifies in addition to SCLC component contains non-SCLC component as well and as such is considered as infrequent SCLC variant (Ebisu et al., 2018). The 5 year survival rate of people taking all lung cancer types into account is only 22%. The defined survival is relatively more in women (25%) than in men (18%) (Yu et al., 2022). While the 5 year rate of survival for SCLC is only 7%, it is 26% for NSCLC indicating that SCLC is relatively more lethal over NSCLC (Elkbuli et al., 2020; Siegel et al., 2022). Further, this survival varies in case of NSCLC depending on whether this cancer is localized, regional or metastatic. Localized indicates that cancer is within the confines of lungs and has not gone outside, regional signifies that cancer has gained access to the surrounding lymph nodes while metastatic means that cancer has gained access to distant body parts. The aforesaid survival rate for localized NSCLC is 63%, for regional around 35% and for metastatic NSCLC only 7% (Siegel et al., 2022). Speaking in general the SCLC survival rate (5-year) is only 7%. However, the aforementioned rate for localized SCLC is 27%, for regional form of SCLC is 16% while that of malignant or distant SCLC this parameter

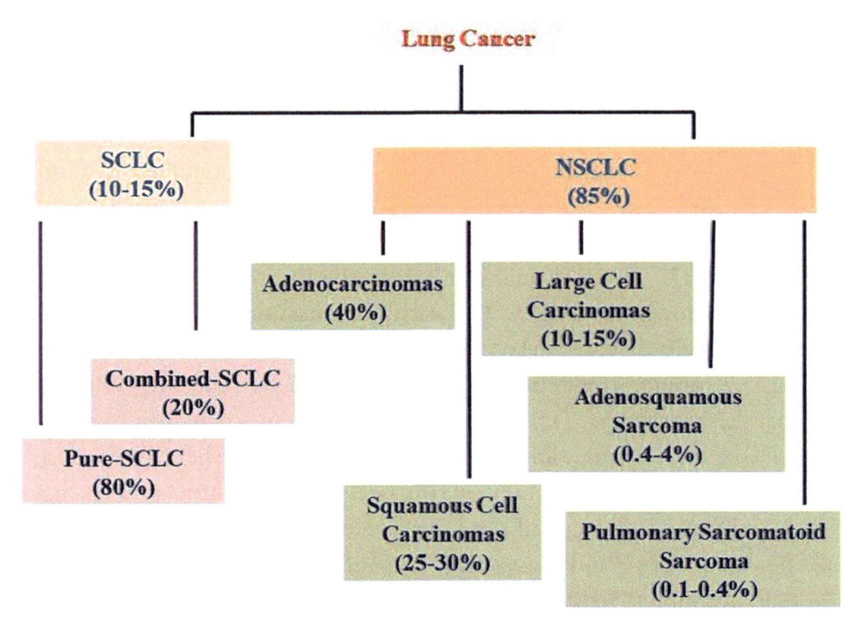

Fig. 1 Overview of different types of lung cancer and their occurrence frequency. Lung cancer has two main subtypes SCLC and NSCLC. While SCLC contributes to 10–15% of all lung cancers, NSCLC constitutes 85% of all lung cancers. Adenocarcinomas form 40% of entire lung cancer cases whereas squamous cell carcinomas contribute 25–30% and large cell carcinomas make up 10–15% of lung cancers. Adenosquamous sarcoma and pulmonary sarcomatoid sarcoma form 0.4–4% and 0.1–0.4% of total lung cancers respectively. 80% of the SCLC are of pure-SCLC type while combined-SCLC constitutes 20% of the whole SCLCs. 41–89% NSCLCs manifest EGFR overexpression. 89% and 41% of squamous tumors and adenocarcinomas show hyperexpression of EGFR respectively. About 31.6% patients of NSCLC portray EGFR mutation with 15% in squamous cell carcinoma and 36.5% in subtype adenocarcinoma. Deletion at 19th position (DEL19) of EGFR forms 39.3% of whole EGFR mutations while lysine to arginine conversion at 858 position (L858R) constitutes 50.7% of total EGFR mutations.

is only 3% (Siegel et al., 2022). According to reports about lung cancer, the survival rate has escalated in USA from 14.5% to 23.7%. While 2 year survival of 35% has been reported in men who were confirmed to have NSCLC in 2014, only 26% survival was estimated in patients who were diagnosed in year 2001. This gives an indication that that the survival of NSCLC patients irrespective of ethnicities/races has significantly become better with the passage of time (Howlader et al., 2020). Mesothelium is simply an epithelial structure and is regarded as embryonic mesoderm derivative. It covers and provides protection to majority of internal organs including lungs and heart. Cancer developing from mesothelium is termed as mesothelioma.

Multiple types of mesothelioma have been reported encompassing pleural mesothelioma, pericardial mesothelioma and peritoneal mesothelioma. Pericardial mesothelioma primarily grows in pericardium (heart lining) (Mott, 2012), Pleural mesothelioma affects pleura and accounts for 80% of entire mesotheliomas. Exposure to asbestos is considered as the premier cause of pleural mesothelioma (Thomas, Karakattu, Cagle, & Hoskere, 2021). It has been estimated that latency period of malignant pleural mesothelioma ranges from 15 to 60 years. It is noteworthy that pleural mesothelioma is not considered lung cancer as it does not occur inside lungs but involves pleura, the serous membrane (Frost, 2013).

3. Conventional and other lung cancer therapies and their significant limitations

The conventional therapies target various molecular targets involved in initiation and promotion of lung cancer in one way or the other. Epidermal growth factor receptor (EGFR), a well-established glycoprotein (transmembrane) acts as receptor for epidermal growth factor (EGF) and other ligands. EGFR consists of a single chain that has several functionally important regions. These regions encompass extracellular domain that acts as binding site for EGF followed by a transmembrane sequence which is short in length, cytoplasmic region encompassing phosphorylation domain at C-terminal and tyrosine kinase domain (Lee, Hazlett, & Koland, 2006). On binding to ligand this surface protein undergoes dimerization and subsequent autophosphorylation on tyrosine residue culminating in cell proliferation (Sabbah, Hajjo, & Sweidan, 2020). Atypical expression of EGFR has strong involvement in NSCLC pathogenesis (Ohsaki et al., 2000). Further, its expression in defined cancer has implications in decreased survival, low chemosensitivity and recurrent metastasis to lymph nodes (Fontanini et al., 1998). Activation of EGFR has link with facilitation of angiogenesis in tumors and this angiogenesis in turn favors proliferation and metastasis of these cells (Hung et al., 2016). Expression of this transmembrane protein in cancer cells regulates the generation of vascular endothelial growth factor (VEGF) and interleukin 8 (IL-8) (Ferrara, Gerber, & LeCouter, 2003; Waugh & Wilson, 2008). EGFR overexpression ranges from 41% to 89% in NSCLC. It has been estimated that only 41% of adenocarcinomas but 89% of squamous tumors manifest overexpression of this transmembrane receptor (Al Olayan, Al Hussaini, & Jazieh, 2012; Lynch et al., 2004). Most of the mutations of EGFR observed in NSCLC are spaced in its tyrosine kinase domain exons (18–21) (Hsu, Yang, Mok, & Loong, 2018). One of the study performed in 2019 demonstrated that 31.6% of patients having NSCLC carry EGFR mutation. While 15% of these patients were having squamous cell carcinoma, 36.5% manifested adenocarcinoma. L858R and DEL19 mutations contributed 50.7% and 39.3% of entire EGFR mutations (Kumari et al., 2019). Mutations in EGFR are hardly observed in SCLC unlike NSCLC. Frequency of EGFR mutations in SCLC is 0.2–4% (Bhuiyan, Siddiqui, Zirkiyeva, Agladze, & Bashir, 2021; Varghese et al., 2014).

Inhibitors of EGFR have been used for countering NSCLC but these inhibitors are effectual where tumors are strongly reliant on signaling elicited by this receptor. Among the presently available EGFR inhibitors lapatinib, gefitinib and erlotinib are prominent (Yuan, Huang, Chen, Wu, & Xu, 2019). These tyrosine kinase inhibitors are given through oral route and as indicated by their names modulate the tyrosine kinase domain of EGFR. On the other hand certain monoclonal antibodies encompassing panitumumab and cetuximab are injected through intravenous mode (Freeman et al., 2009; Pirker & Filipits, 2012). However, unlike tyrosine kinase inhibitors these antibodies disrupt extracellular ligand docking (binding) (Baselga, 2001). While the cetuximab has demonstrated enhanced therapeutic benefit in concert with other chemotherapeutic agents, tyrosine kinase inhibitors of EGFR manifested no advantage in collaboration with standard cytotoxic agents (Pirker & Filipits, 2012). However, tyrosine kinase inhibitors as individual agents have been observed to extend survival in aggressive NSCLC. Following use of first generation inhibitors of tyrosine kinase it has been reported that patients develop resistance to these inhibitors (Lee et al., 2013). This resistance has been predominantly ascribed to mutation of tyrosine kinase component of EGFR and has been confirmed to be substitution of threonine at position 790 by methionine (Suda, Onozato, Yatabe, & Mitsudomi, 2009; Xu, Xie, Ni, & Liu, 2015). Further, amplification of mesenchymal epithelial transition *(MET)* gene in addition to mutation of RAS also contributes to defined resistance (Engelman et al., 2007; Pao et al., 2005; Shanker, Willcutts, Roth, & Ramesh, 2010). However, the above mentioned mutation of tyrosine kinase has been noticed in a subset of patients which have not all been subjected to treatment involving the first generation tyrosine kinase inhibitors.

Osimertinib considered as third generation inhibitor of tyrosine kinase of above defined growth factor receptor has demonstrated substantially better progression free survival over erlotinib and gefitinib both being first generation inhibitors (Ramalingam et al., 2020; Takeda & Nakagawa, 2019). RAS family of genes has HRAS, KRAS and NRAS under its umbrella and these genes encode GTPases which in turn modulate cell proliferation *via* distinct pathways. KRAS of RAS family is the most predominant RAS mutated in lung cancer (Slebos & Rodenhuis, 1992). About 20–40% lung adenocarcinomas show mutations in KRAS (Adderley, Blackhall, & Lindsay, 2019; Ricciuti et al., 2016). Sorafenib, a multikinase inhibitor administered through oral route targets RAF, PDGFR, VEGFR-2, and VEGFR-3 (Wilhelm et al., 2006). Mutation in EGFR (C797S) was confirmed to induce resistance toward third generation EFGR-tyrosine kinase inhibitors including HM61713 and AZD9291 (Wang, Tsui, Liu, Song, & Liu, 2016). Proliferation as well as migration of lung cancer cells bearing EGFR mutation is augmented by amplification of MET and thus targeting this amplification may improve therapeutic advantage (Baldacci et al., 2018). From these grim facts certain things are evident that EGFR inhibitors in conjunction with crizotinib (MET inhibitor) or AMG510 (KRAS G12C inhibitor) may result in increased clinical outcome (Huang, Guo, Wang, & Fu, 2021; Nwizu et al., 2011).

Activation of JAK-STAT pathways has implications in multiple solid tumors and NSCLC is not an exception (Qureshy, Johnson, & Grandis, 2020). STAT3, an extensively explored member of STAT family plays a critical role in multiple malignancies (Dutta, Sabri, Li, & Li, 2014). Also STAT pathways modulate apoptotic genes, restrain growth of cells and make the EGFR-inhibitor therapy effectual. Coupling EGFR-tyrosine kinase inhibiting molecules with inhibitors of JAK/STAT pathways may augment the anticancer activity and circumvent drug resistance encountered in NSCLC. Indeed a study in which ruxolitinib (JAK2) inhibitor was used in association with cisplatin concluded that the former sensitizes the cisplatin-resistant cells to cisplatin (Hu et al., 2014). Another JAK2 inhibitor CYT387 when given in combination with cetuximab exerted enhanced cytotoxic effects toward resistant NSCLC lines including H1975 (Hu, Dong, Liu, Liu, & Chen, 2016). Further, this augmented anticancer effect was also recorded in, *in vivo* models. An interesting study delineated that lower expression of suppressors of cytokine signaling 3 (SOCS3) but enhanced miR-410 occurs in A549 (lung cancer)-cells over non-tumorigenic BEAS-2B cells. Transfection of either anti-miR-410 or SOCS3 in lung cancer cells markedly induced apoptosis in these cells while considerably declining the phosphorylated levels of JAK1/JAK2 and STAT3.

These findings suggest miR-410 may emerge as a propitious molecular target for overcoming lung cancer (Li, Zheng, & Yuan, 2018). Human epidermal growth factor receptor 3 (HER3) is the product of *ERBB3*. This receptor has connection with resistance of NSCLC cells against EGFR-tyrosine kinase targeting molecules. This resistance has been accredited to compensatory HER3 upregulation. Monoclonal antibodies targeting this receptor including lumretuzumab are currently being evaluated in early clinical studies (Cejalvo et al., 2016). Immunotherapy against lung cancer involving targeting of programmed cell death ligand-1 (PD-L1), cytotoxic T-lymphocyte-associated antigen-4 (CTLA-4) have received green signals for their use against lung cancer and many other cancer types. On 24th of October 2016, Food and Drug Administration (FDA) gave approval to pembrolizumab also known by its brand name as Keytruda for treating the patients carrying metastatic-NSCLC expressing PD-L1. Pembrolizumab, the selective monoclonal antibody targeting programmed cell death protein 1 was approved as first line treatment for those patients of metastatic-NSCLC demonstrating high PD-L1 expression on minimum 50% of the tumor cells (tumor proportion score (TPS) equal to or more than 50%) but without EGFR or anaplastic lymphoma kinase (ALK) aberrations (Khoja, Butler, Kang, Ebbinghaus, & Joshua, 2015; Pai-Scherf et al., 2017). Further these patients should not have undergone systemic chemotherapy for this malignancy. Additionally, pembrolizumab has been given green signal as a second-line therapy for metastatic-NSCLC patients whose disease has advanced following platinum-based therapy and which manifest expression of PD-L1 minimum on 1% (1%) of tumor cells. On 17th of June 2019 this programmed cell death protein 1 inhibitor was conferred accelerated acceptance (approval) by FDA for the treatment of metastatic-SCLC as well with certain conditions (Pai-Scherf et al., 2017). Despite these efforts, due to immune microenvironment intricacy of tumors, immunotherapy is not effective against major proportion of cancer subjects (Herbst et al., 2016; Rittmeyer et al., 2017; Yuan et al., 2019).

4. Surgery and radiation therapy for lung cancer

Both these modalities have significant importance in treating cancers including lung cancer. Surgery is also opted especially against lung cancer upto certain stages only. This procedure is hardly done for SCLC whereas patients having non-SCLC at early stage are subjected to surgery. It is very unfortunate that patients of NSCLC when diagnosed for this type of lung cancer have already crossed the stage where only surgical resection can serve the purpose. Due to this reason patients of NSCLC are subjected to multidisciplinary treatment. Advanced methods of surgery involve approaches which are invasive to a lesser degree thereby reducing the sufferings associated with lesser advanced methods (Lackey & Donington, 2013). Stages I as well as II are considered as early phase of NSCLC and is primarily treated through surgery. During surgery of lung cancer either the part of lung is removed or the entire of it. In wedge resection a triangular piece of lung containing tumor and small part of typical tissue surrounding it is removed. Segmental resection also termed as segmentectomy involves removal of portion of lung smaller than a full lobe but bigger than a wedge section. Both these methods are infact the subtypes of sub-lobar type resection. While pneumonectomy involves removal of full lung, the term lobectomy is used when a single lobe among the five lobes of two lungs is removed (Lackey & Donington, 2013). Conventional radiation therapy is not considered as superior substitute to surgery for early phase lung cancer due to its discouraging outcome. This involves a machine from which radiations emerge and are aimed at abnormal (cancer) area. Majority of machines used in radiation therapy employ photon beams (Parashar, Arora, & Wernicke, 2013). For patients who are in the early stage of NSCLC and are medically incurable, an advanced method known as stereotactic body radiation therapy (SBRT) is opted. Just after some days following SBRT, 90% of the lung cancer patients revert to normalcy. SBRT makes it possible to deliver the high dose of radiation to abnormal tissue (tumor) in very targeted fashion, therefore typical tissues get extremely small radiation exposure and are thus spared (Andruska et al., 2021). On the whole, modalities beyond the immune check point, EGFR, KRAS, JAK2, miR-410, HER3 inhibition and conventional radiation therapy are urgently required to address the drawbacks associated with ongoing approved anti-lung cancer treatments.

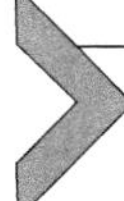

5. Histone deacetylases (HDACs) in lung cancer pathogenesis

Many enzymes regulate gene expression by modifying specific residues of histone proteins through covalent addition of certain groups. However, certain regulate the DNA-methylome *via* methylation and demethylation of cytosine. Such enzymes are referred by a common term as epigenetic modulation/regulatory enzymes (Jin, Liu, Luo, Liu, & Liu, 2022). These regulator enzymes are mostly writers and erasers. HDACs come under the boundary of erasers as they remove covalently linked acetyl groups from DNA-integrated histone proteins. These enzymes are meant to switch off transcription following epigenetic route. They function contradictory to histone acetyltransferases (HATs) which are popular for covalent linking of acetyl groups to basic lysine residues of DNA-wrapped histones (Richters & Koehler, 2017). Typical HAT-HDAC function always keeps histone acetylome in equilibrium and any irregularity in usual HDAC functioning imbalances this acetylome inducing epigenetic-mediated deregulation of transcription and in the long run cancer. Broadly HDACs consist of two types of enzyme families namely Sirtuins and Classical HDACs. While the members of former enzyme family (SIR1-SIR7) use NAD+ for their functioning, HDAC classes (Class I-II and IV) within the latter family rely on zinc to perform deacetylase activity (Seto & Yoshida, 2014). This classification has been put in simplified form in Fig. 2 (A) along with the critical catalytic domains of Classical HDACs. Certain HDACs occurs as units of complexes formed of multiple proteins. As an example Sin3 complex consists of six core subunits including HDAC1/2, SAP30, SAP18, Sin3A/Sin3B, RbAp48, and RbAp46. Sin3 complex is functional at promoter as well as at transcribed regions of the genome (Glass & Rosenfeld, 2000; Zhang, Iratni, Erdjument-Bromage, Tempst, & Reinberg, 1997). Another complex containing HDAC1/HDAC2 as one of the component is Co-REST complex. First recognized as corepressor of RE-1 Silencing Transcription Factor (REST) and afterwards was found to be a part of HDAC1/2 possessing complexes. In addition to HDAC1/2, Co-REST complex possesses lysine-specific demethylase 1 (LSD1), zinc-finger protein 217, Sox-like protein and p80 (You, Tong, Grozinger, & Schreiber, 2001). It has been demonstrated that for interaction with Co-REST only one zinc figure motif is critical. Further, the mutations in this motif capable of preventing binding also nullify repression (Andrés et al., 1999). Methyl-CpG binding protein 2 (MECP2), as the name indicates binds to methylated cytosine residues of DNA following which it recruits the corepressor complex having HDAC1, HDAC2 in addition to Sin3A as its components. MECP2 consists of two functionally important domains. One domain namely MECP2 binding domain (MBD) binds to methylated CpG regions while transcription repressor domain (TRD) recruits the above mentioned

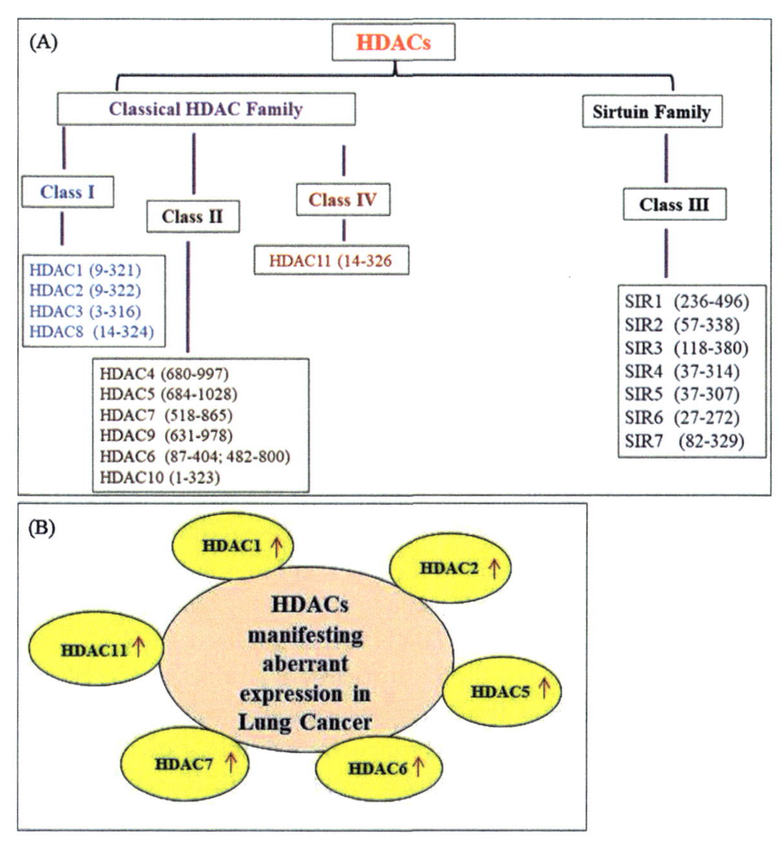

Fig. 2 Overview of HDAC classification (A) and different isozymes overexpressed in lung cancer (B). Two families namely classical HDAC family and sirtuin family come under HDACs. Classical HDACs need zinc for deacetylation function while sirtuins use NAD+ for normal functioning. The critical catalytic region residues of all HDACs are mentioned in brackets. It is evident that isozymes of all the classes of classical HDACs demonstrate elevated expression in lung cancer.

co-repressor complex (Kavalali, Nelson, & Monteggia, 2011). As this complex contains HDAC1 and HDAC2 which deacetylate nucleosomal histones the entire process results in chromatin condensation and ultimately transcriptional silencing (Nan et al., 1998).

Emerging findings have unbossomed the role of chromatin modification regulating enzymes in lung cancer (Ganai, 2020b; Langevin, Kratzke, & Kelsey, 2015). Various zinc-dependent HDACs demonstrate atypical expression in lung and other cancer types (Ganai, 2020d). As HDACs alter histone acetylation which comes under epigenetic signatures thus there is no controversy in mentioning them as epigenetic or epi-targets. Adenocarcinoma manifested greater expression of HDAC1 than squamous cell carcinoma. Negative correlation was observed between the expression of this HDAC at transcript or at protein level and lung cancer patient survival (Cao et al., 2017). Cell lines of lung adenocarcinoma showed hyperexpression of Class IIb HDAC by name HDAC6 and its overexpression facilitated proliferation of these cells that positively coincided with its deacetylase function. Overrepresentation of this HDAC impaired gefitinib sensitivity through EGFR stabilization (Wang et al., 2016). Plakoglobin also termed as γ-catenin obstructs lung cancer. Ectopic expression of Class IIa isozyme HDAC7 in lung cancer cells inhibited its degree of expression. This HDAC inhibits the aforesaid catenin by attaching to its promoter region. Proliferation as well as migration got facilitated by overexpression of HDAC7 in lung cancer cells. From these findings it is highly obvious that high expression of HDAC7 triggers proliferation and migration of these cells through indirect inhibition of plakoglobin (Sang et al., 2019). From lung cancer tissues abnormal expression of HDAC10 has been confirmed. This HDAC is localized mainly in the cytoplasmic part of lung cancer cells but in nearby typical (normal) cells occurs within the nuclear confines. In a variety of lung cancer cell lines, the cytoplasmic spacing of HDAC10 has been certified. Importantly, growth of lung cancer cells was found to be heightened by HDAC10. Knockdown of this Class IIb isozyme invoked cell cycle blockade and programmed cell death (Yang et al., 2014). Unusual HDAC2 expression has also been recorded in lung cancer. Inactivation of this HDAC reduced growth of tumor cells and induction of apoptosis through inhibition of Bcl2 and stimulation of Bax and p53 (Jung et al., 2012). HDAC7, the member of Class IIa facilitates lung cancer by way of deacetylating STAT3. HDAC7 forms direct interaction with STAT3 and this finally results in deacetylation of latter. Depletion of this enzyme in tumor cells substantially elevated the acetylation of STAT3 and its tyrosine phosphorylation (Lei et al., 2017). Higher degree of expression of only member of Class IV HDAC (HDAC11) coincided with low patient outcome and its depleted levels markedly alleviated cancer stem cells auto renewal. Besides depletion or obstruction of this HDAC declined expression of Sox2 and this transcription is critical for continuance of cancer stem cells. This suggests that HDAC11 is a crucial molecular target for vanquishing lung cancer (Bora-Singhal et al., 2020). RT-qPCR and analysis of western blots revealed that HDAC5 is markedly upregulated in lung cancer cells and tissues.

Its overexpression in lung cancer cells markedly augmented proliferation and restrained their apoptosis (Zhong et al., 2018). Elevated expression of HDAC7 triggered burgeoning and metastasis of lung cancer cells. HDAC7 hyperexpression hypoacetylated β-catenin by deacetylating it at lysine residue located at 49th position in primary sequence. Additionally in cancer cells over-representing this HDAC substantial site specific decline in phosphorylation (Ser45) of β-catenin was noticed. HDAC7 overexpression favored nuclear shuttling of above mentioned catenin where it promoted the expression of fibroblast growth factor 18 (FGF18) (Guo et al., 2022). Conclusively, anomalous expression of classical HDACs is potentially involved in onset and facilitation of lung cancer through modulation of various molecular players. Thus, for restoring normalcy molecules with the potential to block the activity of these zinc-dependent HDACs are increasingly surfacing as novel and promising anti-lung cancer therapeutics (Fig. 2(B).

6. Current scenario and chemistry of HDAC inhibitors

Certain molecules occlude the aberrant deacetylase function of eraser HDACs and by doing so reset the usual acetylome. These molecules have become popular as histone deacetylase inhibitors (HDACi). Basically HDACi come under the banner of epigenetic drugs or epi-weapons as they target transcriptional co-repressors (HDACs) which are significantly involved in suppression of gene expression following epigenetic mechanism (Ganai, 2020c). Chemically HDACi may be macrocyclic, hydroxamates, short chain fatty acids, benzamide derivatives, isothiocyanates, flavonoids, stilbenes or organosulfur compounds. Major proportion of HDACi target HDACs across diverse classes, some target entirely a particular HDAC while certain target only some or all members restricted to a single class (Ganai, 2019; Maccallini et al., 2022; Mottamal, Zheng, Huang, & Wang, 2015). Due to their highly impressive therapeutic effect against cancer four of these inhibitors have been bestowed approval by U.S FDA and some have been granted orphan approval for treating different types of blood malignancies such as multiple myeloma, cutaneous T-cell lymphoma and peripheral T-cell lymphoma (Bondarev et al., 2021; Boumber, Younes, & Garcia-Manero, 2011; Kim, Ali, Oh, & Seo, 2022). HDACi exert pleiotropic effects on cancer cells encompassing derepression of epigenetically suppressed pro-apoptotic genes, restrain angiogenesis, elicit differentiation, cell cycle arrest and obstruct conversion of epithelial phenotype to mesenchymal one (Mamdani & Jalal, 2020). As the theme of the article is HDACi in the context of lung cancer thus the downstream paragraphs will focus sequentially on these inhibitors as single agents and in combinatorial form with other chemotherapeutic agents against this cancer.

7. HDAC inhibitor-based monotherapy toward lung cancer

Significant research has been done to explore the cytotoxic effects of HDACi in different models of lung cancer. While testing the hydroxamate trichostatin A (TSA) on SCLC cell line DMS53, it was found that this inhibitor induces growth inhibition in these cells. TSA treatment caused morphological alterations in these cells and increased p21 and p27 levels. Expectedly this inhibitor raised PARP cleavage and concurrently depleted B-cell lymphoma 2 (*Bcl-2*) (Platta, Greenblatt, Kunnimalaiyaan, & Chen, 2007). As TSA is pan-HDAC inhibitor it targets a range of HDACs such as HDAC1-4, 6, 7 and 9 (Cacabelos & Torrellas, 2016). In A549 cells, TSA elicited apoptotic death through modulation of several molecular targets including cyclooxygenase-2 (COX-2). This apoptosis was accompanied with inhibition of B-Cell Leukemia/Lymphoma 2 (Bcl-2) and pro-apoptotic protein Bcl-2-associated X protein (Bax) upregulation. Caspase-3 and 9 activation and eventual poly-ADP ribose polymerase (PARP) degradation was noticed in treated cells. While substantial decline in cyclooxygenase-2 (COX-2) was quantified at transcript and at protein stage no such considerable effect was noticed on degree of cyclooxygenase-1 (COX-1) (Choi, 2005). Entinostat, selective to Class I HDACs has portrayed encouraging outcome against NSCLC (Ruiz, Raez, & Rolfo, 2015). This inhibitor selectively blocks the function of HDAC1 and HDAC3 (Truong et al., 2021). Sal-like protein 4 (SALL4), the crucial stem cell factor has a role in upkeeping embryonic stem cell pluripotency. Atypical expression of this factor has been confirmed in a wide range of cancers including lung cancer (Kobayashi, Kuribayashi, Tanaka, & Watanabe, 2011; Yang et al., 2008). Knockdown of SALL4 markedly collapsed the tumor growth over control group clearly indicating its expression is important for facilitating lung tumor dimensions. Insulin like growth factor 1 receptor (IGF1R) and EGFR expression is enhanced by SALL4 *via* Casitas B-lineage lymphoma proto-oncogene-b (CBL-B) ubiquitin ligase downregulation. Broad range of lung cancer lines (17) differing in extent of SALL4 expression were treated with HDAC inhibitor entinostat. H661 and PC-9 lines with higher SALL4 expression proved to be more sensitive to entinostat exposure over other lines with relatively lower expression of this protein. Entinostat treatment lowered the expression of SALL4 but elevated the levels of CBL-B in lung cancer (H661)-line. This occurred due to attenuation of SALL4 binding to promoter of *CBLB* following treatment with entinostat (10 μM) (Yong et al., 2016). Panobinostat, a known inhibitor of hydroxamate type demonstrated antiproliferative effect on multiple non-small cell lung cancer lines such as SKMES, H838, Calu-1, A549, and H226. Being pan-inhibitor, it targets all the 11 isozymes of Classical HDACs (HDAC1-HDAC11) (Singh, Patel, Jain, Patel, & Rajak, 2016). Experiments on xenograft mice (NOD-SCID) revealed that in comparison to control panobinostat markedly attenuated disease progression by 54% (Wang et al., 2018). This establishes that panobinostat has potent activity against NSCLC and needs further evaluation for certifying its use against this cancer.

Sodium 4-phenylbutyrate falls within the boundary of short chain fatty acid group HDACi. This inhibitor has the potential to induce growth arrest, differentiation and cytotoxicity in diverse cancer models. Multiple HDACs including HDAC1-5 and HDAC7-9 are inhibited by this short chain fatty acid (Cacabelos & Torrellas, 2016). While sodium 4-phenylbutyrate invoked apoptosis in NCI-H1792 and NCI-H460 lines it proved neutral toward HBE4-E6/E7 (bronchial epithelial line). The apoptotic effect was associated with stimulation of Janus Kinase (JAK) and extracellular signal-regulated kinase (ERK) post-treatment with said inhibitor. In a couple of lung cancer lines SK-LU-1 and A549, sodium 4-phenylbutyrate elicited growth arrest and this arrest was found to have cross-talk with ERK activation as no Janus kinase stimulation was noticed in these two lines. While sodium 4-phenylbutyrate-triggered apoptosis was rescued by JNK RNA interference or JNK inhibitor, inhibition of MEK failed to do so (Zhang, Wei, Yang, & Yu, 2004). This means that sodium 4-phenylbutyrate-stimulated apoptosis is mediated through Janus kinase activation. Mocetinostat, a selective inhibitor of benzamide derivative group inhibiting HDACs (HDAC1-3, 8 and 11) has shown promising activity against preclinical cancer models (Hahn et al., 2015). This class selective HDAC inhibitor was evaluated on a couple of NSCLC lines namely LX-1 and A-549, the former line being squamous cell carcinoma and the hind one adenocarcinoma line. Mocetinostat reduced cell survival following a concentration dependent trend. Lower concentrations of CI-994 induced growth arrest while higher concentration (160 μM) provoked apoptosis in NSCLC cells (Loprevite et al., 2005). It has been observed that large cell carcinoma and squamous cell carcinoma manifest greater HDAC activities over normal as well as adenocarcinoma tissues. Additionally, it has been confirmed that NCI-H1299, Calu-6 and SCLC cells exhibit elevated HDAC activities than normal and the remaining lung cancer cells. Growth of lung cancer cells was reduced by vorinostat treatment whereas no growth alteration was seen in normal lung cells. Among the tested lung cancer lines Calu-6 showed maximum sensitivity to vorinostat as indicated by the lowest IC_{50} (5 μM) (You, Han, & Park, 2017). Lung cancer like other cancer types involves not only genetic alterations but also epigenetic dysregulation. Both these deviations result in silencing of tumor suppressor genes culminating in cancer. While testing vorinostat on various NSCLC cell lines the result obtained was that this inhibitor suppressed growth of these cells. Induction of G0-G1 arrest and p21WAF1 was noticed in vorinostat exposure cells (Komatsu et al., 2006). Vorinostat, the FDA stamped HDAC inhibitor for CTCL has demonstrated significant activity against large cell lung carcinoma (LCC). This approved HDAC inhibitor manifested anti-proliferative effect against NCI-H460 cell line, the typical example of LCC. Vorinostat stimulated apoptosis and cell cycle arrest in this line and these effects were accompanied with reduced phosphorylation of AKT and ERK. Besides, the expression profile of hypoxia-inducible factor 1-alpha (HIF-1α) and VEGF (pro-angiogenic proteins) was inhibited by this marvelous inhibitor of HDACs. In xenograft model bearing a tumor of NCI-H460 cells, vorinostat inhibited tumor advancement (Zhao et al., 2014). The conclusion is that vorinostat is credible for tacking growth of both LCC cells under *in vitro* conditions and LCC tumors under *in vivo* conditions. Using HDACi individually against solid tumors results only in moderate therapeutic benefits. This is due to multiple resistance mechanisms elicited in cancer cells in reciprocation to these inhibitors. This has significantly minimized the evaluation and use of these inhibitors as single agents and has concurrently accentuated their use in collaboration with standard cancer chemotherapeutics (Fig. 3).

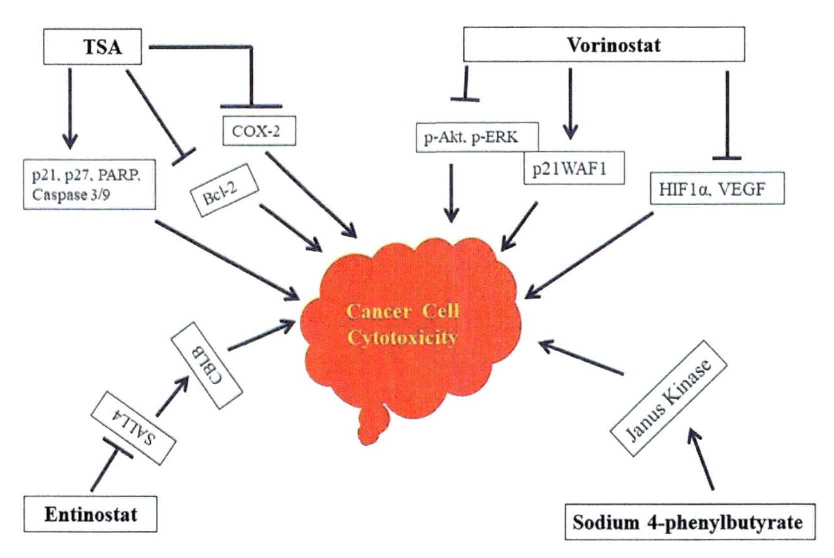

Fig. 3 Targets inhibited or facilitated by structurally diverse HDACi for kindling cytotoxicity in lung cancer models. Sodium-4-phenylbutyrate stimulates Janus Kinase pathway for inducing cytotoxicity while entinostat inhibits SALL4 due to which CBLB gets activated triggering death signaling. Vorinostat inhibits HIF1α and VEGF, p-AKT, and p-ERK but facilitates p21WAF1 for invoking cytotoxic effect in lung cancer cells. TSA on the other hand inhibits Bcl-2 and COX-2 but activates p21, p27, caspase 3/9, and PARP.

8. HDACi together with other chemotherapeutics against lung cancer

The modest antineoplastic benefits gained from single agent use of HDACi have been improved to a greater level by coupling these glorious molecules with traditional or conventional drugs (Hontecillas-Prieto et al., 2020). This combined procedure has strikingly raised the therapeutic advantage of these inhibitors and has synchronously abated their toxic effects toward typical (non-cancerous) cells of the body (Suraweera, O'Byrne, & Richard, 2018). Most notably, the combinatorial scheme narrows the opportunity of cancer cells to develop insensitivity toward HDACi (Ganai, 2020a).

Vorinostat has been assessed on small cell lung cancer cells either in association with cisplatin only or in conjunction with etoposide and cisplatin (etoposide/cisplatin). The rationale of selecting this combination was that due to chromatin decompaction induced by vorinostat cisplatin may gain enhanced access to DNA to intensify its damage. The above mentioned triplet and doublet combinations were studied on H146 and H209 SCLC cells. Further, the vorinostat in association with cisplatin was evaluated on nude mice bearing a tumor of H209 cells. Substantial depletion of cell viability and greater cytotoxic effect was induced by triplet combination. Combination of this HDAC inhibitor and cisplatin facilitated cell cycle arrest and induction of apoptosis. Compared to cells treated only with vorinostat, the combined regimens instigated greater α-tubulin and histone H3 acetylation. While vorinostat alleviated the thymidylate synthase expression this effect was maintained by cisplatin coupling. Vorinostat together with cisplatin substantially shrinked tumor growth in above mentioned mice model suggesting that this combination works even under *in vivo* environment (Pan et al., 2016). Vorinostat has also been tested against lung adenocarcinoma cells in combination with gefitinib (EGFR-tyrosine kinase blocker) or in association with sorafenib targeting multiple kinases. Coupling vorinostat with EGFR-TK inhibitor exerted synergistic inhibition on growth of above mentioned adenocarcinoma cells. This combination kindled apoptosis in these cells by way of obstructing IGF-1R/AKT-backed signaling. Notably vorinostat-gefitinib dual use potentially collapsed tumor growth in mice carrying lung adenocarcinoma tumor. Contrarily, the HDAC inhibitor-sorafenib collaboration showed no additional benefits over single treatment involving vorinostat (Jeannot et al., 2016). In another study on lung cancer patients, combination of programmed cell death protein 1 blocker (pembrolizumab) and vorinostat co-treatment confirmed better tolerability of the duo (Gray et al., 2019).

NSCLC associated with KRAS mutation respond poorly to inhibitors directed to EGFR. Panobinostat although approved for multiple myeloma has shown activity against NSCLC. Panobinostat has been examined on NSCLC cells resistant to gefitinib. This HDAC inhibitor circumvents gefitinib-resistance in NSCLC cells bearing canonical EGFR but mutated KRAS. The dual combination exhibited synergistic growth inhibitory effect on NSCLC even under *in vivo* settings. Panobinostat, but not gefitinib was individually able to inhibit the transcript of transcriptional co-activator with PDZ-binding motif abbreviated as TAZ. Combining the duo demonstrated synergistic effect on inhibition of TAZ, EGFR ligand and EGFR, the two hind proteins being the downstream TAZ targets. Panobinostat-induced inhibition of TAZ sensitized these cells to cytotoxic effects of gefitinib. This effect was mediated by disruption of AKT/mTOR (mammalian target of rapamycin) signaling cascade (Lee et al., 2017). From these results it is highly obvious that using panobinostat in concert with gefitinib elicits cytotoxicity even in NSCLC cells resistant to the latter. The ability of panobinostat to circumvent the hypoxia-generated cisplatin resistance in A549 and H23 cells was investigated. Cisplatin and panobinostat were synchronously delivered to these lines and it was found that in comparison to single therapy with cisplatin, co-treatment markedly reduced viability of H23 cells (Fischer et al., 2015).

It has been observed that high survivin expression coincides with low prognosis, drug resistance and recurrence of tumor. Same applies to non-small cell lung cancer also. The molecular mechanism through which survivin triggers the above mentioned processes has been explored and additionally the combined benefits of Class I selective HDAC inhibitor namely entinostat together with paclitaxel have been examined. The experimental findings certified that entinostat potentiates paclitaxel-induced antiproliferative effects toward NSCLC cells. Entinostat induced the expression of miR-203 which in turn alleviated the survivin expression. Samples taken from patients of this cancer showed downregulated levels of miR-203. Further higher expression of survivin as well as DNA methyltransferase I (DNMT1) was quantified and this expression markedly corresponded with lower expression of miR-203 (Wang et al., 2016). Valproic acid, the member of short chain fatty acid group of HDACi obstructs Class I HDACs (HDAC1-3,8) more effectively over Class IIa (HDAC 4-5, 7 and 9) has demonstrated cisplatin sensitizing activity on NSCLC cells. This HDAC inhibitor significantly downregulated the expression status of ATP-binding cassette transporter ABCA1 at both levels (Chen et al., 2017; Ni et al., 2017). VPA has the potential to decline the promoter activity of *ABCA1* in H358 and A549 cells. Phosphorylation level of Specificity protein 1(Sp1), a well-known transcription factor was decreased following valproic acid exposure. Phosphorylation of this transcription factor promotes its binding to promoter region of this gene resulting in its transcription (Tan & Khachigian, 2009). VPA-elicited inhibition of ABCA1 was hampered by HDAC2 (isozyme of Class I) overexpression. Further, overexpression of this HDAC also alleviated the sensitivity induced by valproic acid in NSCLC cells (Chen et al., 2011). It is strongly discernible that valproic acid sensitizes NSCLC cells to conventional drug cisplatin through downregulation of ABCA1 that is mediated through HDAC2 inhibition and depletion in phosphorylated levels of Sp1. Another inhibitor of short chain fatty acid group sodium butyrate has been investigated on lung adenocarcinoma

(A549) cell line. Among the glioma-associated oncogene homolog (GLI) proteins, GLI1 (glioma-associated oncogene homolog 1) predominantly enhances NSCLC proliferation. The other two GLI2 and GLI3 had moderate and slight effect on proliferation of these cells respectively (Bermudez, Hennen, Koch, Lindner, & Eickelberg, 2013). Docetaxel and sodium butyrate as individual agents blocked proliferation of A549 cells and facilitated their commitment toward apoptosis. Combination of these agents demonstrated substantially improved effects over either of the individual drugs. This benefit was recorded even in H1299 cell line, another lung adenocarcinoma line. Sodium butyrate-docetaxel combination evinced additive effect in inhibiting the GLI1 expression. While the combined treatment escalated the protein levels of p21, cyclin A and Bax, decline in degree of CDK1, Ki-67, cyclin D1, survivin, and Bcl-2 was quite prominent (Chen et al., 2020). These additive effects were counteracted by GLI1 overexpression suggesting that the improved benefits of dual combination are interceded by deepened inhibition of this transcription factor. Another interesting study explored the tendency of sodium phenylbutyrate to sensitize the NSCLC lines to erlotinib, gefitinib, and cisplatin. This HDAC inhibitor dose dependently inhibited the growth of three NSCLC lines including H1650, Calu1, and A549. The maximum inhibition was noticed in case of H1650 as lowest IC_{50} (4.53 mM) was quantified in this case followed by Calu1 (8.53) and 8.53 (10 mM). Addition of defined HDAC inhibitor to gefitinib, cisplatin or erlotinib strengthened their anti-proliferative effects. All the combinations manifested synergistic growth inhibitory effect against all the NSCLC lines tested (Al-Keilani, Alzoubi, & Jaradat, 2018). This proves that sodium phenylbutyrate potentiates the antiproliferative effects of standard anticancer drugs toward NSCLC cells.

Survivin, an antiapoptotic protein facilitates cell survival and as such its inhibitors have been developed and are tested against various cancer types. Inhibitor of this protein namely YM-155 has been evaluated against NSCLC (Guo et al., 2015). This survivin inhibitor has proven anti-lung cancer activity. However, its inhibitory effect on survivin does not last long and is not enough for achieving preferred therapeutic benefit. HDAC inhibitor entinostat has marked anticancer activity and compared to YM-155 lasts for longer duration. While exploring whether this HDAC inhibitor can augment the anticancer effect of survivin inhibitor in lung adenocarcinoma lines it was found that entinostat strongly enhances the antiproliferative activity of YM-155 in A549 cells. Entinostat increased the degree of H3 acetylation, facilitated DNMTs degradation and escalated the expression of miR-195 and miR-138 by inhibiting the methylation of genes encoding these miRs. Most importantly, miR-195 and miR-138 displayed synergistic effect in combination with YM-155 as they lower the expression of survivin by binding to its three untranslated region (Wang, Zhu, et al., 2016). From these results one can infer that using Class I selective inhibitor entinostat together with survivin inhibitor YM-155 greatly increases their anti-proliferative effect on cells of lung adenocarcinoma and thus this combination deserves further evaluation through clinical studies. SGI-110 is actually a prodrug which on metabolism yields decitabine, the famous DNMT inhibitor. Like zebularine and azacytidine, decitabine is also a cytosine nucleoside analogue (Griffiths et al., 2013).

While this prodrug depleted tumor burden by 35% in a rat models bearing an engrafted tumor at usual place composed of Calu6 cells, its conjunction with entinostat decreased the defined burden by 56% (Tellez et al., 2014). Combination of erlotinib and entinostat has been checked on NCI-H1993 cells bearing MET amplification and NCI-H1975 cells possessing EGFR-L858R/T790M mutation. In A549 xenograft (E-cadherin positive) in nude mice, entinostat alone induced 70% tumor growth inhibition, cetuximab by 65% and the dual combination showed 88% inhibition. On the other hand H460 xenografts (E-cadherin negative) demonstrated 75%, 64% and 3% inhibition of tumor growth on administration of cetuximab-entinostat, entinostat and cetuximab respectively. Entinostat in conjunction with erlotinib induced synergistic cytotoxic effect in or NCI-H1993 and NCI-H1975 lines as combination index below 1 was obtained (Witta et al., 2012). In nutshell, entinostat combination with cetuximab or erlotinib showed highly enhanced pharmacological benefit against herculean lung cancer models.

Belinostat, a hydroxamate restraining isozymes of all the three classes of Classical HDACs, sensitized lung cancer cells (platinum resistant) through modulation of ABCC2 (Sawas, Radeski, & O'Connor, 2015; To, Tong, & Fu, 2017). Synergistic cytotoxicity was noticed in these resistant lung lines on combined treatment with belinostat and cisplatin. This effect was noticed only when the duo was added concomitantly or belinostat was bath applied prior to cisplatin. The dual combination in aforesaid modes enhanced the intracellular accumulation of conventional drug cisplatin and formation of adducts of DNA-platinum. Combined exposure lowered the expression of ABCC2 functioning as efflux transporter and also the gene encoding excision repair cross-complementing group 1 (ERCC1) protein critically participating in nucleotide excision repair. Belinostat-induced

ABCC2 downregulation was accompanied with elevated binding of transcriptional repressor to its promoter area and depleted attachment of TFIIB to the defined area (To et al., 2017). This signifies that belinostat has the property of sensitizing cisplatin-resistant lung cancer cells and this sensitization is mediated through downregulation of drug efflux protein ABCC2 and the gene the product of which is a protein crucial for DNA repair. Phase I study involving belinostat with two standard drugs cisplatin and etoposide revealed that the maximum tolerated dose of this HDAC inhibitor is 500 mg/m/24 h while that of cisplatin and etoposide is 60 mg/m and 80 mg/m respectively (Balasubramaniam et al., 2018). This triplet combination was found to be safe and quite working in SCLC subjects. Vorinostat was evaluated in phase II study in concert with paclitaxel and carboplatin in patients having NSCLC in advanced stage. The efficacy of these conventional therapeutics was augmented by hydroxamate vorinostat in the said patients (Ramalingam et al., 2010). Lung squamous cell carcinoma (SCC) cells displayed MAPK pathway downregulation and apoptosis on treatment with belinostat. Cancer cells resistant to cisplatin portrayed anomalous activation of this pathway and applying belinostat in combination with cisplatin obstructed cisplatin-elicited ERK phosphorylation and exerted potent cytotoxic effect in these cells which was found to be synergistic. At molecular level belinostat facilitated proteasomal degradation of son of sevenless (SOS) indirectly. This protein acts upstream to MAPK pathway and as such regulates this pathway. Belinostat-induced transcriptional upregulation of FBXW10 and FBXO3 (F-box proteins) which in turn target SOS directly. Importantly, bortezomib (proteasome inhibitor) or knockdown of *FBXO3/FBXW10 using siRNA rescued the belinostat-induced effects further corroborating the above statement* (Kong et al., 2017). Somatic mutations in the EGFR-kinase domain make the NSCLC cells resistant to inhibitors targeting this domain. Thus, it is an ongoing process to search such agents that can sensitize these therapeutically challenging cells to EGFR-tyrosine kinase blocking agents. Belinostat has been used in collaboration with Hsp90 inhibitor (17-DMAG) on 12 NSCLC lines. These inhibitors when applied as single agents inhibited growth of almost all lines. EGFR expression and phosphorylated-Akt levels were declined following belinostat treatment. The dual combination showed synergic antiproliferative effect in above mentioned cells (Sudo et al., 2013). These agents separately as well as in combination thwarted tumor formation (tyrosine kinase inhibitor-resistant) in xenograft model. Vorinostat has been assessed against NSCLC cell lines in association with bortezomib, a well-established proteasome inhibitor. This combination proved to be synergistic in inducing apoptosis in these cells. Apoptosis induced by this dual drug exposure was found to be partially mediated through enhanced ROS production and this conclusion was derived because N-acetylcysteine inhibited the vorinostat-bortezomib stimulated apoptosis. Further, this combination strongly triggered caspase-3 activation and Beclin-1 breakdown (Park et al., 2010). Enhancer of Zeste homolog 2 (EZH2) overexpression has implications in NSCLC and inhibitor of this methyltransferase (3-deazaneplanocin A) has anti-proliferative activity against NSCLC. Vorinostat was examined along with 3-deazaneplanocin A against NSCLC cells and it was recorded that the dual administration synergistically inhibited the proliferation of under consideration NSCLC lines irrespective of their EGFR profile. Synergistic antiproliferative effect induced by co-treatment was accompanied with lowering of H3K27me3, moderately raised levels of histone acetylation and decrease in EZH2 besides other PRC2 (polycomb repressive complex 2) proteins. Vorinostat in combination with 3-deazaneplanocin A resulted in p27Kip1 accumulation, inhibition of cyclin A in treated cells. Importantly, this co-treatment inhibited EGFR signaling both in EGFR-mutated and EGFR canonical cells predominantly through dephosphorylation of this receptor. Most importantly, this combinatorial tactics inhibited tumor growth under *in vivo* conditions more efficiently than vorinostat or EZH2 inhibitor without imparting discernible toxicity (Takashina et al., 2016). The outcome entices to speculate that coupling histone methyltransferase inhibitors with HDAC inhibitors may be a promising method of tackling EGFR mutated NSCLC. Entinostat combined with DNMT inhibitor azacytidine (vidaza) has shown positive results in, *in vivo* lung adenocarcinomas. Growth of lung adenocarcinomas (K-ras/p53 mutant) was noticeably affected by this co-treatment. Decrease in tumor growth was also observed in EGFR or K-ras mutant tumors upon combined exposure. These effects were mediated through enhanced proapoptotic gene and p21 expression (Belinsky et al., 2011). These significant results indicated that coupling DNMT inhibitor with benzamide derivative HDAC inhibitor offers encouraging results against lung adenocarcinomas possessing mutations in diverse proteins. Sodium butyrate and bortezomib combination was explored in NSCLC cells including H460, H358, and A549. The cells were subjected to sequential treatment first with sodium butyrate followed by bortezomib. Sodium butyrate enhanced the transcriptional activity of NFκB (nuclear factor κB) fourfold over control in these lines. Sodium butyrate-driven NFκB activation was reverted to basal level by application of bortezomib and this proteasome inhibitor also stabilized p53 and p21 proteins. While individual use of sodium butyrate was able to induce only slightest apoptosis, its ordered combination with bortezomib heightened apoptosis substantially (three to fourfold) (Denlinger, Keller, Mayo, Broad, & Jones, 2004). The kernel of these findings is that bortezomib potentiates sodium butyrate induced apoptosis predominantly through NFκB suppression and to certain extent by inducing p21/p53 stabilization. A mind blowing study involved belinostat along with CYC202 (CDK inhibitor) in A549 cells bearing canonical p53. The combined approach induced marked anti-proliferative effect in comparison to singlet belinostat treatment. Synchronous application of these drugs considerably induced apoptosis in these cells. Variety of changes was noticed in cells that were given the co-treatment. These changes encompass cleavage of certain caspases (caspase-8/3), cleavage of PARP, p53 induction and BID activation. Protein such as XIAP and Mcl-1 underwent depletion following concurrent exposure. Although silencing of p53 failed to rescind enhanced cytotoxicity, caspase inhibitors markedly negated the defined enhancement. Most notably, belinostat-CYC202 collaboration proved to be effective even against H2444 cells having p53 null status (Ong et al., 2016) (Table 1). The core of these findings is that belinostat and CYC202 joint therapy engenders apoptosis in NSCLC cells regardless of their p53 status.

Table 1 Protein targets modulated in various lung cancer lines following combinatorial treatment.

Drug combination	Lung cancer lines employed	Cellular targets inhibited	Molecular targets promoted
Vorinostat/Cisplatin or Vorinostat/Cisplatin/Etoposide	H209, H146	Thymidylate synthase	α-tubulin acetylation, histone H3 acetylation
Vorinostat/Gefitinib	H1719, H322, A549, H358, H1299	IGF-1R/AKT signaling	–
Panobinostat/Gefitinib	H441, A549, H460	TAZ, EGFR ligand, EGFR, AKT/mTOR signaling	–
Entinostat/Paclitaxel	H460, A549	Survivin	miR-203
Valproic acid/Cisplatin	A549, H358	ABCA1, p-Sp1, HDAC2	–

Table 1 Protein targets modulated in various lung cancer lines following combinatorial treatment.—cont'd

Drug combination	Lung cancer lines employed	Cellular targets inhibited	Molecular targets promoted
Sodium butyrate/ Docetaxel	A549, H1299	GLI1, CDK1, Bcl-2, Ki-67, survivin, cyclin D1	p21, Bax, cyclin A
Entinostat/YM-155	*A549*	*DNMTs*	*H3 acetylation,* miR-195, miR-138
Belinostat/Cisplatin		ABCC2, ERCC1	
Cisplatin/Belinostat	H520, H226, H2170, H596	ERK phosphorylation, SOS, FBXW10, FBXO3	–
Belinostat/17-DMAG	H1975, HCC827, H520, H1666, H1299, H460, PC9, H1650	EGFR, p-Akt	–
3-deazaneplanocin A/Vorinostat	H1975, H1299, A549	H3K27me3, EZH2, cyclin A, EGFR signaling	–
Entinostat/ Azacytidine	Calu-6	–	p21
Sodium butyrate/ Bortezomib	H358, H460, A549	NFκB	p21/p53
Belinostat/CYC202	A549	XIAP, Mcl-1	p53, BID activation

9. Clinical studies of HDACi with immune checkpoint inhibitors toward lung cancer

HDACi have been observed to augment the efficacy of immune checkpoint inhibitors against lung cancer in clinical studies. Pembrolizumab, a PD-1 inhibitor was tested in conjunction with vorinostat in phase I/Ib trial against advanced form of metastatic-NSCLC. Patients were subjected to 200 or 400 mg of vorinostat per day orally and 200 mg of PD-1 inhibitor (intravenously) every 3 weeks. Among the treated patients 13 were chosen for phase I and 20 were selected for phase Ib forming a total of 33 patients. Patients enrolled for phase I study included were both pembrolizumab-naïve and pembrolizumab-pretreated to verify dose limiting toxicities (DLTs). The combination was nicely tolerated and manifested anti-tumor activity in spite of the advancement on previous pembrolizumab treatment (Gray et al., 2019). Belinostat was tested as intravenous infusion in collaboration with two standard anticancer molecules cisplatin and etoposide against SCLC. While maximum tolerated dose of belinostat was found to be 500 mg/m^2/24 h, the defined dose for etoposide and cisplatin was estimated to be 80 and 60 mg/m^2/24 h respectively. The combination was found to be active and safe against this cancer and frequently observed toxicities were hematological (Balasubramaniam et al., 2018).

10. Conclusion and future suggestions

HDACi of diverse types have manifested encouraging preclinical outcome in cell and animal models of lung cancer. These inhibitors engender cytotoxic effects in lung cancer models through modulation of proapoptotic, anti-apoptotic, cell cycle regulating, HDACs and other proteins. Although HDAC inhibitor based monotherapy against lung cancer has exhibited significant cytotoxicity, the desired therapeutic effect is not acquired through this approach. Monotherapy involving HDACi has another flip side that is toxicity toward typical body cells is also noticed at several occasions and this toxicity has been ascribed to higher doses of HDACi required for exerting cytotoxic effect. Combined approach involving these inhibitors in concert with other chemotherapeutic molecules has solved the limitations encountered with these inhibitors to a greater level. This strategy was found to generate markedly raised cytotoxic effects even against cisplatin and other conventional drug resistant lung cancer models. Further, these beneficial effects were recorded at low dose combinations of two or more therapeutics including HDACi. Thus, through combinatorial approach no overt toxicity is observed toward normal body cells. The improved cytotoxicity over individual agents was found to be either synergistic or additive. Although free combination of synthetic HDACi and other drugs have shown advantages over single agent therapy, it has nowadays become common to encapsulate HDACi singly or together with other therapeutics in nanoparticles for further enhancement in tumor specific drug delivery, bioavailability and cytotoxicity. However, this nanocombinatorial scheme is still undone especially in the context of HDACi with other therapeutics toward lung cancer. Most importantly, no significant clinical trials are available that will corroborate the results attained in cell and orthotopic rodent models of lung cancer. Conclusively, codelivery of HDACi with other anticancer agents through tumor directed nanoparticles is still an unmet field and use of these epigenetic route targeting inhibitors requires substantial clinical trials for elevating them for treating lung cancer in human beings.

Acknowledgments

Ganai SA acknowledges his family members for cooperation during the preparation of this article.

References

Adderley, H., Blackhall, F. H., & Lindsay, C. R. (2019). KRAS-mutant non-small cell lung cancer: Converging small molecules and immune checkpoint inhibition. *eBioMedicine*, *41*, 711–716. https://doi.org/10.1016/j.ebiom.2019.02.049.

Al Olayan, A., Al Hussaini, H., & Jazieh, A. R. (2012). The roles of epidermal growth factor receptor (EGFR) inhibitors in the management of lung cancer. *Journal of Infection and Public Health*, *5*(Suppl 1), S50–S60. https://doi.org/10.1016/j.jiph.2012.09.004.

Al-Keilani, M. S., Alzoubi, K. H., & Jaradat, S. A. (2018). The effect of combined treatment with sodium phenylbutyrate and cisplatin, erlotinib, or gefitinib on resistant NSCLC cells. *Clinical Pharmacology*, *10*, 135–140. https://doi.org/10.2147/cpaa.s174074.

Andrés, M. E., Burger, C., Peral-Rubio, M. J., Battaglioli, E., Anderson, M. E., Grimes, J., et al. (1999). CoREST: A functional corepressor required for regulation of neural-specific gene expression. *Proceedings of the National Academy of Sciences of the United States of America*, *96*(17), 9873–9878. https://doi.org/10.1073/pnas.96.17.9873.

Andruska, N., Stowe, H. B., Crockett, C., Liu, W., Palma, D., Faivre-Finn, C., et al. (2021). Stereotactic radiation for Lung cancer: A practical approach to challenging scenarios. *Journal of Thoracic Oncology*, *16*(7), 1075–1085. https://doi.org/10.1016/j.jtho.2021.04.002.

Balasubramaniam, S., Redon, C. E., Peer, C. J., Bryla, C., Lee, M. J., Trepel, J. B., et al. (2018). Phase I trial of belinostat with cisplatin and etoposide in advanced solid tumors, with a focus on neuroendocrine and small cell cancers of the lung. *Anti-Cancer Drugs*, *29*(5), 457–465. https://doi.org/10.1097/cad.0000000000000596.

Baldacci, S., Kherrouche, Z., Cockenpot, V., Stoven, L., Copin, M. C., Werkmeister, E., et al. (2018). MET amplification increases the metastatic spread of EGFR-mutated NSCLC. *Lung Cancer*, *125*, 57–67. https://doi.org/10.1016/j.lungcan.2018.09.008.

Baselga, J. (2001). The EGFR as a target for anticancer therapy—Focus on cetuximab. *European Journal of Cancer*, *37*(Suppl 4), S16–S22. https://doi.org/10.1016/s0959-8049(01)00233-7.

Belinsky, S. A., Grimes, M. J., Picchi, M. A., Mitchell, H. D., Stidley, C. A., Tesfaigzi, Y., et al. (2011). Combination therapy with vidaza and entinostat suppresses tumor growth and reprograms the epigenome in an orthotopic lung cancer model. *Cancer Research*, *71*(2), 454–462. https://doi.org/10.1158/0008-5472.can-10-3184.

Bermudez, O., Hennen, E., Koch, I., Lindner, M., & Eickelberg, O. (2013). Gli 1 mediates lung cancer cell proliferation and sonic hedgehog-dependent mesenchymal cell activation. *PLoS One*, *8*(5), e63226. https://doi.org/10.1371/journal.pone.0063226.

Bhuiyan, S., Siddiqui, R. S., Zirkiyeva, M., Agladze, M., & Bashir, T. (2021). A rare case of small cell Lung Cancer with an epidermal growth factor receptor mutation and its response to Osimertinib. *Cureus*, *13*(5), e15136. https://doi.org/10.7759/cureus.15136.

Bondarev, A. D., Attwood, M. M., Jonsson, J., Chubarev, V. N., Tarasov, V. V., & Schiöth, H. B. (2021). Recent developments of HDAC inhibitors: Emerging indications and novel molecules. *British Journal of Clinical Pharmacology*, *87*(12), 4577–4597. https://doi.org/10.1111/bcp.14889.

Bora-Singhal, N., Mohankumar, D., Saha, B., Colin, C. M., Lee, J. Y., Martin, M. W., et al. (2020). Novel HDAC11 inhibitors suppress lung adenocarcinoma stem cell self-renewal and overcome drug resistance by suppressing Sox2. *Scientific Reports*, *10*(1), 4722. https://doi.org/10.1038/s41598-020-61295-6.

Boumber, Y., Younes, A., & Garcia-Manero, G. (2011). Mocetinostat (MGCD0103): A review of an isotype-specific histone deacetylase inhibitor. *Expert Opinion on Investigational Drugs*, *20*(6), 823–829. https://doi.org/10.1517/13543784.2011.577737.

Cacabelos, R., & Torrellas, C. (2016). Chapter 32—Pharmacoepigenomics. In T. O. Tollefsbol (Ed.), *Medical Epigenetics* (pp. 585–617). Academic Press. https://doi.org/10.1016/B978-0-12-803239-8.00032-6.

Cao, L.-L., Song, X., Pei, L., Liu, L., Wang, H., & Jia, M. (2017). Histone deacetylase HDAC1 expression correlates with the progression and prognosis of lung cancer: A meta-analysis. *Medicine*, *96*(31), e7663. https://doi.org/10.1097/MD.0000000000007663.

Cejalvo, J. M., Fleitas, T., Felip, E., Mendivil, A. N., Martinez-Garcia, M., Taus, A., et al. (2016). A phase Ib study of lumretuzumab, a glycoengineered monoclonal antibody targeting HER3, in combination with carboplatin and paclitaxel as 1st-line treatment in patients with squamous non-small cell lung cancer. *Annals of Oncology*, *27*, vi118. https://doi.org/10.1093/annonc/mdw368.15.

Chan, B. A., & Hughes, B. G. (2015). Targeted therapy for non-small cell lung cancer: Current standards and the promise of the future. *Translational Lung Cancer Research*, *4*(1), 36–54. https://doi.org/10.3978/j.issn.2218-6751.2014.05.01.

Chen, M., Jiang, W., Xiao, C., Yang, W., Qin, Q., Mao, A., et al. (2020). Sodium butyrate combined with docetaxel for the treatment of Lung adenocarcinoma A549 cells by targeting Gli1. *Oncotargets and Therapy*, *13*, 8861–8875. https://doi.org/10.2147/ott.s252323.

Chen, X., Zhao, Y., Guo, Z., Zhou, L., Okoro, E. U., & Yang, H. (2011). Transcriptional regulation of ATP-binding cassette transporter A1 expression by a novel signaling pathway. *The Journal of Biological Chemistry*, *286*(11), 8917–8923. https://doi.org/10.1074/jbc.M110.214429.

Chen, J. H., Zheng, Y. L., Xu, C. Q., Gu, L. Z., Ding, Z. L., Qin, L., et al. (2017). Valproic acid (VPA) enhances cisplatin sensitivity of non-small cell lung cancer cells via HDAC2 mediated down regulation of ABCA1. *Biological Chemistry*, *398*(7), 785–792. https://doi.org/10.1515/hsz-2016-0307.

Choi, Y. H. (2005). Induction of apoptosis by trichostatin A, a histone deacetylase inhibitor, is associated with inhibition of cyclooxygenase-2 activity in human non-small cell lung cancer cells. *International Journal of Oncology*, *27*(2), 473–479.

Denlinger, C. E., Keller, M. D., Mayo, M. W., Broad, R. M., & Jones, D. R. (2004). Combined proteasome and histone deacetylase inhibition in non–small cell lung cancer. *The Journal of Thoracic and Cardiovascular Surgery*, *127*(4), 1078–1086. https://doi.org/10.1016/S0022-5223(03)01321-7.

Dutta, P., Sabri, N., Li, J., & Li, W. X. (2014). Role of STAT3 in lung cancer. *Jakstat*, *3*(4), e999503. https://doi.org/10.1080/21623996.2014.999503.

Ebisu, Y., Ishida, M., Saito, T., Murakawa, T., Uemura, Y., & Tsuta, K. (2018). Combined small cell carcinoma with giant cell carcinoma component of the lung: A case successfully diagnosed by computed tomography-guided fine-needle aspiration cytology. *Oncology Letters*, *15*(2), 1907–1911. https://doi.org/10.3892/ol.2017.7448.

Elkbuli, A., Byrne, M. M., Zhao, W., Sutherland, M., McKenney, M., Godinez, Y., et al. (2020). Gender disparities in lung cancer survival from an enriched Florida population-based cancer registry. *Annals of Medicine and Surgery*, *60*, 680–685. https://doi.org/10.1016/j.amsu.2020.11.081.

Engelman, J. A., Zejnullahu, K., Mitsudomi, T., Song, Y., Hyland, C., Park, J. O., et al. (2007). MET amplification leads to gefitinib resistance in lung cancer by activating ERBB3 signaling. *Science*, *316*(5827), 1039–1043. https://doi.org/10.1126/science.1141478.

Ferrara, N., Gerber, H. P., & LeCouter, J. (2003). The biology of VEGF and its receptors. *Nature Medicine*, *9*(6), 669–676. https://doi.org/10.1038/nm0603-669.

Fischer, C., Leithner, K., Wohlkoenig, C., Quehenberger, F., Bertsch, A., Olschewski, A., et al. (2015). Panobinostat reduces hypoxia-induced cisplatin resistance of non-small cell lung carcinoma cells via HIF-1α destabilization. *Molecular Cancer*, *14*, 4. https://doi.org/10.1186/1476-4598-14-4.

Fontanini, G., De Laurentiis, M., Vignati, S., Chinè, S., Lucchi, M., Silvestri, V., et al. (1998). Evaluation of epidermal growth factor-related growth factors and receptors and of neoangiogenesis in completely resected stage I-IIIA non-small-cell lung cancer: Amphiregulin and microvessel count are independent prognostic indicators of survival. *Clinical Cancer Research*, *4*(1), 241–249.

Freeman, D. J., Bush, T., Ogbagabriel, S., Belmontes, B., Juan, T., Plewa, C., et al. (2009). Activity of panitumumab alone or with chemotherapy in non-small cell lung carcinoma cell lines expressing mutant epidermal growth factor receptor. *Molecular Cancer Therapeutics*, *8*(6), 1536–1546. https://doi.org/10.1158/1535-7163.mct-08-0978.

Frost, G. (2013). The latency period of mesothelioma among a cohort of British asbestos workers (1978-2005). *British Journal of Cancer*, *109*(7), 1965–1973. https://doi.org/10.1038/bjc.2013.514.

Ganai, S. A. (2019). Different groups of HDAC inhibitors based on various classifications. In S. A. Ganai (Ed.), *Histone Deacetylase Inhibitors—Epidrugs for Neurological Disorders* (pp. 33–38). Springer Singapore. https://doi.org/10.1007/978-981-13-8019-8_5.

Ganai, S. A. (2020a). Combining histone deacetylase inhibitors with other anticancer agents as a novel strategy for circumventing limited therapeutic efficacy and mitigating toxicity. In S. A. Ganai (Ed.), *Histone Deacetylase Inhibitors in Combinatorial Anticancer Therapy* (pp. 203–239). Springer Singapore. https://doi.org/10.1007/978-981-15-8179-3_10.

Ganai, S. A. (2020b). Epigenetic regulator enzymes and their implications in distinct malignancies. In *Histone Deacetylase Inhibitors in Combinatorial Anticancer Therapy* (pp. 35–65). Springer Singapore. https://doi.org/10.1007/978-981-15-8179-3_2.

Ganai, S. A. (2020c). Singlet anticancer therapy through epi-weapons histone deacetylase inhibitors and its shortcomings. In *Histone Deacetylase Inhibitors in Combinatorial Anticancer Therapy* (pp. 173–201). Springer Singapore. https://doi.org/10.1007/978-981-15-8179-3_9.

Ganai, S. A. (2020d). Strong involvement of classical histone deacetylases and mechanistically distinct Sirtuins in bellicose cancers. In *Histone Deacetylase Inhibitors in Combinatorial Anticancer Therapy* (pp. 75–95). Springer Singapore. https://doi.org/10.1007/978-981-15-8179-3_4.

Glass, C. K., & Rosenfeld, M. G. (2000). The coregulator exchange in transcriptional functions of nuclear receptors. *Genes & Development*, *14*(2), 121–141.

Gray, J. E., Saltos, A., Tanvetyanon, T., Haura, E. B., Creelan, B., Antonia, S. J., et al. (2019). Phase I/Ib study of Pembrolizumab plus Vorinostat in advanced/metastatic non-small cell Lung Cancer. *Clinical Cancer Research*, *25*(22), 6623–6632. https://doi.org/10.1158/1078-0432.ccr-19-1305.

Griffiths, E. A., Choy, G., Redkar, S., Taverna, P., Azab, M., & Karpf, A. R. (2013). SGI-110: DNA methyltransferase inhibitor oncolytic. *Drugs of the Future*, *38*(8), 535–543. https://pubmed.ncbi.nlm.nih.gov/26190889.

Guo, K., Huang, P., Xu, N., Xu, P., Kaku, H., Zheng, S., et al. (2015). A combination of YM-155, a small molecule survivin inhibitor, and IL-2 potently suppresses renal cell carcinoma in murine model. *Oncotarget*, *6*(25), 21137–21147. https://doi.org/10.18632/oncotarget.4121.

Guo, K., Ma, Z., Zhang, Y., Han, L., Shao, C., Feng, Y., et al. (2022). HDAC7 promotes NSCLC proliferation and metastasis via stabilization by deubiquitinase USP10 and activation of β-catenin-FGF18 pathway. *Journal of Experimental & Clinical Cancer Research*, *41*(1), 91. https://doi.org/10.1186/s13046-022-02266-9.

Hahn, N. M., Picus, J., Bambury, R. M., Pal, S. K., Hart, L. L., Grivas, P., et al. (2015). A phase 2 study of the histone deacetylase (HDAC) inhibitor mocetinostat in patients with urothelial carcinoma (UC) and inactivating alterations of acetyltransferase genes. *Journal of Clinical Oncology*, *33*(15_suppl), TPS4575. https://doi.org/10.1200/jco.2015.33.15_suppl.tps4575.

Herbst, R. S., Baas, P., Kim, D. W., Felip, E., Pérez-Gracia, J. L., Han, J. Y., et al. (2016). Pembrolizumab versus docetaxel for previously treated, PD-L1-positive, advanced non-small-cell lung cancer (KEYNOTE-010): A randomised controlled trial. *Lancet*, *387*(10027), 1540–1550. https://doi.org/10.1016/s0140-6736(15)01281-7.

Hontecillas-Prieto, L., Flores-Campos, R., Silver, A., de Álava, E., Hajji, N., & García-Domínguez, D. J. (2020). Synergistic enhancement of Cancer therapy using HDAC inhibitors: Opportunity for clinical trials [review]. *Frontiers in Genetics*, *11*, 578011. https://doi.org/10.3389/fgene.2020.578011.

Howlader, N., Forjaz, G., Mooradian, M. J., Meza, R., Kong, C. Y., Cronin, K. A., et al. (2020). The effect of advances in Lung-Cancer treatment on population mortality. *The New England Journal of Medicine*, *383*(7), 640–649. https://doi.org/10.1056/NEJMoa1916623.

Hsu, W. H., Yang, J. C., Mok, T. S., & Loong, H. H. (2018). Overview of current systemic management of EGFR-mutant NSCLC. *Annals of Oncology*, *29*(suppl_1), i3–i9. https://doi.org/10.1093/annonc/mdx702.

Hu, Y., Dong, X. Z., Liu, X., Liu, P., & Chen, Y. B. (2016). Enhanced antitumor activity of Cetuximab in combination with the Jak inhibitor CYT387 against non-small-cell Lung Cancer with various genotypes. *Molecular Pharmaceutics*, *13*(2), 689–697. https://doi.org/10.1021/acs.molpharmaceut.5b00927.

Hu, Y., Hong, Y., Xu, Y., Liu, P., Guo, D. H., & Chen, Y. (2014). Inhibition of the JAK/STAT pathway with ruxolitinib overcomes cisplatin resistance in non-small-cell lung cancer NSCLC. *Apoptosis*, *19*(11), 1627–1636. https://doi.org/10.1007/s10495-014-1030-z.

Huang, L., Guo, Z., Wang, F., & Fu, L. (2021). KRAS mutation: From undruggable to druggable in cancer. *Signal Transduction and Targeted Therapy*, *6*(1), 386. https://doi.org/10.1038/s41392-021-00780-4.

Huang, S. Y., Shen, S. J., & Li, X. Y. (2013). Pulmonary sarcomatoid carcinoma: A clinicopathologic study and prognostic analysis of 51 cases. *World Journal of Surgical Oncology*, *11*, 252. https://doi.org/10.1186/1477-7819-11-252.

Hung, M. S., Chen, I. C., Lin, P. Y., Lung, J. H., Li, Y. C., Lin, Y. C., et al. (2016). Epidermal growth factor receptor mutation enhances expression of vascular endothelial growth factor in lung cancer. *Oncology Letters*, *12*(6), 4598–4604. https://doi.org/10.3892/ol.2016.5287.

Jeannot, V., Busser, B., Vanwonterghem, L., Michallet, S., Ferroudj, S., Cokol, M., et al. (2016). Synergistic activity of vorinostat combined with gefitinib but not with sorafenib in mutant KRAS human non-small cell lung cancers and hepatocarcinoma. *Oncotargets and Therapy*, *9*, 6843–6855. https://doi.org/10.2147/OTT.S117743.

Jin, Y., Liu, T., Luo, H., Liu, Y., & Liu, D. (2022). Targeting epigenetic regulatory enzymes for Cancer therapeutics: Novel small-molecule Epidrug development. *Frontiers in Oncology*, *12*, 848221. https://doi.org/10.3389/fonc.2022.848221.

Jung, K. H., Noh, J. H., Kim, J. K., Eun, J. W., Bae, H. J., Xie, H. J., et al. (2012). HDAC2 overexpression confers oncogenic potential to human lung cancer cells by deregulating expression of apoptosis and cell cycle proteins. *Journal of Cellular Biochemistry, 113*(6), 2167–2177. https://doi.org/10.1002/jcb.24090.

Karim, N. A., Schuster, J., Eldessouki, I., Gaber, O., Namad, T., Wang, J., et al. (2018). Pulmonary sarcomatoid carcinoma: University of Cincinnati experience. *Oncotarget, 9*(3), 4102–4108. https://doi.org/10.18632/oncotarget.23468.

Kavalali, E. T., Nelson, E. D., & Monteggia, L. M. (2011). Role of MeCP2, DNA methylation, and HDACs in regulating synapse function. *Journal of Neurodevelopmental Disorders, 3*(3), 250–256. https://doi.org/10.1007/s11689-011-9078-3.

Khoja, L., Butler, M. O., Kang, S. P., Ebbinghaus, S., & Joshua, A. M. (2015). Pembrolizumab. *Journal for Immunotherapy of Cancer, 3*, 36. https://doi.org/10.1186/s40425-015-0078-9.

Kim, J. H., Ali, K. H., Oh, Y. J., & Seo, Y. H. (2022). Design, synthesis, and biological evaluation of histone deacetylase inhibitor with novel salicylamide zinc binding group. *Medicine, 101*(17), e29049. https://doi:10.1097/MD.0000000000029049.

Kobayashi, D., Kuribayashi, K., Tanaka, M., & Watanabe, N. (2011). Overexpression of SALL4 in lung cancer and its importance in cell proliferation. *Oncology Reports, 26*(4), 965–970. https://doi.org/10.3892/or.2011.1374.

Komatsu, N., Kawamata, N., Takeuchi, S., Yin, D., Chien, W., Miller, C. W., et al. (2006). SAHA, a HDAC inhibitor, has profound anti-growth activity against non-small cell lung cancer cells. *Oncology Reports, 15*(1), 187–191.

Kong, L. R., Tan, T. Z., Ong, W. R., Bi, C., Huynh, H., Lee, S. C., et al. (2017). Belinostat exerts antitumor cytotoxicity through the ubiquitin-proteasome pathway in lung squamous cell carcinoma. *Molecular Oncology, 11*(8), 965–980. https://doi.org/10.1002/1878-0261.12064.

Kumari, N., Singh, S., Haloi, D., Mishra, S. K., Krishnani, N., Nath, A., et al. (2019). Epidermal growth factor receptor mutation frequency in squamous cell carcinoma and its diagnostic performance in cytological samples: A molecular and Immunohistochemical study. *World Journal of Oncology, 10*(3), 142–150. https://doi.org/10.14740/wjon1204.

Lackey, A., & Donington, J. S. (2013). Surgical management of lung cancer. *Seminars in Interventional Radiology, 30*(2), 133–140. https://doi.org/10.1055/s-0033-1342954.

Langevin, S. M., Kratzke, R. A., & Kelsey, K. T. (2015). Epigenetics of lung cancer. *Translational Research, 165*(1), 74–90. https://doi.org/10.1016/j.trsl.2014.03.001.

Lee, W.-Y., Chen, P.-C., Wu, W.-S., Wu, H.-C., Lan, C.-H., Huang, Y.-H., et al. (2017). Panobinostat sensitizes KRAS-mutant non-small-cell lung cancer to gefitinib by targeting TAZ. *International Journal of Cancer, 141*(9), 1921–1931. https://doi.org/10.1002/ijc.30888.

Lee, N. Y., Hazlett, T. L., & Koland, J. G. (2006). Structure and dynamics of the epidermal growth factor receptor C-terminal phosphorylation domain. *Protein Science, 15*(5), 1142–1152. https://doi.org/10.1110/ps.052045306.

Lee, J. K., Shin, J. Y., Kim, S., Lee, S., Park, C., Kim, J. Y., et al. (2013). Primary resistance to epidermal growth factor receptor (EGFR) tyrosine kinase inhibitors (TKIs) in patients with non-small-cell lung cancer harboring TKI-sensitive EGFR mutations: An exploratory study. *Annals of Oncology, 24*(8), 2080–2087. https://doi.org/10.1093/annonc/mdt127.

Lei, Y., Liu, L., Zhang, S., Guo, S., Li, X., Wang, J., et al. (2017). Hdac 7 promotes lung tumorigenesis by inhibiting Stat3 activation. *Molecular Cancer, 16*(1), 170. https://doi.org/10.1186/s12943-017-0736-2.

Li, C., & Lu, H. (2018). Adenosquamous carcinoma of the lung. *Oncotargets and Therapy, 11*, 4829–4835. https://doi.org/10.2147/ott.s164574.

Li, M., Zheng, R., & Yuan, F. L. (2018). MiR-410 affects the proliferation and apoptosis of lung cancer A549 cells through regulation of SOCS3/JAK-STAT signaling pathway. *European Review for Medical and Pharmacological Sciences*, *22*, 5987–5993. https://doi.org/10.26355/eurrev_201809_15933.

Loprevite, M., Tiseo, M., Grossi, F., Scolaro, T., Semino, C., Pandolfi, A., et al. (2005). In vitro study of CI-994, a histone deacetylase inhibitor, in non-small cell lung cancer cell lines. *Oncology Research*, *15*(1), 39–48. https://doi.org/10.3727/096504005775082066.

Lynch, T. J., Bell, D. W., Sordella, R., Gurubhagavatula, S., Okimoto, R. A., Brannigan, B. W., et al. (2004). Activating mutations in the epidermal growth factor receptor underlying responsiveness of non-small-cell lung cancer to gefitinib. *The New England Journal of Medicine*, *350*(21), 2129–2139. https://doi.org/10.1056/NEJMoa040938.

Maccallini, C., Ammazzalorso, A., De Filippis, B., Fantacuzzi, M., Giampietro, L., & Amoroso, R. (2022). HDAC inhibitors for the therapy of triple negative breast Cancer. *Pharmaceuticals (Basel, Switzerland)*, *15*(6), 667. https://doi.org/10.3390/ph15060667.

Mamdani, H., & Jalal, S. I. (2020). Histone deacetylase inhibition in non-small cell Lung Cancer: Hype or Hope? *Frontiers in Cell and Development Biology*, *8*, 582370. https://doi.org/10.3389/fcell.2020.582370.

Mott, F. E. (2012). Mesothelioma: A review. *The Ochsner Journal*, *12*(1), 70–79.

Mottamal, M., Zheng, S., Huang, T. L., & Wang, G. (2015). Histone deacetylase inhibitors in clinical studies as templates for new anticancer agents. *Molecules*, *20*(3), 3898–3941. https://doi.org/10.3390/molecules20033898.

Nan, X., Ng, H. H., Johnson, C. A., Laherty, C. D., Turner, B. M., Eisenman, R. N., et al. (1998). Transcriptional repression by the methyl-CpG-binding protein MeCP2 involves a histone deacetylase complex. *Nature*, *393*(6683), 386–389. https://doi.org/10.1038/30764.

Ni, L., Wang, L., Yao, C., Ni, Z., Liu, F., Gong, C., et al. (2017). The histone deacetylase inhibitor valproic acid inhibits NKG2D expression in natural killer cells through suppression of STAT3 and HDAC3. *Scientific Reports*, 7(1), 45266. https://doi.org/10.1038/srep45266.

Nwizu, T., Kanteti, R., Kawada, I., Rolle, C., Vokes, E. E., & Salgia, R. (2011). Crizotinib (PF02341066) as a ALK/MET inhibitor- special emphasis as a therapeutic drug against Lung cancer. *Drugs of the Future*, *36*(2), 91–99. https://doi.org/10.1358/dof.2011.036.02.1584112.

Ohsaki, Y., Tanno, S., Fujita, Y., Toyoshima, E., Fujiuchi, S., Nishigaki, Y., et al. (2000). Epidermal growth factor receptor expression correlates with poor prognosis in non-small cell lung cancer patients with p53 overexpression. *Oncology Reports*, 7(3), 603–607. https://doi.org/10.3892/or.7.3.603.

Ong, P. S., Wang, L., Chia, D. M., Seah, J. Y., Kong, L. R., Thuya, W. L., et al. (2016). A novel combinatorial strategy using Seliciclib(®) and Belinostat(®) for eradication of non-small cell lung cancer via apoptosis induction and BID activation. *Cancer Letters*, *381*(1), 49–57. https://doi.org/10.1016/j.canlet.2016.07.023.

Pai-Scherf, L., Blumenthal, G. M., Li, H., Subramaniam, S., Mishra-Kalyani, P. S., He, K., et al. (2017). FDA approval summary: Pembrolizumab for treatment of metastatic non-small cell Lung Cancer: First-line therapy and beyond. *The Oncologist*, *22*(11), 1392–1399. https://doi.org/10.1634/theoncologist.2017-0078.

Pan, C.-H., Chang, Y.-F., Lee, M.-S., Wen, B. C., Ko, J.-C., Liang, S.-K., et al. (2016). Vorinostat enhances the cisplatin-mediated anticancer effects in small cell lung cancer cells. *BMC Cancer*, *16*(1), 857. https://doi.org/10.1186/s12885-016-2888-7.

Pao, W., Wang, T. Y., Riely, G. J., Miller, V. A., Pan, Q., Ladanyi, M., et al. (2005). KRAS mutations and primary resistance of lung adenocarcinomas to gefitinib or erlotinib. *PLoS Medicine*, *2*(1), e17. https://doi.org/10.1371/journal.pmed.0020017.

Parashar, B., Arora, S., & Wernicke, A. G. (2013). Radiation therapy for early stage lung cancer. *Seminars in Interventional Radiology, 30*(2), 185–190. https://doi.org/10.1055/s-0033-1342960.

Park, M.-Y., Kim, D. R., Park, J.-S., Cho, Y.-J., Lee, S.-W., Yoon, H.-I., et al. (2010). Combination therapy of Bortezomib (PS-341) and SAHA (Vorinostat) in non-small cell Lung Cancer cell lines. *Journal of Lung Cancer, 9*(2), 77–84. https://doi.org/10.6058/jlc.2010.9.2.77.

Pirker, R., & Filipits, M. (2012). Cetuximab in non-small-cell lung cancer. *Translational Lung Cancer Research, 1*(1), 54–60. https://doi.org/10.3978/j.issn.2218-6751.11.01.

Platta, C. S., Greenblatt, D. Y., Kunnimalaiyaan, M., & Chen, H. (2007). The HDAC inhibitor trichostatin A inhibits growth of small cell lung cancer cells. *The Journal of Surgical Research, 142*(2), 219–226. https://doi.org/10.1016/j.jss.2006.12.555.

Qin, J., & Lu, H. (2018). Combined small-cell lung carcinoma. *Oncotargets and Therapy, 11*, 3505–3511. https://doi.org/10.2147/ott.s159057.

Qureshy, Z., Johnson, D. E., & Grandis, J. R. (2020). Targeting the JAK/STAT pathway in solid tumors. *Journal of Cancer Metastasis and Treatment, 6*, 27.

Ramalingam, S. S., Maitland, M. L., Frankel, P., Argiris, A. E., Koczywas, M., Gitlitz, B., et al. (2010). Carboplatin and paclitaxel in combination with either vorinostat or placebo for first-line therapy of advanced non-small-cell lung cancer. *Journal of Clinical Oncology, 28*(1), 56–62. https://doi.org/10.1200/jco.2009.24.9094.

Ramalingam, S. S., Vansteenkiste, J., Planchard, D., Cho, B. C., Gray, J. E., Ohe, Y., et al. (2020). Overall survival with Osimertinib in untreated, EGFR-mutated advanced NSCLC. *The New England Journal of Medicine, 382*(1), 41–50. https://doi.org/10.1056/NEJMoa1913662.

Ricciuti, B., Leonardi, G. C., Metro, G., Grignani, F., Paglialunga, L., Bellezza, G., et al. (2016). Targeting the KRAS variant for treatment of non-small cell lung cancer: Potential therapeutic applications. *Expert Review of Respiratory Medicine, 10*(1), 53–68. https://doi.org/10.1586/17476348.2016.1115349.

Richters, A., & Koehler, A. N. (2017). Epigenetic modulation using small molecules—Targeting histone acetyltransferases in disease. *Current Medicinal Chemistry, 24*(37), 4121–4150. https://doi.org/10.2174/0929867324666170223153115.

Rittmeyer, A., Barlesi, F., Waterkamp, D., Park, K., Ciardiello, F., von Pawel, J., et al. (2017). Atezolizumab versus docetaxel in patients with previously treated non-small-cell lung cancer (OAK): A phase 3, open-label, multicentre randomised controlled trial. *Lancet, 389*(10066), 255–265. https://doi.org/10.1016/s0140-6736(16)32517-x.

Rudin, C. M., Brambilla, E., Faivre-Finn, C., & Sage, J. (2021). Small-cell lung cancer. *Nature Reviews. Disease Primers*, 7(1), 3. https://doi.org/10.1038/s41572-020-00235-0.

Ruiz, R., Raez, L. E., & Rolfo, C. (2015). Entinostat (SNDX-275) for the treatment of non-small cell lung cancer. *Expert Opinion on Investigational Drugs, 24*(8), 1101–1109. https://doi.org/10.1517/13543784.2015.1056779.

Sabbah, D., Hajjo, R., & Sweidan, K. (2020). Review on epidermal growth factor receptor (EGFR) structure, signaling pathways, interactions, and recent updates of EGFR inhibitors. *Current Topics in Medicinal Chemistry, 20*, 815–834. https://doi.org/10.2174/1568026620666200303123102.

Sang, Y., Sun, L., Wu, Y., Yuan, W., Liu, Y., & Li, S.-W. (2019). Histone deacetylase 7 inhibits plakoglobin expression to promote lung cancer cell growth and metastasis. *International Journal of Oncology, 54*(3), 1112–1122. https://doi.org/10.3892/ijo.2019.4682.

Sawas, A., Radeski, D., & O'Connor, O. A. (2015). Belinostat in patients with refractory or relapsed peripheral T-cell lymphoma: A perspective review. *Therapeutic Advances in Hematology, 6*(4), 202–208. https://doi.org/10.1177/2040620715592567.

Seto, E., & Yoshida, M. (2014). Erasers of histone acetylation: The histone deacetylase enzymes. *Cold Spring Harbor Perspectives in Biology*, *6*(4), a018713. https://doi.org/10.1101/cshperspect.a018713.

Shanker, M., Willcutts, D., Roth, J. A., & Ramesh, R. (2010). Drug resistance in lung cancer. *Lung Cancer (Auckl)*, *1*, 23–36.

Siegel, R. L., Miller, K. D., Fuchs, H. E., & Jemal, A. (2022). Cancer statistics, 2022. *CA: a Cancer Journal for Clinicians*, *72*(1), 7–33. https://doi.org/10.3322/caac.21708.

Singh, A., Patel, V. K., Jain, D. K., Patel, P., & Rajak, H. (2016). Panobinostat as Pan-deacetylase inhibitor for the treatment of pancreatic Cancer: Recent Progress and future prospects. *Oncotargets and Therapy*, *4*(1), 73–89. https://doi.org/10.3322/caac.21708.

Slebos, R. J., & Rodenhuis, S. (1992). The ras gene family in human non-small-cell lung cancer. *Journal of the National Cancer Institute. Monographs*, (13), 23–29.

Suda, K., Onozato, R., Yatabe, Y., & Mitsudomi, T. (2009). EGFR T790M mutation: A double role in lung cancer cell survival? *Journal of Thoracic Oncology*, *4*(1), 1–4. https://doi.org/10.1097/JTO.0b013e3181913c9f.

Sudo, M., Chin, T. M., Mori, S., Doan, N. B., Said, J. W., Akashi, M., et al. (2013). Inhibiting proliferation of gefitinib-resistant, non-small cell lung cancer. *Cancer Chemotherapy and Pharmacology*, *71*(5), 1325–1334. https://doi.org/10.1007/s00280-013-2132-y.

Suraweera, A., O'Byrne, K. J., & Richard, D. J. (2018). Combination therapy with histone deacetylase inhibitors (HDACi) for the treatment of Cancer: Achieving the full therapeutic potential of HDACi. *Frontiers in Oncology*, *8*, 92. https://doi.org/10.3389/fonc.2018.00092.

Takashina, T., Kinoshita, I., Kikuchi, J., Shimizu, Y., Sakakibara-Konishi, J., Oizumi, S., et al. (2016). Combined inhibition of EZH2 and histone deacetylases as a potential epigenetic therapy for non-small-cell lung cancer cells. *Cancer Science*, *107*(7), 955–962. https://doi.org/10.1111/cas.12957.

Takeda, M., & Nakagawa, K. (2019). First- and second-generation EGFR-TKIs are all replaced to Osimertinib in chemo-naive EGFR mutation-positive non-small cell Lung Cancer? *International Journal of Molecular Sciences*, *20*(1), 146. https://doi.org/10.3390/ijms20010146.

Tan, N. Y., & Khachigian, L. M. (2009). Sp1 phosphorylation and its regulation of gene transcription. *Molecular and Cellular Biology*, *29*(10), 2483–2488. https://doi.org/10.1128/mcb.01828-08.

Tellez, C. S., Grimes, M. J., Picchi, M. A., Liu, Y., March, T. H., Reed, M. D., et al. (2014). SGI-110 and entinostat therapy reduces lung tumor burden and reprograms the epigenome. *International Journal of Cancer*, *135*(9), 2223–2231. https://doi.org/10.1002/ijc.28865.

Thomas, A., Karakattu, S., Cagle, J., & Hoskere, G. (2021). Malignant pleural mesothelioma epidemiology in the United States from 2000 to 2016. *Cureus*, *13*(4), e14605. https://doi.org/10.7759/cureus.14605.

To, K. K. W., Tong, W.-S., & Fu, L.-W. (2017). Reversal of platinum drug resistance by the histone deacetylase inhibitor belinostat. *Lung Cancer*, *103*, 58–65. https://doi.org/10.1016/j.lungcan.2016.11.019.

Travis, W. D., Brambilla, E., Burke, A. P., Marx, A., & Nicholson, A. G. (2015). Introduction to the 2015 World Health Organization classification of tumors of the lung, pleura, thymus, and heart. *Journal of Thoracic Oncology*, *10*(9), 1240–1242. https://doi.org/10.1097/jto.0000000000000663.

Truong, A. S., Zhou, M., Krishnan, B., Utsumi, T., Manocha, U., Stewart, K. G., et al. (2021). Entinostat induces antitumor immune responses through immune editing of tumor neoantigens. *The Journal of Clinical Investigation*, *131*(16), e138560. https://doi.org/10.1172/jci138560.

Ung, M., Rouquette, I., Filleron, T., Taillandy, K., Brouchet, L., Bennouna, J., et al. (2016). Characteristics and clinical outcomes of Sarcomatoid carcinoma of the Lung. *Clinical Lung Cancer, 17*(5), 391–397. https://doi.org/10.1016/j.cllc.2016.03.001.

Varghese, A. M., Zakowski, M. F., Yu, H. A., Won, H. H., Riely, G. J., Krug, L. M., et al. (2014). Small-cell lung cancers in patients who never smoked cigarettes. *Journal of Thoracic Oncology, 9*(6), 892–896. https://doi.org/10.1097/jto.0000000000000142.

Wang, L., Syn, N. L., Subhash, V. V., Any, Y., Thuya, W. L., Cheow, E. S. H., et al. (2018). Pan-HDAC inhibition by panobinostat mediates chemosensitization to carboplatin in non-small cell lung cancer via attenuation of EGFR signaling. *Cancer Letters, 417*, 152–160. https://doi.org/10.1016/j.canlet.2017.12.030.

Wang, Z., Tang, F., Hu, P., Wang, Y., Gong, J., Sun, S., et al. (2016). HDAC6 promotes cell proliferation and confers resistance to gefitinib in lung adenocarcinoma. *Oncology Reports, 36*, 589–597. https://doi.org/10.3892/or.2016.4811.

Wang, S., Tsui, S. T., Liu, C., Song, Y., & Liu, D. (2016). EGFR C797S mutation mediates resistance to third-generation inhibitors in T790M-positive non-small cell lung cancer. *Journal of Hematology & Oncology, 9*(1), 59. https://doi.org/10.1186/s13045-016-0290-1.

Wang, S., Zhu, L., Zuo, W., Zeng, Z., Huang, L., Lin, F., et al. (2016). MicroRNA-mediated epigenetic targeting of Survivin significantly enhances the antitumor activity of paclitaxel against non-small cell lung cancer. *Oncotarget*, 7(25), 37693–37713. https://doi.org/10.18632/oncotarget.9264.

Waugh, D. J., & Wilson, C. (2008). The interleukin-8 pathway in cancer. *Clinical Cancer Research, 14*(21), 6735–6741. https://doi.org/10.1158/1078-0432.ccr-07-4843.

Wilhelm, S., Carter, C., Lynch, M., Lowinger, T., Dumas, J., Smith, R. A., et al. (2006). Discovery and development of sorafenib: A multikinase inhibitor for treating cancer. *Nature Reviews Drug Discovery, 5*(10), 835–844. https://doi.org/10.1038/nrd2130.

Witta, S. E., Chan, D., Girard, L., Zheng, D., Franekova, V., Peyton, M., et al. (2012). Effects of entinostat onresistance to cetuximab and EGFR TKIs in non-small cell lung cancer. *Journal of Clinical Oncology, 30*(15_suppl), e18077. https://doi.org/10.1200/jco.2012.30.15_suppl.e18077.

Xia, C., Dong, X., Li, H., Cao, M., Sun, D., He, S., et al. (2022). Cancer statistics in China and United States, 2022: Profiles, trends, and determinants. *Chinese Medical Journal, 135*(5), 584–590. https://doi.org/10.1097/cm9.0000000000002108.

Xu, M., Xie, Y., Ni, S., & Liu, H. (2015). The latest therapeutic strategies after resistance to first generation epidermal growth factor receptor tyrosine kinase inhibitors (EGFR TKIs) in patients with non-small cell lung cancer (NSCLC). *Annals of Translational Medicine, 3*(7), 96. https://doi.org/10.3978/j.issn.2305-5839.2015.03.60.

Yang, J., Chai, L., Fowles, T. C., Alipio, Z., Xu, D., Fink, L. M., et al. (2008). Genome-wide analysis reveals Sall4 to be a major regulator of pluripotency in murine-embryonic stem cells. *Proceedings of the National Academy of Sciences of the United States of America, 105*(50), 19756–19761. https://doi.org/10.1073/pnas.0809321105.

Yang, Y., Huang, Y., Wang, Z., Wang, H.-T., Duan, B., Ye, D., et al. (2014). HDAC10 promotes lung cancer proliferation via AKT phosphorylation. *Oncotarget*, 7, 59388–59401. https://doi.org/10.18632/oncotarget.10673.

Yong, K. J., Li, A., Ou, W.-B., Hong, C. K. Y., Zhao, W., Wang, F., et al. (2016). Targeting SALL4 by entinostat in lung cancer. *Oncotarget*, 7(46), 75425–75440. https://www.oncotarget.com/article/12251/text/.

You, B. R., Han, B. R., & Park, W. H. (2017). Suberoylanilide hydroxamic acid increases anti-cancer effect of tumor necrosis factor-α through up-regulation of TNF receptor 1 in lung cancer cells. *Oncotarget, 8*(11), 17726–17737. https://doi.org/10.18632/oncotarget.14628.

You, A., Tong, J. K., Grozinger, C. M., & Schreiber, S. L. (2001). CoREST is an integral component of the CoREST- human histone deacetylase complex. *Proceedings of the National Academy of Sciences of the United States of America, 98*(4), 1454–1458. https://doi.org/10.1073/pnas.98.4.1454.

Yu, X. Q., Yap, M. L., Cheng, E. S., Ngo, P. J., Vaneckova, P., Karikios, D., et al. (2022). Evaluating prognostic factors for sex differences in Lung Cancer survival: Findings from a large Australian cohort. *Journal of Thoracic Oncology, 17*(5), 688–699. https://doi.org/10.1016/j.jtho.2022.01.016.

Yuan, M., Huang, L.-L., Chen, J.-H., Wu, J., & Xu, Q. (2019). The emerging treatment landscape of targeted therapy in non-small-cell lung cancer. *Signal Transduction and Targeted Therapy, 4*(1), 61. https://doi.org/10.1038/s41392-019-0099-9.

Zappa, C., & Mousa, S. A. (2016). Non-small cell lung cancer: Current treatment and future advances. *Translational Lung Cancer Research, 5*(3), 288–300. https://doi.org/10.21037/tlcr.2016.06.07.

Zhang, Y., Iratni, R., Erdjument-Bromage, H., Tempst, P., & Reinberg, D. (1997). Histone deacetylases and SAP18, a novel polypeptide, are components of a human Sin3 complex. *Cell, 89*(3), 357–364. https://doi.org/10.1016/s0092-8674(00)80216-0.

Zhang, X., Wei, L., Yang, Y., & Yu, Q. (2004). Sodium 4-phenylbutyrate induces apoptosis of human lung carcinoma cells through activating JNK pathway. *Journal of Cellular Biochemistry, 93*(4), 819–829. https://doi.org/10.1002/jcb.20173.

Zhao, Y., Yu, D., Wu, H., Liu, H., Zhou, H., Gu, R., et al. (2014). Anticancer activity of SAHA, a potent histone deacetylase inhibitor, in NCI-H460 human large-cell lung carcinoma cells in vitro and in vivo. *International Journal of Oncology, 44*(2), 451–458. https://doi.org/10.3892/ijo.2013.2193.

Zhong, L., Sun, S., Yao, S., Han, X., Gu, M., & Shi, J. (2018). Histone deacetylase 5 promotes the proliferation and invasion of lung cancer cells. *Oncology Reports, 40*(4), 2224–2232. https://doi.org/10.3892/or.2018.6591.

CHAPTER FIVE

From ecology to oncology: To understand cancer stem cell dormancy, ask a Brine shrimp (*Artemia*)

Christopher R. Wood[a,*], Wen-Tao Wu[a], Yao-Shun Yang[a], Jin-Shu Yang[a], Yongmei Xi[b], and Wei-Jun Yang[a,*]

[a]MOE Laboratory of Biosystem Homeostasis and Protection, College of Life Sciences, Zhejiang University, Hangzhou, Zhejiang, China

[b]The Women's Hospital, and Institute of Genetics, Zhejiang University School of Medicine, Zhejiang Provincial Key Laboratory of Genetic & Developmental Disorders, Hangzhou, Zhejiang, China

*Corresponding authors: e-mail address: johnsamuel@zju.edu.cn; w_jyang@cls.zju.edu.cn

Contents

Advances in Cancer Research, Volume 158
ISSN 0065-230X
https://doi.org/10.1016/bs.acr.2022.12.004

Abstract

The brine shrimp (*Artemia*), releases embryos that can remain dormant for up to a decade. Molecular and cellular level controlling factors of dormancy in *Artemia* are now being recognized or applied as active controllers of dormancy (quiescence) in cancers. Most notably, the epigenetic regulation by SET domain-containing protein 4 (SETD4), is revealed as highly conserved and the primary control factor governing the maintenance of cellular dormancy from *Artemia* embryonic cells to cancer stem cells (CSCs). Conversely, DEK, has recently emerged as the primary factor in the control of dormancy exit/reactivation, in both cases. The latter has been now successfully applied to the reactivation of quiescent CSCs, negating their resistance to therapy and leading to their subsequent destruction in mouse models of breast cancer, without recurrence or metastasis potential. In this review, we introduce the many mechanisms of dormancy from *Artemia* ecology that have been translated into cancer biology, and herald *Artemia*'s arrival on the model organism stage. We show how *Artemia* studies have unlocked the mechanisms of the maintenance and termination of cellular dormancy. We then discuss how the antagonistic balance of SETD4 and DEK fundamentally controls chromatin structure and consequently governs CSCs function, chemo/radiotherapy resistance, and dormancy in cancers. Many key stages from transcription factors to small RNAs, tRNA trafficking, molecular chaperones, ion channels, and links with various pathways and aspects of signaling are also noted, all of which link studies in *Artemia* to those of cancer on a molecular and/or cellular level. We particularly emphasize that the application of such emerging factors as SETD4 and DEK may open new and clear avenues for the treatment for various human cancers.

1. Introduction

The brine shrimp (*Artemia*) is one of the most extreme examples of animal dormancy. Upon the approach of winter, the *Artemia* female, in an increasingly saline environment, with lower temperatures (exposure to 2.2 K or over 1 h at 103 °C), food scarcity, and shortened photoperiod, releases embryos that are both encased in a protective cyst shell and in a deeply dormant diapause state that can remain as such for between 3 months and 18 years (Clegg, 2002; Dai et al., 2017; Drinkwater & Clegg, 2018). As conditions yet worsen, triggered by environmental factors such as temperature or dehydration, diapause is replaced by a post-diapause state. This state, while still representing dormancy, leaves the protected embryos poised to hatch out quickly upon the resumption of suitable conditions (Fig. 1). It is understood that on a cellular level this highly tolerant dormancy state is characterized by a G_0 state of cellular quiescence (Clegg, 2002; Dai et al., 2017).

Consider now a tumor undergoing its own chemo-radiotherapeutic barrage. Such an attack, lethal for the most part to active cancer cells, can be

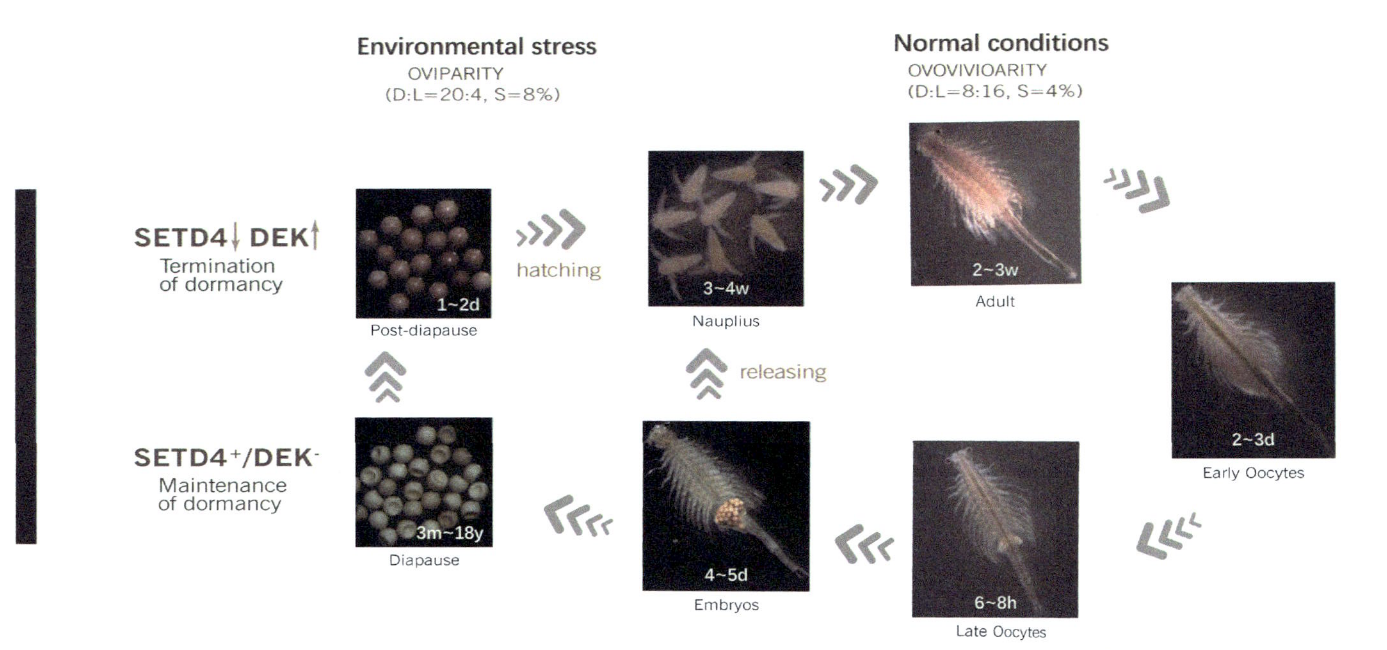

Fig. 1 *Artemia* alternative reproductive strategies as triggered by environmental conditions. (A) Ovoviviparous (directly developing, [right]) and oviparous (diapause-destined, [left]) reproductive pathways of *Artemia*. In the Lab, the ovoviviparous reproductive cycle is maintained under favorable conditions (16 h light/8 h dark cycle and 4% [w/v] salinity). Conversely, the oviparous pathway is triggered by unfavorable conditions (4 h light/20 h dark cycle and 8% salinity) where chitinous shell encased embryos are released and remain in a semi-perpetual dormant state unless activation factors are applied. Duration required for between adjacent phases is marked: d, day; w, week; m, month; y, year. (B) Cell cycle progression during diapause formation, post-diapause and active nauplius life stages.

weathered by a subset of CSCs hunkered down in a potentially comparable strategy of dormancy. Such a dormancy, is also characterized CSCs adopting state of G_0 cellular quiescence. By thus removing themselves from the cell cycle they render themselves extremely difficult to detect, track and target, and facilitate immune evasion. Such a state therefore characterizes tolerance to cancer drugs. When opportunistic environment is resumed, such as beyond the completion of treatments or therapies that target the tumor cell bulk, the subset of dormant cells can then reactivate, resume proliferation, and metastasis can then occur (Jia et al., 2019; Recasens & Munoz, 2019) (Table 1).

Table 1 Comparing the characteristics of *Artemia* diapause/dormancy and CSCs dormancy.

Items	*Artemia* diapause/domancy	CSCs domancy	References
Similarities	Stress-tolerance	Stress-tolerance	Recasens and Munoz (2019) and Drinkwater and Clegg (2018)
	Dormant state of cell cycle arrest at G_0 phase	Adopting state of cell cycle arrest at G_0 phase	Jia et al. (2019) and Recasens and Munoz (2019)
	Maintenance/ reactivation of quiescence controlled by SETD4/DEK	Maintenance/ reactivation of quiescence controlled by SETD4/DEK	Ye et al. (2019) and Yang et al. (2022)
Difference	Triggered by environmental stress (temperature, salinity, photoperiod, etc.)	Triggered by chemo-radiotherapeutic stress	Clegg (2002), Recasens and Munoz (2019), and Dai et al. (2017)
	Normal development	Disease condition	Abatzopoulos, Beardmore, Clegg, and Sorgeloos (2002) and Marusyk, Janiszewska, and Polyak (2020)

Vastly differing scale and context aside, this seems to be a remarkably similar strategy to that of *Artemia*, one perhaps not simply coincidental or analogous. Could the same cell-cycle related molecular triggers, newly discovered as active in the maintenance and termination of this profound state of cellular dormancy/quiescence and likely related to environmental resistance in *Artemia* embryos, be likewise active in the control of cancer dormancy, providing cells similar chemo/radiotherapy tolerance and facilitating metastasis beyond their activation? If so, could such dormancy initiation, maintenance, and termination (reactivation) factors be actively manipulated and applied to "domesticate" cancer cells to stand up or lie down upon request? Encouraging answers to the affirmative are emerging.

As we will detail more fully below, *Artemia* studies ultimately uncovered the key controlling factor of the first, dormancy maintenance, by revealing that it is epigenetic regulation by the histone lysine methyltransferase SETD4 that primarily governs the maintenance of a quiescent state (Recasens & Munoz, 2019). Then, equally significantly, came the discovery of its counterpoint in *Artemia* in the shape of the nuclear protein DEK, as the specific and primary activation factor for quiescent cells (Dai et al., 2017). This latter discovery opened the opportunity to follow an "awakening strategy" to reactivate dormant cells, thus removing their treatment tolerance and increasing their susceptibility to anti-proliferative drugs (Damen, van Rheenen, & Scheele, 2021; Jia et al., 2019; Recasens & Munoz, 2019). This was finally applied in vivo to a murine breast cancer model and shown to successfully facilitate the exit/reactivation of these formerly therapy-tolerant cancer cells. The negation of this tolerance to therapy, their subsequent destruction without cancer recurrence and/or metastasis, was then demonstrated (Yang et al., 2022).

In identifying many of the regulatory factors of cellular quiescence related to *Artemia* dormancy, many other novel conserved mechanisms have also been discovered as applicable to human and animal-model cancers and may also hold considerable potential for breakthroughs in the treatment of cancers, particularly in terms of chemo-radiotherapy resistance and metastasis. This review in particular links the emerging understanding of the role of CSC quiescence in the process of cancers to an ideal model organism for its study, *Artemia*. We therefore present this humble shrimp as a robust and accessible model for the study of cellular dormancy and its related conserved molecular mechanisms, particularly in cancer, but also in somatic/homeostatic contexts.

2. Dormancy, diapause, and quiescence

2.1 Dormancy, diapause, and quiescence at organismal level

Ecologically, dormancy ranges from seasonal metabolic depressions of hibernating hedgehogs or the deciduous autumnal leaf drop, to that of animal diapause, which can represent high levels of stress tolerance or metabolic suppression, often within or beyond adverse stimulus. A trigger for dormancy termination and the resumption of full activity is then required. Conversely, in the context of the study of human cancer, dormancy is usually defined in a tumor context as quiescence and diapause is rarely in focus.

Diapause ranges in context from the Dauer of *C. elegans* through to temperature-induced diapause of the annual killifish (Hu & Brunet, 2018) to the examples of mammalian embryonic diapause where embryo retention in seals (Deng et al., 2018), possums (Oates, Bradshaw, Bradshaw, Stead-Richardson, & Philippe, 2007), wallabies (Fenelon & Renfree, 2018; Flint & Renfree, 1982), nematode (Angelo & Van Gilst, 2009), fruit flies (Kučerová et al., 2016), silk warm (Yamashita, 1996), even giant pandas (Zhang, Li, Wang, & Hull, 2009), optimize their birth periods to environmental conditions. Overall, diapause has been defined as a pre-programmed, genetically determined state, noted as mediated by neurohormones. It is often a seasonal phenomenon; incorporating an arrested state of development; often linked to varying levels and associated aspects of metabolic depression or cell-cycle arrest; and may correspond to tolerance to extreme environmental conditions. It is usually associated with just one developmental stage of an organism, most typically embryonic. While control of diapause occurs via endogenous factors, it is exogenous environmental cues that signal the entry and termination of the diapause state. Differing aspects of diapause are noted as highly varied across species and species strains (Diniz, de Albuquerque, Oliva, de Melo-Santos, & Ayres, 2017; Hand, Denlinger, Podrabsky, & Roy, 2016).

At the organismal/ecological level another form dormancy, quiescence, is also of note. Quiescence, in this context, is suggested to be imposed more directly by unfavorable conditions (e.g., anoxia, temperature or humidity), not pre-programmed in the same way as diapause, often triggered more abruptly, occurring more rapidly, and less predictably and/or less seasonally. When the adverse stimulus ceases, physiological activity is then restored (Diniz et al., 2017). This latter organism level description of quiescence clearly equates to the post-diapause state in *Artemia* (post-diapause and quiescence being so synonymous on an organism level as to be in some instances amalgamated into the term "post-diapause quiescence" (Diniz et al., 2017)) and describing a state primed for reactivation or resumption of organismal development. Overall, while many aspects of dormancy can occur in relatively mild forms, *Artemia* takes it to the other extreme, displaying the most profound degree of metabolic depression ever measured during diapause (Hand et al., 2016). Though a small number of other species such as *C. elegans* (Padilla & Ladage, 2012) or Tardigrades (Guidetti, Boschini, Altiero, Bertolani, & Rebecchi, 2008) show comparable traits, *Artemia*'s extreme dormancy, plus their encasing in an indestructible cyst shell, has certainly attracted much attention (Table 2).

2.2 Cellular quiescence and its regulation mechanisms

On a cellular level, and in the context of the study of human physiology and disease, dormancy is often defined as quiescence and is a particular hot topic relating to stem cells (Bankaitis, Ha, Kuo, & Magness, 2018) or CSCs (Ye et al., 2019). In this context the definitions of cellular quiescence are not gleaned from the ecological state of the organism, but defined more directly by observing the state of the cells. Upon noting a non-proliferative state where cells such as CSCs appear to have exited the cell cycle into a G_0 state, the term cellular quiescence is often then applied. Cellular quiescence occurs during *Artemia* diapause with cells existing in a state of G_0 (Table 1). For clarity, though not all publications follow this definition, in this paper we use the term 'quiescence' to indicate a dormancy where a reversable exit of the cell-cycle into a G_0 state of non-proliferation is confirmed or strongly suspected. More general "dormancy" can be characterized also by a state of cell cycle arrest at the G1 or G2 phase (Li & Clevers, 2010). This brings us back to the occasions where the diapause context clearly overlaps with observations of mechanisms that relate to cell-cycle arrest or quiescence (Hand et al., 2016).

Quiescence, as an expressly non-permanent and reversible state, needs to be actively maintained in the cell. Quiescent cells are therefore able to be reactivated to re-enter the cell cycle in response to various aspects of signaling. This can occur as readily in a positive manner, as in healthy hemostatic processes of repair, as much as in the negative cancer cell focus we are outlining here. This often includes subsets of cells occurring as quiescent in an otherwise active organism, thus representing a step away from the often whole-organism embryonic observations of the diapause context. The reactivation of such fundamental "sleep-like" quiescent cells such as fibroblasts, lymphocytes, and stem cells into proliferation is being increasingly

Table 2 A comparison of the main mechanisms of dormancy in the brine shrimp to the analogous mechanisms of dormancy in other model species.

Species (animal model)	Life span	Developmental stage of diapause entry	Duration of diapause	Trigger of diapause termination	Potential for genetic manipulation	References
Brine shrimp [*Artemia franciscana*]	2–3 months	Embryo	(a decade +)	Low temperature treatment	High	Clegg (2002), Dai et al. (2017), and Drinkwater and Clegg (2018)
Daphia [*Daphnia magna*]	1–2 months	Embryo	(a decade +)	Resumption of optimal environmental conditions	High	Abatzopoulos et al. (2002)
Fruit fly [*Drosophila melanogaster*]	2–2.5 months	Reproductive diapause	(days–weeks)	Amino acid; temperature and photoperiod	High	Kučerová et al. (2016)
Silkworm [*Bombyx mori*]	4–6 weeks	Embryo	(months to 1 year +)	Low temperature	High	Yamashita (1996)
Mosquito [*Numerous species*]	1 year	Embryo, larvae, adult	(6 months) (1 month); (4–8 months)	Resumption of optimal environmental conditions	High	Diniz et al. (2017)
Giant panda [*Ailuropoda melanoleuca*]	20–30 years	Embryo	(17 days)	Resumption of optimal environmental conditions	None	Fenelon and Renfree (2018)

Honey possum [*Tarsipes rostratus*]	1–3 years	Embryo	(<2 weeks)	Shortening day length	None	Oates et al. (2007)
Wallabies [*Notamacropus eugenii*]	12–15 years	Embryo	(11 months)	Increase oestradiol level	None	Flint and Renfree (1982)
Arabidopsis [*Arabidopsis thaliana*]	1 year	Embryo	a decade +	Resumption of optimal environmental conditions	High	Footitt, Douterelo, Clay, and Finch-Savage (2011)
Killifish [*Nothobranchius furzeri*]	4–6 months	Embryo	(months to years)	Resumption of optimal environmental conditions	High	Hu and Brunet (2018)
Nematode [*Caenorhabditis elegans*]	2–3 weeks	Embryo, larvae, adult germline	(Several days) (2–4 weeks); (1 month)	Resumption of optimal environmental conditions	High	Angelo and Van Gilst (2009) and Padilla and Ladage (2012)

understood as important for normal homeostasis such as in processes of tissue repair and regeneration and in other developmental aspects that are vital for the growth and health of higher multicellular organisms. For example, playing important role in neurogenesis, *Drosophila* embryonic neuroblasts remain quiescent at the post-embryonic stage and then are reactivated to resume asymmetric division from first to third larval stages (Holguera & Desplan, 2018). The reservoir of neural stem cells (NSCs) in the murine brain is another good example. Protected from full depletion by an increase in quiescence, these quiescent stem cells, once activated, then reveal similar proliferation and differentiation abilities to those of progenitor cells (Kalamakis et al., 2019). Likewise, quiescent Bmi1-expressing intestinal stem cells compensate for the depletion of Lgr5-expressing intestinal cells due to normal abrasion or specific injury, thus serving as an alternative stem cell pool to permit regeneration (Bankaitis et al., 2018).

In such processes, ongoing studies are now beginning to implicate these same *Artemia* discovered/studied factors as the governors of the entry, maintenance and exit of the quiescent cellular state. For example, related to SETD4, from the discovery of its role in quiescence maintenance in *Artemia*, SETD4 has now been revealed to control a quiescent subset of pancreatic cells which then activate to generate newborn acinar cells in response to cerulein-induced pancreatitis (Tian et al., 2021); SETD4 controlled quiescent c-Kit+ cells are seen to contribute to cardiac neovascularization of capillaries beyond their activation in the heart, (Tian et al., 2021; Xing et al., 2021); SETD4 control of cellular quiescence facilitates maintenance of NSCs and their subsequent regenerative functions in response to brain injury and aging (Cai et al., 2022); and quiescent SETD4+ cells in the intestinal tract are seen to survive radiation exposure and then activate to produce Sca-1-expressing cell types to restore and regenerate the epithelial wall (Huang et al., 2022), etc. In this way SETD4 (and its counterpart DEK—see later) are being revealed though their initial studies in *Artemia* as fundamental conserved epigenetic histone modifiers acting at the chromosome level in the control of quiescence/dormancy for both somatic and cancer cells/stem cell contexts (Table 1).

2.3 Cancer stem cell quiescence and chemo-radiotherapeutic tolerance

The relative success of a combined chemo-radiotherapeutic attack on the tumor bulk has been oft contrasted with its ineffectiveness upon a subset of therapy-resistant cells, which then reassert in relapse or metastasis

(Marusyk et al., 2020). Cellular dormancy is becoming recognized as central to this mechanism (Goddard, Bozic, Riddell, & Ghajar, 2018; Phan & Croucher, 2020). Dormancy within the heterogeneous cancer stem cell population, first noted in 2010 (Pece et al., 2010), was then promoted as a, possibly the, key factor in the mediation of chemoresistance for tumor-propagating CSCs. This brought a new opportunity to target CSC entry/exit into this reversible, quiescent, chemo-resistant state (Brown et al., 2017).

Similarly, disseminated tumor cells (DTCs) may lie dormant (quiescent) for up to a decade prior their activation (Gomis & Gawrzak, 2017), or even be maintained asymptomatically without activation for the lifetime of an organism (Recasens & Munoz, 2019). In either case their depressed metabolic state essentially renders them in "stealth mode" in terms of the immune system itself with attempts to detect, track and or target such cells remaining highly problematic. No treatment or molecular marker for targeting DTCs prior to metastasis is currently available. As we will go on to detail in Section 4, these mechanisms relate strongly to chromatin and histone proteins around which DNA is wound with condensed heterochromatin being a primary characteristic of cellular dormancy, and euchromatin's loose or uncondensed form, characterizing dormancy exit, reactivation, and proliferation (Boonsanay et al., 2016; Liu et al., 2013). Just as the focus upon the epigenetic regulation of the transitions between the two was becoming a prime focus for general aspects of cellular transcription (Liau et al., 2017; Puig et al., 2018; Saito et al., 2010), and its silencing (Berger, 2007; Bierhoff et al., 2014; Kirmizis et al., 2007), it was exactly this being studied relating to Artemia's own dormant/active transitions, where the molecular controllers of this process were being uncovered. These controllers were then demonstrated to be particularly conserved features and translatable to cancer biology and other homeostatic aspects of quiescence.

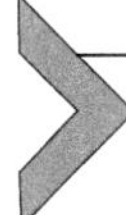

3. *Artemia* used as an animal model to study cancer dormancy

3.1 *Artemia* as an experimental animal

A quarter of a century has now passed since the Brine shrimp, *Artemia*, an "aquatic *Drosophila*," first emerged as a potential model organism. Ease of rearing and experimental manipulation, low economic cost, ability to synchronize larvae (nauplii) in the lab, an established genomic database (https://antagen.kopri.re.kr/project/genome_ info_iframe.php?Code=AF01) plus a second in development (https://www.ugent.be/bw/ asae/en/research/aquaculture/research), an established RNA interference (RNAi) system (Chen, Lin, et al., 2016) and a clear background of organism's biochemistry, represent a strong practical foundations (Marco, Garesse, Cruces, & Renart, 2018). *Artemia* labs now exist in the United States, Belgium, United Kingdom, Canada, China, Spain, Greece, Iran, and Chile. Through these, *Artemia* research has been applied into many areas beyond that of dormancy, including environmental toxicity, agricultural systems, extremophile research, the cyst shell as a bio-material, and age-related diseases (Abatzopoulos et al., 2002). However, it as a robust and accessible model for the nature of diapause and cellular quiescence that *Artemia* arguably provides the highest potential.

Artemia is also one of the most important live food items used in larval aquaculture worldwide. Although *A. franciscana* is the most usual for both aquaculture and scientific research, a variety of geographically separated *Artemia* strains are globally distributed over in saline water ecosystems including six bisexually reproductive *Artemia* species, *A. franciscana, A. persimilis, A. salina, A. sinica, A. tibetiana,* and *A. urmiana*, plus a number of asexual (parthenogenetic) geographic strains. All release ovoviviparous nauplii or oviparous diapause cysts, under favorable or unfavorable conditions, respectively (Abatzopoulos et al., 2002).

3.2 *Artemia* life cycle and the corresponding cellular and molecular characteristics

In ideal environments *Artemia* release live young via an oviviparous pathway. However, in the lab, under hyper-salinity (8%) with short light photoperiod (4h light, 20h dark), they become oviparous with early diapause embryos released in a protective cyst shell. In such a state extreme resilience toward environmental stress is displayed to a rarely rivaled extent (Clegg, 1993; Clegg, 2002; Drinkwater & Clegg, 2018). The cycle continues with triggers, such a temperature or dehydration, then leading to a post-diapause state of dormancy, a state more primed for reactivation. Activation then occurs upon the resumption of suitable environmental conditions (Fig. 1).

For *Artemia*, whether in the wild finding themselves in a naturally occurring adverse environment, or in the lab where an artificially triggered adverse environment is imposed through high salinity, exposure to radiation, or other stresses, the noted response is largely identical. Within the protective cyst shells, the cells of the early-stage embryos are protected from DNA damage by a profound state of quiescence as maintained by SETD4. The SETD4 counterpoint DEK is then applied, either simply triggered by the end of the negative stimulus and the resumption of a favorable environment, or where DEK is deliberately and artificially induced in a lab setting. In either case DEK then controls the activation of these formerly quiescent cells, which leads to both the resumption of cellular synthesis, and also the removal their former tolerance properties (Yang et al., 2022).

Initial attention was given to explore the cellular state of such diapause and post-diapause embryos in Artemia, examining to what extent and in what capacity cell-cycle arrest or quiescence was in evidence. Western blot revealed strong inhibition of CDK6, cyclin D3, phosphorylated Rb at Thr356, and phosphorylated histone H3 at Ser^{10}, in both diapause and post-diapause stages. Low expressions of these mitosis markers indicated that the cells were in a relatively non-active or quiescent state as compared to pre-diapause or larval stages in which cells were actively dividing (Ye et al., 2019). To distinguish the specific cell cycle phases for these diapause and post-diapause embryos, DNA content was analyzed by flow cytometry. More than 90% of cells of diapause embryos were in the G_0/G1 phase (quiescence) whereas more than 85% of cells were in the G2/M phase of cell cycle arrest for post-diapause embryos. These observations were confirmed via the incorporation of 5-bromo-2′-deoxyuridine (BrdU). The absence of BrdU signals in diapause embryos confirmed cell cycle arrest prior to the S phase for diapause embryos, whereas the detection of same signal in post-diapause embryos indicated progression to the G2/M phase (Ye et al., 2019).

3.3 Toward cancer CSCs application

Upon the discovery of such states of dormancy, what controlled this "handbrake" of the cell-cycle became a huge question. A number of the key puzzle pieces for the regulation of cellular quiescence have now been uncovered using *Artemia*, some of which are already being applied into cancer biology. While many of these will be summarized here, the consideration will be maintained that the most fundamental and conserved discovery of potentially universal application lies in the chromosome level epigenetic modification linked to the transition of chromatin structure that is governed by the balance between SETD4 and DEK. Detailing these and other noted factors, here we aim to provide a platform for a more unified approach and promote *Artemia* as a model for the highly important emerging understanding of the role of quiescence in cancers, particularly related to CSCs, chemo/radiotherapy resistance, dormancy, remission, and metastasis.

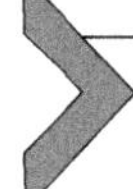

4. Epigenetic regulation of chromatin structure as the central mechanism to the cellular quiescence transition

4.1 The centrality of histone modification

The nucleosome, the basic subunit of chromatin, is a protein core of histone proteins (primarily H3, H4, H2A, and H2B6) around which DNA is wound. It is highly dynamic, undergoing aspects of assembly and disassembly, elongation, and contraction, all functioning to open or close nuclear DNA for business (Fig. 2). Such changes render DNA either accessible or inaccessible to various factors (i.e., non-coding RNAs or transcription factors), which can function to ensure either efficient mRNA synthesis or, conversely, in its effective shutdown. In such a way transcription, replication, the various phases of the cell cycle, and the maintenance of embryonic stem cells, are all regulated (Venkatesh & Workman, 2015).

Enzymes that epigenetically add histone post-translational modifications, such as acetylation, phosphorylation, methylation, ubiquitylation, and sumoylation, together with the involvement of various chromatin remodeling factors and histone chaperones, promote the required changes in histones and overall nucleosome composition that regulate this DNA accessibility by either providing the scaffolding upon which effector proteins may bind, or denying protein access and/or binding to chromatin. In this way they act to facilitate or repress cellular processes, respectively (Venkatesh & Workman, 2015). Such changes in chromatin structure, particularly the transition between the condensed form, heterochromatin, and the loose or open form, euchromatin, represent the essential mechanism of transfer between the potentially long-term dormant state of G_0 (quiescence) and that of actively cycling cells in *Artemia*. In discovering the identities of the fundamental controllers of this process, many were then shown to be highly conserved and revealed as not simply controlling quiescence/dormancy entry, maintenance and exit in an *Artemia* context, but also functioning in the corresponding maintenance and termination of cellular quiescence in CSCs (Fig. 2).

4.2 H3K56 acetylation (H3K56ac) initiates cellular quiescence

The first indication that chromatin/histone modification may be central to cellular quiescence was that in yeast, acetylation of histone H3 lysine 56 (H3K56ac) occurs during normal replication-coupled nucleosome assembly and is balanced by histone acetyltransferase and histone deacetylases

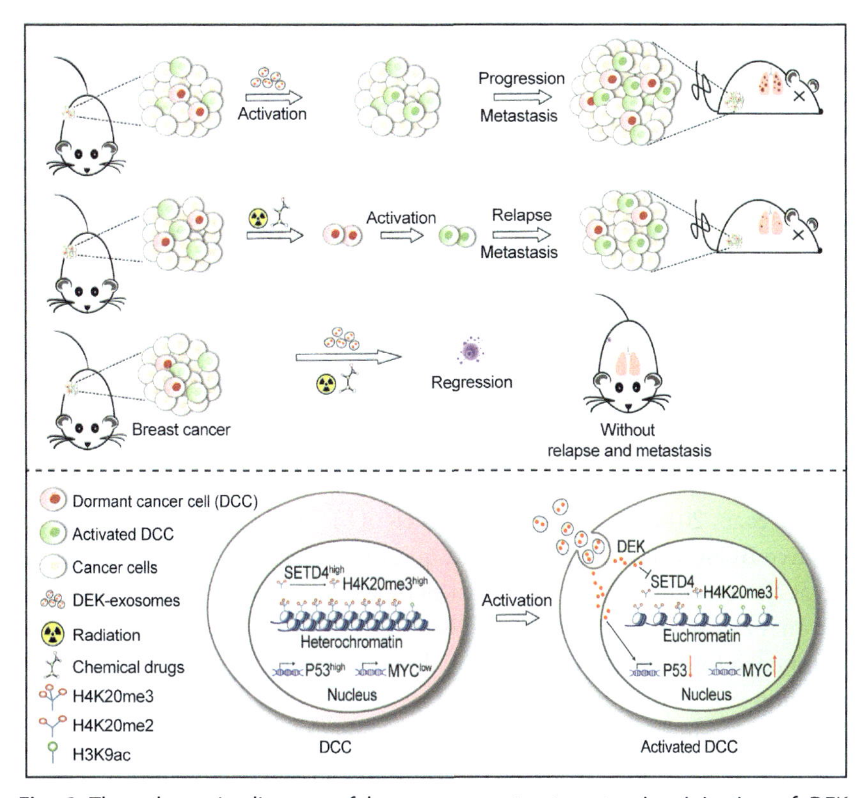

Fig. 2 The schematic diagram of breast cancer treatment using injection of DEK-containing exosomes combined with chemoradiotherapy, and the mechanism of DEK-induced quiescent cancer stem cell activation. Exosomal DEK activated qCSCs to promote tumor progression and metastasis. The qCSCs were resistant to chemoradiotherapy and could re-activate to induce tumor relapse and metastasis. Exosomal DEK activated qCSCs depriving them of their therapy-tolerance. Chemoradiotherapy could then eradicate the CSCs resulting in treatment without relapse and metastasis. Exosomal DEK downregulated the level of SETD4 and H4K20me3, facilitating the transformation of heterochromatin to euchromatin. This epigenetic regulation caused changes in gene expression levels, such as the downregulation of *p53* and the upregulation of *MYC*.

in the regulation of the cell cycle responses to DNA damage repair (Li et al., 2008; Masumoto, Hawke, Kobayashi, & Verreault, 2005). For *Artemia*, increased H3K56ac was observed on chromatin upon diapause initiation, and subsequently decreased upon diapause termination. It was thus implicated in cell cycle arrest, G_0 entry/exit, quiescence, and associated processes (Zhou et al., 2013). This *Artemia* study then turned to HeLa cells where a block in deacetylation using the histone deacetylase inhibitor nicotinamide resulted in similar increases of chromatin H3K56ac levels and the induction of an artificial cell cycle arrest (Zhou et al., 2013) (Fig. 2).

4.3 SETD4/H4K20me3 maintains cellular quiescence

However, a novel and more fundamental histone transition was later observed in *Artemia* in the regulation of the trimethylation of H4K20 (H4K20me3), already an established marker for heterochromatin, by SETD4. Dai et al. discovered abundantly expressed SETD4 facilitating H4K20me3 catalysis *Artemia* diapause embryos (Dai et al., 2017). In contrast to controls, SETD4 knockdown reduced H4K20me3 resulting in embryos unable to enter or sustain a diapause state or enter the G_0 phase.

SETD4 was then found to control breast cancer stem cell quiescence in a similar manner, facilitating heterochromatin formation via H4K20me3 catalysis (Ye et al., 2019). High expression levels of SETD4 were seen in quiescent vs active BCSCs and SETD4-defined quiescent CSCs (qCSCs) were identified as clustered together as a distinct cell type within the heterogeneous population of CSCs in breast, lung, liver, cervical, and gastric cancers and proposed to form the basis for chemoradiotherapy tolerance, dormancy under remission, and as the originator cells of recurrence. This discovery may open the door to new approaches to cancer treatment. This master control process for the onset and maintenance of cellular dormancy was therefore shown to be conserved from *Artemia* to cancers (Fig. 2) (Yang et al., 2022). Interestingly, in contrast to the fundamental nuclear functions of SETD4 above, another fundamental action of mammalian SETD4 has been recently identified in the cytoplasm (Wang et al., 2022) with SETD4 mediated methylation of KU70, a non-histone substrate of SETD4, occurring in the cytoplasm resulting in the suppression of apoptosis. This cytoplasmic function may complement our highlighted nuclear function where the nuclear function is active in the triggering of quiescence, and where its cytoplasmic function correspondingly acts to prevent apoptosis that would otherwise occur. Both functions could be understood as fundamentally protective on a cellular level for a potentially long-term quiescent state.

4.4 DEK terminates cellular quiescence

The nuclear protein DEK, initially cloned in the early 1990s, is considered as highly conserved, though absent in lower eukaryotes such as yeast or *C. elegans* (Jia et al., 2019; von Lindern et al., 1992). While active in aspects of heterochromatin integrity, replication, cellular differentiation, apoptosis, senescence etc., and beyond cellular penetration functioning via intracellular trafficking/nuclear translocation (Saha et al., 2013), most prior studies have focused upon in its relationship to tumorigenesis and chemoresistance as a promoter of tumor growth/survival (Riveiro-Falkenbach & Soengas, 2010). In *Artemia*, DEK was noted as downregulated in quiescent cells of diapause embryos and subsequently enriched upon post-diapause reactivation (Fig. 1). It was then actively confirmed, as the counterpoint to SETD4, to function to terminate diapause and reactivate quiescent cells via its epigenetic action in histone modification, leading to an increase of the more uncoiled and genetically active euchromatin and decrease of the more condensed and genetically inactive heterochromatin. Knockdown of DEK resulted in diapause embryos where control animals produced free-swimming nauplii (Jia et al., 2019).

In cancers, DEK has previously been noted for cellular growth, invasion, and mammosphere formation (Privette Vinnedge et al., 2011), and suggested to confer stem cell-like qualities related to chemotherapy resistance (Privette Vinnedge, Kappes, Nassar, & Wells, 2013). Cellular quiescence in an *Artemia* DEK knockdown context was demonstrated as applicable to MCF-7 and MKN45 human cancer cell lines. DEK knockdown caused brief period of cellular quiescence, with increased environmental stress tolerance, followed by apoptosis. DEK depletion was linked to the upregulation of p53 and down-regulation of AURKA and Wnt signaling pathways in both *Artemia* and MCF-7 cells (Jia et al., 2019), Ongoing *Artemia* linked studies were initiated. These increasingly suggested that DEK may be the main controller in the reactivation of cancer cells from their dormant state.

The prime focus emerging at this stage was the emerging understanding that SETD4, acting in the initiation and maintenance of the dormant/quiescent state, occurs primarily via its regulation of the euchromatin to heterochromatin transition, this representing a transition from open and genetically active DNA to a more tightly coiled and genetically inactive form. This then led on to the discovery of DEK as its counterpoint, implicated in the reactivation of such cells, the suppression of SETD4, via the transition in the opposing direction, heterochromatin to euchromatin (Fig. 2). This was then applied to an impressively successfully cancer treatment model (see below).

4.5 A step toward clinical application: DEK-reactivated quiescence vs SETD4-maintained chemoradiotherapy tolerance in cancer models

Former studies from the same team had heighted DEK, as a counterpoint to SETD4, acting in the activation of quiescent cells in *Artemia* (Jia et al., 2019). Their studies then proceeded to uncover the mechanism of breast quiescent CSC activation, as triggered by exogenous DEK, and to demonstrate that the application of DEK leads to the activation of such quiescent cells, thus switching them from a state of resistance to one of susceptibility to chemoradiotherapy treatment. The role of DEK brought with its potential novel clinical opportunities for removal of quiescence-linked therapy resistance. This was directly applied in the study of our lab (Yang et al., 2022).

In this study, assays of chemoradiotherapy resistant selection were initially used to isolate small populations of chemoradiotherapy-resistant cancer cells from two types of mouse breast tumors and one type of human breast tumor in cancer cell line derived xenograft mice. These quiescent therapy-resistant cells were confirmed with low expressions of proliferative markers and maintained themselves without proliferation over a 6-month chemical-drug treatment period. However, upon completion of such a treatment, and after the removal of chemical drugs, the activation of these cells was then confirmed with corresponding increases of Ki67, H3S10ph, and PCNA, as they began to proliferate to form tumorspheres in vitro. Notably, all three types of breast tumors showed abundant expression of SETD4 but lacked DEK, however once reactivated, active CSCs demonstrated the reverse—enriched DEK but a lack of SETD4.

The use of exogenous DEK to activate qCSCs was then confirmed. The supplementation of DEK-GFP proteins into qCSCs in vitro resulted in the activation of qCSCs and the corresponding suppression of SETD4. Drug treatment was then initiated contrasting these DEK supplemented group where qCSCs had been activated, with untreated control cells still containing cells remaining in their quiescent state. Nearly all such DEK activated cells died when treatment was concurrent with 30 h of DEK-GFP supplementation. However, 98.5% of the three types of control qCSCs that had not been activated survived the same treatment. In addition, no subsequent recurrence or metastasis was observed in the mice of the DEK + drug treatment group. This was in direct contrast to controls where metastasis and recurrence occurred post drug treatment in all cases. The team concluded that an exciting new option for cancer therapy was emerging where exogenous DEK could be used to trigger activation of qCSCs, thus switching them from resistance to susceptibility to chemotherapy (Yang et al., 2022).

Mechanistically, DEK's integral relationship with chromatin structure during its function in the activation of qCSCs was examined using TEM analysis. Analysis of markers confirmed pronounced decreases in condensed heterochromatin and corresponding increases euchromatin. Studies of

MCF7 qCSCs further confirmed changes of chromatin structure as triggered by the application of exogenous DEK characterized by higher chromatin accessibility (open structure) in active aCSCs relative to quiescent qCSCs. The ability of DEK to trigger the quiescence exit of CSCs by directly binding to chromatin, facilitating its genome-wide accessibility was concluded. The upregulation of 25 genes involved in MYC targets and downregulation of 30 genes related with the p53 signaling pathway were similarly noted along with significant decreases in p53, p21, and PUMA, and increases of MYC after qCSCs activation.

To supplement the above observations, clinical data obtained from nearly 4000 breast cancer patients showed that a major differentiation exists between the SETD4 high expressing cohorts, with a median recurrence time of 46 months, and the low SETD4 low expression cohorts, showing median recurrence of over 1 year later (58 months). When the patients were divided into different clinical stage (stages I, II, and III), the number of SETD4-marked qCSCs in stage III was obviously higher than that in stage I and II. Twinned with mouse derived data, this showed SETD4-marked qCSCs with notably higher representation in post-chemotherapy tumors. The team concluded that high SETD4 expression is characteristic of accelerated tumor recurrence and malignancy, probably attributable to the activation of surviving SETD4-marked qCSCs post-therapy. Conversely, detection of SETD4+ cells showed that chemoradiotherapy combined with DEK-exosomes could almost eliminate qCSCs in 4T1- and EMT6-tumors from their corresponding CDX mice. In vitro studies then confirmed that therapy-resistant cancer cells isolated from breast tumors of two clinical patients could similarly be activated using DEK-exosomes, and then destroyed (Yang et al., 2022).

5. Broader factors and signaling pathways conserved between *Artemia*, mouse, and human cancers

Prior to the focus upon SETD4 and DEK, *Artemia* studies had discovered many other novel functions, many with potential applications for cancer mechanism/treatment. These examples move from transcription, to small RNA post-transcription factors, tRNA trafficking, and translation, RNA and Protein chaperons, chloride ion channel, to cell cycle regulation and to metabolic factors (Table 3, and below in detail).

Among these, the p8 gene encodes a nuclear and cytoplasmic stress protein that provides rapid robust and transient responses to various stress factors

Table 3 A summary of broader factors and signaling pathways conserved between *Artemia*, mouse, and human cancers.

Factors in *Artemia*	Factors in human	Function in Artemia (diapause/quiescence)	Molecular pathway(s)	Application to animal/human cancer cells	References
SETD4	SETD4	Maintenance of quiescence	Heterochromatin formation via H4K20me3	Breast, lung, liver, cervical, and gastric cancers	Ye et al. (2019) and Yang et al. (2022)
DEK	DEK	Reactivation of quiescence	Transition of heterochromatin to euchromatin, linking to P53, MYC targets Wnt pathways	Mouse breast tumor and human breast cancer cells	Privette Vinnedge et al. (2013) and Jia et al. (2019)
p8	p8	Stress response	Autophagy	MKN45, HEK293T, HT1080, and MCF7 cell lines	Jia et al. (2016) and Lin et al. (2016)
miR-34	miR-34	Cell proliferation	PLK1, linking to cyclin K, MEK-ERK-Rsk2 pathways	Breast cancer	Achari, Winslow, Ceder, and Larsson (2014) and Zhao et al. (2015)
miR-100	miR-100	Cell apoptosis	mTor pathway	Glioblastoma mouse models, esophageal squamous cell carcinoma	Alrfaei, Vemuganti, and Kuo (2013), Ambros (2004), and Shi et al. (2010)
PLK1	PLK1	Regulation of quiescence	Cyclin K and MEK-ERK-Rsk2 pathways	HeLa cells and human gastric CSC-like cells	Li et al. (2012) and Cholewa, Liu, and Ahmad (2013)
Rsk1	Rsk1	Initiation of quiescence	PLK1 and MEK-ERK-Rsk2 pathways	Lung cancer, breast cancer mouse models	Duan et al. (2014) and Zhu et al. (2017)

Rsk2	Rsk2	Inhibition of quiescence	MEK-ERK pathway	Lung cancer, breast cancer mouse models	Lara et al. (2011) and Stratford et al. (2012)
LARP	LARP	Inducing cell cycle arrest	tRNA trafficking	HeLa, HT1080, and MKN45 cell lines	Bayfield, Yang, and Maraia (2010) and Chen, Lin, et al. (2016)
Artemin	N/A	Stress-tolerance	Protection against RNA degradation	Prevention of denaturation for mammalian cells	Warner, Brunet, MacRae, and Clegg (2004) and King, Toxopeus, and MacRae (2014)
p26	N/A	Stress-tolerance	Protection against protein degradation	N/A	Malitan, Cohen, and MacRae (2019)
CFTR	CFTR	Cell membrane potential maintenance	Linking to Wnt, AURKA and p53 pathway	Mouse intestinal tumors, MCF-7 cell line	Li et al. (2019)
Trehalase	N/A	Stress-tolerance	mTOR-independent autophagy	Human keratinocytes, cancer cell lines	Yang et al. (2013) and Chen, Li, et al. (2016)
PHB1	PHB1 (or PHB)	Inhibition of DNA replication	Ras-Raf signaling pathway	Mouse cervical tumors	Mishra, Murphy, and Murphy (2006), Chiu et al. (2013), and Polier et al. (2012)

(Malicet et al., 2006). p8 elevation was noted just prior to *Artemia* diapause, but absent in cyst-encased diapause embryos. The note of its regulation of autophagy, acting via the targeting of the endoplasmic reticulum during diapause initiation (Jia et al., 2016; Lin et al., 2016), links with its corresponding role in cancer cells (Chen, Hu, Wang, Lou, & Zhou, 2015), with its overexpression in many human cancers (Mohammad, Seachrist, Quirk, & Nilson, 2004; Su et al., 2001). Ectopic expression of *Artemia* p8 in human gastric cancer cells (MKN45), embryonic kidney cells (HEK293T), fibrosarcomas (HT1080), and breast cancer cells (MCF7), induced AMPK/TOR-independent autophagy (Lin et al., 2016). The p8-mediated autophagy system during *Artemia* diapause may provide a useful model to investigate its mechanism in cancers.

Differential expression profiles of micro RNAs (miRNAs) between the stages of Artemia diapause highlighted miR-100 and miR-34, for their respective and opposing roles in cell cycle progression with both miRs regulating the cell cycle during diapause entry by targeting of polo-like kinase 1 (PLK1), and by activation the MEK-ERK-RSK2 pathway and cyclin K, leading to the suppression of RNA polymerase II (RNAP II) (Alrfaei et al., 2013). miR-34 was active in cell cycle arrest by silencing cyclin K expression, resulting in the downregulation of RNAP II activity during diapause entry (Zhao et al., 2015). miR-34-regulation of cyclin K was also noted as a potential therapeutic target to reduce cell proliferation and improve survival in animal models of glioblastomas, and in breast CSCs and other cancer cells (Achari et al., 2014; Zhao et al., 2015). Upregulation of miR-34 expression, with links to the p53 pathway, induced apoptosis, cell cycle arrest, and cell death (Okada et al., 2014). Contrasting to miR-34, miR-100 was noted as important for cell division, the silence of which considered critical for quiescence in the early *Artemia* embryos (Zhao et al., 2015). miR-100 is also positively proliferation-related in cancers and noted as aberrantly expressed (Alrfaei et al., 2013; Ambros, 2004; Shi et al., 2010).

Artemia studies then suggested a clear mechanism of mitosis shut down and clarified PLK1 upstream control by aurora A and with links between PLK1 and ribosomal S6 kinases 1 (RSK1) (Li et al., 2012). PLK1 has a particularly strong relationship with carcinogenesis and the loss of cell-cycle checkpoint control (Cholewa et al., 2013; Chun et al., 2010). RNAi inhibition of Plk1 has been linked to growth arrest or apoptosis in HeLa cells and human gastric CSC-like cell populations (Gumireddy et al., 2005; Li et al., 2012; Liu & Erikson, 2003). Plk1 specific inhibitors resulted in activation of

RSK1, also then inducing quiescence (Zhu et al., 2017). Overexpression of Plk1 in Hela cells resulted in suppression of RSK1, with upstream control via the MEK-ERK pathway. *Artemia* RSK2 was revealed to be a negative regulator of cellular quiescence, its knockdown leading to mitotic arrest and decreases in cyclin D3/phosphorylated histone H3 and the creation of pseudo-diapause (diapause-like) cysts (Duan et al., 2014). The negative regulation of mitotic arrest by *Artemia* Rsk2 was noted to promote cell proliferation and was linked to tumorigenesis. General RSK inhibition has also proved effective in blocking both invasion and metastasis in cancer models (Lara et al., 2011; Stratford et al., 2012). Contrasts between the two *Artemia* isoforms (RSK1 in the promotion of, and RSK2 the inhibition of, quiescence), may help clarify relationships between the differing human RSK isoforms.

Artemia La-related protein (LARP) was demonstrated to act as an upstream signal for tRNA trafficking, regulating cell cycle progression responses to environmental stresses (Berger, 2007). LARP is particularly active in genotoxic stress responses where its regulation of tRNA trafficking provides a physiological adaptation integral to DNA damage responses at the cell cycle checkpoint level (Bayfield et al., 2010). Accumulation of LARP was identified as one of the key triggers of cell cycle arrest and initiator of *Artemia* diapause via its binding of tRNA in the nucleus, preventing export to cytoplasm and leading to the nuclear accumulation of tRNA (Chen, Lin, et al., 2016). In a xenograft mouse tumor model, LARP exogenous gene expression could suppress growth in HeLa and HT1080 cell lines and induce cell cycle arrest in HeLa and MKN45 cell lines. Knockdown of LARP resulted in a depleted cyclin D3 and CDK6 expression and the block in Rb and histone H3 phosphorylation (Chen, Lin, et al., 2016).

The diapause-specific molecular chaperone Artemin was estimated with a high percentage of 7.2% of the soluble protein in *Artemia* cysts. It was linked to the equally abundant heat shock protein p26, extending stress-tolerance properties to the cyst-encased embryo (King et al., 2014; Warner et al., 2004). A decrease in Artemin corresponded to a significant decrease of freezing and desiccation tolerance by preventing RNA degradation, while p26 acts to prevent protein degradation (Warner et al., 2004). p26 RNAi knockdown further highlighted p26 as integral to *Artemia* diapause, strongly modifying a wide range of proteins within the diapause proteome, also affecting embryo development and stress tolerance (Malitan et al., 2019). Artemin has also been observed to convey similar stress-tolerance and prevention of denaturation formammalian cells (Warner et al., 2004). While potentially useful in aspects of maintenance of quiescence, Artemin remains a poor candidate for any trigger or controlling factor.

In *Artemia* diapause embryos the cystic fibrosis transmembrane conductance regulator (CFTR) was noted as undergoing fundamental transition. The expression of CFTR remained silenced with membrane potential hyperpolarized (Vmem+) and intracellular chloride concentration high in diapause embryos. Upon the resumption of CFTR expression in post diapause embryos, the Vmem became depolarized (Vmem−) and the intracellular chloride concentration reduced, indicating a relationship between cell membrane potential and quiescence (Li et al., 2019). CFTR deficiency in CFTR mutant mice leading to intestinal tumors in >60% of mice tested (Than et al., 2016). Combined analysis of CFTR with RNA sequencing (RNA-Seq) of GlyH-101-treated MCF-7 cells concluded that CFTR inhibition down-regulated Wnt and AURKA and upregulated p53 signaling pathways (Li et al., 2019).

Artemia trehalose is noted as amply stored in cyst-encased diapause embryos and considered vital to both their extraordinary stress tolerance and in the provision of a post-diapause energy source (Yang et al., 2013). Transcriptional regulation of trehalase occurs via *Artemia* trehalase-associated protein (ArTAP), expressed in diapause destined embryos but absent from ovivivoparus (live young) pathways. ArTAP-knockdown caused increases in trehalase and induction of apoptosis in *Artemia* cysts. Trehalose treatment also provided a strong cytoprotective effect on cells exposed to UVB radiation and had inhibitory effects upon cellular proliferation in human keratinocytes, also leading to mTOR-independent autophagy in various cancer related cell lines (Chen, Li, et al., 2016).

Prohibitin 1 (PHB1) is a known regulator of aspects of the cell-cycle and transcription, the latter in its association with E2F, Rb, and the chromatin-remodeling complexes Brg-1 and Brm (Fusaro, Dasgupta, Rastogi, Joshi, & Chellappan, 2003; Mishra et al., 2006), and as an inhibitor of DNA replication via its interaction with MCM2–7 (a mini chromosome maintenance complex of proteins) (Artal-Sanz & Tavernarakis, 2009). In *Artemia*, PHB1 is highlighted for its role in the metabolic re-initiation stage, promoting post-diapause cell cycle progression and shown to be critical for subsequent mitochondrial maturation and yolk degradation during development (Ye et al., 2014). PHB1 is also a recognized target in clinical oncology. The phosphorylation of PHB1 activates the Ras-Raf signaling pathway, regulating epithelial cell adhesion, migration (Rajalingam et al., 2005), and metastasis in cervical tumors (Chiu et al., 2013). Rocaglamide or other anti-cancer compounds, inhibit the Ras-Raf signaling pathway via their binding to PHB in promising anti-cancer approaches (Polier et al., 2012).

More direct applications from *Artemia* may be forthcoming. As shown in Table 3, many pathways and connections and the various upstream/downstream linking factors and mechanisms still await discovery. Nevertheless, we believe each observation/study remains valid, important, interesting, and represents many key stages from transcription factors, to small RNAs, tRNA trafficking, molecular chaperones, ion channels, and links with various pathways and aspects of signaling, all of which link studies in *Artemia* to those of cancer on a molecular and/or cellular level.

However, it must be noted that the above observations were viewed more secondarily, considered more as stepping stones leading to the understanding of the primacy of the SETD4/DEK relationship in epigenetic regulation of chromatin modification and as controlling the dormant/activation transition in both somatic and cancer stem cell contexts. It was a growing realization throughout the two decades covering all of the above studies that finally led to the discovery that this transition, as controlled via exosomal DEK application, could represent the removal of chemoradiotherapy tolerance in cancer stem cells and the subsequent negation of recurrence, or relapse.

6. Conclusion and future perspectives

Overall, the discovery of qCSCs have been relatively recent and the wider recognition that quiescence may strongly underlie many aspects of cancer biology is strongly emerging. However, there has been a struggle to find good animal models for quiescence and direct investigation of quiescence factors in cancers are in their relative infancy. In *Artemia*, direct investigations of specific triggers of cellular quiescence maintenance and termination are ongoing. While some are novel, others have already been associated with wider and more general aspects of proliferation, metastasis, recurrence, or resistance in cancers. Initial investigations, more generally associated to characteristics of the whole tumor, cancer cell or cancer stem cell population, are now beginning to lead to more focused studies of the potential positive/negative regulators of quiescence or of the control of dormancy in cancer cell and/or qCSC sub-populations. With *Artemia* comes opportunity for a model for this much tighter focus. Many related aspects of the mechanisms regulating the quiescence/dormancy maintenance and exit/activation for cancer cells, and indeed other somatic cells, have been previously highlighted at a cellular level. However, few show such a potentially clear, consistent, and evolutionarily conserved mechanism that seems to be represented by the SETD4/DEK fluctuation as it provides epigenetic modulation of chromatin structure relating to the transition from a dormant and resistant state of quiescence, and that of active proliferation. It remains highly notable that this mechanism was first discovered in *Artemia*.

In two of the three strategies against dormant cancer cells, those of (i) a "sleeping strategy" to maintain cancer cells in the dormant state, thus preventing reactivation and metastasis and (ii) an "awakening strategy" to reactivate dormant cells to increase their susceptibility to anti-proliferative drugs (Recasens & Munoz, 2019), it is clear that *Artemia* studies have made fundamental contributions to both, the first primarily via SETD4 and second principally via DEK. In the second they have led to a particularly defining step. In this "awakening strategy" endogenous application of DEK via exosomes was indeed seen to remove chemoradiotherapy resistance by triggering quiescence exit of breast cancer stem cells which were then subsequently eradicated using traditional cancer therapies in both murine in vivo and in in vitro models of human breast cancer. A clear route to clinical application is therefore presented for this strategy and is now being explored. This strategy goes a long way beyond simply targeting the tumor bulk, but incorporates the deeper goal of the removal of relapse and metastasis potential. While the use of exosomes as drug delivery vehicles is also not a new concept in the treatment of cancer (Johnsen et al., 2014), the novel use of this concept as an effectively utilized delivery system for DEK in the activation of breast cancer qCSCs to their subsequent eradication, should open the door for similar attempts for many other cancer types.

For the first "sleeping strategy" *Artemia* studies may have also moved the field forwards. Triggers to turn the whole cancer population quiescent and therefore inactive, thereafter maintaining the cancer cell population in that dormant state, are certainly interesting alternatives to strategies which are often unsuccessfully focused upon the entire eradication of the cancer cell population (Klement, 2016). In one of the key discoveries that could confer such a practical potential toward such a strategy comes the recognition of SETD4 as a fundamental master controller for the maintenance of cellular dormancy. Importantly SETD4 has enabled labelling and targeting of quiescent cells in a manner not previously possible, largely negating their former "stealth mode" status and enabling the identification of subsets of qCSCs within an otherwise active cancer cell and CSC populations. Again, let it not be forgotten that SETD4 was first discovered as a master regulator of dormancy in *Artemia*. Differential expression of SETD4 was noted in quiescent vs active BCSCs and SETD4-defined qCSCs and this then led to the discovery of SETD4-defined qCSCs clustered together as a distinct cell type within the CSCs population of breast, lung, liver, cervical, and gastric cancers. As such SETD4 was then boldly proposed to form the basis for chemoradiotherapy resistance, dormancy under remission, and as the originator cells of recurrence.

However, the potential clinical use of SETD4 to maintain a perpetual sleeping strategy for already dormant cells, or to apply to active cancer cells to force a transition into non-proliferative perpetual dormancy, still represents a goal to be achieved. How to prevent SETD4 suppression occurring during the activation process is a big issue. Despite SETD4 firmly holding this subset of cells quiescent during treatment regimes, the completion of treatment is often characterized by the awakening from quiescence into metastasis/reactivation. How to prevent this awakening remains unanswered. Similarly, the possibility to force and then maintain cancer cells, whether in a tumor setting or as disseminated cells, into a perpetual state of quiescence/dormancy, remains fundamentally problematic due to lack of appropriate delivery, specific targeting system and unknown duration of dormancy that any such potential application of SETD4 could promote. Despite this, SETD4 perhaps remains as the most promising target for the sleeping strategy. Possibly, it would be more realistic to firstly attempt to apply SETD4 to simply delay cancer progression, rather than attempting to confer full and perpetual cancer dormancy. Could SETD4 be then applied to provide conditions unattractive to reactivation factors that promote tumor regrowth or that promote the seed cells of quiescence (such as disseminated cancer cells) to activate into metastasis? Could SETD4 also provide an opportunity to track, locate, and/or target dormant CSCs? If so, how could this be tested and applied in pre-clinical trials and studies? These are key questions for further application. Until such answers are forthcoming DEK's application to the awakening/irradiation strategy may remain *Artemia*'s primary contribution toward a feasible and translatable cancer treatment.

Other *Artemia* linked factors have also been implicated in the key stages of the initiation of quiescence in cancers (by PLK1 and RSK1 pathways, and H3K56), negative control of mitotic arrest (by RSK2), reactivation beyond quiescence or the prevention of quiescence (by PHB1, and CFTR) (Table 3), many with potential applications in areas of targeting, chemoradiotherapy resistance, metastasis, dormancy, and subsequent recurrence. The realization that this dormant/quiescent state requires not just triggering but, somewhat paradoxically, active maintenance is another key consideration. Where the trigger for quiescence may have been applied, the conditions/mechanisms for the maintenance of quiescence nevertheless must remain or apoptosis or activation will likely result. Options to negate the opportunity for cells to remain in such a dormant state may be multiple. Despite this, the direct activation of quiescent/dormant CSCs linked to the eradication of the whole cancer population via the application of DEK combined with traditional treatments remains the most oven-ready option that has stemmed from *Artemia* studies thus far. The move toward clinical for multiple cancer types should be soon forthcoming. These findings, as obtained from *Artemia* studies, have achieved the ability to identify, monitor, activate, and then eliminate qCSCs using human cancer cell lines and in vivo tumor models, the *Artemia* model has already more than proved itself. The potential for new treatments for cancer through approaches that target epigenetic modification may end up owing quite a debt of gratitude to the humble brine-shrimp.

Acknowledgment

This work was supported by the National Natural Science Foundation of China (31730084).

References

Abatzopoulos, T., Beardmore, J., Clegg, J., & Sorgeloos, P. (2002). *Artemia: Basic and Applied Biology* (p. 286). Dordrecht, The Netherlands: Kluwer Academic Publishers.

Achari, C., Winslow, S., Ceder, Y., & Larsson, C. (2014). Expression of miR-34c induces G2/M cell cycle arrest in breast cancer cells. *BMC Cancer*, *14*, 538.

Alrfaei, B. M., Vemuganti, R., & Kuo, J. S. (2013). microRNA-100 targets SMRT/NCOR2, reduces proliferation, and improves survival in glioblastoma animal models. *PLoS One*, *8*, e80865.

Ambros, V. (2004). The functions of animal microRNAs. *Nature*, *431*, 350–355.

Angelo, G., & Van Gilst, M. R. (2009). Starvation protects germline stem cells and extends reproductive longevity in C. elegans. *Science*, *326*, 954–958.

Artal-Sanz, M., & Tavernarakis, N. (2009). Prohibitin and mitochondrial biology. *Trends in Endocrinology and Metabolism*, *20*, 394–401.

Bankaitis, E. D., Ha, A., Kuo, C. J., & Magness, S. T. (2018). Reserve stem cells in intestinal homeostasis and injury. *Gastroenterology*, *155*, 1348–1361.

Bayfield, M. A., Yang, R., & Maraia, R. J. (2010). Conserved and divergent features of the structure and function of La and La-related proteins (LARPs). *Biochimica et Biophysica Acta*, *1799*, 365–378.

Berger, S. L. (2007). The complex language of chromatin regulation during transcription. *Nature*, *447*, 407–412.

Bierhoff, H., Dammert, M. A., Brocks, D., Dambacher, S., Schotta, G., & Grummt, I. (2014). Quiescence-induced LncRNAs trigger H4K20 trimethylation and transcriptional silencing. *Molecular Cell*, *54*, 675–682.

Boonsanay, V., Zhang, T., Georgieva, A., Kostin, S., Qi, H., Yuan, X., et al. (2016). Regulation of skeletal muscle stem cell quiescence by Suv4-20h1-dependent facultative heterochromatin formation. *Cell Stem Cell, 18*, 229–242.
Brown, J. A., Yonekubo, Y., Hanson, N., Sastre-Perona, A., Basin, A., Rytlewski, J. A., et al. (2017). TGF-β-induced quiescence mediates chemoresistance of tumor-propagating cells in squamous cell carcinoma. *Cell Stem Cell, 21*, 650–664.e658.
Cai, S. L., Yang, Y. S., Ding, Y. F., Yang, S. H., Jia, X. Z., Gu, Y. W., et al. (2022). SETD4 cells contribute to brain development and maintain adult stem cell reservoir for neurogenesis. *Stem Cell Reports, 17*, 2081–2096.
Chen, S. S., Hu, W., Wang, Z., Lou, X. E., & Zhou, H. J. (2015). p8 attenuates the apoptosis induced by dihydroartemisinin in cancer cells through promoting autophagy. *Cancer Biology & Therapy, 16*, 770–779.
Chen, X., Li, M., Li, L., Xu, S., Huang, D., Ju, M., et al. (2016). Trehalose, sucrose and raffinose are novel activators of autophagy in human keratinocytes through an mTOR-independent pathway. *Scientific Reports, 6*, 28423.
Chen, D. F., Lin, C., Wang, H. L., Zhang, L., Dai, L., Jia, S. N., et al. (2016). An La-related protein controls cell cycle arrest by nuclear retrograde transport of tRNAs during diapause formation in Artemia. *BMC Biology, 14*, 16.
Chiu, C. F., Ho, M. Y., Peng, J. M., Hung, S. W., Lee, W. H., Liang, C. M., et al. (2013). Raf activation by Ras and promotion of cellular metastasis require phosphorylation of prohibitin in the raft domain of the plasma membrane. *Oncogene, 32*, 777–787.
Cholewa, B. D., Liu, X., & Ahmad, N. (2013). The role of polo-like kinase 1 in carcinogenesis: Cause or consequence? *Cancer Research, 73*, 6848–6855.
Chun, G., Bae, D., Nickens, K., O'Brien, T. J., Patierno, S. R., & Ceryak, S. (2010). Polo-like kinase 1 enhances survival and mutagenesis after genotoxic stress in normal cells through cell cycle checkpoint bypass. *Carcinogenesis, 31*, 785–793.
Clegg, J. S. (1993). Respiration of Artemia franciscana embryos after continuous anoxia over 1-year period. *Journal of Comparative Physiology B, 163*, 48–51.
Clegg, J. C. T. (2002). *Physiology and biochemical aspects of Artemia ecology*. Patrick Sorgeloos, Clive NA Trotman: Artemia Biology Edited ByRobert A Browne.
Dai, L., Ye, S., Li, H. W., Chen, D. F., Wang, H. L., Jia, S. N., et al. (2017). SETD4 regulates cell quiescence and catalyzes the trimethylation of H4K20 during diapause formation in artemia. *Molecular and Cellular Biology, 37*.
Damen, M. P. F., van Rheenen, J., & Scheele, C. (2021). Targeting dormant tumor cells to prevent cancer recurrence. *The FEBS Journal, 288*, 6286–6303.
Deng, L., Li, C., Chen, L., Liu, Y., Hou, R., & Zhou, X. (2018). Research advances on embryonic diapause in mammals. *Animal Reproduction Science, 198*, 1–10.
Diniz, D. F. A., de Albuquerque, C. M. R., Oliva, L. O., de Melo-Santos, M. A. V., & Ayres, C. F. J. (2017). Diapause and quiescence: Dormancy mechanisms that contribute to the geographical expansion of mosquitoes and their evolutionary success. *Parasites & Vectors, 10*, 310.
Drinkwater, L., & Clegg, J. (2018). *Experimental biology of cyst diapause* (pp. 93–118).
Duan, R. B., Zhang, L., Chen, D. F., Yang, F., Yang, J. S., & Yang, W. J. (2014). Two p90 ribosomal S6 kinase isoforms are involved in the regulation of mitotic and meiotic arrest in Artemia. *The Journal of Biological Chemistry, 289*, 16006–16015.
Fenelon, J. C., & Renfree, M. B. (2018). The history of the discovery of embryonic diapause in mammals. *Biology of Reproduction, 99*, 242–251.
Flint, A. P., & Renfree, M. B. (1982). Oestradiol-17 beta in the blood during seasonal reactivation of the diapausing blastocyst in a wild population of tammar wallabies. *The Journal of Endocrinology, 95*, 293–300.
Footitt, S., Douterelo, I., Clay, H., & Finch-Savage, W. (2011). Dormancy cycling in Arabidopsis seeds is controlled by seasonally distinct hormone-signaling pathways. *Proceedings of the National Academy of Sciences of the United States of America, 108*, 20236–20241. https://doi.org/10.1073/pnas.1116325108.

Fusaro, G., Dasgupta, P., Rastogi, S., Joshi, B., & Chellappan, S. (2003). Prohibitin induces the transcriptional activity of p53 and is exported from the nucleus upon apoptotic signaling. *The Journal of Biological Chemistry*, *278*, 47853–47861.

Goddard, E. T., Bozic, I., Riddell, S. R., & Ghajar, C. M. (2018). Dormant tumour cells, their niches and the influence of immunity. *Nature Cell Biology*, *20*, 1240–1249.

Gomis, R. R., & Gawrzak, S. (2017). Tumor cell dormancy. *Molecular Oncology*, *11*, 62–78.

Guidetti, R., Boschini, D., Altiero, T., Bertolani, R., & Rebecchi, L. (2008). Diapause in tardigrades: A study of factors involved in encystment. *The Journal of Experimental Biology*, *211*, 2296–2302.

Gumireddy, K., Reddy, M. V., Cosenza, S. C., Boominathan, R., Baker, S. J., Papathi, N., et al. (2005). ON01910, a non-ATP-competitive small molecule inhibitor of Plk1, is a potent anticancer agent. *Cancer Cell*, *7*, 275–286.

Hand, S. C., Denlinger, D. L., Podrabsky, J. E., & Roy, R. (2016). Mechanisms of animal diapause: Recent developments from nematodes, crustaceans, insects, and fish. *American Journal of Physiology. Regulatory, Integrative and Comparative Physiology*, *310*, R1193–R1211.

Holguera, I., & Desplan, C. (2018). Neuronal specification in space and time. *Science*, *362*, 176–180.

Hu, C. K., & Brunet, A. (2018). The African turquoise killifish: A research organism to study vertebrate aging and diapause. *Aging Cell*, *17*, e12757.

Huang, X. T., Li, T., Li, T., Xing, S., Tian, J. Z., Ding, Y. F., et al. (2022). Embryogenic stem cell-derived intestinal crypt fission directs de novo crypt genesis. *Cell Reports*, *41*, 111796. https://doi.org/10.1016/j.celrep.2022.111796. PMID: 36516755.

Jia, W. H., Li, A. Q., Feng, J. Y., Ding, Y. F., Ye, S., Yang, J. S., et al. (2019). DEK terminates diapause by activation of quiescent cells in the crustacean Artemia. *The Biochemical Journal*, *476*, 1753–1769.

Jia, S. N., Lin, C., Chen, D. F., Li, A. Q., Dai, L., Zhang, L., et al. (2016). The transcription factor p8 regulates autophagy in response to palmitic acid stress via a mammalian target of rapamycin (mTOR)-independent signaling pathway. *The Journal of Biological Chemistry*, *291*, 4462–4472.

Johnsen, K. B., Gudbergsson, J. M., Skov, M. N., Pilgaard, L., Moos, T., & Duroux, M. (2014). A comprehensive overview of exosomes as drug delivery vehicles—Endogenous nanocarriers for targeted cancer therapy. *Biochimica et Biophysica Acta*, *1846*, 75–87.

Kalamakis, G., Brüne, D., Ravichandran, S., Bolz, J., Fan, W., Ziebell, F., et al. (2019). Quiescence modulates stem cell maintenance and regenerative capacity in the aging brain. *Cell*, *176*, 1407–1419.e1414.

King, A. M., Toxopeus, J., & MacRae, T. H. (2014). Artemin, a diapause-specific chaperone, contributes to the stress tolerance of Artemia franciscana cysts and influences their release from females. *The Journal of Experimental Biology*, *217*, 1719–1724.

Kirmizis, A., Santos-Rosa, H., Penkett, C. J., Singer, M. A., Vermeulen, M., Mann, M., et al. (2007). Arginine methylation at histone H3R2 controls deposition of H3K4 trimethylation. *Nature*, *449*, 928–932.

Klement, G. L. (2016). Eco-evolution of cancer resistance. *Science Translational Medicine*, *8*, 327fs325.

Kučerová, L., Kubrak, O. I., Bengtsson, J. M., Strnad, H., Nylin, S., Theopold, U., et al. (2016). Slowed aging during reproductive dormancy is reflected in genome-wide transcriptome changes in Drosophila melanogaster. *BMC Genomics*, *17*, 50.

Lara, R., Mauri, F. A., Taylor, H., Derua, R., Shia, A., Gray, C., et al. (2011). An siRNA screen identifies RSK1 as a key modulator of lung cancer metastasis. *Oncogene*, *30*, 3513–3521.

Li, R., Chen, D. F., Zhou, R., Jia, S. N., Yang, J. S., Clegg, J. S., et al. (2012). Involvement of polo-like kinase 1 (Plk1) in mitotic arrest by inhibition of mitogen-activated protein kinase-extracellular signal-regulated kinase-ribosomal S6 kinase 1 (MEK-ERK-RSK1) cascade. *The Journal of Biological Chemistry*, *287*, 15923–15934.

Li, L., & Clevers, H. (2010). Coexistence of quiescent and active adult stem cells in mammals. *Science*, *327*, 542–545.

Li, A. Q., Sun, Z. P., Liu, X., Yang, J. S., Jin, F., Zhu, L., et al. (2019). The chloride channel cystic fibrosis transmembrane conductance regulator (CFTR) controls cellular quiescence by hyperpolarizing the cell membrane during diapause in the crustacean Artemia. *The Journal of Biological Chemistry*, *294*, 6598–6611.

Li, Q., Zhou, H., Wurtele, H., Davies, B., Horazdovsky, B., Verreault, A., et al. (2008). Acetylation of histone H3 lysine 56 regulates replication-coupled nucleosome assembly. *Cell*, *134*, 244–255.

Liau, B. B., Sievers, C., Donohue, L. K., Gillespie, S. M., Flavahan, W. A., Miller, T. E., et al. (2017). Adaptive chromatin remodeling drives glioblastoma stem cell plasticity and drug tolerance. *Cell Stem Cell*, *20*, 233–246.e237.

Lin, C., Jia, S. N., Yang, F., Jia, W. H., Yu, X. J., Yang, J. S., et al. (2016). The transcription factor p8 regulates autophagy during diapause embryo formation in Artemia parthenogenetica. *Cell Stress & Chaperones*, *21*, 665–675.

Liu, L., Cheung, T. H., Charville, G. W., Hurgo, B. M., Leavitt, T., Shih, J., et al. (2013). Chromatin modifications as determinants of muscle stem cell quiescence and chronological aging. *Cell Reports*, *4*, 189–204.

Liu, X., & Erikson, R. L. (2003). Polo-like kinase 1 in the life and death of cancer cells. *Cell Cycle*, *2*, 424–425.

Malicet, C., Giroux, V., Vasseur, S., Dagorn, J. C., Neira, J. L., & Iovanna, J. L. (2006). Regulation of apoptosis by the p8/prothymosin alpha complex. *Proceedings of the National Academy of Sciences of the United States of America*, *103*, 2671–2676.

Malitan, H. S., Cohen, A. M., & MacRae, T. H. (2019). Knockdown of the small heat-shock protein p26 by RNA interference modifies the diapause proteome of Artemia franciscana. *Biochemistry and Cell Biology*, *97*, 471–479.

Marco, R., Garesse, R., Cruces, J., & Renart, J. (2018). *Artemia* molecular genetics. *Artemia biology* (pp. 1–20). CRC Press.

Marusyk, A., Janiszewska, M., & Polyak, K. (2020). Intratumor heterogeneity: The rosetta stone of therapy resistance. *Cancer Cell*, *37*, 471–484.

Masumoto, H., Hawke, D., Kobayashi, R., & Verreault, A. (2005). A role for cell-cycle-regulated histone H3 lysine 56 acetylation in the DNA damage response. *Nature*, *436*, 294–298.

Mishra, S., Murphy, L. C., & Murphy, L. J. (2006). The Prohibitins: Emerging roles in diverse functions. *Journal of Cellular and Molecular Medicine*, *10*, 353–363.

Mohammad, H. P., Seachrist, D. D., Quirk, C. C., & Nilson, J. H. (2004). Reexpression of p8 contributes to tumorigenic properties of pituitary cells and appears in a subset of prolactinomas in transgenic mice that hypersecrete luteinizing hormone. *Molecular Endocrinology*, *18*, 2583–2593.

Oates, J. E., Bradshaw, F. J., Bradshaw, S. D., Stead-Richardson, E. J., & Philippe, D. L. (2007). Reproduction and embryonic diapause in a marsupial: Insights from captive female Honey possums, Tarsipes rostratus (Tarsipedidae). *General and Comparative Endocrinology*, *150*, 445–461.

Okada, N., Lin, C. P., Ribeiro, M. C., Biton, A., Lai, G., He, X., et al. (2014). A positive feedback between p53 and miR-34 miRNAs mediates tumor suppression. *Genes & Development*, *28*, 438–450.

Padilla, P. A., & Ladage, M. L. (2012). Suspended animation, diapause and quiescence: Arresting the cell cycle in C. elegans. *Cell Cycle*, *11*, 1672–1679.

Pece, S., Tosoni, D., Confalonieri, S., Mazzarol, G., Vecchi, M., Ronzoni, S., et al. (2010). Biological and molecular heterogeneity of breast cancers correlates with their cancer stem cell content. *Cell*, *140*, 62–73.
Phan, T. G., & Croucher, P. I. (2020). The dormant cancer cell life cycle. *Nature Reviews. Cancer*, *20*, 398–411.
Polier, G., Neumann, J., Thuaud, F., Ribeiro, N., Gelhaus, C., Schmidt, H., et al. (2012). The natural anticancer compounds rocaglamides inhibit the Raf-MEK-ERK pathway by targeting prohibitin 1 and 2. *Chemistry & Biology*, *19*, 1093–1104.
Privette Vinnedge, L. M., Kappes, F., Nassar, N., & Wells, S. I. (2013). Stacking the DEK: From chromatin topology to cancer stem cells. *Cell Cycle*, *12*, 51–66.
Privette Vinnedge, L. M., McClaine, R., Wagh, P. K., Wikenheiser-Brokamp, K. A., Waltz, S. E., & Wells, S. I. (2011). The human DEK oncogene stimulates β-catenin signaling, invasion and mammosphere formation in breast cancer. *Oncogene*, *30*, 2741–2752.
Puig, I., Tenbaum, S. P., Chicote, I., Arqués, O., Martínez-Quintanilla, J., Cuesta-Borrás, E., et al. (2018). TET2 controls chemoresistant slow-cycling cancer cell survival and tumor recurrence. *The Journal of Clinical Investigation*, *128*, 3887–3905.
Rajalingam, K., Wunder, C., Brinkmann, V., Churin, Y., Hekman, M., Sievers, C., et al. (2005). Prohibitin is required for Ras-induced Raf-MEK-ERK activation and epithelial cell migration. *Nature Cell Biology*, 7, 837–843.
Recasens, A., & Munoz, L. (2019). Targeting cancer cell dormancy. *Trends in Pharmacological Sciences*, *40*, 128–141.
Riveiro-Falkenbach, E., & Soengas, M. S. (2010). Control of tumorigenesis and chemoresistance by the DEK oncogene. *Clinical Cancer Research*, *16*, 2932–2938.
Saha, A. K., Kappes, F., Mundade, A., Deutzmann, A., Rosmarin, D. M., Legendre, M., et al. (2013). Intercellular trafficking of the nuclear oncoprotein DEK. *Proceedings of the National Academy of Sciences of the United States of America*, *110*, 6847–6852.
Saito, Y., Kitamura, H., Hijikata, A., Tomizawa-Murasawa, M., Tanaka, S., Takagi, S., et al. (2010). Identification of therapeutic targets for quiescent, chemotherapy-resistant human leukemia stem cells. *Science Translational Medicine*, *2*, 17ra19.
Shi, W., Alajez, N. M., Bastianutto, C., Hui, A. B., Mocanu, J. D., Ito, E., et al. (2010). Significance of Plk1 regulation by miR-100 in human nasopharyngeal cancer. *International Journal of Cancer*, *126*, 2036–2048.
Stratford, A. L., Reipas, K., Hu, K., Fotovati, A., Brough, R., Frankum, J., et al. (2012). Targeting p90 ribosomal S6 kinase eliminates tumor-initiating cells by inactivating Y-box binding protein-1 in triple-negative breast cancers. *Stem Cells*, *30*, 1338–1348.
Su, S. B., Motoo, Y., Iovanna, J. L., Xie, M. J., Mouri, H., Ohtsubo, K., et al. (2001). Expression of p8 in human pancreatic cancer. *Clinical Cancer Research*, 7, 309–313.
Than, B. L., Linnekamp, J. F., Starr, T. K., Largaespada, D. A., Rod, A., Zhang, Y., et al. (2016). CFTR is a tumor suppressor gene in murine and human intestinal cancer. *Oncogene*, *35*, 4179–4187.
Tian, J. Z., Xing, S., Feng, J. Y., Yang, S. H., Ding, Y. F., Huang, X. T., et al. (2021). SETD4-expressing cells contribute to pancreatic development and response to cerulein induced pancreatitis injury. *Scientific Reports*, *11*, 12614. https://doi.org/10.1038/s41598-021-92075-5. PMID: 34131249. PMCID: PMC8206148.
Venkatesh, S., & Workman, J. L. (2015). Histone exchange, chromatin structure and the regulation of transcription. *Nature Reviews. Molecular Cell Biology*, *16*, 178–189.
von Lindern, M., Fornerod, M., Soekarman, N., van Baal, S., Jaegle, M., Hagemeijer, A., et al. (1992). Translocation t(6;9) in acute non-lymphocytic leukaemia results in the formation of a DEK-CAN fusion gene. *Baillière's Clinical Haematology*, *5*, 857–879.
Wang, Y., Liu, B., Lu, H., Liu, J., Romanienko, P. J., Montelione, G. T., et al. (2022). SETD4-mediated KU70 methylation suppresses apoptosis. *Cell Reports*, *39*, 110794.

Warner, A. H., Brunet, R. T., MacRae, T. H., & Clegg, J. S. (2004). Artemin is an RNA-binding protein with high thermal stability and potential RNA chaperone activity. *Archives of Biochemistry and Biophysics*, *424*, 189–200.

Xing, S., Tian, J. Z., Yang, S. H., Huang, X. T., Ding, Y. F., Lu, Q. Y., et al. (2021). Setd4 controlled quiescent c-Kit(+) cells contribute to cardiac neovascularization of capillaries beyond activation. *Scientific Reports*, *11*, 11603.

Yamashita, O. (1996). Diapause hormone of the silkworm, Bombyx mori: Structure, gene expression and function. *Journal of Insect Physiology*, *42*, 669–679.

Yang, F., Chen, S., Dai, Z. M., Chen, D. F., Duan, R. B., Wang, H. L., et al. (2013). Regulation of trehalase expression inhibits apoptosis in diapause cysts of Artemia. *The Biochemical Journal*, *456*, 185–194.

Yang, Y. S., Jia, X. Z., Lu, Q. Y., Cai, S. L., Huang, X. T., Yang, S. H., et al. (2022). Exosomal DEK removes chemoradiotherapy resistance by triggering quiescence exit of breast cancer stem cells. *Oncogene*, *41*, 2624–2637.

Ye, S., Ding, Y. F., Jia, W. H., Liu, X. L., Feng, J. Y., Zhu, Q., et al. (2019). SET domain-containing protein 4 epigenetically controls breast cancer stem cell quiescence. *Cancer Research*, *79*, 4729–4743.

Ye, X., Zhao, Y., Zhao, L. L., Sun, Y. X., Yang, J. S., & Yang, W. J. (2014). Characterization of PHB1 and its role in mitochondrial maturation and yolk platelet degradation during development of Artemia embryos. *PLoS One*, *9*, e109152.

Zhang, H., Li, D., Wang, C., & Hull, V. (2009). Delayed implantation in giant pandas: The first comprehensive empirical evidence. *Reproduction*, *138*, 979–986.

Zhao, L. L., Jin, F., Ye, X., Zhu, L., Yang, J. S., & Yang, W. J. (2015). Expression profiles of miRNAs and involvement of miR-100 and miR-34 in regulation of cell cycle arrest in Artemia. *The Biochemical Journal*, *470*, 223–231.

Zhou, R., Yang, F., Chen, D. F., Sun, Y. X., Yang, J. S., & Yang, W. J. (2013). Acetylation of chromatin-associated histone H3 lysine 56 inhibits the development of encysted Artemia embryos. *PLoS One*, *8*, e68374.

Zhu, L., Xing, S., Zhang, L., Yu, J. M., Lin, C., & Yang, W. J. (2017). Involvement of Polo-like kinase 1 (Plk1) in quiescence regulation of cancer stem-like cells of the gastric cancer cell lines. *Oncotarget*, *8*, 37633–37645.

CHAPTER SIX

Multi-CpG linear regression models to accurately predict paclitaxel and docetaxel activity in cancer cell lines

Manny D. Bacolod[a,*], Paul B. Fisher[b,c,d], and Francis Barany[a]

[a]Department of Microbiology and Immunology, Weill Cornell Medicine, New York, NY, United States
[b]Department of Human and Molecular Genetics, Virginia Commonwealth University, School of Medicine, Richmond, VA, United States
[c]VCU Institute of Molecular Medicine, Virginia Commonwealth University, School of Medicine, Richmond, VA, United States
[d]VCU Massey Cancer Center, Virginia Commonwealth University, School of Medicine, Richmond, VA, United States
*Corresponding author: e-mail address: mdb2005@med.cornell.edu

Contents

Advances in Cancer Research, Volume 158
ISSN 0065-230X
https://doi.org/10.1016/bs.acr.2022.12.005

Abstract

The microtubule-targeting paclitaxel (PTX) and docetaxel (DTX) are widely used chemotherapeutic agents. However, the dysregulation of apoptotic processes, microtubule-binding proteins, and multi-drug resistance efflux and influx proteins can alter the efficacy of taxane drugs. In this review, we have created multi-CpG linear regression models to predict the activities of PTX and DTX drugs through the integration of publicly available pharmacological and genome-wide molecular profiling datasets generated using hundreds of cancer cell lines of diverse tissue of origin. Our findings indicate that linear regression models based on CpG methylation levels can predict PTX and DTX activities (log-fold change in viability relative to DMSO) with high precision. For example, a 287-CpG model predicts PTX activity at R^2 of 0.985 among 399 cell lines. Just as precise ($R^2 = 0.996$) is a 342-CpG model for predicting DTX activity in 390 cell lines. However, our predictive models, which employ a combination of mRNA expression and mutation as input variables, are less accurate compared to the CpG-based models. While a 290 mRNA/mutation model was able to predict PTX activity with R^2 of 0.830 (for 546 cell lines), a 236 mRNA/mutation model could calculate DTX activity at R^2 of 0.751 (for 531 cell lines). The CpG-based models restricted to lung cancer cell lines were also highly predictive ($R^2 \geq 0.980$) for PTX (74 CpGs, 88 cell lines) and DTX (58 CpGs, 83 cell lines). The underlying molecular biology behind taxane activity/resistance is evident in these models. Indeed, many of the genes represented in PTX or DTX CpG-based models have functionalities related to apoptosis (e.g., *ACIN1, TP73, TNFRSF10B, DNASE1, DFFB, CREB1, BNIP3*), and mitosis/microtubules (e.g., *MAD1L1, ANAPC2, EML4, PARP3, CCT6A, JAKMIP1*). Also represented are genes involved in epigenetic regulation (*HDAC4, DNMT3B*, and histone demethylases *KDM4B, KDM4C, KDM2B*, and *KDM7A*), and those that have never been previously linked to taxane activity (*DIP2C, PTPRN2, TTC23, SHANK2*). In summary, it is possible to accurately predict taxane activity in cell lines based entirely on methylation at multiple CpG sites.

Abbreviations

DTX docetaxel
PTX paclitaxel

1. Introduction

Despite rapid progress in developing molecularly-targeted therapeutics and immunotherapy, classical (or standard) chemotherapy drugs still constitute the bulk of anti-cancer treatment regimens that are considered essential by most countries (Robertson, Barr, Shulman, Forte, & Magrini, 2016). Moreover, there are recent FDA approvals (https://www.fda.gov/drugs/) (i.e., 2017–22) that involve targeted therapeutics or immunotherapy in combination with one or more standard chemotherapy drugs, which can kill cancer cells through various mechanisms, including disruption of DNA replication by acting as DNA-damaging agents (alkylators, methylators, cross-linkers, intercalators) or inactivation of topoisomerases, disruption of nucleic acids metabolism (anti-metabolites), and disruption of mitosis by targeting microtubules (Chabner & Longo, 2011). Among the most widely used anti-mitotic drugs are the taxanes paclitaxel (PTX) and docetaxel (DTX). These drugs prevent mitosis by binding to GDP-bound β tubulin (which would prevent the depolymerization of microtubules) (Xiao et al., 2006). PTX has been approved for breast cancer (alone or with doxorubicin), non-small cell lung cancer (NSCLC) (with cisplatin), ovarian cancer (alone or with cisplatin), and AIDS-related Kaposi sarcoma (as second-line therapy). In addition, FDA has approved DTX as a treatment against breast cancer (with doxorubicin and cyclophosphamide), NSCLC (alone or with cisplatin), prostate cancer (castration-resistant), squamous cell carcinoma of the head and neck (with cisplatin and 5-fluorouracil), stomach cancer or gastroesophageal junction adenocarcinoma (alone) (see www.cancer.gov).

There is extensive knowledge of the various factors and mechanisms influencing taxanes' anti-cancer activities (Mosca, Ilari, Fazi, Assaraf, & Colotti, 2021). As expected, these factors include proteins that directly interact with microtubules. The kinesins (such as KIF14, KIF23, KIF18B, KIF24, and KIFC1), whose movements along microtubules power mitosis and meiosis, are over-expressed in taxane-resistant cell lines (Galletti et al., 2020). The microtubule-associated proteins MAP4 and MAPT (tau protein), which, when bound to microtubules, can hinder PTX's access to its target, are upregulated in taxane-resistant breast cancer lines (Rouzier et al., 2005). Akin to many anti-cancer drugs, taxane effectiveness can be modulated by both efflux and influx pump proteins in the membrane. Elevated expression of the multi-drug resistance (MDR) factor and efflux pump ABCB1 (Das et al., 2021) and down-regulation of influx transporter OATP1B3 (or SLCO1B3) (Takano et al., 2009) are contributing factors to reduced taxane activity. As with many other anti-cancer therapeutics, apoptotic processes can also modulate taxane efficacy (Zhao, Tang, Wang, & Najafi, 2022). Overexpression of the anti-apoptotic Bcl-2 and Bcl-xL proteins, downregulation of pro-apoptotic Bax protein, or modulation of Fas/Fas ligand expression can all impede apoptosis, resulting in cancer cells that are resistant to PTX (Huang et al., 1997). It has also been demonstrated that N-terminal serine phosphorylation of p53, which stabilizes the protein from proteasomal degradation, can lead to EGFR transactivation, and eventual evasion of cancer cells from PTX activity (Gan, Wang, Xu, & Yang, 2011). Other transcription factors such as FOXC2, Twist, Slug, and Snail that are involved in epithelial-to-mesenchymal transition (EMT) can pivot cancer cells toward taxane resistance (Rosano et al., 2011; Zhou et al., 2015). Metabolic degradation of taxanes through hepatic enzymes, such as CYP3A4, and CYP2C8, is another factor influencing resistance against these drugs (de Weger, Beijnen, & Schellens, 2014; van Eijk, Boosman, Schinkel, Huitema, & Beijnen, 2019).

It is now accepted as almost axiomatic that a given transcript's expression level results from numerous competing and complementary gene regulatory forces, many of which are epigenetic. One of the significant factors in epigenetic regulation is the methylation status of CpG islands in the gene promoter regions (Ottaviano et al., 1994). For specific genes, reduced methylation at the promoter regions indicates higher transcriptional activity (and vice versa). This is indeed the case for many genes that play essential roles in cancer drug resistance, including *MGMT* (*O6-methylguanine- DNA methyltransferase*) (Bacolod & Barany, 2020), which is the primary resistance factor against the methylator/alkylator Temozolomide and Carmustine; the MDR gene *ABCB1* (Bacolod, 2021), and *GSTM1* (*glutathione S-transferase mu 1*) (Bacolod, 2021), which can deactivate many chemotherapeutic drugs. This inverse relationship between expression and promoter methylation was previously observed on the gene *CHFR* (*checkpoint with forkhead and ring finger domains*), whose function as a mitotic checkpoint protein influences taxane activity (Wang et al., 2011; Yun et al., 2015). As previously reported by Bacolod, Fisher and Barany (Bacolod et al., 2015; Bacolod & Barany, 2020; Bacolod, Barany, Pilones, Fisher, & de Castro, 2019), a hypomethylated promoter region often exhibits a localized open chromatin structure that is transcriptionally active. This is evident in the enrichment of histone marks H3K4m1, H3K4m2, H3K4m3, and H3K9ac (all indicators of a transcriptionally active promoter (Campbell & Turner, 2013; Ernst et al., 2011; Ghirlando et al., 2012)) within that hypomethylated promoter region, upon integration of ChIP-Seq-generated ENCODE (2004) data into our analysis. However, the promoter's CpG methylation and histone marks are not the only epigenetic factors that may influence transcription. We previously reported that methylation of CpG sites in the gene body is positively (and highly) correlated with MGMT expression in tissues and cell lines (Bacolod & Barany, 2020). It is also important to note that there are many genes in the entire genome wherein the relationship between CpG methylation (promoter, or anywhere else) and expression is not straightforward (the authors have extensively studied the integrated methylation and expression datasets from The Cancer Genome Atlas or TCGA (Vazquez & Boehm, 2020) and Cancer Cell Line Encyclopedia or CCLE (Barretina et al., 2012) projects).

Having established how a cancer cell responds to a drug (such as taxane) is likely influenced by a discordant expression of many genes and that some of these genes may be epigenetically regulated at varying degrees of complexities, we also "hypothesize" that accurate models of the relationships between drug activity and the molecular profiles (CpG methylation, expression) of the cell lines prior to exposure to the drug can be constructed. This can be accomplished by integrating publicly available pharmacological and molecular profiling data for hundreds of cancer cell lines. This information will then allow us to conduct additional bioinformatic analyses to assess the biological relevance of the markers relating to how they may contribute to the overall activities of PTX and DTX.

2. Datasets and methods used to develop predictive multi-marker linear regression models

2.1 Publicly available genomic datasets

The modeling we describe is based on integrated analyses of publicly available genomic and pharmacological datasets described below.

2.1.1 Pharmacological data from the PRISM project

The PRISM (Profiling Relative Inhibition Simultaneously in Mixtures) dataset is the output of a sequencing-based high-throughput drug screening (Corsello et al., 2020). The PRISM approach entails the drug treatment of pools of molecularly barcoded cancer cell lines. What served as a measure of viability (V) post-exposure to the drug is the abundance of these barcodes (relative to cells treated with DMSO) (calculated as fold-change). The 19Q4 version of the dataset (4518 drugs, most of which at 2.5 μM dosage, against 568 cell lines) was downloaded from the Cancer Dependency Map (DepMap) Project portal (https://depmap.org/portal/download/). For our modeling, we pulled the viability data for two drugs: PTX (BRD-K62008436), and DTX (BRD-K30577245).

2.1.2 Genome-wide methylation data from GEO

The dataset GSE68379 (Iorio et al., 2016) was generated through genome-wide methylation profiling (Illumina 450K array) of genomic DNA extracted from 1001 cancer cell lines. It was downloaded from NCBI Gene Expression Omnibus (GEO) data repository (https://www.ncbi.nlm.nih.gov/geo/).

2.1.3 Expression (total mRNA) dataset

Also obtained from the DepMap portal is the RNASeq-generated total mRNA (22Q1 version) dataset from the Cancer Cell Line Encyclopedia (CCLE) project (Ghandi et al., 2019). Each gene expression value in the total mRNA dataset is presented as log2(TPM +1). The downloaded dataset (19,177 unique genes X 1393 cell lines matrix) is quantile normalized prior to further analysis.

2.1.4 Mutational dataset

Downloaded from the DepMap portal is a table of every mutation call (in Mutation Annotation Format or MAF format) for 1393 cell lines, also generated via the CCLE Project. The table (22Q1 version) was then transformed into an array format (row = 19,177 genes, column = 1393 cell lines), with each cell representing a non-silent mutation count for a given gene. The table was then re-coded into binary form (presence and absence of non-silent mutation are "1" and "0," respectively).

2.1.5 Cell line annotations

The clinicopathological information for every cell line was also downloaded from the DepMap data portal. This information includes the primary disease of the tissue of origin, tissue subtype, and source age. The cell lines represented the following cancer types: bile duct cancer, bladder cancer, bone cancer, brain cancer, breast cancer, colon/colorectal cancer, endometrial/uterine cancer, esophageal cancer, gallbladder cancer, gastric cancer, head and neck cancer, kidney cancer, liver cancer, lung cancer, neuroblastoma, ovarian cancer, pancreatic cancer, prostate cancer, rhabdoid, sarcoma, skin cancer, and thyroid cancer.

2.2 Bioinformatic and statistical tools used to develop predictive multi-CpG linear regression models

2.2.1 Genomics data processing

Most of the data processing was accomplished using commercial JMP Pro 13 and JMP Genomics software (SAS, Cary, NC), and Broad Institute's web-ware Morpheus (https://software.broadinstitute.org/morpheus/) (Broad Institute, Cambridge, MA). These programs were utilized for various processes, including dataset normalization, visualization, re-coding, data merging, and other needed data manipulation tasks.

2.2.2 Statistical analyses

The JMP software was used to accomplish various statistical calculations, including correlation analysis, distribution analysis, and linear regression model fitting.

2.2.3 Biomarker annotations

Necessary biomarker annotations were obtained from online sources. These include the following annotations: (a) *Illumina 450K methylation array annotations*. This annotation file ("HumanMethylation450 v1.2 manifest file") was downloaded from the Illumina website (https://support.illumina.com/). Included in the annotation files are CpG sites' genomic coordinates (Genome Build 37), RefSeq Gene Name, RefSeq Gene Group, and relative location to CpG Island, information that was aligned to the genomic coordinates via the UCSC Genome Brower (https://genome.ucsc.edu). The RefSeq Gene Group indicates whether the CpG site's position is relative to the region in the gene loci (e.g., TSS1500, TSS200, Body, 5′utr, 3′utr). If the CpG site is within or near the CpG island, it is indicated as either "Island," "N Shore," "N Shelf," "S Shore," or "S Shelf." (b) *Gene Annotations from ENSEMBL*. Through the ENSEMBL Biomart website (https://useast.ensembl.org/info/data/biomart/index.html), gene annotations were obtained. Attributes include stable gene IDs and associated GO terms originating from the Gene Ontology (GO) website (http://geneontology.org/) (Gene Ontology, 2021).

2.2.4 Associated molecular pathways and processes

Identifying pathways related to a subset of genes was performed through the Reactome website (https://reactome.org/) (Jassal et al., 2020). The built-in program can generate a list of over-represented pathways associated with any given list of genes. The output includes the following: (a) Reactome pathway (R), (b) the number of identifiers (or genes) submitted (or found) (F) in the analysis, (c) the total (T) number of genes curated to belong to pathway R, (d) the associated probability score (*P*), calculated using Binomial Test, and (e) false discovery rate (FDR) which estimates the false positives through Benjamini-Hochberg procedure (Fabregat et al., 2017).

2.2.5 Genome browsers

The UCSC Genome Browser (https://genome.ucsc.edu/) and the Integrative Genomics Viewer (IGV) (https://software.broadinstitute.org/software/igv/) were used to visualize the genomic locations of the CpG markers and their positions relative to various regions of the gene locus (e.g., promoter, exons, introns, CpG islands).

3. Developing the predictive models

3.1 CpG methylation-based linear regression models for predicting paclitaxel and docetaxel activities in cancer cell lines

The first step in the analysis is identifying the cell lines with matched genome-wide methylation profiling and PTX (or DTX) viability data. The resulting data subsets for 399 and 390 cell lines for PTX/methylation and DTX/methylation, respectively, were subjected to further analyses. First, we calculated the correlation coefficient (R, Pearson) between each CpG marker's methylation coefficient (β) (which ranges from 0 = unmethylated to 1 = fully methylated) and the resulting viability change (log fold change relative to DMSO) after treatment with PTX or DTX. The CpG markers were then ranked according to absolute R values. The top 500 markers for each drug were then used as initial input variables for standard least squares linear regression modeling. This results in initial regression model output, which can be augmented by removing or adding new variables upon further inspection. In general, individual variables (i.e., CpG sites) exhibiting low LogWorth values [−log (base 10) of p values; indicator of the variable's contribution to the model] are removed. Through this approach, we were able to generate three predictive models for PTX activities in all cancer cell lines: models "PTX-287-CpG-PanOnc" (Fig. 1A, *left*), "PTX-265-CpG-PanOnc" (Fig. 1A, *center*), and "PTX-249-CpG-PanOnc" (Fig. 1A, *right*). As the name implies, the 287 CpG marker-model is a linear regression model for the calculation of predicted viability (V_p) after PTX treatment of all cancer cell lines ("PanOnc" = Pan Oncology), based on the methylation of 287 CpG markers (Eq. 1). The terms β_i and c_i refer to the methylation value (0 to 1) and the coefficient (or estimate), respectively, for the CpG marker i.

$$V_p(paclitaxel) = -58.858 + \sum_{i=1}^{287} c_i\beta_i \quad (1)$$

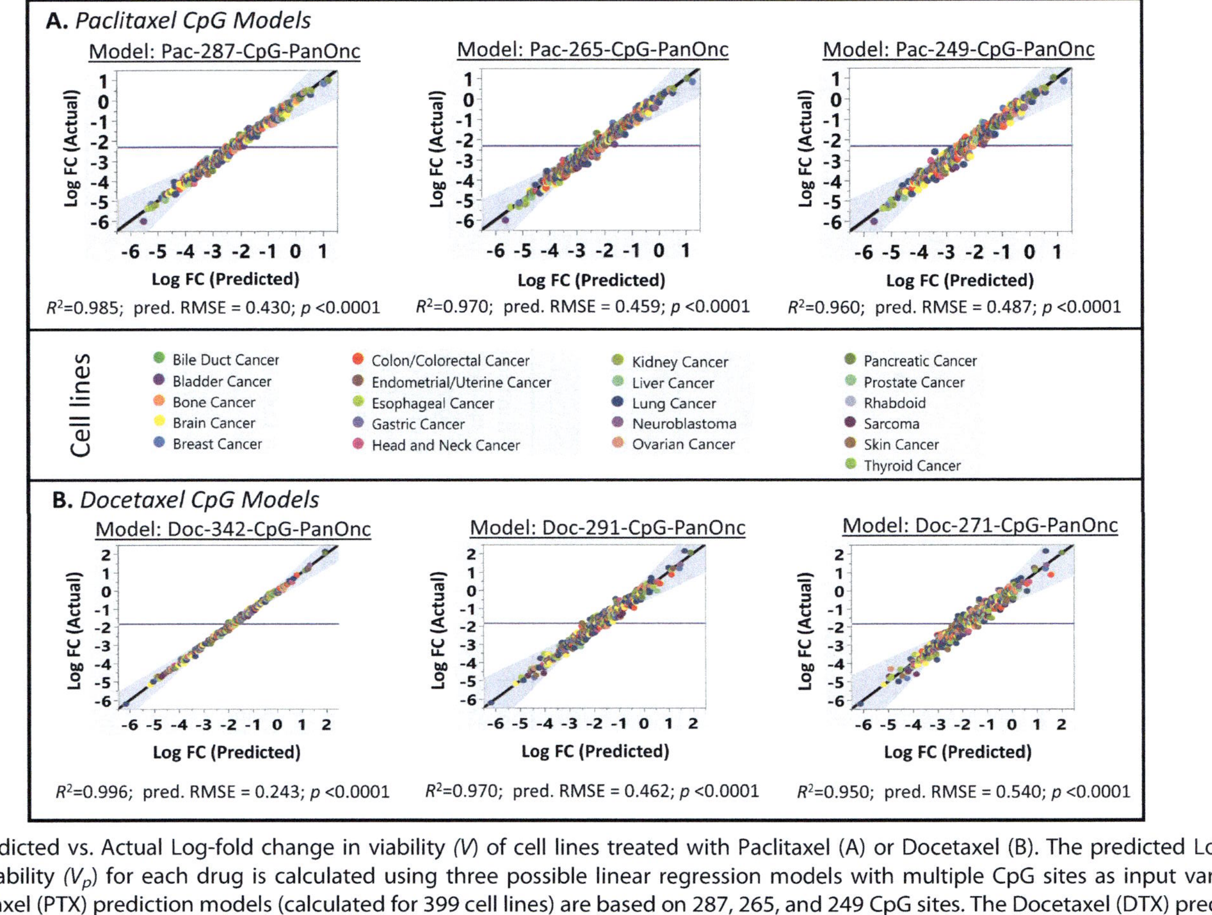

Fig. 1 Predicted vs. Actual Log-fold change in viability *(V)* of cell lines treated with Paclitaxel (A) or Docetaxel (B). The predicted Log-fold change viability (V_p) for each drug is calculated using three possible linear regression models with multiple CpG sites as input variables. The Paclitaxel (PTX) prediction models (calculated for 399 cell lines) are based on 287, 265, and 249 CpG sites. The Docetaxel (DTX) prediction models (calculated for 390 cell lines) are based on 342, 291, and 271 CpG sites. RMSE = root mean square error.

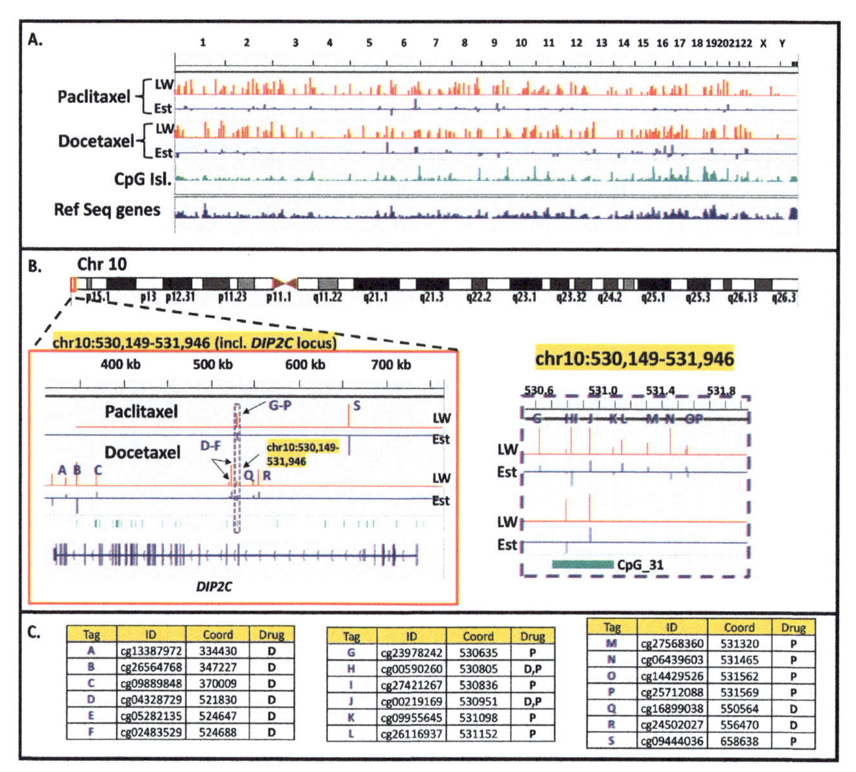

Fig. 2 (A) The location of the CpG markers included in the paclitaxel and docetaxel predictive models (PTX-287-CpG-PanOnc and DTX-342-CpG-PanOnc, respectively). The markers are indicated by the location of matched LogWorth (LW) and Estimate (Est) bars for each CpG marker. Also shown are the locations of CpG Islands in the genome. (B) The left panel is a zoomed-in region of chromosome 10 focusing on the *DIP2C* locus. Indicated are the positions of CpG markers (labeled A-S and marked by LW and Est bars) within the *DIP2C* locus that are included in the two predictive models. Shown on the right panel is a higher resolution of the region encompassing markers G-P (including the two markers H and J, within the CpG island CpG_31). (C) List of the DIP2C CpG markers, marked in B, along with their coordinates (GRCh37).

The correlation between the predicted (V_p) and actual (V_a) viability for PTX is very high ($R^2=0.985$). The models with 265 CpGs (PTX-265-CpG-PanOnc) and 249 CpGs (PTX-249-CpG-PanOnc) have R^2 values of 0.970 and 0.960, respectively. The linear regression models for DTX activity depicted in Fig. 2B are just as accurate. The 342 CpG model (DTX-342-CpG-PanOnc) (Eq. 2) exhibits an R^2 of 0.996. The 291 and 271 CpG models rendered R^2 of 0.97 and 0.95, respectively.

$$V_p(docetaxel) = -22.845 + \sum_{i=1}^{342} c_i\beta_i \quad (2)$$

We also inspected the possible correlation of the input variables. In the 287 CpG model for PTX, there are 41,041 unique pairings [i.e., (287 × 286)/2], and only 301 (0.73%) exhibited $R \geq 0.60$. For the 342 CpG DTX model, out of the 58,311 possible pairings [i.e., (342 × 341)/2], only 34 (which is 0.058%) had $R \geq 0.60$. Thus, it is safe to say that there is a high degree of independence among the variables (CpG sites) in the two models.

3.2 The biological relevance of the genes represented by the CpG markers: Known associations with apoptosis, mitosis, microtubules, epigenetics, and taxane-resistance

The CpG sites represented in PTX and DTX predictive models are spread throughout the genome (Fig. 2A). They can be located within or in the proximity of the promoter region or within the body (introns, exons) of a protein-coding gene (See Table 1 for a partial list). Also included in the models are CpG sites found in intergenic regions whose regulatory roles are not yet clear. Our subsequent analyses of the CpG sites indicate that many of them are part of genes associated with apoptotic and mitotic processes. This was accomplished by simply matching the keywords "apoptotic," "mitotic," "microtubule," "transporter," and "epigenetic" to the Gene Ontology (GO) terms associated with the CpG sites' Gene IDs. The results can be visualized in the volcano plots shown in Figs. 3 and 4. The LogWorth for each CpG site in those graphs is plotted against its corresponding estimate value (coefficient). The datapoints (CpG) sites that match the keywords above are highlighted in distinct colors.

3.2.1 Apoptosis-related markers

In the PTX-287-CpG-PanOnc model, several CpG cites are located in loci of the following apoptosis-related genes (Fig. 3A): *ACIN1* (*apoptotic chromatin condensation inducer 1*), *CADM1* (*cell adhesion molecule 1*), *CD44* (*CD44 molecule, Indian blood group*), *CREB1* (*cAMP responsive element binding protein 1*), *HDAC4* (*histone deacetylase 4*), *LRP2* (*LDL receptor-related protein 2*), *MCF2L* (*MCF.2 cell line derived transforming sequence*), *NME2* (*NME/NM23 nucleoside diphosphate kinase 2*), *PPP2R2B* (*protein phosphatase 2 regulatory subunit B, beta*), *PRKDC* (*protein kinase, DNA-activated, catalytic subunit*), *RPS6KA2* (*ribosomal protein S6 kinase A2*), *SYNGAP1* (*synaptic Ras GTPase activating protein 1*), *TLR4* (*toll-like receptor 4*), *TNFRSF10B* (*TNF receptor superfamily member 10b*), *TOX3* (*TOX high mobility group box family*

Table 1 A partial list of the CpG markers included in the predictive models for activities of paclitaxel (PTX-287-CpG-PanOnc) and docetaxel (DTX-342-CpG-PanOnc) in cancer cell lines.

CpG ID	Gene ID	Chr.	Coord. (hg19)	UCSC Ref Gene Group	Rel. to CpG Is	Log Worth	Estimate (Abs)	Estimate (sign)	GO Tag	Drug
cg15641872	*ZBTB2*	6	151,687,416	Body		15.37	64.81	(+)	NNNN	PTX
cg01867419		6	30,421,515		S Shore	13.01	30.82	(−)	NNNN	PTX
cg16052317	*TTC23*	15	99,789,855	TSS200	N Shore	15.12	29.91	(+)	NNNN	PTX
cg26163234	*TTC23*	15	99,789,622	1stExon; 5′UTR	N Shore	8.19	26.73	(−)	NNNN	PTX
cg25385940	*TTC23*	15	99,789,637	1stExon; 5′UTR	N Shore	11.48	21.53	(−)	NNNN	PTX
cg23761332	*CREB1*	2	208,393,956	TSS1500	Island	7.99	18.72	(+)	YNNN	PTX
ch.9.84366407R		9	85,176,587	undefined	undefined	8.29	18.68	(+)	NNNN	PTX
cg11300341	*RAI1*	17	17,689,562	5′UTR	S Shelf	12.88	18.47	(−)	NNNN	PTX
cg06894146	*ZNF512B*	20	62,595,859	Body	S Shore	12.72	18.09	(+)	NNNN	PTX
cg26886259		5	151,486,992			11.89	17.18	(+)	NNNN	PTX
cg27122614	*PTPRN2*	7	157,913,741	Body		11.26	16.1	(+)	NNNN	PTX
cg24576995	*PTPRN2*	7	157,854,008	Body		10.36	14.77	(−)	NNNN	PTX
cg26308909	*MAD1L1*	7	2,233,090	Body		6.71	14.2	(+)	NYNN	PTX
cg01802713	*ZBTB47*	3	42,695,307	1stExon; 5′UTR	N Shore	13.63	14.19	(−)	NNNN	PTX

cg02971219	*C17orf28*	17	72,954,508	Body		11.59	13.55	(+)	NNNN	PTX
cg06399164	*ZBTB47*	3	42,695,035	TSS200	N Shore	13.55	12.83	(+)	NNNN	PTX
cg08954353	*TNFRSF10B*	8	22,925,656	Body	Island	11.2	12.81	(+)	YNNN	PTX
cg09444036	*DIP2C*	10	658,638	Body		13.85	11.96	(−)	NNNN	PTX
cg27421267	*DIP2C*	10	530,836	Body	Island	10.86	11.89	(−)	NNNN	PTX
cg13401703	*TTC23*	15	99,789,777	1stExon; 5′UTR	N Shore	4.29	11.7	(+)	NNNN	PTX
cg22932220	*NME2*	17	49,244,634	Body	Island	4.29	11.67	(+)	YNNN	PTX
cg03106056	*MARCHF4*	2	217,237,937	TSS1500	S Shore	13.73	11.61	(−)	NNNN	PTX
cg20987283		8	135,863,314			15.69	11.29	(−)	NNNN	PTX
cg15637420	*SMPD4*	2	130,938,337	Body; TSS1500	N Shore	11.29	11.15	(−)	NNNN	PTX
cg18703227	*LEMD2*	6	33,757,854	TSS1500	S Shore	12.13	10.82	(+)	NNNN	PTX
cg18384681		1	55,462,126		N Shore	7.84	10.76	(+)	NNNN	PTX
cg10723020	*CREB1*	2	208,393,829	TSS1500	N Shore	9.76	10.55	(+)	YNNN	PTX
cg09409457	*PTPRN2*	7	157,954,722	Body	N Shelf	10.82	10.53	(−)	NNNN	PTX
cg10975798		8	94,567,610			14.69	10.39	(−)	NNNN	PTX
cg12251075	*NME2*	17	49,244,717	Body	S Shore	9.18	10.36	(−)	YNNN	PTX
cg01778939		15	93,118,381	undefined	undefined	12.67	10.2	(−)	NNNN	PTX

Continued

Table 1 A partial list of the CpG markers included in the predictive models for activities of paclitaxel (PTX-287-CpG-PanOnc) and docetaxel (DTX-342-CpG-PanOnc) in cancer cell lines.—cont'd

CpG ID	Gene ID	Chr.	Coord. (hg19)	UCSC Ref Gene Group	Rel. to CpG Is	Log Worth	Estimate (Abs)	Estimate (sign)	GO Tag	Drug
cg01916724	*PARP3*	3	51,975,220	TSS1500; Body	N Shore	13.93	10.13	(−)	NYNN	PTX
cg14149337	*TP53RK, SLC13A3*	20	45,313,167	3′UTR; TSS200		14.44	9.86	(−)	NNNN	PTX
cg00591949	*UTS2D, CCDC50*	3	191,048,439	TSS200; Body	S Shore	15.62	9.83	(+)	NNNN	PTX
cg13420089	*LOC100129066*	9	92,253,441	TSS1500	undefined	9.31	9.56	(−)	NNNN	PTX
cg16453926		6	28,829,764		N Shore	14.12	9.48	(−)	NNNN	PTX
cg10917619	*NRXN1*	2	51,255,627	5′UTR	S Shore	9.73	9.07	(+)	NNNN	PTX
cg00219169	*DIP2C*	10	530,951	Body	Island	10.68	8.97	(+)	NNNN	PTX
cg18770052		1	5,661,886			12.86	8.74	(−)	NNNN	PTX
cg07430967	*C17orf28*	17	72,954,350	Body		9.17	8.53	(−)	NNNN	PTX
cg01725297	*CLCN6*	1	11,888,644	Body		20.2	54.16	(−)	NNYN	DTX
cg09399405	*CHMP1A*	16	89,712,825	Body	N Shore	22.08	48.1	(+)	NYNN	DTX
cg10462529	*RPTOR*	17	78,880,748	Body	S Shore	16.34	44.22	(−)	NYNN	DTX
cg09876202	*LYRM4*	6	5,145,019	Body		19.19	43.88	(+)	NNNN	DTX
cg06260964	*CHST12*	7	2,473,003	Body	Island	23.1	40.83	(−)	NNNN	DTX

cg12128149	*TECPR1*	7	97,852,022	Body		17.01	39.72	(+)	NNNN	DTX
cg06260789	*ZC3H7B*	22	41,741,977	Body	Island	16.28	36.48	(+)	NNNN	DTX
cg09605818	*TRPV1*	17	3,489,147	Body		21.02	36.42	(+)	YNNN	DTX
cg12218193	*CROCCL1*	1	16,955,163	Body		18.48	35.92	(−)	NNNN	DTX
cg02059950	*BANP*	16	88,094,335	Body		21.38	33.37	(−)	NNNN	DTX
cg01294717	*HDAC4*	2	240,018,022	Body		22.17	32.84	(−)	YNNY	DTX
cg00302030	*C7orf50*	7	1,071,633	Body	S Shelf	23.51	32.82	(+)	NNNN	DTX
cg05142023	*MTHFSD*	16	86,570,343	Body		19.41	32.33	(−)	NNNN	DTX
cg24249695	*ZFAND3*	6	37,896,180	Body		23.83	31.53	(+)	NNNN	DTX
cg09018739	*CPNE2,*	16	57,180,107	Body		18.89	31.17	(+)	NNNN	DTX
cg15134482	*TRRAP*	7	98,581,089	Body	S Shore	21.02	31.16	(−)	NNNN	DTX
cg24198678	*DNASE1*	16	3,703,304	5′UTR; 1stExon		18.85	27.94	(−)	YNNN	DTX
cg17091733		15	41,778,394		Island	12.79	27.13	(+)	NNNN	DTX
cg08032476	*RAB34*	17	27,045,176	Body; TSS1500; 1stExon	N Shore	21.53	26.8	(+)	NNNN	DTX
cg00948047	*BANP*	16	88,029,877	Body	N Shelf	15.33	26.6	(−)	NNNN	DTX
cg25662053		12	122,216,080	3′UTR	undefined	15.12	25.76	(+)	NNNN	DTX

Continued

Table 1 A partial list of the CpG markers included in the predictive models for activities of paclitaxel (PTX-287-CpG-PanOnc) and docetaxel (DTX-342-CpG-PanOnc) in cancer cell lines.—cont'd

CpG ID	Gene ID	Chr.	Coord. (hg19)	UCSC Ref Gene Group	Rel. to CpG Is	Log Worth	Estimate (Abs)	Estimate (sign)	GO Tag	Drug
	RHOF, TMEM120B									
cg14148108	*PFKL*	21	45,744,314	Body	Island	19.1	25.57	(−)	NNNN	DTX
cg22803868	*RAB34*	17	27,045,164	Body; TSS1500; 1stExon	N Shore	22.01	25.44	(−)	NNNN	DTX
cg05253151	*SYNJ2*	6	158,506,818	Body	N Shore	11.54	24.28	(+)	NNNN	DTX
cg15641872	*ZBTB2*	6	151,687,416	Body		11.61	24.1	(+)	NNNN	DTX
cg10164264	*KIAA1530*	4	1,373,477	Body		19.57	24.01	(+)	NNNN	DTX
cg01445100	*BANP*	16	88,103,339	Body	S Shore	18.28	23.53	(+)	NNNN	DTX
cg11867012	*RPS19,*	19	42,364,667	5′UTR	Island	12.74	23.26	(+)	NNNN	DTX
cg09526129	*C7orf50*	7	1,103,262	Body	N Shore	13.88	21.93	(−)	NNNN	DTX
cg05282135	*DIP2C*	10	524,647	Body	N Shore	23.57	21.6	(+)	NNNN	DTX
ch.14.270578R	*NPAS3*	14	34,102,682	Body	undefined	13.43	21.28	(−)	NNNN	DTX
cg01778790	*MYH11*	16	15,815,346	Body		16.26	21.28	(+)	NNNN	DTX
cg09233191	*ULK1*	12	132,404,856	Body		7.98	21.08	(−)	NNNN	DTX
cg25312876	*KDM4B*	19	5,151,581	3′UTR	Island	7.65	20.99	(+)	NNNN	DTX

cg00367863	*MVP*	16	29,854,122	Body	N Shore	22	20.92	(+)	NNNN	DTX
cg15922228	*LSM7*	19	2,321,726	Body	Island	22.4	20.74	(+)	NNNN	DTX
cg13676816	*SEMA4D*	9	91,976,972	3′UTR	undefined	15.96	20.55	(−)	YNNN	DTX
cg00398759		7	155,574,330		S Shore	12.7	20.44	(+)	NNNN	DTX
cg20929407	*MYO1C*	17	1,375,459	Body	Island	10.31	20.3	(−)	NNNY	DTX
cg04110164	*TRIM26*	6	30,158,249	Body	undefined	8.63	20.12	(−)	NNNN	DTX

Shown are markers with the 40 highest absolute estimate values (for each model). The GO (Gene Ontology) tag refers to "*apoptosis*," "*microtubule or mitosis*," "*transporter*," and "*epigenetic*" tags in order. For example, the tag "YNNN" indicates that the marker gene is associated with "apoptosis" and not with the other three terms.

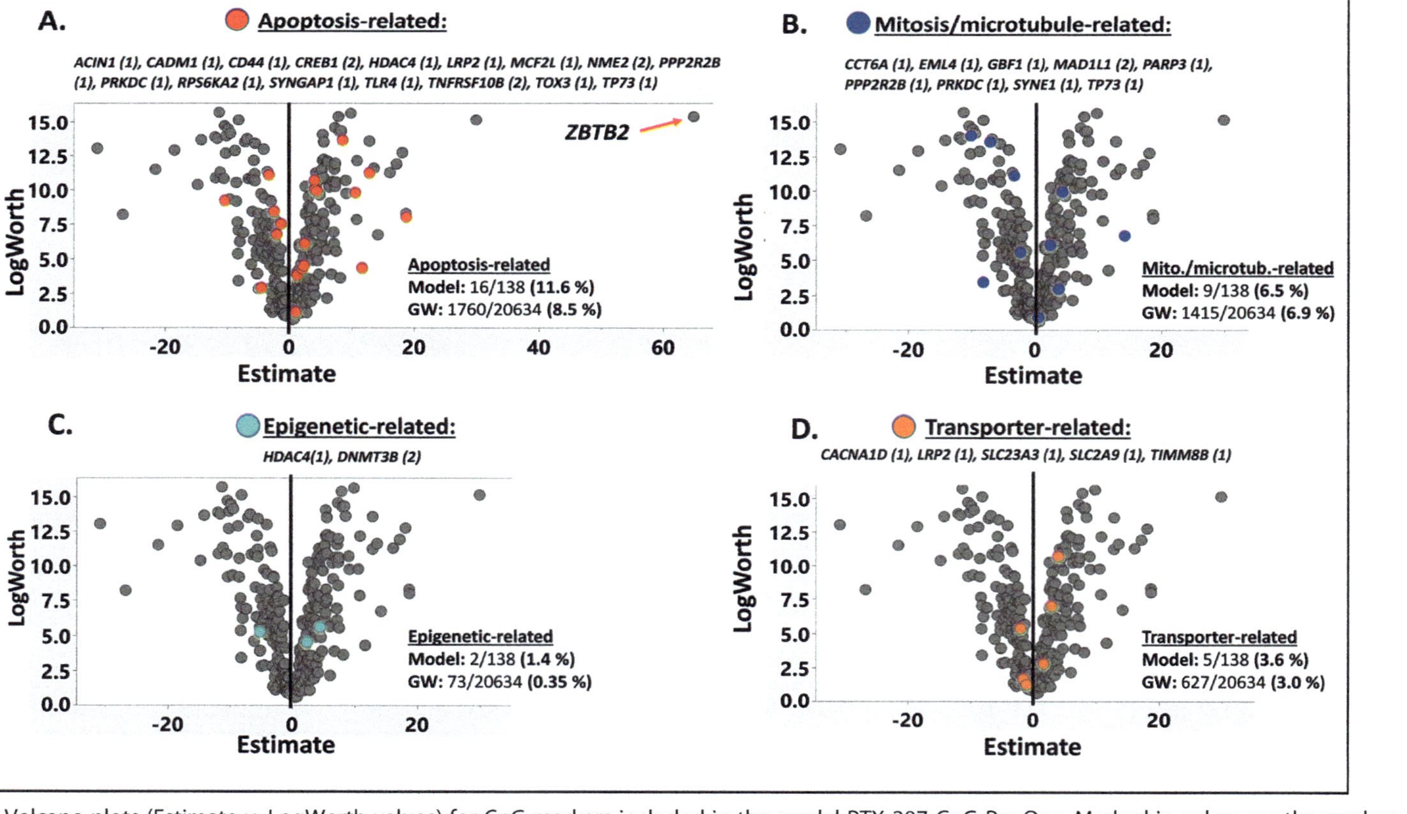

Fig. 3 Volcano plots (Estimate v. LogWorth values) for CpG markers included in the model PTX-287-CpG-PanOnc. Marked in colors are the markers whose genes are associated with Gene Ontology (GO) terms "apoptosis" (A), "mitosis," or "microtubules" (B), "epigenetics" (C), and "transporter" (D). An outlier marker (CpG site for the gene *ZBTB2*) is indicated in the first panel. Each panel also indicates the proportion of genes associated with each corresponding GO term(s) that are either represented in the *Model* or the entire Illumina 450K array (i.e., genome-wide or *GW*).

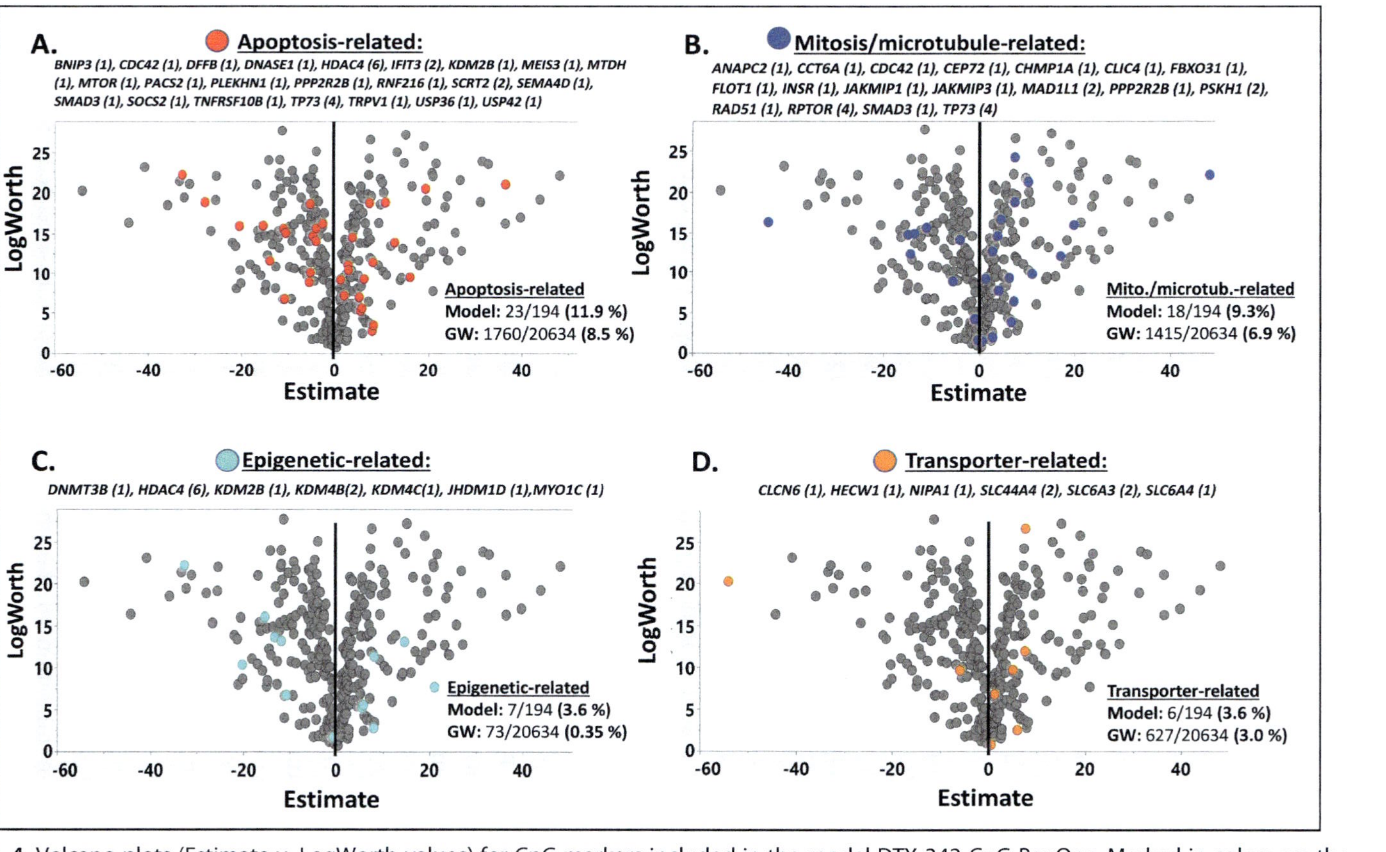

Fig. 4 Volcano plots (Estimate v. LogWorth values) for CpG markers included in the model DTX-342-CpG-PanOnc. Marked in colors are the markers whose genes are associated with Gene Ontology (GO) terms "apoptosis" (A), "mitosis," or "microtubules" (B), "epigenetics" (C), and "transporter" (D). Each panel also indicates the proportion of genes associated with each corresponding GO term(s) that are either represented in the *Model* or the entire Illumina 450K array (i.e., genome-wide or *GW*).

member 3), and *TP73* (*tumor protein p73*). Similarly, the predictive model DTX-342-CpG-PanOnc includes CpG sites from the following apoptosis-associated genes (Fig. 4A): *BNIP3* (*BCL2 interacting protein 3*), *CDC42* (*cell division cycle 42*), *DFFB* (*DNA fragmentation factor subunit beta*), *DNASE1* (*deoxyribonuclease 1*), *HDAC4* (*histone deacetylase 4*), *IFIT2* (*interferon induced protein with tetratricopeptide repeats 2*), *KDM2B* (*lysine demethylase 2B*), *MEIS3* (*Meis homeobox 3*), *MTDH* (*metadherin*), *MTOR* (*mechanistic target of rapamycin kinase*), *PACS2* (*phosphofurin acidic cluster sorting protein 2*), *PLEKHN1* (*pleckstrin homology domain containing N1*), *PPP2R2B* (*protein phosphatase 2 regulatory subunit Bbeta*), *RNF216* (*ring finger protein 216*), *SCRT2* (*scratch family transcriptional repressor 2*), *SEMA4D* (*semaphorin 4D*), *SMAD3* (*SMAD family member 3*), *SOCS2* (*suppressor of cytokine signaling 2*), *TNFRSF10B* (*TNF receptor superfamily member 10b*), *TP73* (*tumor protein p73*), *TRPV1* (*transient receptor potential cation channel subfamily V member 1*), *USP36* (*ubiquitin specific peptidase 36*), and *USP42* (*ubiquitin specific peptidase 42*). Upon closer look at the genes listed above, we can recognize that their roles are spread out in multiple but interconnected facets that make up the apoptotic process. *TNFRSF10B* codes for one of the TRAIL death receptors (DR5) and is crucial in initiating the extrinsic pathway (Wang & El-Deiry, 2003). The gene *PACS2* has been reported to be an essential effector of TRAIL for apoptosis promotion (Guicciardi et al., 2014). TP73 (a member of the TP53 family) is a transcription factor involved in the transactivation of various pro-apoptotic genes such as *FAS* (part of the extrinsic pathway), as well as *BAX* and *PUMA* (Melino et al., 2004; Terrasson et al., 2005). BAX and PUMA are members of the Bcl-2 family proteins, which modulate mitochondrial membrane permeability during the mitochondrial or intrinsic pathway. SCRT2 is a transcriptional repressor that regulates PUMA expression (Rodriguez-Aznar & Nieto, 2011). PLEKHN1 functions as a positive apoptosis regulator by enhancing Bax-Bak oligomerization (Kuriyama, Tsuji, Sakuma, Yamamoto, & Tanaka, 2018). Also listed above is BNIP3, an anti-apoptotic member of the Bcl-2 family (Mellor & Harris, 2007). DFFB (Liu et al., 1998), DNASE1 (Nagahama et al., 2008), and ACIN1 (Sahara et al., 1999) are directly involved in chromatin condensation and DNA fragmentation, which occur during programmed cell death. Evidence points to the role of TLR4 signaling in caspase-3 activation, another essential component of programmed cell death (Hsu, Lin, Hu, Shu, & Lu, 2018). Apoptosis can be induced upon inhibition of the phosphatidylinositol kinase-related kinase MTOR (Chen & Zhou, 2020), the transcription factor CREB1 (Zhao, Zhang, Wang, Wang, & He, 2021), and the regulatory protein MTDH

(Rajasekaran et al., 2015). The kinase PRKDC is an essential mediator of DNA-damage-induced apoptosis (Norbury & Zhivotovsky, 2004). The cell adhesion molecule CADM1 also plays a role in apoptosis promotion and inhibition of proliferation (Fisser et al., 2015). A recent report has demonstrated that the nucleoside diphosphate kinase NME2 can act as a negative regulator of apoptosis by promoting the transcription of various anti-apoptotic factors, including mir-200 (Gong, Yang, Wang, Wang, & Zhang, 2020). We also separately subjected the genes represented in the PTX-287-CpG-PanOnc and DTX-342-CpG-PanOnc predictive models to Reactome analysis (See Table 2).

Table 2 The top Reactome pathways associated with genes represented in the CpG-based paclitaxel (PTX-287-CpG-PanOnc) and docetaxel (DTX-342-CpG-PanOnc) predictive models.

Drug	Reactome ID	Pathway name	Entities *P* value	Entities FDR	Submitted entities found
PTX	R-HSA-428359	Insulin-like Growth Factor-2 mRNA Binding Proteins (IGF2BPs/IMPs/VICKZs) bind RNA	5.64E-07	3.22E-04	*CD44*
PTX	R-HSA-6803211	TP53 Regulates Transcription of Death Receptors and Ligands	6.70E-05	1.91E-02	*TNFRSF10B; TP73*
PTX	R-HSA-9707616	Heme signaling	1.50E-03	1.58E-01	*RAI1; CREB1; CLOCK; TLR4*
PTX	R-HSA-75158	TRAIL signaling	4.09E-03	1.58E-01	*TNFRSF10B*
PTX	R-HSA-199920	CREB phosphorylation	5.14E-03	1.58E-01	*CREB1; RPS6KA2*
PTX	R-HSA-112316	Neuronal System	5.32E-03	1.58E-01	*KCNH4; CREB1; RPS6KA2; NRXN1; KCNQ3; CACNA2D3; GRIK4; HTR3A; APBA1; COMT; SYN2; SHANK2*

Continued

Table 2 The top Reactome pathways associated with genes represented in the CpG-based paclitaxel (PTX-287-CpG-PanOnc) and docetaxel (DTX-342-CpG-PanOnc) predictive models.—cont'd

Drug	Reactome ID	Pathway name	Entities *P* value	Entities FDR	Submitted entities found
PTX	R-HSA-6794361	Neurexins and neuroligins	5.66E-03	1.58E-01	*NRXN1; APBA1; SHANK2*
PTX	R-HSA-9022707	MECP2 regulates transcription factors	6.29E-03	1.58E-01	*CREB1*
PTX	R-HSA-156842	Eukaryotic Translation Elongation	7.31E-03	1.58E-01	*EEF1D; RPS18; RPL22L1; RPL39*
PTX	R-HSA-450282	MAPK targets/ Nuclear events mediated by MAP kinases	8.32E-03	1.58E-01	*CREB1; RPS6KA2; DUSP7*
PTX	R-HSA-390450	Folding of actin by CCT/TriC	1.04E-02	1.97E-01	*CCT6A*
PTX	R-HSA-5633008	TP53 Regulates Transcription of Cell Death Genes	1.69E-02	3.04E-01	*TNFRSF10B; TP73*
DTX	R-HSA-6803211	TP53 Regulates Transcription of Death Receptors and Ligands	1.83E-04	1.16E-01	*TNFRSF10B; TP73*
DTX	R-HSA-3214842	HDMs demethylate histones	1.38E-03	4.38E-01	*KDM4B; KDM4C; KDM2B; JHDM1D*
DTX	R-HSA-75158	TRAIL signaling	6.79E-03	5.27E-01	*TNFRSF10B*
DTX	R-HSA-446205	Synthesis of GDP-mannose	1.25E-02	5.27E-01	*PMM2*
DTX	R-HSA-525793	Myogenesis	1.32E-02	5.27E-01	*CDC42; CTNNA2; NTN3*
DTX	R-HSA-6794362	Protein-protein interactions at synapses	1.44E-02	5.27E-01	*PPFIBP2; NRXN1; FLOT1; SHANK2*
DTX	R-HSA-112311	Neurotransmitter clearance	1.68E-02	5.27E-01	*COMT; SLC6A3; SLC6A4*

Results indicate that the highest-rated pathways (i.e., lowest entities *P* value) include "*TP53 Regulates Transcription of Death Receptors and Ligands*" and "*TRAIL Signaling.*" The genes TP73 and TNFRSF10B are mapped in these pathways, which are commonly represented in both PTX and DTX predictive models.

Reports linked taxane activity (or resistance) to some of the genes described above. Recently, PRKDC inhibition was described as a possible approach to overcome taxane resistance (Chao & Goodman Jr., 2021). Also, the over-expression of TLR4 was observed in PTX- resistant prostate cancer cells (Che et al., 2022). The inhibition of TNFRSF10B was reported to enhance PTX-induced apoptosis in non-small-cell lung cancer (Odoux, Albers, Amoscato, Lotze, & Wong, 2002). Knocking down the *MTDH* gene led to the attenuation of PTX resistance in breast cancer (in vivo and in vitro) (Yang et al., 2018).

3.2.2 Mitosis or microtubule-related markers

Incorporated in the PTX-287-CpG-PanOnc model are CpG sites belonging to mitosis or microtubule-related genes (Fig. 3B). Among these genes are *CCT6A* (*chaperonin containing TCP1 subunit 6A*), *EML4* (*EMAP like 4*), *GBF1* (*Golgi brefeldin A resistant guanine nucleotide exchange factor 1*), *MAD1L1* (*mitotic arrest deficient 1 like* 1), *PARP3* (*poly(ADP-ribose) polymerase family member 3*), *PPP2R2B* (*protein phosphatase 2 regulatory subunit B, beta*), *PRKDC, SYNE1* (*spectrin repeat containing nuclear envelope protein 1*), and *TP73*. The mitosis- or microtubule-associated genes (Fig. 4B) represented in the DTX-342-CpG-PanOnc model include *ANAPC2* (*anaphase-promoting complex subunit 2*), *CCT6A* (*chaperonin containing TCP1 subunit 6A)*, *CDC42* (*cell division cycle 42*), *CEP72* (*centrosomal protein 72*), *CHMP1A* (*charged multivesicular body protein 1A*), *CLIC4* (*chloride intracellular channel 4*), *FBXO31* (*F-box protein 31*), *FLOT1* (*flotillin 1*), *INSR* (*insulin receptor*), JAKMIP1 (*Janus kinase and microtubule interacting protein 1*), *JAKMIP3* (*Janus kinase and microtubule interacting protein 3*), *MAD1L1* (*mitotic arrest deficient 1 like 1*), *PPP2R2B* (*protein phosphatase 2 regulatory subunit Bbeta*), *PSKH1* (*protein serine kinase H1*), *RAD51* (*RAD51 recombinase*), *RPTOR* (*regulatory associated protein of MTOR complex 1*), *SMAD3* (*SMAD family member 3*), and *TP73*. Common among the two models are the genes *PPP2R2B, CCT6A, MAD1L1,* and *TP73*. Genes that are directly involved in the mitotic assembly include *MAD1L1* (a checkpoint protein that makes sure chromosomes are correctly aligned at the metaphase plate before anaphase) (Akera & Watanabe, 2016), *ANAPC2* (a component of the anaphase-promoting complex, which promotes the transition from metaphase to anaphase) (Schrock, Stromberg, Scarberry, & Summers, 2020),

EML4 (a microtubule-associated protein, that may be involved in microtubule formation) (Pollmann et al., 2006), *CCT6A* (a component of chaperonin complex TCP1 responsible for folding actins and tubulins) (Sternlicht et al., 1993), *JAKMIP1* and *JAKMIP3* (likely role in Jak signaling and rearrangement of microtubules) (Steindler et al., 2004), *GBF1* (predicted to be associated with microtubule) (Walch et al., 2018) and *CEP72* (leucine-rich-repeat protein localized in microtubule-organizing centrosome) (Oshimori, Li, Ohsugi, & Yamamoto, 2009). The elevated expression of MAD1L1 was reported to be associated with poor prognosis and PTX-resistance in breast cancer (Kivilaakso, 1985). PARP proteins (such as PARP3) are essential regulators of mitotic functions (including centrosome function and assembly of mitotic spindles) (Slade, 2019). The recombinase RAD51 is considered an essential safeguard of genomic integrity and protects DNA synthesis through its DNA repair function (Wassing et al., 2021). Another gene critical during the mitotic process is *CLIC4*, a regulator of cortical cytoskeleton stability during cell division (Peterman, Valius, & Prekeris, 2020).

3.2.3 Epigenetic-related markers

The PTX-287-CpG-PanOnc model includes CpG sites in *HDAC4* (*histone deacetylase 4*) and *DNMT3B* (*DNA methyltransferase 3 beta*) (Fig. 3C). The same two genes, plus *MYO1C* (*myosin IC*), are represented in the DTX-342-CpG-PanOnc model (Fig. 4C). Our Reactome analysis has identified the pathway "*HDMs demethylate histones*" as one of the top significant pathways among the genes represented in the DTX predictive model. Mapped in the pathway are the histone demethylase genes *KDM4B* (*lysine demethylase 4B*), *KDM4C* (*lysine demethylase 4C*), *KDM2B* (*lysine demethylase 2B*), and *JHDM1D* (*KDM7A, lysine demethylase 7A*). Reports contend that genes coding for proteins that play crucial roles in epigenetic regulation are primary drivers of cancer drug resistance. Hence, inhibition of DNMT3B has been shown to reverse resistance against sorafenib in liver carcinoma (Lai et al., 2019) and enzalutamide in prostate cancer (Farah et al., 2022). The contributions of HDAC4 to resistance against Pt Drugs (Stronach et al., 2011), 5-fluorouracil (Yu et al., 2013), and etoposide (etoposide) have been reported in the literature. Equally important epigenetic regulators are the histone lysine demethylases (such as KDM2B, KDM4B, KDM4C, and KDM7A/JHDM1D), which function by removing the methylation marks on histone tails (Sterling, Menezes, Abbassi, & Munoz, 2020). We can only assume that these factors may regulate many genes related to cancer drug efficacy and resistance. Recent publications describe the roles of

KDM4B, KDM4C, and KDM7A in regulating androgen receptor (AR) (Gao et al., 2021; Lee et al., 2020; Tang et al., 2020).

3.2.4 Transporter-related markers

The transporter-related markers for the PTX-287-CpG-PanOnc model (Fig. 3D) include CpG within the loci of genes *CACNA1D* (*calcium voltage-gated channel subunit alpha1 D*), *LRP2* (*LDL receptor-related protein 2*), *SLC23A3* (*solute carrier family 23 member 3*), *SLC2A9* (*solute carrier family 2 member 9*), *TIMM8B* (*translocase of inner mitochondrial membrane 8 homolog B*). For the DTX-342-CpG-PanOnc model (Fig. 4D), similar genes that are covered include *CLCN6* (*chloride voltage-gated channel 6*), *HECW1* (*HECT, C2, and WW domain containing E3 ubiquitin-protein ligase 1*), *NIPA1* (*NIPA magnesium transporter 1*), *SLC44A4* (*solute carrier family 44 members 4*), *SLC6A3* (solute carrier family 6 member), and *SLC6A4* (solute carrier family 6 member 4). There is no assumption that any of the aforementioned transporter-related genes function in drug influx or efflux, thus influencing drug activity. Nevertheless, recent reports point to the upregulation of SLC2A9 in a cytostatic-resistant ovarian cancer cell line (Januchowski et al., 2014) and polyploid cancer resistant to a combination of cisplatin, PTXl, and DTX (Wang, Lu, & Lan, 2017).

3.2.5 Other markers

As shown in Fig. 5, certain genes are highly represented in the models. These include the genes *DIP2C* (*disco interacting protein 2 homolog C*) (See Fig. 2B and C for detail about the marker locations), *PTPRN2* (*protein tyrosine phosphatase receptor type N2*), *SHANK2* (*SH3 and multiple ankyrin repeat domains 2*), and *TTC23* (*tetratricopeptide repeat domain 23*) for the PTX model (Fig. 5A), and genes *DIP2C*, *NRG2* (*neuregulin 2*), *SHANK2*, and *RAB34* (*RAB34, member RAS oncogene family*) for the DTX model (Fig. 5B). Our speculation of how the abovementioned genes influence taxane activity is not as highly supported in the literature as the apoptosis and mitosis-related genes (discussed). An earlier report has demonstrated that aberrant expression of PTPRN2 can lead to apoptosis resistance in cancer cells (Sorokin et al., 2015). Mutations in the gene *SHANK2* (a molecular scaffold protein) have been shown to impair apoptosis (Unsicker et al., 2021). RAB34, a Golgi-bound GTPase, regulates breast cancer cell adhesion, migration, and invasion (Sun et al., 2018). NRG2 has been shown to play a role in glioma cell migration (Zhao, Yi, Ou, & Qiao, 2021). A report indicates that the knockout of the gene *DIP2C* triggered EMT in cancer cells

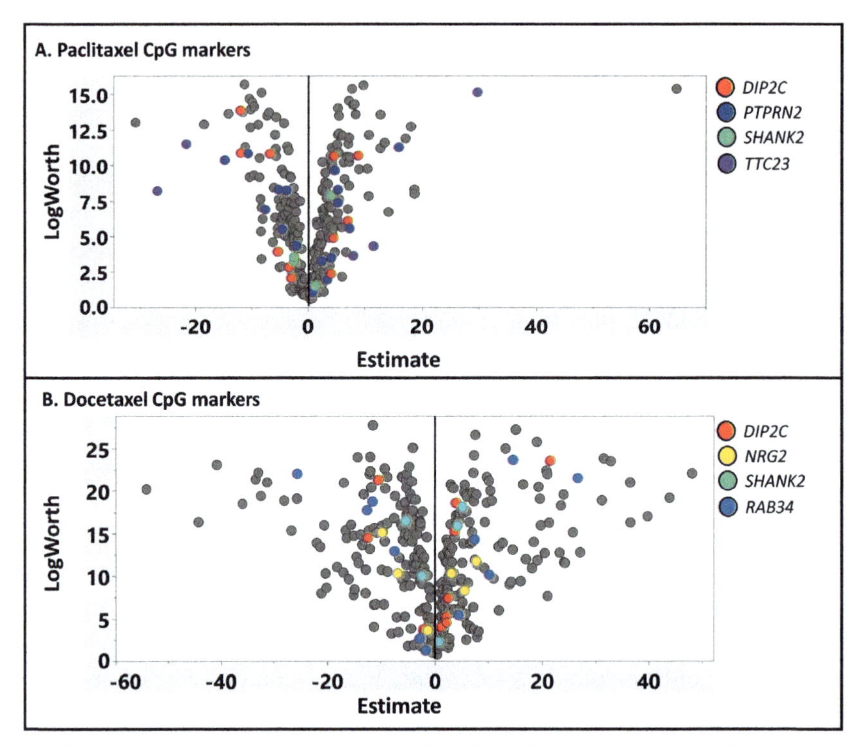

Fig. 5 Volcano plots (Estimate v. LogWorth values) for CpG markers included in the model PTX-287-CpG-PanOnc (A) and DTX-342-CpG-PanOnc (B) Marked in colors are the CpG sites highly represented genes *DIP2C, PTPRN2, SHANK2,* and *TTC23* for the paclitaxel model, and the genes *DIP2C, NRG2, SHANK2,* and *RAB34* for the docetaxel model.

(Larsson et al., 2017). There are few biological studies concerning the gene *TTC23*, although its structural motif (tetratricopeptide repeat) is found in many proteins involved in protein-protein interactions (Perez-Riba & Itzhaki, 2019). Also represented in the DTX predictive model is VIM (vimentin), a type III intermediate filament protein that is a well-known EMT marker (Satelli & Li, 2011). The phosphatase DUSP7 (dual specificity phosphatase 7) can regulate ERK2 activity toward proper chromosome alignment during cell division and thus may be necessary for mitosis (Guo et al., 2021). A recent study has demonstrated that DUSP7 can mediate paclitaxel resistance (Yang et al., 2020). The gene *GPC6* (glypican 6) codes for a proteoglycan which can play a role in cell growth and division. Moreover, GPC6 was upregulated in the PTX-resistant ovarian cancer cell line (Januchowski, Zawierucha, Rucinski, Nowicki, & Zabel, 2014). CCR6 (C-C motif chemokine receptor 6) has been similarly implicated in PTX resistance in ovarian cancer (Chen et al., 2021). The inhibition of TP53RK (TP53 regulating kinase) has been shown to augment taxane activity (Peterson et al., 2010). The authors described TP53RK expression as marker of molecular marker for response to anti-mitotic agents. The gene *ZBTB2* (*zinc finger and BTB domain containing 2*), marked as an outlier in Fig. 3A, codes for a transcription factor, whose inhibition has been demonstrated to inhibit proliferation and induce apoptosis (Yang et al., 2019).

3.3 Paclitaxel and docetaxel predictive models based on mRNA expression and mutation markers

It is also possible to generate PTX and DTX activity predictive models with transcriptional and mutational data as input variables. This was accomplished by first integrating the PTX and DTX pharmacological data with genome-wide expression and mutational data for the cancer cell lines. Forty-six cancer lines match PTX viability, mRNA, and mutational data. On the other hand, DTX viability, mRNA, and mutational data were integrated for 531 cell lines. Similar to constructing CpG-based models (See Section 3.1), we had to identify the short list of mRNA and mutation markers that could be entered as initial inputs for linear regression model calculations. First, the R_m values (correlation coefficient) between drug viability and genome-wide mRNA expression were calculated, followed by ranking the genes according to absolute R_m values. Then, we identified the genes with at least a 5% mutation rate in the CCLE mutational dataset (which totals 1523 genes). The mutation class (1 = mutant, 0 = wild type) was temporarily classified as numeric values to calculate absolute R_{mut} values between viability and mutation. The top 500 mRNAs (based on absolute R_m) and the top 100 mutations (based on absolute R_{mut}) markers were initially used as input variables in the modeling. Just like using the CpG approach, manual inspection improved the models. An additional consideration was that the number of markers in the models should be roughly similar to those used in CpG models. Fig. 6A illustrates the PTX predictive models PTX-290-mRNA-mut-PanOnc (consists of 277 mRNA and 13 mutational markers) and PTX-181-mRNA-mut-PanOnc (171 mRNA and 10 mutational markers) exhibited R^2 values of 0.83 and 0.70, respectively. On the other hand, the DTX predictive models (Fig. 6B) DTX-236-mRNA-mut-PanOnc (200 mRNA and 36 mut) and DTX-181-mRNA-mut-PanOnc (157 mRNA and 24 mutational markers) models resulted in R^2 values of 0.82 and 0.77 respectively. The resulting regression equations for the PTX-290-mRNA-mut-PanOnc and DTX-236-mRNA-mut-PanOnc are shown in

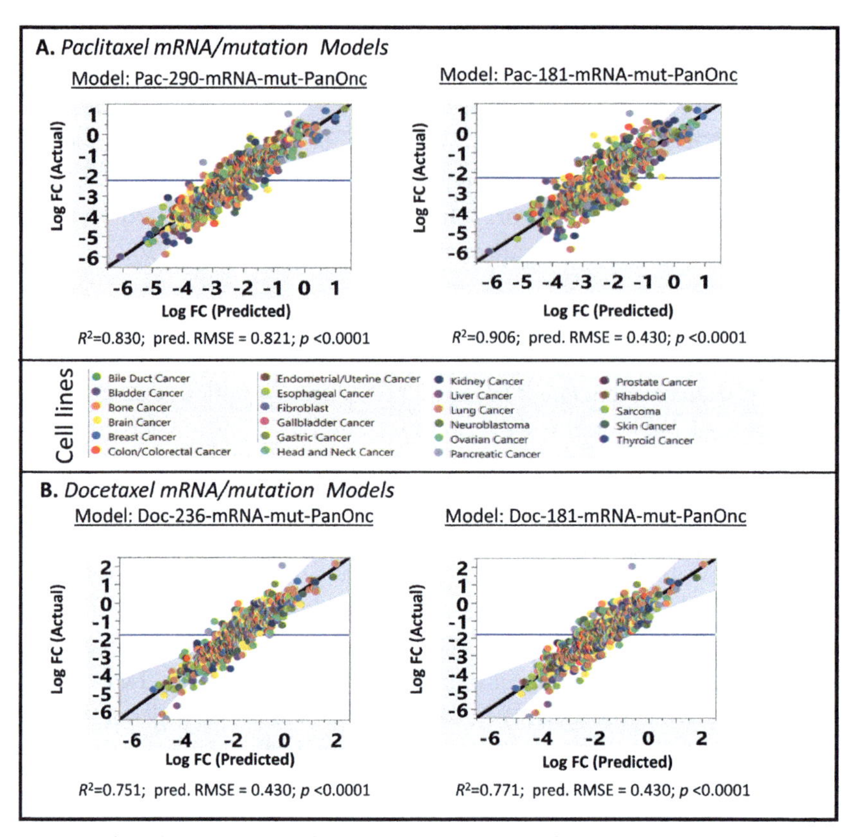

Fig. 6 Predicted vs. Actual Log-fold change viability *(V)* of cell lines that are treated with Paclitaxel (A) or Docetaxel (B). The predicted Log-fold change viability (V_p) for each drug is calculated using two possible linear regression models with multiple mRNAs and mutational markers as input variables. The numbers included in the name of each predictive model refer to the number of markers used in the model. The Paclitaxel and Docetaxel models were constructed using data for 546 and 532 cell lines, respectively. RMSE = root mean square error.

Eq. (3) and Eq. (4), respectively. In these equations, M_i and c_i refer to mRNA expression value and estimate for marker i. For the mutational gene markers (j), their contributions to the models are the constant values d_j and $-d_j$ for the absence and presence of mutation, respectively.

$$V_p(paclitaxel) = -40.374 + \sum_{i=1}^{277} c_i M_i + \sum_{j=1}^{13} \begin{pmatrix} wt \rightarrow d_j \\ mut \rightarrow -d_j \end{pmatrix} \tag{3}$$

$$V_p(docetaxel) = -9.307 + \sum_{i=1}^{200} c_i M_i + \sum_{j=1}^{36} \begin{pmatrix} wt \rightarrow d_j \\ mut \rightarrow -d_j \end{pmatrix} \tag{4}$$

3.4 Many genes represented in the mRNA/mutation-based predictive models have functionalities associated with mitosis, microtubules, epigenetic regulation, and apoptosis

As observed in the CpG-based predictive models, the mRNA/mutation-based models are represented by genes with apoptotic, mitotic, transporter, and epigenetic functionalities (Figs. 7 and 8 and Table 3). In the paclitaxel model PTX-290-mRNA-mut-PanOnc, the apoptosis-related genes (Fig. 7A) include the mRNA markers *AATF* (*apoptosis antagonizing transcription factor*) (Passananti, Floridi, & Fanciulli, 2007), *GADD45G* (*growth arrest and DNA damage inducible gamma*) (Salvador, Brown-Clay, & Fornace Jr., 2013), *MOAP1* (*modulator of apoptosis 1*) (Wu et al., 2015), *TP63* (*tumor protein p63*) (Wang, Lu, Wang, Li, & Chen, 2022), and *TRAF3* (*TNF receptor-associated factor 3*) (Baker & Reddy, 1996). The microtubule/mitosis-related

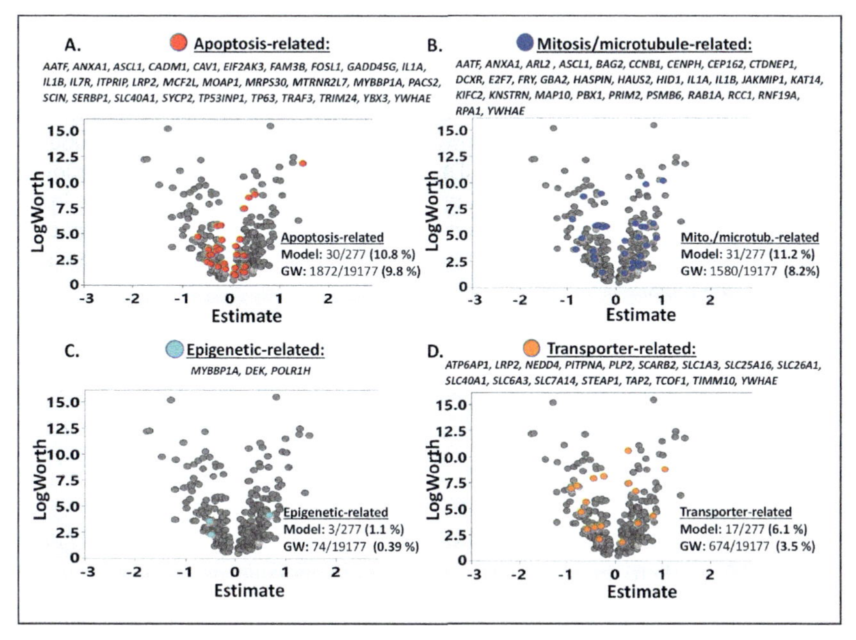

Fig. 7 Volcano plots (Estimate v. LogWorth values) for mRNA markers included in the model PTX-290-mRNA-mut-PanOnc. Marked in colors are the markers whose genes are associated with Gene Ontology (GO) terms "apoptosis" (A), "mitosis," or "microtubules" (B), "epigenetics" (C), and "transporter" (D). Each panel also indicates the proportion of genes associated with each corresponding GO term(s) that are either represented in the *Model* or the entire CCLE RNASeq dataset (i.e., genome-wide or *GW*). In the predictive models, mutation markers are not associated with estimates (See Eq. 3) and are not represented in the plots.

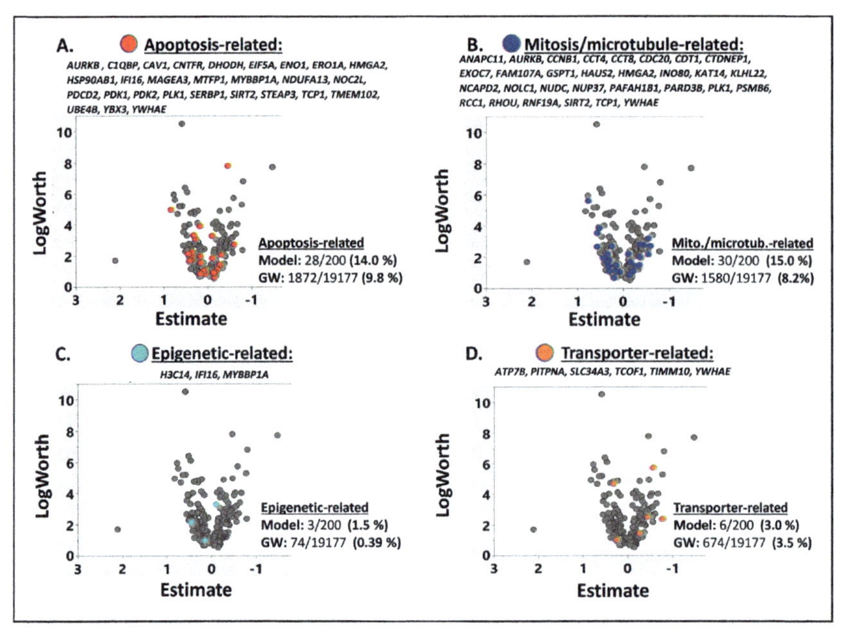

Fig. 8 Volcano plots (Estimate v. LogWorth values) for mRNA markers included in the model DTX-236-mRNA-mut-PanOnc. Marked in colors are the markers whose genes are associated with Gene Ontology (GO) terms "apoptosis" (A), "mitosis," or "microtubules" (B), "epigenetics" (C), and "transporter" (D). Each panel also indicates the proportion of genes associated with each corresponding GO term(s) that are either represented in the *Model* or the entire CCLE RNASeq dataset (i.e., genome-wide or *GW*). In the predictive models, mutation markers are not associated with estimates (See Eq. 4) and are not represented in the plots.

genes (Fig. 7B) include *CCNB1* (*cyclin B1*) (Runion & Pipa, 1970), *CENPH* (*centromere protein H*) (Amor, Kalitsis, Sumer, & Choo, 2004), *CEP162* (*centrosomal protein 162*) (Wang et al., 2013), *FRY* (*FRY microtubule-binding protein*) (Nagai, Ikeda, Chiba, Kanno, & Mizuno, 2013), *JAKMIP1* (*Janus kinase and microtubule interacting protein 1*) (Steindler et al., 2004), KIFC2 (*kinesin family member C2*) (Galletti et al., 2020), *MAP10* (*microtubule-associated protein 10*) (Fong et al., 2013), *RCC1* (*regulator of chromosome condensation 1*) (Zhang, Furuta, Arnaoutov, & Dasso, 2018), and *RPA1* (*replication protein A1*) (Qin et al., 2020). The transporter-coding genes (Fig. 7D) include the mRNA markers solute carrier family members *SLC1A3*, *SLC25A16*, *SLC26A1*, *SLC40A1*, *SLC6A3*, and *SLC7A14*, *SCARB2* (*scavenger receptor class B member 2*), and *TAP2* (*transporter 2, ATP binding cassette subfamily B member*) (Golin & Ambudkar, 2015). Also classified as a transporter gene is the

Table 3 A partial list of the mRNA and mutation markers included in the predictive models for activities of paclitaxel (PTX-290-mRNA-mut-PanOnc) and docetaxel (DTX-236-mRNA-mut-PanOnc) in cancer cell lines.

Gene ID	Type	NCBI ID	Chr	Karyotype band	NCBI gene description	Log Worth	Estimate (Abs)	Estimate (sign)	GO Tag	Drug
MTRNR2L7	mRNA	100288485	10	p11.21	MT-RNR2 like 7	1.203	76.6673	(+)	YNNN	PTX
BSND	mRNA	7809	1	p32.3	barttin CLCNK type accessory beta subunit	9.8352	9.0932	(−)	NNNN	PTX
H2BW2	mRNA	286,436	X	q22.2	H2B histone family member M	10.0691	5.9521	(+)	NNNN	PTX
PLEKHD1	mRNA	400224	14	q24.1	pleckstrin homology and coiled-coil domain containing D1	12.1956	1.748	(−)	NNNN	PTX
FAM98B	mRNA	283742	15	q14	family with sequence similarity 98 member B	12.2377	1.6937	(−)	NNNN	PTX
SERBP1	mRNA	26135	1	p31.3	SERPINE1 mRNA binding protein 1	11.8009	1.4726	(+)	YNNN	PTX
NOL7	mRNA	51406	6	p23	nucleolar protein 7	9.749	1.4536	(−)	NNNN	PTX
LRFN2	mRNA	57497	6	p21.1	leucine rich repeat and fibronectin type III domain containing 2	6.2767	1.3886	(+)	NNNN	PTX
FSCN2	mRNA	25794	17	q25.3	fascin actin-bundling protein 2, retinal	12.4454	1.2879	(+)	NNNN	PTX

Continued

Table 3 A partial list of the mRNA and mutation markers included in the predictive models for activities of paclitaxel (PTX-290-mRNA-mut-PanOnc) and docetaxel (DTX-236-mRNA-mut-PanOnc) in cancer cell lines.—cont'd

Gene ID	Type	NCBI ID	Chr	Karyotype band	NCBI gene description	Log Worth	Estimate (Abs)	Estimate (sign)	GO Tag	Drug
CCDC28A	mRNA	25901	6	q24.1	coiled-coil domain containing 28A	15.1876	1.2865	(−)	NNNN	PTX
HMGN4	mRNA	10473	6	p22.2	high mobility group nucleosomal binding domain 4	11.9093	1.2744	(+)	NNNN	PTX
GEMIN4	mRNA	50628	17	p13.3	*gem* nuclear organelle associated protein 4	6.5838	1.2083	(−)	NNNN	PTX
UNC79	mRNA	57578	14	q32.12	unc-79 homolog, NALCN channel complex subunit	3.6911	1.203	(−)	NNNN	PTX
SAMD3	mRNA	154075	6	q23.1	sterile alpha motif domain containing 3	9.4761	1.1928	(−)	NNNN	PTX
TCOF1	mRNA	6949	5	q32	treacle ribosome biogenesis factor 1	8.7548	1.062	(+)	NNYN	PTX
ZNF701	mRNA	55762	19	q13.41	zinc finger protein 701	11.12	1.0484	(−)	NNNN	PTX
EIF3I	mRNA	8668	1	p35.2	eukaryotic translation initiation factor 3 subunit I	6.7599	1.037	(−)	NNNN	PTX
DCXR	mRNA	51181	17	q25.3	dicarbonyl and L-xylulose reductase	10.1604	1.0204	(+)	NYNN	PTX

ARL8A	mRNA	127829	1	q32.1	ADP ribosylation factor like GTPase 8A	11.487	0.9999	(+)	NNNN	PTX
ASH1L	mRNA	55870	1	q22	ASH1 like histone lysine methyltransferase	5.3656	0.9946	(–)	NNNN	PTX
UTP4	mRNA	84916	16	q22.1	UTP4 small subunit processome component	4.7943	0.99	(–)	NNNN	PTX
PGAM1	mRNA	5223	10	q24.1	phosphoglycerate mutase 1	8.8485	0.9733	(–)	NNNN	PTX
PRRC1	mRNA	133,619	5	q23.2	proline rich coiled-coil 1	6.8304	0.9564	(–)	NNNN	PTX
NUDT9	mRNA	53343	4	q22.1	nudix hydrolase 9	10.2543	0.9432	(+)	NNNN	PTX
PPIL1	mRNA	51645	6	p21.2	peptidylprolyl isomerase like 1	5.6712	0.9227	(–)	NNNN	PTX
MRPS22	mRNA	56945	3	q23	mitochondrial ribosomal protein S22	7.0506	0.9213	(+)	NNNN	PTX
COX10	mRNA	1352	17	p12	cytochrome *c* oxidase assembly factor heme A: farnesyltransferase COX10	7.0878	0.9	(–)	NNNN	PTX
TIMM10	mRNA	26519	11	q12.1	translocase of inner mitochondrial membrane 10	6.8821	0.8919	(–)	NNYN	PTX
PSMB6	mRNA	5694	17	p13.2	proteasome subunit beta 6	4.6758	0.8842	(+)	NYNN	PTX

Continued

Table 3 A partial list of the mRNA and mutation markers included in the predictive models for activities of paclitaxel (PTX-290-mRNA-mut-PanOnc) and docetaxel (DTX-236-mRNA-mut-PanOnc) in cancer cell lines.—cont'd

Gene ID	Type	NCBI ID	Chr	Karyotype band	NCBI gene description	Log Worth	Estimate (Abs)	Estimate (sign)	GO Tag	Drug
KNSTRN	mRNA	90417	15	q15.1	kinetochore localized astrin (SPAG5) binding protein	6.4478	0.8813	(−)	NYNN	PTX
PAK1IP1	mRNA	55003	6	p24.2	PAK1 interacting protein 1	4.9897	0.8688	(+)	NNNN	PTX
EIF2AK4	mRNA	440275	15	q15.1	eukaryotic translation initiation factor 2 alpha kinase 4	7.5037	0.8653	(+)	NNNN	PTX
CYTH2	mRNA	9266	19	q13.33	cytohesin 2	6.8313	0.8618	(+)	NNNN	PTX
SCO1	mRNA	6341	17	p13.1	SCO cytochrome *c* oxidase assembly protein 1	5.7301	0.8539	(+)	NNNN	PTX
VN1R1	mRNA	57191	19	q13.43	vomeronasal 1 receptor 1	3.5837	0.8533	(−)	NNNN	PTX
EXOC2	mRNA	55770	6	p25.3	exocyst complex component 2	7.709	0.8466	(−)	NNNN	PTX
TARS1	mRNA	6897	5	p13.3	threonyl-tRNA synthetase	9.0232	0.8438	(−)	NNNN	PTX
MRPL18	mRNA	29074	6	q25.3	mitochondrial ribosomal protein L18	3.7039	0.8311	(+)	NNNN	PTX
LRRC37A3	mRNA	374819	17	q24.1	leucine rich repeat containing 37 member A3	15.4672	0.8188	(+)	NNNN	PTX

RCC1	mRNA	1104	1	p35.3	regulator of chromosome condensation 1	6.047	0.8146	(+)	NYNN	PTX
ODF3	mRNA	113746	11	p15.5	outer dense fiber of sperm tails 3	1.9215	5.8143	(+)	NNNN	DTX
BSND	mRNA	7809	1	p32.3	barttin CLCNK type accessory beta subunit	1.6856	2.1157	(+)	NNNN	DTX
CRYBB3	mRNA	1417	22	q11.23	crystallin beta B3	7.7004	1.473	(−)	NNNN	DTX
C1QBP	mRNA	708	17	p13.2	complement C1q binding protein	4.9354	0.8245	(+)	YNNN	DTX
FAM98B	mRNA	283742	15	q14	family with sequence similarity 98 member B	6.7971	0.8089	(−)	NNNN	DTX
SLC34A3	mRNA	142,680	9	q34.3	solute carrier family 34 member 3	2.3557	0.794	(−)	NNYN	DTX
USP10	mRNA	9100	16	q24.1	ubiquitin specific peptidase 10	5.2825	0.7791	(−)	NNNN	DTX
WRAP53	mRNA	55135	17	p13.1	WD repeat containing antisense to TP53	5.9458	0.7702	(+)	NNNN	DTX
EXOC7	mRNA	23265	17	q25.1	exocyst complex component 7	5.5923	0.7439	(+)	NYNN	DTX
UTP4	mRNA	84916	16	q22.1	UTP4 small subunit processome component	4.6813	0.6816	(+)	NNNN	DTX

Continued

Table 3 A partial list of the mRNA and mutation markers included in the predictive models for activities of paclitaxel (PTX-290-mRNA-mut-PanOnc) and docetaxel (DTX-236-mRNA-mut-PanOnc) in cancer cell lines.—cont'd

Gene ID	Type	NCBI ID	Chr	Karyotype band	NCBI gene description	Log Worth	Estimate (Abs)	Estimate (sign)	GO Tag	Drug
EIF3I	mRNA	8668	1	p35.2	eukaryotic translation initiation factor 3 subunit I	4.0164	0.6759	(−)	NNNN	DTX
GEMIN4	mRNA	50628	17	p13.3	*gem* nuclear organelle associated protein 4	3.3958	0.6433	(−)	NNNN	DTX
RAB5B	mRNA	5869	12	q13.2	RAB5B, member RAS oncogene family	5.1783	0.6398	(+)	NNNN	DTX
TCP1	mRNA	6950	6	q25.3	t-complex 1	2.7105	0.6148	(−)	YYNN	DTX
TIMM10	mRNA	26519	11	q12.1	translocase of inner mitochondrial membrane 10	5.7208	0.5884	(−)	NNYN	DTX
DNAJB9	mRNA	4189	7	q31.1	DnaJ heat shock protein family (Hsp40) member B9	10.4993	0.5827	(+)	NNNN	DTX
WDR36	mRNA	134430	5	q22.1	WD repeat domain 36	5.1951	0.5769	(+)	NNNN	DTX
NOLC1	mRNA	9221	10	q24.32	nucleolar and coiled-body phosphoprotein 1	3.1843	0.5683	(−)	NYNN	DTX
CTDNEP1	mRNA	23399	17	p13.1	CTD nuclear envelope phosphatase 1	2.1708	0.5679	(−)	NYNN	DTX
DHX33	mRNA	56919	17	p13.2	DEAH-box helicase 33	2.5383	0.561	(−)	NNNN	DTX

INO80	mRNA	54617	15	q15.1	INO80 complex ATPase subunit	2.6539	0.553	(+)	NYNN	DTX
NUDC	mRNA	10726	1	p36.11	nuclear distribution C, dynein complex regulator	3.5242	0.5485	(+)	NYNN	DTX
ELOA	mRNA	6924	1	p36.11	elongin A	4.0275	0.5364	(−)	NNNN	DTX
TIPIN	mRNA	54962	15	q22.31	TIMELESS interacting protein	3.9006	0.5277	(+)	NNNN	DTX
VTA1	mRNA	51534	6	q24.1	vesicle trafficking 1	3.1044	0.5256	(−)	NNNN	DTX
TLCD3A	mRNA	79850	17	p13.3	TLC domain containing 3A	6.3904	0.51	(+)	NNNN	DTX
TPI1	mRNA	7167	12	p13.31	triosephosphate isomerase 1	3.8646	0.5089	(−)	NNNN	DTX
CCT8	mRNA	10694	21	q21.3	chaperonin containing TCP1 subunit 8	3.8445	0.5037	(+)	NYNN	DTX
PRPF8	mRNA	10594	17	p13.3	pre-mRNA processing factor 8	2.2662	0.5032	(+)	NNNN	DTX
NYNRIN	Mut	57523	14	q12	NYN domain and retroviral integrase containing	8.5276	0.5009	Wt (−), Mut(+)	NNNN	DTX
CACNA2D1	Mut	781	7	q21.11	calcium voltage-gated channel auxiliary subunit alpha2delta 1	5.9926	0.4913	Wt (−), Mut(+)	NNNN	DTX
LTV1	mRNA	84946	6	q24.2	LTV1 ribosome biogenesis factor	2.3906	0.4782	(+)	NNNN	DTX

Continued

Table 3 A partial list of the mRNA and mutation markers included in the predictive models for activities of paclitaxel (PTX-290-mRNA-mut-PanOnc) and docetaxel (DTX-236-mRNA-mut-PanOnc) in cancer cell lines.—cont'd

Gene ID	Type	NCBI ID	Chr	Karyotype band	NCBI gene description	Log Worth	Estimate (Abs)	Estimate (sign)	GO Tag	Drug
ZNHIT6	mRNA	54680	1	p22.3	zinc finger HIT-type containing 6	4.9519	0.4759	(−)	NNNN	DTX
GSPT1	mRNA	2935	16	p13.13	G1 to S phase transition 1	2.7644	0.4691	(−)	NYNN	DTX
GTF2H3	mRNA	2967	12	q24.31	general transcription factor IIH subunit 3	3.294	0.4679	(+)	NNNN	DTX
MTFP1	mRNA	51537	22	q12.2	mitochondrial fission process 1	7.783	0.4654	(−)	YNNN	DTX
ZNF613	mRNA	79898	19	q13.41	zinc finger protein 613	6.0968	0.4621	(+)	NNNN	DTX
PITPNA	mRNA	5306	17	p13.3	phosphatidylinositol transfer protein alpha	2.485	0.4567	(−)	NNYN	DTX
WDR33	Mut	55339	2	q14.3	WD repeat domain 33	4.3308	0.4523	Wt (+), Mut(−)	NNNN	DTX
RUVBL2	mRNA	10856	19	q13.33	RuvB like AAA ATPase 2	3.0049	0.4458	(−)	NNNN	DTX

Shown are markers with the 40 highest absolute estimate values (for each model). The GO (Gene Ontology) tag refers to "*apoptosis*," "*microtubule or mitosis*," "*transporter*," and "*epigenetic*" tags in order. For example, the tag "YNNN" indicates that the marker gene is associated with "apoptosis" and not with the other three terms.

mutational marker *CFTR* (CF *transmembrane conductance regulator*) (Keppler, 2011) which is mutated in 7% of the cell lines and contributes to V_p (paclitaxel) the values of −0.31 and 0.31 if mutated and wild type, respectively. That can be interpreted as mutated *CFTR* positively contributing to PTX efficacy. Previously identified as one of the epigenetic-related markers in the PTX CpG model, *HDAC4* (Stronach et al., 2011; Yu et al., 2013) is also part of the PTX mRNA/mutation model as a mutational marker. If mutated, *HDAC4* (6% among CCLE cell lines) adds positively 0.37 V_p (paclitaxel), thus contributing to PTX resistance. Other mutational markers contributing to the PTX model are *PCDHA11* (*protocadherin alpha 11*), and LRRIQ3 (*leucine-rich repeats and IQ motif containing 3*). Both genes are mutated at 5% rate and contribute to V_p (paclitaxel), the values of −0.68 and −0.59, respectively, if mutated. There are very few publications regarding those genes.

Among the apoptosis-related genes (Fig. 8A) represented in the DTX-236-mRNA-mut-PanOnc model are the *AURKB* (*aurora kinase B*) (as mRNA) (Yoon et al., 2012), *CREBBP* (*CREB binding protein*) (as mutation) (Bordonaro & Lazarova, 2015), *MTFP1* (*mitochondrial fission process 1*) (as mRNA) (Morita et al., 2017), *PDCD2* (*programmed cell death 2*) (as mRNA) (Zhang et al., 2015), *PLK1* (*polo-like kinase 1*) (as mRNA) (Iliaki, Beyaert, & Afonina, 2021), and *EPB41L3* (*erythrocyte membrane protein band 4.1 like 3*) (as mutation) (Tuerxun et al., 2022). The mutation rate of *CREBBP* is relatively high (at 12%) and contributes 0.23 toward DTX resistance, while *EPB41L3* is mutated at a 7% rate and contributes 0.35 toward DTX efficacy. Also covered in the DTX model are genes associated with mitotic or microtubule functionalities (Fig. 8B). These include the mRNA markers *ANAPC11* (*anaphase-promoting complex subunit 11*) (Schrock et al., 2020), *CDC20* (*cell division cycle 20*) (Greil, Engelhardt, & Wasch, 2022), *CDT1* (*chromatin licensing and DNA replication factor 1*) (Fujita, 2006), *CENPF* (*centromere protein F*) (Amor et al., 2004), *PLK1* (*polo-like kinase 1*), and *RCC1* (*regulator of chromosome condensation 1*) (Zhang et al., 2018). RADIL (Rap associating with DIL domain) (Ahmed et al., 2012) is a microtubule component and contributes a value of 0.21 toward DTX efficacy. Other mutational markers that are part of the DTX predictive model are *SLC34A3 (6% mutation; subtracts 0.22 if mutated), NYNRIN* (*NYN domain and retroviral integrase containing*) (7% mutation, adds 0.5 if mutated), and *CACNA2D1* (*calcium voltage-gated channel auxiliary subunit alpha2delta 1*) (5% mutation rate, adds 0.49 if mutated). In melanoma cells, silencing of NYNRIN led to increased invasive activity, which is expected given its role as a mediator of the metastasis suppressor function of NME1 and NME2 (Leonard et al., 2021). Silencing CACNA2D1 seemingly has the opposite effect, as the invasion of gastric cancer cells was suppressed when the gene was knocked out (Inoue et al., 2022). Another mutational marker in the DTX model is the gene *WDR33* (*WD repeat domain 33*) (5% mutation rate; subtracts −0.45 if mutated). There are only few publications regarding this gene's potential role in cancer.

According to our "Reactome" analysis, RNA processing pathways are enriched among the genes represented in the PTX-290-mRNA-mut-PanOnc model. Among these are the pathways "*RNA Polymerase III Transcription Initiation From Type 3 Promoter*" and "*rRNA modification in the nucleus and cytosol*" (See Table 4). However, the apoptotic pathway "*TP53*

Table 4 The top Reactome pathways associated with genes represented in the mRNA- and mutation-based paclitaxel (PTX-290-mRNA-mut-PanOnc) and docetaxel (DTX-236-mRNA-mut-PanOnc) predictive models.

Drug	Reactome ID	Pathway name	Entities *P* value	Entities FDR	Submitted entities found
DTX	R-HSA-69278	Cell Cycle, Mitotic	4.67E-08	4.15E-05	*YWHAE; HSP90AB1; AURKB; H2AC18; ANAPC11; PSMB6; CDC20; CCNB1; H3C14; RCC1; RFC5; RANBP2; CDT1; NUDC; H2BC6; PLK1; HAUS2; SIRT2; H2BC18; NUP93; CTDNEP1; CENPF; NCAPD2; NUP37; PAFAH1B1*
DTX	R-HSA-1640170	Cell Cycle	1.80E-07	8.00E-05	*YWHAE; HSP90AB1; AURKB; H2AC18; ANAPC11; PSMB6; CDC20; CCNB1; H3C14; RUVBL2; TMEM107; RCC1; RFC5; RANBP2; CDT1; NUDC; H2BC6; PLK1; HAUS2; SIRT2; H2BC18; NUP93; CTDNEP1; CENPF; NCAPD2; WRAP53; NUP37; PAFAH1B1*

Table 4 The top Reactome pathways associated with genes represented in the mRNA- and mutation-based paclitaxel (PTX-290-mRNA-mut-PanOnc) and docetaxel (DTX-236-mRNA-mut-PanOnc) predictive models.—cont'd

Drug	Reactome ID	Pathway name	Entities *P* value	Entities FDR	Submitted entities found
DTX	R-HSA-68886	M Phase	2.49E-06	7.38E-04	*YWHAE; RANBP2; NUDC; H2BC6; PLK1; AURKB; HAUS2; SIRT2; H2BC18; ANAPC11; H2AC18; CTDNEP1; PSMB6; NUP93; CDC20; CENPF; CCNB1; H3C14; RCC1; NCAPD2; NUP37; PAFAH1B1*
DTX	R-HSA-156711	Polo-like kinase mediated events	4.63E-06	1.03E-03	*CCNB1; CENPF; PLK1*
DTX	R-HSA-69620	Cell Cycle Checkpoints	1.58E-05	2.80E-03	*YWHAE; RFC5; RANBP2; NUDC; H2BC6; PLK1; AURKB; H2BC18; ANAPC11; PSMB6; CDC20; CENPF; CCNB1; NUP37; PAFAH1B1*
DTX	R-HSA-3214815	HDACs deacetylate histones	2.28E-05	3.37E-03	*H3C14; H2BC6; GPS2; CHD4; H2BC18; H2AC18*
DTX	R-HSA-68875	Mitotic Prophase	4.16E-05	5.29E-03	*CTDNEP1; NUP93; RANBP2; CCNB1; H3C14; H2BC6; PLK1; H2BC18; NUP37; H2AC18*
DTX	R-HSA-2299718	Condensation of Prophase Chromosomes	7.30E-05	8.10E-03	*CCNB1; H3C14; H2BC6; PLK1; H2BC18; H2AC18*
DTX	R-HSA-5625740	RHO GTPases activate PKNs	1.19E-04	1.13E-02	*YWHAE; H3C14; H2BC6; MYH10; H2BC18; PDK1; H2AC18*
DTX	R-HSA-5578749	Transcriptional regulation by small RNAs	1.29E-04	1.13E-02	*NUP93; RANBP2; H3C14; H2BC6; H2BC18; NUP37; H2AC18*

Continued

Table 4 The top Reactome pathways associated with genes represented in the mRNA- and mutation-based paclitaxel (PTX-290-mRNA-mut-PanOnc) and docetaxel (DTX-236-mRNA-mut-PanOnc) predictive models.—cont'd

Drug	Reactome ID	Pathway name	Entities *P* value	Entities FDR	Submitted entities found
DTX	R-HSA-157579	Telomere Maintenance	1.88E-04	1.26E-02	*RFC5; H2BC6; RUVBL2; TMEM107; H2BC18; WRAP53; H2AC18*
DTX	R-HSA-3214847	HATs acetylate histones	2.00E-04	1.26E-02	*CREBBP; H3C14; H2BC6; RUVBL2; KAT14; H2BC18; H2AC18*
DTX	R-HSA-68882	Mitotic Anaphase	2.02E-04	1.26E-02	*RANBP2; NUDC; PLK1; AURKB; SIRT2; ANAPC11; PSMB6; NUP93; CDC20; CENPF; CCNB1; RCC1; NUP37; PAFAH1B1*
DTX	R-HSA-2555396	Mitotic Metaphase and Anaphase	2.10E-04	1.26E-02	*RANBP2; NUDC; PLK1; AURKB; SIRT2; ANAPC11; PSMB6; NUP93; CDC20; CENPF; CCNB1; RCC1; NUP37; PAFAH1B1*
DTX	R-HSA-69618	Mitotic Spindle Checkpoint	2.14E-04	1.26E-02	*CDC20; RANBP2; NUDC; CENPF; PLK1; AURKB; PAFAH1B1; NUP37; ANAPC11*
DTX	R-HSA-427389	ERCC6 (CSB) and EHMT2 (G9a) positively regulate rRNA expression	2.63E-04	1.44E-02	*H3C14; H2BC6; CHD4; H2BC18; H2AC18*
DTX	R-HSA-5619507	Activation of HOX genes during differentiation	2.95E-04	1.44E-02	*CREBBP; H3C14; H2BC6; HOXA3; H2BC18; H2AC18*
DTX	R-HSA-5617472	Activation of anterior HOX genes in hindbrain development during early embryogenesis	2.95E-04	1.44E-02	*CREBBP; H3C14; H2BC6; HOXA3; H2BC18; H2AC18*

Table 4 The top Reactome pathways associated with genes represented in the mRNA- and mutation-based paclitaxel (PTX-290-mRNA-mut-PanOnc) and docetaxel (DTX-236-mRNA-mut-PanOnc) predictive models.—cont'd

Drug	Reactome ID	Pathway name	Entities *P* value	Entities FDR	Submitted entities found
DTX	R-HSA-195258	RHO GTPase Effectors	3.27E-04	1.46E-02	*YWHAE; RANBP2; NUDC; H2BC6; PLK1; AURKB; H2BC18; H2AC18; CDC20; CENPF; H3C14; MYH10; PDK1; NUP37; PAFAH1B1*
DTX	R-HSA-141424	Amplification of signal from the kinetochores	3.48E-04	1.46E-02	*CDC20; RANBP2; NUDC; CENPF; PLK1; AURKB; PAFAH1B1; NUP37*
DTX	R-HSA-141444	Amplification of signal from unattached kinetochores via a MAD2 inhibitory signal	3.48E-04	1.46E-02	*CDC20; RANBP2; NUDC; CENPF; PLK1; AURKB; PAFAH1B1; NUP37*
DTX	R-HSA-73728	RNA Polymerase I Promoter Opening	4.15E-04	1.66E-02	*H3C14; H2BC6; H2BC18; H2AC18*
DTX	R-HSA-176417	Phosphorylation of Emi1	4.40E-04	1.67E-02	*CDC20; CCNB1; PLK1*
DTX	R-HSA-68877	Mitotic Prometaphase	5.15E-04	1.83E-02	*YWHAE; CDC20; RANBP2; NUDC; CENPF; CCNB1; PLK1; NCAPD2; AURKB; HAUS2; NUP37; PAFAH1B1*
DTX	R-HSA-69275	G2/M Transition	5.37E-04	1.83E-02	*YWHAE; PSMB6; CCNB1; CENPF; HSP90AB1; PLK1; AURKB; HAUS2; PAFAH1B1*
DTX	R-HSA-68616	Assembly of the ORC complex at the origin of replication	8.58E-04	2.38E-02	*H3C14; H2BC6; H2BC18; H2AC18*
DTX	R-HSA-73886	Chromosome Maintenance	9.07E-04	2.38E-02	*RFC5; H2BC6; RUVBL2; TMEM107; H2BC18; WRAP53; H2AC18*

Continued

Table 4 The top Reactome pathways associated with genes represented in the mRNA- and mutation-based paclitaxel (PTX-290-mRNA-mut-PanOnc) and docetaxel (DTX-236-mRNA-mut-PanOnc) predictive models.—cont'd

Drug	Reactome ID	Pathway name	Entities *P* value	Entities FDR	Submitted entities found
DTX	R-HSA-73772	RNA Polymerase I Promoter Escape	9.15E-04	2.38E-02	*H3C14; H2BC6; GTF2H3; H2BC18; H2AC18*
DTX	R-HSA-73854	RNA Polymerase I Promoter Clearance	9.77E-04	2.48E-02	*H3C14; H2BC6; CHD4; GTF2H3; H2BC18; H2AC18*
DTX	R-HSA-5250924	B-WICH complex positively regulates rRNA expression	9.94E-04	2.48E-02	*MYBBP1A; H3C14; H2BC6; H2BC18; H2AC18*
PTX	R-HSA-5660668	CLEC7A/ inflammasome pathway	7.83E-04	6.02E-01	*IL1B*
PTX	R-HSA-76071	RNA Polymerase III Transcription Initiation From Type 3 Promoter	3.59E-03	6.02E-01	*SNAPC1; POLR3B; POLR3G*
PTX	R-HSA-6790901	rRNA modification in the nucleus and cytosol	5.54E-03	6.02E-01	*WDR36; UTP4; CRY2; NAT10; UTP14A; WDR46*
PTX	R-HSA-76046	RNA Polymerase III Transcription Initiation	8.58E-03	6.02E-01	*SNAPC1; POLR3B; POLR3G*
PTX	R-HSA-73780	RNA Polymerase III Chain Elongation	8.79E-03	6.02E-01	*POLR3B; POLR3G*
PTX	R-HSA-72312	rRNA processing	8.97E-03	6.02E-01	*LTV1; WDR36; UTP4; NOB1; ELAC2; NIP7; CRY2; RPP40; NAT10; TFB1M; UTP14A; WDR46*
PTX	R-HSA-73980	RNA Polymerase III Transcription Termination	1.46E-02	6.02E-01	*POLR3B; POLR3G*
PTX	R-HSA-749476	RNA Polymerase III Abortive and Retractive Initiation	1.55E-02	6.02E-01	*SNAPC1; POLR3B; POLR3G*
PTX	R-HSA-74158	RNA Polymerase III Transcription	1.55E-02	6.02E-01	*SNAPC1; POLR3B; POLR3G*
PTX	R-HSA-8948216	Collagen chain trimerization	1.67E-02	6.02E-01	*COL24A1; COL7A1*

Table 4 The top Reactome pathways associated with genes represented in the mRNA- and mutation-based paclitaxel (PTX-290-mRNA-mut-PanOnc) and docetaxel (DTX-236-mRNA-mut-PanOnc) predictive models.—cont'd

Drug	Reactome ID	Pathway name	Entities *P* value	Entities FDR	Submitted entities found
PTX	R-HSA-210746	Regulation of gene expression in endocrine-committed (NEUROG3+) progenitor cells	1.70E-02	6.02E-01	*INSM1*
PTX	R-HSA-8868773	rRNA processing in the nucleus and cytosol	1.78E-02	6.02E-01	*LTV1; WDR36; UTP4; NOB1; NIP7; CRY2; RPP40; NAT10; UTP14A; WDR46*
PTX	R-HSA-6784531	tRNA processing in the nucleus	2.00E-02	6.02E-01	*CLP1; ELAC2; RPP40; CSTF2; FAM98B*
PTX	R-HSA-76066	RNA Polymerase III Transcription Initiation from Type 2 Promoter	2.22E-02	6.02E-01	*POLR3B; POLR3G*
PTX	R-HSA-76061	RNA Polymerase III Transcription Initiation from Type 1 Promoter	2.44E-02	6.02E-01	*POLR3B; POLR3G*
PTX	R-HSA-6791226	Major pathway of rRNA processing in the nucleolus and cytosol	2.51E-02	6.02E-01	*LTV1; WDR36; UTP4; NOB1; NIP7; CRY2; RPP40; UTP14A; WDR46*
PTX	R-HSA-448706	Interleukin-1 processing	2.90E-02	6.02E-01	*IL1A; IL1B*
PTX	R-HSA-445095	Interaction between L1 and Ankyrins	3.69E-02	6.02E-01	*SCN3A*
PTX	R-HSA-6803204	TP53 Regulates Transcription of Genes Involved in Cytochrome C Release	3.69E-02	6.02E-01	*TP53INP1; TP63*
PTX	R-HSA-5576892	Phase 0—rapid depolarization	3.97E-02	6.02E-01	*SCN3A*

Regulates Transcription of Genes Involved in Cytochrome C Release" was also included in the list owing to two-component genes (*TP53INP1, TP63*) being mapped in the pathway. The "Reactome" pathways over-represented in the DTX model are primarily related to mitosis, microtubules, and cell cycle.

These include the pathways: *"Cell Cycle," "Mitotic," "M Phase," "Mitotic Anaphase," "Mitotic Spindle Checkpoint," "G2/M Transition,"* and *"Assembly of the ORC complex at the origin of replication."*

3.5 Lung cancer-specific CpG-based models for taxane activity prediction

Given that both PTX and DTX are used in lung cancer treatment, it would be interesting to generate predictive models exclusive to the lung cancer cell lines. There are 88 lung cancer lines with matching PTX viability and methylation data, while 83 had matching DTX viability and methylation data. We then employed the same approach in the earlier models comprising all the cell lines. As illustrated in Fig. 9, we presented 74-CpG (PTX-74-CpG-LungCan) and 57-CpG (DTX-58-CpG-LungCan) models for PTX and DTX prediction, respectively. Both models resulted in an R^2 value of 0.98. The marker overlap between the pan-oncology and lung cancer models consists of just a few CpG sites. For the PTX models, these common markers are cg16549809 (within a non-coding region), cg22686253 (*C19orf25*), cg10399946 (*RPS18; ribosomal protein S18*), and cg00182421 (*RPL22L1; ribosomal protein L22 like 1*). A single CpG marker is common

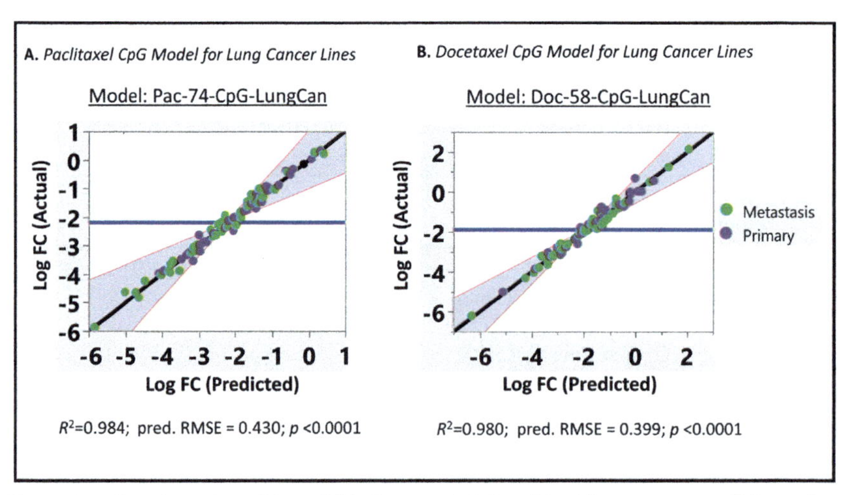

Fig. 9 Predicted vs. Actual Log-fold change viability *(V)* of lung cancer cell lines that are treated with Paclitaxel (PTX) (A) or Docetaxel (DTX) (B). The predicted Log-fold change viability (V_p) for each drug is calculated using a linear regression model with multiple CpG sites. The PTX model was designed using 74-CpG methylation data from 88 cell lines, while the DTX model was constructed using 58-CpG methylation data from 58 cell lines. RMSE = root mean square error.

in lung cancer and pan-oncology DTX models, which is the CpG site cg00398759 (also within a non-coding region). However, six unique genes in the PTX lung cancer model are also represented in the PTX pan-oncology model. These are *SNORA69* (*small nucleolar RNA, H/ACA box 69*), *RPS18* (*ribosomal protein S18*), *RPL22L1* (*ribosomal protein L22 like 1*), *PTPRN2* (*protein tyrosine phosphatase receptor type N2*), *PLEKHH3* (*pleckstrin homology, MyTH4 and FERM domain containing H3*), and *C19orf25* (*chromosome 19 open reading frame 25*). A total of 8 unique genes are commonly present in the DTX lung cancer and pan-oncology models. These are *TRRAP* (*transformation/transcription domain associated protein*), *SMAD3* (*SMAD family member 3*), *RPTOR* (*regulatory associated protein of MTOR complex 1*), *MAD1L1* (*mitotic arrest deficient 1 like 1*), *KIAA1530* (*UV stimulated scaffold protein A*), *INTS1* (*integrator complex subunit 1*), *BANP* (*BTG3 associated nuclear protein*), and ANKRD*11* (*ankyrin repeat domain containing 11*).

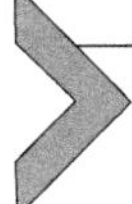

4. Discussion of linear progression models and applications for gene identification and predictors of taxane therapeutic response

In this review, we demonstrate that the anti-cancer activities of two widely used taxane drugs (paclitaxel and docetaxel) can be predicted with high accuracy through linear regression models based entirely on methylation data of cell lines prior to exposure to the drugs. In addition, predictive models which relied on mRNA expression and mutation data were also assembled. However, they exhibited inferior statistical fit relative to the CpG methylation-based models (See Table 5 for a statistical comparison of all the models presented in the review).

As anticipated, many of the markers (CpG, mRNA, mutation) which became part of the predictive models represent genes that are associated with apoptosis, microtubule formation, mitosis, and other processes that influence the activity of the anti-mitotic drugs paclitaxel (PTX) and docetaxel (DTX) (*Please refer to 3. Developing Predictive models sections, where these genes are described in greater detail*). Overall, the proportions of apoptosis-, mitosis/microtubule-, and epigenetic-related genes represented in the predictive models were relatively higher than their proportions in the genome (See Figs. 3, 4, 7, and 8). Among the apoptosis-related genes represented in the CpG-based models are those involved in TRAIL signaling (*TNFRSF10B, PACS2*) (Guicciardi et al., 2014; Wang & El-Deiry,

Table 5 Summarization of the different Paclitaxel (PTX) and Docetaxel (DTX) predictive models depicted in previous tables and figures.

Model Name	Drug	Type	CpG (n)	mRNA (n)	Mut (n)	Cell Lines (n)	R^2 (Actual v. Predicted)	*P* val	ANOVA test (Prob > F)	Ave. LogWorth of individual markers
PTX-287-CpG-PanOnc	Paclitaxel	Pan Oncology	287	0	0	399	0.985	<0.0001	<0.0001	6.436 ± 4.043
PTX-265-CpG-PanOnc	Paclitaxel	Pan Oncology	265	0	0	399	0.970	<0.0001	<0.0001	8.155 ± 5.205
PTX-249-CpG-PanOnc	Paclitaxel	Pan Oncology	249	0	0	399	0.960	<0.0001	<0.0001	8.296 ± 5.474
DTX-342-CpG-PanOnc	Docetaxel	Pan Oncology	342	0	0	390	0.996	<0.0001	<0.0001	12.271 ± 6.533
DTX-291-CpG-PanOnc	Docetaxel	Pan Oncology	291	0	0	390	0.970	<0.0001	<0.0001	9.084 ± 5.587
DTX-271-CpG-PanOnc	Docetaxel	Pan Oncology	271	0	0	390	0.950	<0.0001	<0.0001	6.843 ± 4.518
PTX-74-CpG-LungCan	Paclitaxel	Lung Cancer	74	0	0	88	0.984	<0.0001	<0.0001	3.737 ± 1.192
DTX-58-CpG-LungCan	Docetaxel	Lung Cancer	58	0	0	83	0.980	<0.0001	<0.0001	4.284 ± 2.402
PTX-290-mRNA-mut-PanOnc	Paclitaxel	Pan Oncology	0	277	13	546	0.830	<0.0001	<0.0001	4.205 ± 2.935
PTX-181-mRNA-mut-PanOnc	Paclitaxel	Pan Oncology	0	171	10	546	0.700	<0.0001	<0.0001	2.673 ± 1.666
DTX-236-mRNA-mut-PanOnc	Docetaxel	Pan Oncology	0	200	36	531	0.751	<0.0001	<0.0001	2.254 ± 1.562
DTX-181-mRNA-mut-PanOnc	Docetaxel	Pan Oncology	0	157	24	531	0.770	<0.0001	<0.0001	2.577 ± 1.683

2003), chromatin condensation, and DNA fragmentation processes (*DFFB, DNASE1, ACIN1*) (Liu et al., 1998) (Nagahama et al., 2008; Sahara et al., 1999), members of the *BCL2* family of genes and their regulators (*BNIP3, SCRT2, TP73, PLEKHN1*) (Kuriyama et al., 2018; Melino et al., 2004; Mellor & Harris, 2007; Rodriguez-Aznar & Nieto, 2011; Terrasson et al., 2005), and others whose dysregulations have been associated with apoptosis (*NME1, PRKDC, TLR4, MTDH*) (Chao & Goodman Jr., 2021; Gong et al., 2020; Hsu et al., 2018) (Rajasekaran et al., 2015). The markers included in the mRNA/mutation-based models include regulatory kinases (*AURKB* and *PLK1*) (Iliaki et al., 2021; Yoon et al., 2012), a regulator of BAX (*MOAP1*) (Wu et al., 2015), a mitochondrial protein (*MTFP1*) (Morita et al., 2017), the transcription factor *TP63* (Wang et al., 2022), and the stress response factor *GADD45G* (Salvador et al., 2013). Their roles in apoptotic processes are widely reported in the literature. The CpG-based models are also represented by genes coding for proteins directly involved in microtubule formation and functionality (*EML4, JAKMIP1, JAKMIP3*) (Pollmann et al., 2006; Steindler et al., 2004), as well as the component proteins of either kinetochore plate (*MAD1L1*) (Akera & Watanabe, 2016), the centrosome (*CEP72*), or the anaphase-promoting complex (*ANAPC2*) (Schrock et al., 2020). The mRNA/mutation-based models for either drug also covered genes that are involved in microtubule formation and activation (*JAKMIP1, FRY, MAP10, KIFC*) or are components of the kinetochore (*CENPF, CENPH*), the centrosome (*CEP162*) (Oshimori et al., 2009), and the anaphase-promoting complex (ANAPC11). Several transporter protein-coding genes were included in CpG- and mRNA/mutation-based models. However, *SLC2A9* (a CpG site within its locus is part of the DTX-342-CpG-PanOnc model) was previously reported as a possible factor in drug resistance (Januchowski, Zawierucha, Rucinski, Andrzejewska, et al., 2014; Wang et al., 2017). According to previous reports, other genes prominently represented in the CpG models are related to cancer progression. The genes *DIP2C* (11 and 10 CpG sites represented in PTX and DTX models, respectively) (See Fig. 2B and C), *RAB34* (Sun et al., 2018) (1 and 11 CpG sites in PTX and DTX models, respectively.), and *NRG2* (Zhao, Yi, et al., 2021) (6 CpG sites in DTX model) are involved in either epithelial-mesenchymal transition (EMT), or cancer cell migration. Evidence points to *PTPRN2* (Sorokin et al., 2015) (16 and 1 CpG sites in PTX and DTX models, respectively) and *SHANK2* (Unsicker et al., 2021) (4 and 5 CpG sites in PTX and DTX models, respectively) being associated with the apoptotic process. There are few publications regarding

TTC23 (5 and 1 CpG sites in PTX and DTX models, respectively). However, its main structural domain (tetratricopeptide repeat) is associated with protein-protein interactions (Perez-Riba & Itzhaki, 2019).

Of particular interest is that the predictive models include genes that play essential roles in epigenetic regulation. These include the genes for DNA methyltransferase DNMT3B (for de novo methylation of CpG sites) (Gagliardi, Strazzullo, & Matarazzo, 2018), the histone deacetylase HDAC4 (Wang, Qin, & Zhao, 2014), and the histone lysine demethylases KDM4B, KDM4C, KDM2B, and JHDM1D (or KDM7A) (Sterling et al., 2020). The proteins mentioned above can collectively modulate the transcriptional activities of genes. DNMT3B is represented in both drugs' CpG models. HDAC4 is represented in both drugs' CpG model (as well as in PTX's mRNA/mutation model). The histone lysine demethylases are represented in DTX but not in the PTX CpG model. These observations regarding the contributions of epigenetic regulatory genes to the PTX and DTX predictive models are consistent with a wealth of evidence on the importance of epigenetic processes (which, in addition to CpG methylation and histone modifications, also include miRNAs and RNA methylation) in cancer drug resistance (Bacolod, 2021). Indeed, there are reports in the literature describing how inhibition of DNA methyl transferases (by drugs such as azacytidine, decitabine) and histone deacetylases (by drugs such as belinostat, panobinostat, vorinostat) can modulate cancer drug resistance (Rauscher, Greil, & Geisberger, 2021). The addition of decitabine could reverse enzalutamide resistance in prostate cancer cells (Farah et al., 2022). Sorafenib resistance was reversed in liver carcinoma through RNAi or addition of Nanaomycin A (which is selective DNMT3B inhibitor) (Lai et al., 2019). Inhibition of DNA methyl transferases by miR-200b and miR-200c led to reversal of cisplatin resistance in ovarian cancer cells (Wu, Xie, Shi, & Wang, 2020). In another report, the over-expression of DNMT3B is associated with resistance against pralatrexate, an antifolate approved for relapsed/refractory peripheral T-cell lymphoma (Oiwa et al., 2021). However, one study has demonstrated that adding decitabine can induce PTX- and DTX-resistance in gastric cancer cells by restoring the expression of CHFR (checkpoint with forkhead and ring finger domains), an E3 Ligase that functions in mitotic checkpoint (Satoh et al., 2003). Reports point to the expression of HDAC4 directly contributing to cancer drug resistance (and its inhibition can lead to its reversal). Investigators proved that HDAC4-overexpression could induce 5-fluorouracil (5-FU) resistance in breast cancer cells (Yu et al., 2013). Platinum-resistant ovarian cancer biopsies exhibited over-expression of HDAC4 (Stronach et al., 2011). It has also been demonstrated that HDAC inhibition by sodium butyrate can decrease the expression of various MDR transporter genes *ABCC1*, *ABCC2*, and *ABCC3* in lung and colon cancer cells (Shi et al., 2020). When treated with the HDAC inhibitor belinostat, the chemo-resistant, cancer stem cell-like (CSL) state of ER-positive breast cancer-derived organoids was reversed (Chi et al., 2021). A recent report demonstrated that in pancreatic cancer, HDACs are regulated by the protein MARK2 (Microtubule affinity-regulating kinase 2), and that inhibition of HDACs or MARK2 can lead to the reversal of PTX-resistance (Zeng et al., 2022). Our observation that HDAC4 mutation positively contributes to PTX resistance (in the predictive model PTX −290-mRNA-mut-PanOnc) may be consistent with these reports.

The reports discussed above point to DNA Methyltransferase and histone deacetylase expression as directly correlated with drug resistance. An exception is a report by Satoh and colleagues, who claim the opposite effect of DNMT inhibition (Satoh et al., 2003). We can only assume that it is far more complicated than it seems. It may depend on the type of drug and cellular context. DNMT and HDAC inhibitors can influence the transcription of many genes with various functionalities. For example, the addition of the DNMT inhibitor azacytidine can lead to hypomethylation of the promoter regions (and increased expression) of a gene with pro-apoptotic function and another with anti-apoptotic activity. What further complicates the outcome is that for specific genes, CpG methylation in the gene body (i.e., inner introns, exons) may have the opposite effect on transcription (direct relationship). This direct correlation between gene expression and CpG methylation at the gene body is observed in the gene MGMT, the primary resistance factor against the chemotherapeutic drug temozolomide, part of the usual regimen for the treatment of brain tumors (Bacolod & Barany, 2020). As we present in this review, taxane activity is accurately predicted by linear regression models of multiple CpG markers, representing biomarkers representing a simplification in defining a very complex process. In essence, the CpG markers in the models have varied effects on the expression of genes with wide-ranging roles in taxane activity or resistance. Direct confirmation of functionality of the genes identified using our linear regression models of multiple CpG markers requires careful experimentation involving both gain and loss-of-function genetic approaches.

The potential of CpG methylation in tumor samples as a predictor of drug activity is being explored in numerous studies. There have been reports of CpG markers (specific to a particular gene or group of genes) being utilized as a possible predictor of the efficacy of classical chemotherapeutic drugs, the new generation of targeted therapeutics, and immunotherapy. For example, one report points to the combination of CpG methylation and miRNA data being able to predict how breast cancer patients respond to PTX (Bomane, Goncalves, & Ballester, 2019). Another report indicated that CpG methylation could predict breast cancer patients' response to neoadjuvant chemotherapy (Pedersen et al., 2022). Another recent publication also demonstrated that methylation signatures could predict cetuximab response among colorectal cancer patients (Otsuki et al., 2022). In addition, there are various reports of gene-specific methylation capable of predicting response to chemotherapeutic agents such as methylation of *TMEM240* (breast cancer's response to hormone therapy) (Lin et al., 2022), *LRP12* (NSCLC response to carboplatin) (Grasse et al., 2018), *LAMA3* (ovarian cancer response to chemotherapy) (Feng, Huang, Zhang, & Li, 2021), *RAD51C* (serous ovarian carcinoma response to PARP inhibitors) (Nesic et al., 2021), and *MGMT* (brain tumors' response to temozolomide) (Brandner et al., 2021). Immunotherapy is another area where methylation profiling can be helpful as a predictive tool, as the authors have previously discussed (Bacolod, Barany, & Fisher, 2019). One group in particular has reported that methylation signatures can predict response to anti-PD-1 immune checkpoint inhibitors in sarcoma (Starzer et al., 2021) and head and neck squamous cell carcinoma (Starzer et al., 2022) patients.

It is evident that CpG methylation, as a diagnostic assay marker, has many advantages over other markers. First, the fraction of methylation per CpG site (ranging from 0 to 1) can be the output data generated using various tools, including methylation arrays, bisulfite PCR, and bisulfite sequencing. Moreover, genomic samples are relatively easy to prepare and are more stable (BLUEPRINT_CONSORTIUM, 2016).

5. Future perspectives

This report establishes that linear regression models (especially those using CpG markers as input variables) can serve as potential predictors of taxane activity in cancer cell lines. However, the models based on cancer cell lines are not automatically applicable to tissues acquired from cancer patients. Moreover, the predictive models must be validated using independent but similarly acquired datasets (which, as of this writing, are not yet available). Nevertheless, many of the markers that became components of the models are associated with processes with proven roles in taxane activity/resistance. The approach we presented could also identify genes that previously had not been associated with taxanes. The application of the approaches described in this review is readily amenable to other therapeutic drugs and could be paradigm-shifting in how cancer sensitivity to therapeutic agents is defined prior to therapy and informative in cases where therapeutic resistance occurs.

Acknowledgments

This study was supported by: (a) Weill Cornell Medicine funding through the distribution of royalties from intellectual property generated by the Barany laboratory (MDB, FB), (b) A Sponsored Research Agreement between AcuamarkDx and Weill Cornell Medicine (MDB, FB), (c) NIH/NCI grants 1R01 CA259599 and 1R01 CA244993 (PBF), (d) the Commonwealth Health Research Board (CHRB) (PBF), (e) the National Foundation for Cancer Research (PBF), (f) development funds from the Department of Human and Molecular Genetics, the VCU Institute of Molecular Medicine (VIMM) and the VCU Massey Cancer Center (MCC), and (g) support from Thelma Newmeyer Corman Chair in Cancer Research (PBF) at VCU School of Medicine. We truly appreciate the generosity of many of our colleagues in the cancer research field for sharing their datasets and analytical tools.

References

Ahmed, S. M., Theriault, B. L., Uppalapati, M., Chiu, C. W., Gallie, B. L., Sidhu, S. S., et al. (2012). KIF14 negatively regulates Rap1a-Radil signaling during breast cancer progression. *The Journal of Cell Biology, 199*(6), 951–967.

Akera, T., & Watanabe, Y. (2016). The spindle assembly checkpoint promotes chromosome bi-orientation: A novel Mad1 role in chromosome alignment. *Cell Cycle, 15*(4), 493–497.

Amor, D. J., Kalitsis, P., Sumer, H., & Choo, K. H. (2004). Building the centromere: From foundation proteins to 3D organization. *Trends in Cell Biology, 14*(7), 359–368.

Bacolod, M. D. (2021). The epigenetic factors that drive Cancer drug resistance. *Current Cancer Drug Targets, 21*(4), 269–273.

Bacolod, M. D., & Barany, F. (2020). MGMT epigenetics: The influence of gene body methylation and other insights derived from integrated methylomic, transcriptomic, and chromatin analyses for various cancer types. *Current Cancer Drug Targets, 21(4)*, 360–374.

Bacolod, M. D., Barany, F., & Fisher, P. B. (2019). Can CpG methylation serve as surrogate markers for immune infiltration in cancer? *Advances in Cancer Research, 143*, 351–384.

Bacolod, M. D., Barany, F., Pilones, K., Fisher, P. B., & de Castro, R. J. (2019). Pathways- and epigenetic-based assessment of relative immune infiltration in various types of solid tumors. *Advances in Cancer Research, 142*, 107–143.

Bacolod, M. D., Das, S. K., Sokhi, U. K., Bradley, S., Fenstermacher, D. A., Pellecchia, M., et al. (2015). Examination of epigenetic and other molecular factors associated with mda-9/Syntenin dysregulation in Cancer through integrated analyses of public genomic datasets. *Advances in Cancer Research, 127*, 49–121.

Baker, S. J., & Reddy, E. P. (1996). Transducers of life and death: TNF receptor superfamily and associated proteins. *Oncogene, 12*(1), 1–9.

Barretina, J., Caponigro, G., Stransky, N., Venkatesan, K., Margolin, A. A., Kim, S., et al. (2012). The Cancer cell line Encyclopedia enables predictive modelling of anticancer drug sensitivity. *Nature*, *483*(7391), 603–607.

BLUEPRINT_CONSORTIUM. (2016). Quantitative comparison of DNA methylation assays for biomarker development and clinical applications. *Nature Biotechnology*, *34*(7), 726–737.

Bomane, A., Goncalves, A., & Ballester, P. J. (2019). Paclitaxel response can be predicted with interpretable multi-variate classifiers exploiting DNA-methylation and miRNA data. *Frontiers in Genetics*, *10*, 1041.

Bordonaro, M., & Lazarova, D. L. (2015). CREB-binding protein, p300, butyrate, and Wnt signaling in colorectal cancer. *World Journal of Gastroenterology*, *21*(27), 8238–8248.

Brandner, S., McAleenan, A., Kelly, C., Spiga, F., Cheng, H. Y., Dawson, S., et al. (2021). MGMT promoter methylation testing to predict overall survival in people with glioblastoma treated with temozolomide: A comprehensive meta-analysis based on a Cochrane systematic review. *Neuro-Oncology*, *23*(9), 1457–1469.

Campbell, M. J., & Turner, B. M. (2013). Altered histone modifications in cancer. *Advances in Experimental Medicine and Biology*, *754*, 81–107.

Chabner, B., & Longo, D. L. (2011). *Cancer chemotherapy and biotherapy: Principles and practice*. Philadelphia: Wolters Kluwer Health/Lippincott Williams & Wilkins.

Chao, O. S., & Goodman, O. B., Jr. (2021). DNA-PKc inhibition overcomes taxane resistance by promoting taxane-induced DNA damage in prostate cancer cells. *The Prostate*, *81*(14), 1032–1048.

Che, P., Jiang, S., Zhang, W., Zhu, H., Hu, D., & Wang, D. (2022). A comprehensive gene expression profile analysis of prostate cancer cells resistant to paclitaxel and the potent target to reverse resistance. *Human & Experimental Toxicology*, *41*, 9603271221129854.

Chen, M., Su, J., Feng, C., Liu, Y., Zhao, L., & Tian, Y. (2021). Chemokine CCL20 promotes the paclitaxel resistance of CD44(+)CD117(+) cells via the Notch1 signaling pathway in ovarian cancer. *Molecular Medicine Reports*, *24*(3), 635.

Chen, Y., & Zhou, X. (2020). Research progress of mTOR inhibitors. *European Journal of Medicinal Chemistry*, *208*, 112820.

Chi, F., Liu, J., Brady, S. W., Cosgrove, P. A., Nath, A., McQuerry, J. A., et al. (2021). A 'one-two punch' therapy strategy to target chemoresistance in estrogen receptor positive breast cancer. *Translational Oncology*, *14*(1), 100946.

Corsello, S. M., Nagari, R. T., Spangler, R. D., Rossen, J., Kocak, M., Bryan, J. G., et al. (2020). Discovering the anticancer potential of non-oncology drugs by systemic viability profiling. *Nature Cancer*, *1*(2), 235–248.

Das, T., Anand, U., Pandey, S. K., Ashby, C. R., Jr., Assaraf, Y. G., Chen, Z. S., et al. (2021). Therapeutic strategies to overcome taxane resistance in cancer. *Drug Resistance Updates*, *55*, 100754.

de Weger, V. A., Beijnen, J. H., & Schellens, J. H. (2014). Cellular and clinical pharmacology of the taxanes docetaxel and paclitaxel—A review. *Anti-Cancer Drugs*, *25*(5), 488–494.

ENCODE_Project_Consortium. (2004). The ENCODE (ENCyclopedia of DNA elements) project. *Science*, *306*(5696), 636–640.

Ernst, J., Kheradpour, P., Mikkelsen, T. S., Shoresh, N., Ward, L. D., Epstein, C. B., et al. (2011). Mapping and analysis of chromatin state dynamics in nine human cell types. *Nature*, *473*(7345), 43–49.

Fabregat, A., Sidiropoulos, K., Viteri, G., Forner, O., Marin-Garcia, P., Arnau, V., et al. (2017). Reactome pathway analysis: A high-performance in-memory approach. *BMC Bioinformatics*, *18*(1), 142.

Farah, E., Zhang, Z., Utturkar, S. M., Liu, J., Ratliff, T. L., & Liu, X. (2022). Targeting DNMTs to overcome enzalutamide resistance in prostate Cancer. *Molecular Cancer Therapeutics*, *21*(1), 193–205.

Feng, L. Y., Huang, Y. Z., Zhang, W., & Li, L. (2021). LAMA3 DNA methylation and transcriptome changes associated with chemotherapy resistance in ovarian cancer. *Journal of Ovarian Research*, *14*(1), 67.

Fisser, M. C., Rommer, A., Steinleitner, K., Heller, G., Herbst, F., Wiese, M., et al. (2015). Induction of the proapoptotic tumor suppressor gene cell adhesion molecule 1 by chemotherapeutic agents is repressed in therapy resistant acute myeloid leukemia. *Molecular Carcinogenesis*, *54*(12), 1815–1819.

Fong, K. W., Leung, J. W., Li, Y., Wang, W., Feng, L., Ma, W., et al. (2013). MTR120/KIAA1383, a novel microtubule-associated protein, promotes microtubule stability and ensures cytokinesis. *Journal of Cell Science*, *126*(Pt 3), 825–837.

Fujita, M. (2006). Cdt1 revisited: Complex and tight regulation during the cell cycle and consequences of deregulation in mammalian cells. *Cell Division*, *1*, 22.

Gagliardi, M., Strazzullo, M., & Matarazzo, M. R. (2018). "DNMT3B functions: Novel insights from human disease" front cell. *Developmental Biology*, *6*, 140.

Galletti, G., Zhang, C., Gjyrezi, A., Cleveland, K., Zhang, J., Powell, S., et al. (2020). Microtubule engagement with taxane is altered in taxane-resistant gastric cancer. *Clinical Cancer Research*, *26*(14), 3771–3783.

Gan, L., Wang, J., Xu, H., & Yang, X. (2011). Resistance to docetaxel-induced apoptosis in prostate cancer cells by p38/p53/p21 signaling. *The Prostate*, *71*(11), 1158–1166.

Gao, Y., Liu, Y., Liu, Y., Peng, Y., Yuan, B., Fu, Y., et al. (2021). UHRF1 promotes androgen receptor-regulated CDC6 transcription and anti-androgen receptor drug resistance in prostate cancer through KDM4C-mediated chromatin modifications. *Cancer Letters*, *520*, 172–183.

Gene Ontology, C. (2021). The Gene Ontology resource: Enriching a GOld mine. *Nucleic Acids Research*, *49*(D1), D325–D334.

Ghandi, M., Huang, F. W., Jane-Valbuena, J., Kryukov, G. V., Lo, C. C., McDonald, E. R., 3rd, et al. (2019). Next-generation characterization of the cancer cell line Encyclopedia. *Nature*, *569*(7757), 503–508.

Ghirlando, R., Giles, K., Gowher, H., Xiao, T., Xu, Z., Yao, H., et al. (2012). Chromatin domains, insulators, and the regulation of gene expression. *Biochimica et Biophysica Acta*, *1819*(7), 644–651.

Golin, J., & Ambudkar, S. V. (2015). The multidrug transporter Pdr5 on the 25th anniversary of its discovery: An important model for the study of asymmetric ABC transporters. *The Biochemical Journal*, *467*(3), 353–363.

Gong, Y., Yang, G., Wang, Q., Wang, Y., & Zhang, X. (2020). NME2 is a master suppressor of apoptosis in gastric Cancer cells via transcriptional regulation of miR-100 and other survival factors. *Molecular Cancer Research*, *18*(2), 287–299.

Grasse, S., Lienhard, M., Frese, S., Kerick, M., Steinbach, A., Grimm, C., et al. (2018). Epigenomic profiling of non-small cell lung cancer xenografts uncover LRP12 DNA methylation as predictive biomarker for carboplatin resistance. *Genome Medicine*, *10*(1), 55.

Greil, C., Engelhardt, M., & Wasch, R. (2022). The role of the APC/C and its coactivators Cdh1 and Cdc20 in Cancer development and therapy. *Frontiers in Genetics*, *13*, 941565.

Guicciardi, M. E., Werneburg, N. W., Bronk, S. F., Franke, A., Yagita, H., Thomas, G., et al. (2014). Cellular inhibitor of apoptosis (cIAP)-mediated ubiquitination of phosphofurin acidic cluster sorting protein 2 (PACS-2) negatively regulates tumor necrosis factor-related apoptosis-inducing ligand (TRAIL) cytotoxicity. *PLoS One*, *9*(3), e92124.

Guo, X., Ramirez, I., Garcia, Y. A., Velasquez, E. F., Gholkar, A. A., Cohn, W., et al. (2021). DUSP7 regulates the activity of ERK2 to promote proper chromosome alignment during cell division. *The Journal of Biological Chemistry*, *296*, 100676.

Hsu, H. Y., Lin, T. Y., Hu, C. H., Shu, D. T. F., & Lu, M. K. (2018). Fucoidan upregulates TLR4/CHOP-mediated caspase-3 and PARP activation to enhance cisplatin-induced cytotoxicity in human lung cancer cells. *Cancer Letters*, *432*, 112–120.

Huang, Y., Ibrado, A. M., Reed, J. C., Bullock, G., Ray, S., Tang, C., et al. (1997). Co-expression of several molecular mechanisms of multidrug resistance and their significance for paclitaxel cytotoxicity in human AML HL-60 cells. *Leukemia*, *11*(2), 253–257.

Iliaki, S., Beyaert, R., & Afonina, I. S. (2021). Polo-like kinase 1 (PLK1) signaling in cancer and beyond. *Biochemical Pharmacology*, *193*, 114747.

Inoue, H., Shiozaki, A., Kosuga, T., Shimizu, H., Kudou, M., Ohashi, T., et al. (2022). Functions and clinical significance of CACNA2D1 in gastric Cancer. *Annals of Surgical Oncology*, *29*, 4522–4535.

Iorio, F., Knijnenburg, T. A., Vis, D. J., Bignell, G. R., Menden, M. P., Schubert, M., et al. (2016). A landscape of pharmacogenomic interactions in Cancer. *Cell*, *166*(3), 740–754.

Januchowski, R., Zawierucha, P., Rucinski, M., Andrzejewska, M., Wojtowicz, K., Nowicki, M., et al. (2014). Drug transporter expression profiling in chemoresistant variants of the A2780 ovarian cancer cell line. *Biomedicine & Pharmacotherapy*, *68*(4), 447–453.

Januchowski, R., Zawierucha, P., Rucinski, M., Nowicki, M., & Zabel, M. (2014). Extracellular matrix proteins expression profiling in chemoresistant variants of the A2780 ovarian cancer cell line. *BioMed Research International*, *2014*, 365867.

Jassal, B., Matthews, L., Viteri, G., Gong, C., Lorente, P., Fabregat, A., et al. (2020). The reactome pathway knowledgebase. *Nucleic Acids Research*, *48*(D1), D498–D503.

Keppler, D. (2011). Multidrug resistance proteins (MRPs, ABCCs): Importance for pathophysiology and drug therapy. *Handbook of Experimental Pharmacology*, *201*, 299–323.

Kivilaakso, E. (1985). Pathogenetic mechanisms in experimental gastric stress ulceration. *Scandinavian Journal of Gastroenterology. Supplement*, *110*, 57–62.

Kuriyama, S., Tsuji, T., Sakuma, T., Yamamoto, T., & Tanaka, M. (2018). PLEKHN1 promotes apoptosis by enhancing Bax-Bak hetro-oligomerization through interaction with bid in human colon cancer. *Cell Death Discovery*, *4*, 11.

Lai, S. C., Su, Y. T., Chi, C. C., Kuo, Y. C., Lee, K. F., Wu, Y. C., et al. (2019). DNMT3b/OCT4 expression confers sorafenib resistance and poor prognosis of hepatocellular carcinoma through IL-6/STAT3 regulation. *Journal of Experimental & Clinical Cancer Research*, *38*(1), 474.

Larsson, C., Ali, M. A., Pandzic, T., Lindroth, A. M., He, L., & Sjoblom, T. (2017). Loss of DIP2C in RKO cells stimulates changes in DNA methylation and epithelial-mesenchymal transition. *BMC Cancer*, *17*(1), 487.

Lee, K. H., Kim, B. C., Jeong, S. H., Jeong, C. W., Ku, J. H., Kim, H. H., et al. (2020). Histone demethylase KDM7A regulates androgen receptor activity, and its chemical inhibitor TC-E 5002 overcomes cisplatin-resistance in bladder Cancer cells. *International Journal of Molecular Sciences*, *21*(16), 5658.

Leonard, M. K., Puts, G. S., Pamidimukkala, N., Adhikary, G., Xu, Y., Kwok, E., et al. (2021). Comprehensive molecular profiling of UV-induced metastatic melanoma in Nme1/Nme2-deficient mice reveals novel markers of survival in human patients. *Oncogene*, *40*(45), 6329–6342.

Lin, R. K., Su, C. M., Lin, S. Y., Thi Anh Thu, L., Liew, P. L., Chen, J. Y., et al. (2022). Hypermethylation of TMEM240 predicts poor hormone therapy response and disease progression in breast cancer. *Molecular Medicine*, *28*(1), 67.

Liu, X., Li, P., Widlak, P., Zou, H., Luo, X., Garrard, W. T., et al. (1998). The 40-kDa subunit of DNA fragmentation factor induces DNA fragmentation and chromatin condensation during apoptosis. *Proceedings of the National Academy of Sciences of the United States of America*, *95*(15), 8461–8466.

Melino, G., Bernassola, F., Ranalli, M., Yee, K., Zong, W. X., Corazzari, M., et al. (2004). p73 induces apoptosis via PUMA transactivation and Bax mitochondrial translocation. *The Journal of Biological Chemistry*, *279*(9), 8076–8083.

Mellor, H. R., & Harris, A. L. (2007). The role of the hypoxia-inducible BH3-only proteins BNIP3 and BNIP3L in cancer. *Cancer Metastasis Reviews*, *26*(3–4), 553–566.
Morita, M., Prudent, J., Basu, K., Goyon, V., Katsumura, S., Hulea, L., et al. (2017). mTOR controls mitochondrial dynamics and cell survival via MTFP1. *Molecular Cell*, *67*(6), 922–935.e5.
Mosca, L., Ilari, A., Fazi, F., Assaraf, Y. G., & Colotti, G. (2021). Taxanes in cancer treatment: Activity, chemoresistance and its overcoming. *Drug Resistance Updates*, *54*, 100742.
Nagahama, Y., Ishimaru, M., Osaki, M., Inoue, T., Maeda, A., Nakada, C., et al. (2008). Apoptotic pathway induced by transduction of RUNX3 in the human gastric carcinoma cell line MKN-1. *Cancer Science*, *99*(1), 23–30.
Nagai, T., Ikeda, M., Chiba, S., Kanno, S., & Mizuno, K. (2013). Furry promotes acetylation of microtubules in the mitotic spindle by inhibition of SIRT2 tubulin deacetylase. *Journal of Cell Science*, *126*(Pt 19), 4369–4380.
Nesic, K., Kondrashova, O., Hurley, R. M., McGehee, C. D., Vandenberg, C. J., Ho, G. Y., et al. (2021). Acquired RAD51C promoter methylation loss causes PARP inhibitor resistance in high-grade serous ovarian carcinoma. *Cancer Research*, *81*(18), 4709–4722.
Norbury, C. J., & Zhivotovsky, B. (2004). DNA damage-induced apoptosis. *Oncogene*, *23*(16), 2797–2808.
Odoux, C., Albers, A., Amoscato, A. A., Lotze, M. T., & Wong, M. K. (2002). TRAIL, FasL and a blocking anti-DR5 antibody augment paclitaxel-induced apoptosis in human non-small-cell lung cancer. *International Journal of Cancer*, *97*(4), 458–465.
Oiwa, K., Hosono, N., Nishi, R., Scotto, L., O'Connor, O. A., & Yamauchi, T. (2021). Characterization of newly established Pralatrexate-resistant cell lines and the mechanisms of resistance. *BMC Cancer*, *21*(1), 879.
Oshimori, N., Li, X., Ohsugi, M., & Yamamoto, T. (2009). Cep72 regulates the localization of key centrosomal proteins and proper bipolar spindle formation. *The EMBO Journal*, *28*(14), 2066–2076.
Otsuki, Y., Ouchi, K., Takahashi, S., Sasaki, K., Sakamoto, Y., Okita, A., et al. (2022). Altered gene expression due to aberrant DNA methylation correlates with responsiveness to anti-EGFR antibody treatment. *Cancer Science*, *113*(9), 3221–3233.
Ottaviano, Y. L., Issa, J. P., Parl, F. F., Smith, H. S., Baylin, S. B., & Davidson, N. E. (1994). Methylation of the estrogen receptor gene CpG island marks loss of estrogen receptor expression in human breast cancer cells. *Cancer Research*, *54*(10), 2552–2555.
Passananti, C., Floridi, A., & Fanciulli, M. (2007). Che-1/AATF, a multivalent adaptor connecting transcriptional regulation, checkpoint control, and apoptosis. *Biochemistry and Cell Biology*, *85*(4), 477–483.
Pedersen, C. A., Cao, M. D., Fleischer, T., Rye, M. B., Knappskog, S., Eikesdal, H. P., et al. (2022). DNA methylation changes in response to neoadjuvant chemotherapy are associated with breast cancer survival. *Breast Cancer Research*, *24*(1), 43.
Perez-Riba, A., & Itzhaki, L. S. (2019). The tetratricopeptide-repeat motif is a versatile platform that enables diverse modes of molecular recognition. *Current Opinion in Structural Biology*, *54*, 43–49.
Peterman, E., Valius, M., & Prekeris, R. (2020). CLIC4 is a cytokinetic cleavage furrow protein that regulates cortical cytoskeleton stability during cell division. *Journal of Cell Science*, *133*(9), jcs241117.
Peterson, D., Lee, J., Lei, X. C., Forrest, W. F., Davis, D. P., Jackson, P. K., et al. (2010). A chemosensitization screen identifies TP53RK, a kinase that restrains apoptosis after mitotic stress. *Cancer Research*, *70*(15), 6325–6335.
Pollmann, M., Parwaresch, R., Adam-Klages, S., Kruse, M. L., Buck, F., & Heidebrecht, H. J. (2006). Human EML4, a novel member of the EMAP family, is essential for microtubule formation. *Experimental Cell Research*, *312*(17), 3241–3251.

Qin, Z., Bi, L., Hou, X. M., Zhang, S., Zhang, X., Lu, Y., et al. (2020). Human RPA activates BLM's bidirectional DNA unwinding from a nick. *eLife*, *9*, e54098.

Rajasekaran, D., Srivastava, J., Ebeid, K., Gredler, R., Akiel, M., Jariwala, N., et al. (2015). Combination of nanoparticle-delivered siRNA for astrocyte elevated Gene-1 (AEG-1) and all-trans retinoic acid (ATRA): An effective therapeutic strategy for hepatocellular carcinoma (HCC). *Bioconjugate Chemistry*, *26*(8), 1651–1661.

Rauscher, S., Greil, R., & Geisberger, R. (2021). Re-sensitizing tumor cells to cancer drugs with epigenetic regulators. *Current Cancer Drug Targets*, *21*(4), 353–359.

Robertson, J., Barr, R., Shulman, L. N., Forte, G. B., & Magrini, N. (2016). Essential medicines for cancer: WHO recommendations and national priorities. *Bulletin of the World Health Organization*, *94*(10), 735–742.

Rodriguez-Aznar, E., & Nieto, M. A. (2011). Repression of Puma by scratch2 is required for neuronal survival during embryonic development. *Cell Death and Differentiation*, *18*(7), 1196–1207.

Rosano, L., Cianfrocca, R., Spinella, F., Di Castro, V., Nicotra, M. R., Lucidi, A., et al. (2011). Acquisition of chemoresistance and EMT phenotype is linked with activation of the endothelin A receptor pathway in ovarian carcinoma cells. *Clinical Cancer Research*, *17*(8), 2350–2360.

Rouzier, R., Rajan, R., Wagner, P., Hess, K. R., Gold, D. L., Stec, J., et al. (2005). Microtubule-associated protein tau: A marker of paclitaxel sensitivity in breast cancer. *Proceedings of the National Academy of Sciences of the United States of America*, *102*(23), 8315–8320.

Runion, H. I., & Pipa, R. L. (1970). Electrophysiological and endocrinological correlates during the metamorphic degeneration of a muscle fibre in galleria mellonella (L.) (Lepidoptera). *The Journal of Experimental Biology*, *53*(1), 9–24.

Sahara, S., Aoto, M., Eguchi, Y., Imamoto, N., Yoneda, Y., & Tsujimoto, Y. (1999). Acinus is a caspase-3-activated protein required for apoptotic chromatin condensation. *Nature*, *401*(6749), 168–173.

Salvador, J. M., Brown-Clay, J. D., & Fornace, A. J., Jr. (2013). Gadd45 in stress signaling, cell cycle control, and apoptosis. *Advances in Experimental Medicine and Biology*, *793*, 1–19.

Satelli, A., & Li, S. (2011). Vimentin in cancer and its potential as a molecular target for cancer therapy. *Cellular and Molecular Life Sciences*, *68*(18), 3033–3046.

Satoh, A., Toyota, M., Itoh, F., Sasaki, Y., Suzuki, H., Ogi, K., et al. (2003). Epigenetic inactivation of CHFR and sensitivity to microtubule inhibitors in gastric cancer. *Cancer Research*, *63*(24), 8606–8613.

Schrock, M. S., Stromberg, B. R., Scarberry, L., & Summers, M. K. (2020). APC/C ubiquitin ligase: Functions and mechanisms in tumorigenesis. *Seminars in Cancer Biology*, *67*(Pt 2), 80–91.

Shi, B., Xu, F. F., Xiang, C. P., Jia, R., Yan, C. H., Ma, S. Q., et al. (2020). Effect of sodium butyrate on ABC transporters in lung cancer A549 and colorectal cancer HCT116 cells. *Oncology Letters*, *20*(5), 148.

Slade, D. (2019). Mitotic functions of poly(ADP-ribose) polymerases. *Biochemical Pharmacology*, *167*, 33–43.

Sorokin, A. V., Nair, B. C., Wei, Y., Aziz, K. E., Evdokimova, V., Hung, M. C., et al. (2015). Aberrant expression of proPTPRN2 in Cancer cells confers resistance to apoptosis. *Cancer Research*, *75*(9), 1846–1858.

Starzer, A. M., Berghoff, A. S., Hamacher, R., Tomasich, E., Feldmann, K., Hatziioannou, T., et al. (2021). Tumor DNA methylation profiles correlate with response to anti-PD-1 immune checkpoint inhibitor monotherapy in sarcoma patients. *Journal for Immunotherapy of Cancer*, *9*(3), e001458.

Starzer, A. M., Heller, G., Tomasich, E., Melchardt, T., Feldmann, K., Hatziioannou, T., et al. (2022). DNA methylation profiles differ in responders versus non-responders to anti-PD-1 immune checkpoint inhibitors in patients with advanced and metastatic head and neck squamous cell carcinoma. *Journal for Immunotherapy of Cancer*, *10*(3), e003420.

Steindler, C., Li, Z., Algarte, M., Alcover, A., Libri, V., Ragimbeau, J., et al. (2004). Jamip1 (marlin-1) defines a family of proteins interacting with janus kinases and microtubules. *The Journal of Biological Chemistry*, *279*(41), 43168–43177.

Sterling, J., Menezes, S. V., Abbassi, R. H., & Munoz, L. (2020). Histone lysine demethylases and their functions in cancer. *International Journal of Cancer*, *148 (10)*, 2375–2388.

Sternlicht, H., Farr, G. W., Sternlicht, M. L., Driscoll, J. K., Willison, K., & Yaffe, M. B. (1993). The t-complex polypeptide 1 complex is a chaperonin for tubulin and actin in vivo. *Proceedings of the National Academy of Sciences of the United States of America*, *90*(20), 9422–9426.

Stronach, E. A., Alfraidi, A., Rama, N., Datler, C., Studd, J. B., Agarwal, R., et al. (2011). HDAC4-regulated STAT1 activation mediates platinum resistance in ovarian cancer. *Cancer Research*, *71*(13), 4412–4422.

Sun, L., Xu, X., Chen, Y., Zhou, Y., Tan, R., Qiu, H., et al. (2018). Rab34 regulates adhesion, migration, and invasion of breast cancer cells. *Oncogene*, *37*(27), 3698–3714.

Takano, M., Otani, Y., Tanda, M., Kawami, M., Nagai, J., & Yumoto, R. (2009). Paclitaxel-resistance conferred by altered expression of efflux and influx transporters for paclitaxel in the human hepatoma cell line, HepG2. *Drug Metabolism and Pharmacokinetics*, *24*(5), 418–427.

Tang, D. E., Dai, Y., He, J. X., Lin, L. W., Leng, Q. X., Geng, X. Y., et al. (2020). Targeting the KDM4B-AR-c-Myc axis promotes sensitivity to androgen receptor-targeted therapy in advanced prostate cancer. *The Journal of Pathology*, *252*(2), 101–113.

Terrasson, J., Allart, S., Martin, H., Lule, J., Haddada, H., Caput, D., et al. (2005). p73-dependent apoptosis through death receptor: Impairment by human cytomegalovirus infection. *Cancer Research*, *65*(7), 2787–2794.

Tuerxun, G., Abulimiti, T., Abudurexiti, G., Abuduxikuer, G., Zhang, Y., & Abulizi, G. (2022). Over-expression of EPB41L3 promotes apoptosis of human cervical carcinoma cells through PI3K/AKT signaling. *Acta Biochimica Polonica*, *69*(2), 283–289.

Unsicker, C., Cristian, F. B., von Hahn, M., Eckstein, V., Rappold, G. A., & Berkel, S. (2021). SHANK2 mutations impair apoptosis, proliferation and neurite outgrowth during early neuronal differentiation in SH-SY5Y cells. *Scientific Reports*, *11*(1), 2128.

van Eijk, M., Boosman, R. J., Schinkel, A. H., Huitema, A. D. R., & Beijnen, J. H. (2019). Cytochrome P450 3A4, 3A5, and 2C8 expression in breast, prostate, lung, endometrial, and ovarian tumors: Relevance for resistance to taxanes. *Cancer Chemotherapy and Pharmacology*, *84*(3), 487–499.

Vazquez, F., & Boehm, J. S. (2020). The Cancer dependency map enables drug mechanism-of-action investigations. *Molecular Systems Biology*, *16*(7), e9757.

Walch, L., Pellier, E., Leng, W., Lakisic, G., Gautreau, A., Contremoulins, V., et al. (2018). GBF1 and Arf1 interact with Miro and regulate mitochondrial positioning within cells. *Scientific Reports*, *8*(1), 17121.

Wang, S., & El-Deiry, W. S. (2003). TRAIL and apoptosis induction by TNF-family death receptors. *Oncogene*, *22*(53), 8628–8633.

Wang, Q., Lu, F., & Lan, R. (2017). RNA-sequencing dissects the transcriptome of polyploid cancer cells that are resistant to combined treatments of cisplatin with paclitaxel and docetaxel. *Molecular BioSystems*, *13*(10), 2125–2134.

Wang, Y., Lu, H., Wang, Z., Li, Y., & Chen, X. (2022). "TGF-beta1 promotes autophagy and inhibits apoptosis in breast Cancer by targeting TP63" front. *Oncologia*, *12*, 865067.

Wang, Z., Qin, G., & Zhao, T. C. (2014). HDAC4: Mechanism of regulation and biological functions. *Epigenomics*, *6*(1), 139–150.

Wang, W. J., Tay, H. G., Soni, R., Perumal, G. S., Goll, M. G., Macaluso, F. P., et al. (2013). CEP162 is an axoneme-recognition protein promoting ciliary transition zone assembly at the cilia base. *Nature Cell Biology*, *15*(6), 591–601.

Wang, X., Yang, Y., Xu, C., Xiao, L., Shen, H., Zhang, X., et al. (2011). CHFR suppression by hypermethylation sensitizes endometrial cancer cells to paclitaxel. *International Journal of Gynecological Cancer*, *21*(6), 996–1003.

Wassing, I. E., Graham, E., Saayman, X., Rampazzo, L., Ralf, C., Bassett, A., et al. (2021). The RAD51 recombinase protects mitotic chromatin in human cells. *Nature Communications*, *12*(1), 5380.

Wu, S., Xie, J., Shi, H., & Wang, Z. W. (2020). miR-492 promotes chemoresistance to CDDP and metastasis by targeting inhibiting DNMT3B and induces stemness in gastric cancer. *Bioscience Reports*, *40*(3). BSR20194342.

Wu, T., Chen, W., Kong, D., Li, X., Lu, H., Liu, S., et al. (2015). miR-25 targets the modulator of apoptosis 1 gene in lung cancer. *Carcinogenesis*, *36*(8), 925–935.

Xiao, H., Verdier-Pinard, P., Fernandez-Fuentes, N., Burd, B., Angeletti, R., Fiser, A., et al. (2006). Insights into the mechanism of microtubule stabilization by Taxol. *Proceedings of the National Academy of Sciences of the United States of America*, *103*(27), 10166–10173.

Yang, W., Gong, P., Yang, Y., Yang, C., Yang, B., & Ren, L. (2020). Circ-ABCB10 contributes to paclitaxel resistance in breast Cancer through let-7a-5p/DUSP7 Axis. *Cancer Management and Research*, *12*, 2327–2337.

Yang, Y., Li, H., He, Z., Xie, D., Ni, J., & Lin, X. (2019). MicroRNA-488-3p inhibits proliferation and induces apoptosis by targeting ZBTB2 in esophageal squamous cell carcinoma. *Journal of Cellular Biochemistry*, *120*(11), 18702–18713.

Yang, L., Tian, Y., Leong, W. S., Song, H., Yang, W., Wang, M., et al. (2018). Efficient and tumor-specific knockdown of MTDH gene attenuates paclitaxel resistance of breast cancer cells both in vivo and in vitro. *Breast Cancer Research*, *20*(1), 113.

Yoon, M. J., Park, S. S., Kang, Y. J., Kim, I. Y., Lee, J. A., Lee, J. S., et al. (2012). Aurora B confers cancer cell resistance to TRAIL-induced apoptosis via phosphorylation of survivin. *Carcinogenesis*, *33*(3), 492–500.

Yu, S. L., Lee, D. C., Son, J. W., Park, C. G., Lee, H. Y., & Kang, J. (2013). Histone deacetylase 4 mediates SMAD family member 4 deacetylation and induces 5-fluorouracil resistance in breast cancer cells. *Oncology Reports*, *30*(3), 1293–1300.

Yun, T., Liu, Y., Gao, D., Linghu, E., Brock, M. V., Yin, D., et al. (2015). Methylation of CHFR sensitizes esophageal squamous cell cancer to docetaxel and paclitaxel. *Genes & Cancer*, *6*(1–2), 38–48.

Zeng, Y., Yin, L., Zhou, J., Zeng, R., Xiao, Y., Black, A. R., et al. (2022). MARK2 regulates chemotherapeutic responses through class IIa HDAC-YAP axis in pancreatic cancer. *Oncogene*, *41*(31), 3859–3875.

Zhang, M. S., Furuta, M., Arnaoutov, A., & Dasso, M. (2018). RCC1 regulates inner centromeric composition in a ran-independent fashion. *Cell Cycle*, *17*(6), 739–748.

Zhang, J., Wei, W., Jin, H. C., Ying, R. C., Zhu, A. K., & Zhang, F. J. (2015). Programmed cell death 2 protein induces gastric cancer cell growth arrest at the early S phase of the cell cycle and apoptosis in a p53-dependent manner. *Oncology Reports*, *33*(1), 103–110.

Zhao, S., Tang, Y., Wang, R., & Najafi, M. (2022). Mechanisms of cancer cell death induction by paclitaxel: An updated review. *Apoptosis*, *27*(9–10), 647–667.

Zhao, W. J., Yi, S. J., Ou, G. Y., & Qiao, X. Y. (2021). Neuregulin 2 (NRG2) is expressed in gliomas and promotes migration of human glioma cells. *Folia Neuropathologica*, *59*(2), 189–197.

Zhao, Y. J., Zhang, J., Wang, Y. C., Wang, L., & He, X. Y. (2021). "MiR-450a-5p inhibits gastric Cancer cell proliferation, migration, and invasion and promotes apoptosis via targeting CREB1 and inhibiting AKT/GSK-3beta signaling pathway" front. *Oncologia*, *11*, 633366.

Zhou, Z., Zhang, L., Xie, B., Wang, X., Yang, X., Ding, N., et al. (2015). FOXC2 promotes chemoresistance in nasopharyngeal carcinomas via induction of epithelial mesenchymal transition. *Cancer Letters*, *363*(2), 137–145.

CHAPTER SEVEN

Epigenetic adaptations in drug-tolerant tumor cells

Nilanjana Mani, Ankita Daiya, Rajdeep Chowdhury, Sudeshna Mukherjee*, and Shibasish Chowdhury*

Department of Biological Sciences, Birla Institute of Technology and Science (BITS), Pilani, Rajasthan, India

*Corresponding authors: e-mail address: sudeshna@pilani.bits-pilani.ac.in; shiba@pilani.bits-pilani.ac.in

Contents

Abstract

Traditional chemotherapy against cancer is often severely hampered by acquired resistance to the drug. Epigenetic alterations and other mechanisms like drug efflux, drug metabolism, and engagement of survival pathways are crucial in evading drug pressure. Herein, growing evidence suggests that a subpopulation of tumor cells can often tolerate drug onslaught by entering a "persister" state with minimal proliferation. The molecular features of these persister cells are gradually unraveling. Notably,

Advances in Cancer Research, Volume 158
ISSN 0065-230X
https://doi.org/10.1016/bs.acr.2022.12.006

the "persisters" act as a cache of cells that can eventually re-populate the tumor post-withdrawal drug pressure and contribute to acquiring stable drug-resistant features. This underlines the clinical significance of the tolerant cells. Accumulating evidence highlights the importance of modulation of the epigenome as a critical adaptive strategy for evading drug pressure. Chromatin remodeling, altered DNA methylation, and de-regulation of non-coding RNA expression and function contribute significantly to this persister state. No wonder targeting adaptive epigenetic modifications is increasingly recognized as an appropriate therapeutic strategy to sensitize them and restore drug sensitivity. Furthermore, manipulating the tumor microenvironment and "drug holiday" is also explored to maneuver the epigenome. However, heterogeneity in adaptive strategies and lack of targeted therapies have significantly hindered the translation of epigenetic therapy to the clinics. In this review, we comprehensively analyze the epigenetic alterations adapted by the drug-tolerant cells, the therapeutic strategies employed to date, and their limitations and future prospects.

1. Introduction

Cancer is the second leading cause of mortality worldwide (Sung et al., 2021). While in definition it is simply uncontrollable growth of cells, repercussions of this growth are proved to be fatal. Over the years, the scientific community has extensively studied how and what causes cancer, and many mechanisms have been proposed that can drive this uncontrolled cellular growth. The signatures of cancer are considered as the ability to divide even in the absence of any growth factor stimulation, the ability to divide in the presence of anti-growth signals, resisting apoptosis, the potential to maintain telomere length over repeated cell division, stimulation of angiogenesis, and propensity to invade surrounding tissues and colonize to other parts of the body (Hanahan & Weinberg, 2011). It is also postulated that almost all of these features are acquired and essentially vary across the spectrum of cancers. During the cancer progression, the above hallmarks may not exhibit all at once but take different routes to reach the same destination.

Researchers often describe cancer as a series of atavistic events leading to inter and intra-tumor heterogeneity generated through somatic mutations. Even inactivation of the tumor suppressor genes can also be acquired through various epigenetic alterations (Berdasco & Esteller, 2010; Esteller, 2007; Jones & Baylin, 2007). According to the mathematical model of Goldie–Coldman hypothesis, due to intrinsic genetic instability, random mutations within tumor cells lead to a generation of the drug-resistant tumor cell population (Goldie & Coldman, 1985). The generation of these drug-resistant clones is one major hindrance in how patients respond to therapy. Hence, understanding the molecular mechanism of drug resistance in cancer is crucial. However, resistance is a well-defined concept in cancer therapeutics and is exhaustively being studied. Recently, a new drug tolerance concept has garnered much attention from researchers and clinicians. Herein, drug tolerance is defined as the ability of the cells to withstand a drastic yet transient exposure to a chemotherapeutic agent. While resistance is thought to be more of a permanent or stable phenotype, drug tolerance is a stage that precedes resistance and is more of a temporary suite. The "temporary" nature of these cells makes them a perfect target for therapeutics before the tumor becomes rigid to target. However, there is no clear, well-defined distinction between drug-tolerant and drug-resistant cells. The temporary nature of the drug-tolerant cancer cell population is often attributed to the dynamic epigenetic modulation, which is now considered one of the additional Hanahan and Weinberg hallmarks of cancer cells (Hanahan & Weinberg, 2011). Hence, epidrugs are currently being tested as adjuvants with conventional chemotherapeutic drugs to sensitize cancer cells (Majchrzak-Celińska, Warych, & Szoszkiewicz, 2021).

The molecular progression of tumorigenesis and acquiring drug resistance features of cancer cells require genetic and epigenetic programs to function together to alter gene expression. Many events affect gene regulation in cancer, starting from mutational events in the DNA sequence insertions/deletions, inversions, translocations, single nucleotide variants, copy number variation, and gene fusions. One of the primary contributing factors to cancer progression is the suppression of tumor suppressor genes and/or activation of proto/oncogenes. This activation/suppression is usually synonymous with epigenetic change. DNA methylation acts as a switch between the ON or OFF stage of expression or could be hypermethylation of promoters or modification of histone marks. The chromatin structure can modulate gene expression locally or at a distance through looping, nuclearsmatrix association, and nucleosome positioning (Jiang & Pugh, 2009). Similarly, non-coding RNAs like miRNAs, long non-coding RNAs, and siRNAs can also regulate cancer-promoting gene expression by modulation promoters and/or working in conjunction with other epigenetic factors (Kumar, Gonzalez, Rameshwar, & Etchegaray, 2020). In addition to changes at the chromosome level, variations in non-coding RNAs (ncRNAs) and RNA modifications are also dominant in cancer. Existing databases show myriad miRNAs being aberrantly expressed in almost every type of cancer (Xie, Ding, Han, & Wu, 2013; Xie, Ma, & Zhou, 2013). These ncRNAs are targeted by DNA methylation enzymes, sometimes compounding the already complex gene regulation. And these interactions among different epigenetic mechanisms can either synergistically or antagonistically alter genetic expression.

Most of these epigenetic modifications are believed to be erased with new generations. However, studies provide contradictory results that suggest that epigenetic changes may endure in at least four subsequent generations of organisms. The concept of epigenetic memory is emerging and is an exciting field of research. For example, methylation marks can be inherited from either the maternal (Giuliani et al., 2015) or paternal gamete - the concept of genomic imprinting is well established (Soubry, 2015). To better understand the impact of epigenetics on development, areas of gene methylation in one species are compared with its closest living relatives (Hernando-Herraez et al., 2013). Many of the regions in the human genome that are methylated are transcription factors (TFs) that can exert influence over distant regions by promoting or suppressing genes through looping, and a slight difference in the epigenome surrounding the TFs can result in widely varying phenotypes between individuals of the same species (Weirauch & Hughes, 2010). It has also been noticed that histone-modifying enzymes play a significant role in activating oncogenes or inactivating tumor-suppressing genes. Thus, their expression varies between tumor cells and normal tissues. Notably, the genetic alteration in epigenetic enzymes contributes majorly to these known epigenetic modifications. This again substantiates that the interplay of genetic and non-genetic mechanisms gives rise to resistance. In this chapter, we will try to understand the fundamentals of epigenetics and how epigenetics regulates tumor progression and drug tolerance. We shall have a deeper understanding of drug-tolerant persisters and how they are epigenetically regulated. Finally, we will exploit these epigenetic modulations to target the persister cells effectively.

2. Epigenetic regulations

Different arms of epigenetic regulation of gene expression are schematically shown in Fig. 1, which includes transcription factors, non-coding RNAs, DNA methylation, and chromatin remodeling.

2.1 Transcription factors

Transcription factors (TF) are proteins involved in transcribing information embedded in DNA into RNA. Many TFs functions as "master regulators," controlling cellular processes (Whyte et al., 2013) and controlling specific

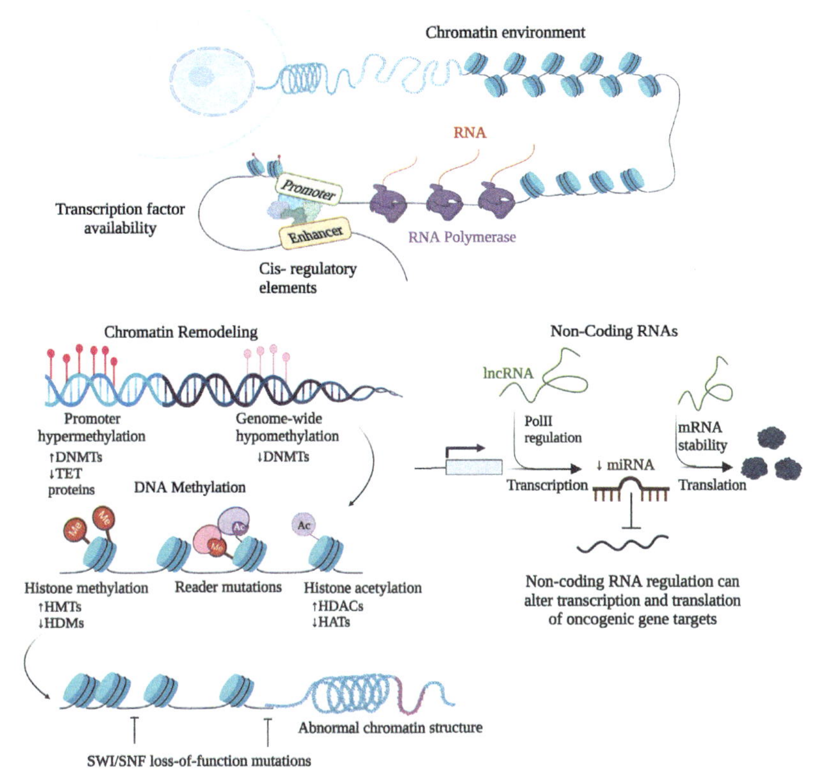

Fig. 1 Schematic diagram showing the different arms of epigenetic regulation in general. Chromatin remodeling entails changes to either histone tails and/or DNA sequences, whereas non-coding RNAs are associated with various chromatin remodeling complexes to change DNA methylation or histone status.

pathways such as immune responses and cell cycle checkpoints. The roles of TFs are often conserved among metazoans (Bejerano et al., 2004), suggesting that there lies a global regulatory network conserved among species. And yet, extensive diversity in TF's turnover in individual cell types is observed (Weirauch & Hughes, 2010), and TFs can duplicate and diverge over longer timescales. The same TF can regulate various genes in different cell types (Geertz, Shore, & Maerkl, 2012), suggesting that these regulatory networks are dynamic even within a single organism. Hence, understanding how TFs recognize binding sites and control transcription could be complex, yet it is crucial to understand their physiological roles in gene expression in organisms.

TF-DNA interaction is the primary determinant for TF functioning, commonly termed motifs. Over the past decades, several in-silico motif search tools like TRANSFAC (Matys et al., 2006) and JASPAR (Mathelier et al., 2016) have been developed to identify TF binding motifs in DNA sequences. It was noticed that TFs could have an even 1000-fold higher preference for specific binding sequences relative to other DNA sequences (Damante et al., 1994; Geertz et al., 2012). It is now known that different histone modifications such as methylation and phosphorylation can epigenetically control the accessibility of promoters to transcription factors, which can further modulate chromatin structure through protein-protein interactions and/or allow competing transcription factors to play that respond to different biological cues (Margueron & Reinberg, 2010). Transcription factors can contribute to epigenetic control of gene expression through the recruitment of regulatory machinery for histone and/or through DNA modifications in which TFs occupy strategic positions in the genome (primarily within promoters) (Esteller, 2007). Among the various DNA modifications, the 5-methylcytosine (5mC) mark is considered a hallmark for DNA methylation, leading to the transcriptionally silencing of cis-regulatory elements (Deniz, Frost, & Branco, 2019). It is observed that CpG islands have TF binding sites (TFBS) that are recognized by TFs through sequence-specific TF–DNA interactions. Herein, transcription factors can either repress or activate the transcription of target genes by recruiting other cofactors. Additionally, several transcription factors directly interact with DNA methyltransferases and/or TET proteins, most likely recruiting these enzymes to particular regions within DNA, thereby forming a feedback loop (Li et al., 2016). Evidence suggests that various ways TFs regulate gene expression, which include assembly of transcripts, regulation of microRNA expression, and controlling the organization of proteins essential for accessibility and transcriptional competency of genomic sequences.

2.2 Non-coding RNAs

Almost all high throughput transcriptomic analyses of the eukaryotic genome, including humans, revealed that although up to 90% genome transcribe, only 1–2% of transcript code for proteins (Feingold & Pachter, 2004). Hence, the majority of transcripts are considered non-coding RNAs (ncRNA). However, compared to mRNAs, most of these putative non-coding RNAs are present at low levels indicating these RNAs may be involved in regulatory roles within the cell. Moreover, the abundance of non-coding RNA somewhat correlates with its level of conservation (Managadze, Rogozin, Chernikova, Shabalina, & Koonin, 2011), which might be a good indicator of its function (Brunet & Doolittle, 2014; Elliott, Linquist, & Gregory, 2014).

Among various ncRNAs, ribosomal, transfer, small nuclear, and small nucleolar RNAs can be classified into infrastructural ncRNAs, whereas microRNAs (miRNAs), Piwi-interacting RNAs (piRNAs), small interfering RNAs (siRNAs), and long non-coding RNAs (lncRNAs) are part of regulatory ncRNAs. At a steady state, a significant fraction of ncRNA human cellular RNA consists of ribosomal RNA (rRNA). However, because of the small size of tRNA, its mole fraction in human cells is higher than that of rRNA. Among other ncRNAs, small nuclear RNAs (snRNA) and small nucleolar RNAs (snoRNAs) are present one to two-fold lower than that of rRNA or tRNA. Among the small regulatory RNAs, miRNAs and piRNAs are more abundant in cells. Though an abundance of small regulatory RNAs is dependent heavily on cell type.

2.2.1 Long non-coding RNAs and their regulatory function

ncRNAs longer than 200 nucleotides with very little or no potential for translation is characterized as "long non-coding RNAs" (lncRNAs). The actual estimation of the number of lncRNAs in humans is difficult because the expression of lncRNAs is cell-specific, development stage-specific, and disease-specific. However, recent genomic studies (Harrow et al., 2012) predict around 15,000 lncRNAs in humans, with even a tiny fraction (less than 1000 lncRNAs) of estimated lncRNAs expressing more than one copy per cell. Although tens of thousands of lncRNAs have been profiled yearly, only a tiny fraction have known specific functions (Djebali et al., 2012; Palazzo & Gregory, 2014). Although many annotated lncRNAs are mRNA-like, they are less in abundance and less evolutionarily conserved (Wu, Yang, & Chen, 2017). lncRNAs are more localized in the nucleus than mRNAs (Quinn & Chang, 2016) and associated with cis elements-protein complexes (Palazzo & Lee, 2018). lncRNAs can act as decoy molecules and scaffold adaptors, among other functions. lncRNAs also bind to transcription factors that can elucidate broader downstream effects on cellular transcriptional programs. Various lncRNA like GAS5 (Hudson et al., 2014; Kino, Hurt, Ichijo, Nader, & Chrousos, 2010), PANDA (Hung et al., 2011), PCGEM1 (Hung et al., 2014), NKILA

(Liu, Liu, Xu, & Li, 2015, CCAT1 (Zhang et al., 2017) are actively involved in gene regulation. Recently discovered lncRNA THOR functions as an oncogene (Hosono et al., 2017).

It has been observed that lncRNAs recruit protein complexes to specific sites (often promoters) and alter gene expression by modifying the chromatin architecture. lncRNAs modulate the chromosome architecture in more than one way, including chromatin remodeling, X-chromosome inactivation, chromosome reshuffling, and histone modification (Loda & Heard, 2019). In this regard, HOTAIR (Imai-Sumida et al., 2020) and INK4 (Meseure et al., 2016) are two prominent lncRNAs. lncRNA HOXA11-AS is known to bind to EZH2 to facilitate histone methylation, which in turn inhibits the transcription of the tumor suppressor gene p21 (Lu, Lv, Qiu, Wang, & Cao, 2017).

Apart from this, lncRNAs like ARLNC1 (Zhang et al., 2018), LincRNA- p21 (Yoon et al., 2012), and MEG3 (Mondal et al., 2015) can directly bind to nucleic acids and alter the normal function of DNA. LncRNA can also function as a "sponge" for miRNAs, the commonly known Let-7. TUG1 has acted as a sponge for miRNAs targeting tumor suppressor PTEN in prostate cancer (Du et al., 2016).

2.2.2 microRNAs as gene expression regulator

miRNAs are short ncRNAs of approximately 19 to 26 nucleotides in length that regulate the expression of other RNAs, primarily mRNAs, through binding between the 5′ end (known as the "seed" region) of the miRNA with complementary sequences in target RNAs. Mature miRNAs more commonly bind to the 3′ untranslated regions (3′UTR) of mRNAs and inhibit their use by either degradation or translational repression (Bartel, 2018; Esquela-kerscher & Slack, 2006). For humans, miRBase v22 shows a total of 2654 mature miRNAs, which could be experimentally verified (Huang, Du, Wen, Lu, & Zhao, 2022). It is now well documented that miRNAs are conserved at their seed region across species, especially within the animal kingdom, which reinforces their critical role in cellular function (Ha, Pang, Agarwal, & Chen, 2008).

A bunch of studies indicates that miRNAs affect the expressions of different components of the epigenetic machinery, while on the other hand, miRNAs can be regulated by epigenetic events (Imai-Sumida et al., 2020). Many studies indicate that miRNAs can epigenetically dysregulate critical enzymes such as DNA Methyltransferases (Meng et al., 2018), TETs, Histone Deacetylases (Saito, Saito, Liang, & Friedman, 2013), and

EZH2. The miRNAs regulating the epigenome are most often termed as epi-miRNAs. Additionally, miRNAs can also regulate histone modification. In-silico analyses have identified histone modifications, including H3K4me1, H3K27me3, H3K27ac, H3K9ac, H3K4me3, and H2AZ that are regulated by miRNAs during mammalian spermatogenesis (Taguchi, 2015).

2.3 DNA methylation as an epigenetic regulator

Since the early 1960s, DNA methylation has been known in prokaryotes. Later in 1977, 5-methylcytosine (5mC) was first identified in eukaryotic organisms, which is now considered a primary epigenetic mechanism (Razin & Cedar, 1977; Razin & Riggs, 1980). The mammalian DNA methylation machinery consists of the enzyme DNA methyltransferases (DNMTs), which establish and maintain DNA methylation patterns (Auclair & Weber, 2012), and the methyl CpG binding proteins (MBDs) that read these methylation marks (Jones & Liang, 2009). In normal cells, DNA methylation usually occurs in repetitive sequences such as satellite DNA, LINES, and SINES (Pappalardo & Barra, 2021). Additionally, CpG islands in promoters remain unmethylated for the underlying genes to be expressed fully (Ndlovu, Denis, & Fuks, 2011). In addition to gene expression regulation, DNA methylation is involved in many biological processes, including RNA splicing (Anastasiadou, Malousi, Maglaveras, & Kouidou, 2011), genome organization (Madakashira & Sadler, 2017), imprinting (Lee et al., 2002) and X chromosome inactivation (XCI) (Pontier & Gribnau, 2011). Hence, DNA methylation is crucial in many biological processes that contribute to human development and diseases, including cancer (Subramaniam, Thombre, Dhar, & Anant, 2014).

2.4 Histone modifications and chromatin remodeling

Since first described by Vincent Allfrey in the early 1960s (Allfrey, Faulkner, & Mirsky, 1964), we know histones are post-translationally modified, and there exists a plethora of modifications. The chromatin structure of eukaryotes can be visualized as an onion- there are a series of organizational layers as you peel right through. The first layer is the DNA sequence that is folded into nucleosomes-fundamental units of chromatin consisting of ~150 bp of DNA wrapped around a histone protein. Four core histones- H2A, H2B, H3, and H4- can be chemically modified. With increasing knowledge of chromatin structure, we now understand how different regulatory elements influence differential gene expression. The positioning of nucleosomes, along with histone variants and modifications, make up the primary structure of chromatin. These modifications affect chromatin structure and recruit remodeling enzymes, which use the energy released by ATP hydrolysis to reposition nucleosomes. These enzymes help maintain nucleosome density and spacing, facilitating adequate gene repression, which helps in preserving histone-DNA contact and facilitating post-translational modifications of histone (Venkatesh & Workman, 2015). The recruitment of proteins and complexes with specific enzymatic activities is now widely accepted as a method by which gene expression is regulated.

Acetylation, methylation, and phosphorylation are three major modifications observed with histone proteins that alter histone function, altering gene expression. The addition of an acetyl group on multiple lysine residues on histone tails leads to the alteration of the positive charge at lysine and the weakening of the interaction between DNA and histone protein (Shahbazian & Grunstein, 2007). Methylation mainly occurs on the side chains of Arginine and/or Lysine residues. Like histone acetylation, the phosphorylation of histones also causes a change in the charge and is known to be a highly dynamic process (Zentner & Henikoff, 2013). The majority of phosphorylation occurs on serine, threonine, and tyrosine residues. In this process, a significant negative charge is added to the histone, influencing the chromatin structure and making it more open (Dou & Gorovsky, 2000). Apart from these significant modifications, a plethora of other alterations collectively referred to as "atypical" histone modifications like histone deamination/citrullination, ubiquitination, ADP-ribosylation, deamination, N6-formylation, and O-GlcNAcylation.

3. Drug tolerant tumor cells (DTPs) and their adaptations

3.1 Origin of DTPs

Gladys Hobby discovered in 1942 that while penicillin could kill most cultured streptococcal cells, just around 1% could withstand penicillin shock (Hobby, Meyer, & Chaffee, 1942). Later, these surviving bacterial cells were given the name "bacterial persisters" by Bigger (1944). It was hypothesized that- the slow cycling of these cells allowed for their survival rather than inherent resistance mechanisms. Recent live imaging experiments had confirmed Bigger's hypothesis, confirming that the bacterial cells that survived antibiotic treatment were not proliferating before the antibiotics were added. Surprisingly, these slow-cycling bacteria revert to a proliferative state upon antibiotic withdrawal, repopulating the sensitive bacterial population. This substantiates that persistence is a temporary phenomenon mediated by phenotypic changes rather than stable genetic alterations.

3.2 Persisters in cancer

Cancer cell populations have been known to show considerable variability in their response to treatment owing to inter and intra-tumor heterogeneity. Most of these phenomena, over the years, have been genetically driven, i.e., through somatic mutations. And if Darwinian forces are at play in such a heterogeneous population, they could drive the resistant clones to repopulate. Typically, mutational and non-mutational adaptations during drug resistance acquisition are thought to be mutually exclusive. While the initial therapy onslaught is overcome through the non-genetic mechanisms, secondary genetic changes could happen in due course of time, leading to disease progression. This population of cells that survive transient exposure to a very high drug concentration is called the drug-tolerant persisters (Sharma et al., 2010). Although these definitions have been well established for microbial species, they have remained ambiguous in cancer and are often used synonymously.

Settleman and colleagues extended the concept of persisters to the cancer field in 2010 to describe cancer cells treated with targeted therapies (Sharma et al., 2010). Using a well-established NSCLC cell line, it was discovered that a small fraction of the original population, approximately 0.3%, survived treatment with a lethal dose of an EGFR inhibitor, erlotinib. The subpopulation of slowly cycling dormant cells was termed the drug-tolerant persisters (DTPs). Over the course, a fraction of these DTPs started proliferating when Erlotinib was reintroduced, giving rise to the second population of cells called drug-tolerant expanded persisters (DTEPs). These are similar to bacterial persisters and give rise to a sensitized cellular population. Other similar studies on different cell lines- melanoma (Sun et al., 2014), colorectal cancer (Russo et al., 2018), breast cancer (Risom et al., 2018), and osteosarcoma (Mukherjee, Dash, Lohitesh, & Chowdhury, 2017) also showed the survival of a small subpopulation of these transient drug-tolerant persister cells. One defining feature of these persisters is their reversibility, i.e., these cells undergo transient changes allowing them to survive, and those changes are majorly epigenetically driven. As already discussed, epigenetics studies change gene functions that are mitotically and/or meiotically heritable and do not involve a difference in the DNA sequence. Epigenetic changes in tumors have long been implicated in cancer development and progression and resistance to chemotherapy (Cheng, He, et al., 2019; Cheng, Sheng, et al., 2019). The tumor epigenome is characterized by global changes associated with DNA methylation, histone modification patterns, and altered expression profiles of chromatin-modifying enzymes. Importantly, epigenetic reprogramming and its potential role in reversing sensitivity have recently been a hot topic. In their pioneering work, Sharma et al. observed the response to epidermal growth factor receptor (EGFR) inhibitors in an EGFR-mutant lung cancer-derived cell line (PC9) that shows exquisite EGFR tyrosine kinase inhibitor sensitivity. It is observed that although the majority of cells were killed upon exposure to an acute drug concentration, a small fraction of slowly growing cells (Drug Tolerant Persisters, aka DTPs) could still be detected 9 days later. Collectively, it is postulated that under selective therapeutic pressure, resistance to treatment can be due to the expansion of these DTPs or the evolution of drug-tolerant cells. However, when these persisters were treated with a histone deacetylase inhibitor, there was a reversal in their sensitivity, suggesting that cancer cell populations can deploy a very dynamic survival strategy in which individual cells can transiently assume a reversibly drug-tolerant state to protect the people from eradication by lethal drug exposures, as shown schematically in Fig. 2.

A better understanding of how persister cells arise may help guide therapeutic strategies to target them and prevent or delay the development of drug resistance, given that they are the bottleneck to acquired drug resistance. While most studies suggest that these persisters are formed due to de novo changes, some studies indicate that drug-tolerant persister populations or cancer cells are already primed to exist before treatment (Hata et al., 2016). To address this dichotomy, Kurppa et al. used DNA barcoding on PC9 cells treated with an EGFR inhibitor, a MEK inhibitor, or both (Kurppa et al., 2020). The findings revealed a high proportion of shared barcodes in cells treated with an EGFR inhibitor alone, implying that cancer cells are being pre-selected. Similarly, Schaffer et al. demonstrated that transcriptional heterogeneity exists at the single-cell level in human BRAF mutant melanoma cells, eventually leading to drug resistance (Shaffer et al., 2017). Another study found that pre-treatment of HER2-overexpressing breast cancer line BT474 parental cells with the GPX4

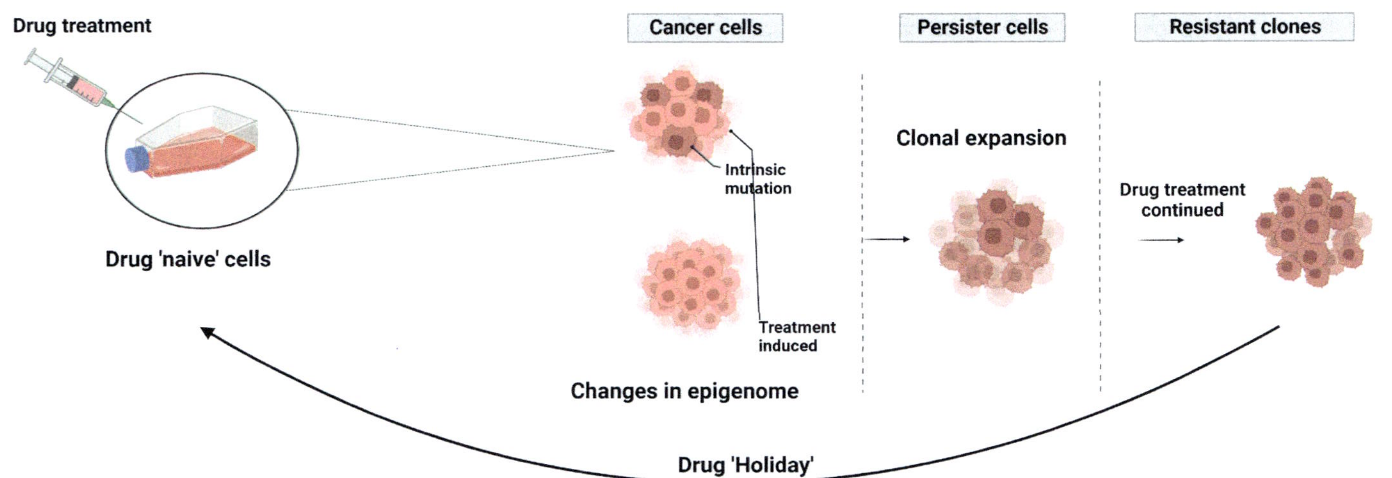

Fig. 2 Illustration showing the evolution of persisters. The population of cells that survive transient exposure to high concentrations of a drug is called the drug-tolerant persisters. These cells eventually proliferate and give rise to resistant clones. This drug tolerance is reversed if the cells are subjected to a "drug holiday," implying the importance of stochastic epigenetic mechanisms favoring their survival.

inhibitor RSL3 reduced the number of persister cells after subsequent treatment with the HER2 inhibitor lapatinib, showing that GPX4-sensitive cells exist (Hangauer et al., 2017).

While the existing heterogeneity might be one cause of the development of persistence, de-novo changes and/or (epigenomic) plasticity might also be a root for the same. When Kurppa et al. performed a combination treatment of EGFR and MEK inhibitors in the same set of DNA barcoding experiments described above, the majority of the resultant barcodes were unique, supporting the notion that this is a stochastic event rather than genetic. Similarly, Risom et al. demonstrated that a drug-tolerant persistent state in breast cancer caused by MEK and PI3K/mTOR inhibitors arise through distinct cell-state transitions involving dynamic remodeling of open chromatin (Risom et al., 2018), as opposed to a simple selection of pre-existing subpopulations. Another study discovered that transient STAT3 activation by IL-6 and FGFR in response to oncogenic kinase inhibition aids in the survival of several oncogene-addicted cancer cell lines (Lee, Zhuang, et al., 2014). All these studies demonstrate that pre-existing and stochastic persister states are not mutually exclusive. While pre-existing clones may contribute to genetic heterogeneity, stochastic events during treatment initiation may drive phenotypic heterogeneity, allowing the fittest progenies of either to survive and proliferate.

3.3 Comparison of DTPs and drug-resistant cells

While much is known about drug resistance, drug persistence is still looming to find newer definitions. The first theme that separates persister cells from resistant clones is that only a tiny population of ~0.1–0.3% cells remain viable despite drug exposure. At the same time, resistance is a more stable state comprising a wider population of cells. Extensive next-generation sequencing systems have allowed us to identify the genetic changes that cause resistance in treatments. Xue et al. performed the selection and subsequent propagation of amplification of the mutant B-Raf proto-oncogene (BRAFamp) in patient-derived tumor xenograft mouse models of both lung cancer and melanoma BRAFV600E mutation (Xue et al., 2017). Consistent with the drug acting as a selection pressure, monotherapy led tumor cells to acquire extensive genetic alterations that could be expanded parallelly.

Contrary to the study by Xue et al. highlighting the genetic perception, Shaffer et al. showed a non-genetic mechanism contributing to the emergence of cells that can evade drug treatment (Shaffer et al., 2017).

The authors took the human melanoma possessing the same BRAF mutation, BRAFV600E, that Xue et al. investigated. Instead, they followed to study mutations in the BRAF gene after giving a BRAF inhibitor Vemurafenib; Vemurafenib only inhibits the mutated BRAFV600E protein. Surprisingly, this treatment nearly eradicated the tumor cells in a population; however, a small subset of cancer cells persisted and developed drug resistance. Intriguingly no mutations in response to drug treatment were found; instead, extreme transcriptional variability at the single-cell level was observed. This variability showed infrequent transcription of several resistance markers at high levels but in a tiny percentage of cells. Adding drug-induced epigenetic reprogramming in these cells is proposed to convert the transient transcriptional state to a stable resistant state. This leads us to define another difference between resistance and tolerance- while drug tolerance is a transient state from which we can return; in contrast, drug resistance is stable for generations, further getting stronger by genetic alterations.

Diverse drug-resistance mechanisms can arise from pre-existing mutations before treatment (extensively studied) and slow-growing persisters after long-term or periodic treatment. Both mechanisms may contribute to drug resistance in clinics. Therefore, tolerance is key to understanding and preventing resistance but is not comprehensively well understood. While most researchers still stress the genetic module of resistance, the stochastic/epigenetic regulation could be the key to not allowing the resistance to even set in the first place.

3.4 Features of DTPs

There is still no clear consensus among the scientific community about the characterization, markers, or even the origin of the various slowly proliferating drug-tolerant phenotypes. The lack of appropriate models or ones that simulate an actual clinical scenario has significantly hindered the identification of suitable markers for tolerance. However, most reports describe tolerant cells as having a slow cell cycle, altered metabolism, an altered epigenetic state, resistance to apoptosis, and immune evasion. In addition, there is increased cellular plasticity and heterogeneity, which allows these cells to evade therapies in multiple ways.

3.4.1 Slow cell cycle

DTPs are frequently compared to quiescent cells due to their reversible proliferation restriction. Quiescence is a physiological cell cycle state (G0) common in adult stem cells from low turnover tissues, but it is also a feature of cancer stem cells (CSCs). While the distinction between CSCs and DTPs is debatable, there are some similarities. Increased tolerance to therapies, for example, is a common feature of CSCs (Zhou, Zhang, Zhang, & Li, 2021), whereas stemness features have been described for DTPs in several cancer subtypes (Raha et al., 2014). Consistent with this premise, cancers with high expression of stemness gene signatures are known to have a poor prognosis, whereas DTPs also resist drug pressure (De Conti, Dias, & Bernards, 2021). While the cellular non-dividing state-quiescence is on one end of the spectrum, senescence, defined as a stable cell cycle arrest, is on the other. Senescence can be induced by telomere shortening, DNA damage, excessive oncogenic signaling, or various stressors (Chen, Hales, & Ozanne, 2007). Because both radio and chemotherapies can induce cancer senescence, which is known as therapy-induced senescence (TIS) (Ewald, Desotelle, Wilding, & Jarrard, 2010), senescent cancer cells share intriguing similarities with normal senescent cells, such as the formation of heterochromatic foci and resistance to apoptotic cell death. Importantly, it has been demonstrated that senescent cancer cells, despite being labeled irreversible, can re-enter the cell cycle in response to changes in the tumor microenvironment or epigenetic alterations, thus reflecting features observed in tolerant cells. Schmitt and colleagues described the enhanced tumor-initiating potential of senescent lymphoma cells. These findings suggest that persister cells induce senescence to avoid therapy slaughter. As previously stated, drug-tolerant cells enter a dormant or diapause state after initial drug exposure. It has previously been proposed that the slow proliferation rates shared by the phenotypes described above provide a survival advantage in the presence of cytotoxic drugs (De Conti et al., 2021). This cell cycle restriction has recently been linked to the upregulation of error-prone DNA repair mechanisms, which can favor the development of complete resistance. According to this scenario, slow proliferating drug tolerance is akin to either quiescence or senescence and is a critical intermediate step towards acquiring genetic resistance.

3.4.2 Metabolism

The persister cells that demonstrate a slow dividing state also affect the metabolism of these cells. Cellular plasticity, which results in changes in chromatin state and transcription modulation, has been shown to slow metabolism, causing it to rely less on glycolysis and more on mitochondrial oxidative phosphorylation (OXPHOS). This pathway is upregulated in models of KRASG12D pancreatic ductal adenocarcinoma (PDAC) (Ju et al., 2017), BRAFV600E melanoma (Shen et al., 2020), as well as Acute Myeloid Leukemia (AML) (Baccelli et al., 2019) and Chronic Myeloid Leukemia (CML) (Kuntz et al., 2017), upon treatment with targeted and/or conventional chemotherapies. Persisters become more oxidatively stressed as OXPHOS activity increases, and they must rely on various antioxidant mechanisms to restore redox homeostasis. In response to oxidative stress, a pathway that gets upregulated is the glutathione-dependent reduction of lipid peroxides. Recent research has identified phospholipid glutathione peroxidase 4 (GPX4) inhibition as a powerful strategy for eradicating DTP cells via ferroptosis (Hangauer et al., 2017).

Similarly, NRF2, a master regulator of redox homeostasis, is also tightly regulated in DTPs to balance oxidative stress (Qin et al., 2022). Interestingly, ALDH, a known stem cell marker, has been shown to protect persister cells from reactive oxygen species (ROS)mediated toxicity (Raha et al., 2014). But what is interesting to note is that increased OXPHOS activity might yet not be a universal DTP feature; in a study on Myc-driven mouse lymphoma model, persister cells tolerant to chemotherapy displayed a very active metabolism, with increased glucose consumption despite exhibiting senescence, thus endorsing the plasticity of these tolerant cells (Dörr et al., 2013). Another feature DTP cells share is the ability to sustain metabolism through alternative nutrient sources such as autophagy or fatty acid oxidation (FAO) pathways (Shen et al., 2020). In this regard, autophagy aids in metabolite recycling derived from macromolecule degradation and, in conjunction with acetyl CoA production in the FAO process, maintains ATP production in the mitochondria. In BRAFV600E melanoma, increased FAO is required by the cells to survive MAPK inhibition before they develop resistance (Shen et al., 2020). Overall, quite little is known about the metabolic properties of persisters, and further research is required to identify probable metabolism-targeting markers.

3.4.3 Immune evasion

It is still not well defined how persisters function at evading response to immunotherapies. On the other hand, a recent study in a murine organotypic ex-vivo model described the emergence of a discrete subpopulation of cells resistant to an immune checkpoint inhibitor (antiPD1 antibody) (Sehgal et al., 2021). These cells expressed Snail and stem cell antigen Sca1 and were resistant to CD8+ T cell-mediated killing. Surprisingly, these tolerant cells were found to be dependent on Birc2/3 anti-apoptotic factors. Inhibiting these factors with antiPD1 therapy resulted in synergistic tumor cell killing in vivo (Meitinger et al., 2020). Furthermore, the interaction of dormant cancer cells and the immune system has been investigated, with immune evasion mechanisms such as the downregulation of MHCI molecules and the upregulation of PDL1 (Saudemont & Quesnel, 2004). Melanoma patients who received anti-PD1 therapy developed resistance through the downregulation of primary histocompatibility complex class 1, which was linked to TGFß activity, SNAI1 upregulation, and a dedifferentiated phenotype (Meitinger et al., 2020). Suppression of MHC expression, interestingly, is also a salient feature of cancer stem cells (CSCs). Not only that but adaptive resistance to immunotherapy is associated with increased expression of alternative immune checkpoint molecules. When we talk about adoptive T cell transfer (ACT), the tumor microenvironment is rich in inflammatory mediators. Particularly, TNFα has been identified to induce a reversible loss of antigens in melanoma cells, preventing them from apoptosis-induced cell death (Landsberg et al., 2012).

Similarly, ACT cleared most of the tumor in another model of squamous cell carcinoma, but a few slow-cycling cells were found looping around. These cells were found to have acquired the expression of CD80, a ligand of the cytotoxic T lymphocyte antigen 4 (CTLA4), thereby decreasing cytotoxic T cell activity. Furthermore, existing research indicates that inflamed stromal cells secrete hepatocyte growth factor (HGF) in response to T cell therapy, favoring the recruitment of immune suppressive neutrophils that limit T cell expansion and function (Landsberg et al., 2012). This suggests that even therapeutic interventions can sometimes cause stochastic changes, favoring persisters' immune evasion.

3.4.4 Apoptosis

Existing studies and our findings show that drug exposure can significantly reduce the overall level of transcripts. In addition, slow-proliferating drug-tolerant cells are also characterized by a low level of mRNA translation. Herein, some studies describe the role of post-transcriptional modifications in enhancing the stability of specific transcribed mRNAs employed in pro-survival regimes. Surprisingly, DTPs from EGFR-mutated NSCLC can survive gefitinib treatment by upregulating the antiapoptotic protein MCL1 via mTORC1-mediated post-transcriptional regulation of mRNA translation (Song, Hosono, Turner, & Jacob, 2018). However, while apoptosis inhibition appears to be a necessary prerequisite for developing eventual drug resistance, how it is instilled and executed requires further investigation. Because the vast majority of drugs in clinical trials target anti-apoptotic proteins, a thorough understanding of how apoptosis works in DTPs is critical.

3.4.5 Cellular plasticity

As mentioned before, a drug onslaught renders the transcriptional machinery to slow down temporarily, followed by comparative restoration of transcript levels as the persisters shun their dormant state and resume proliferation. Therefore, the cells during each of these distinct phases show immense cellular plasticity that underlies their transcriptional control (Niveditha, Sharma, Sahu, et al., 2019). Such plasticity also includes changes in intracellular pathways that may merge into reversible epigenetic changes and transcriptional factor modulation to control cell behavior. DTPs from melanoma cells, for example, exhibited a neural crest stem cell transcriptional state in response to targeted therapy, favoring the development of fully resistant clones (Rambow et al., 2018). There are reports that DTP cells with altered expression of differentiation state markers emerge during MEK and PI3K/mTOR inhibitor treatment in triple-negative and basal-like breast cancers via a cell state transition involving chromatin remodeling, which then plays a critical role in therapy evasion (Risom et al., 2018).

Most often, the reversibility of these DTPs is attributed to changes in chromatin structure, as chromatin is a highly dynamic entity on its own. This process can also be accompanied by a phenotypic switch, such as the epithelial-to-mesenchymal transition (EMT) or the acquisition of stem/progenitor-like phenotypes, as previously discussed. It is thus intriguing to investigate how phenotypic plasticity is linked to drug tolerance and persistence. According to this study, EMT has been linked to the early survival of DTP cells in EGFR-mutated NSCLC after tyrosine kinase inhibition (TKI). To identify EGFR inhibition as a promising therapy to halt DTP survival and expansion, mesenchymal cells from NSCLC biopsies were used in place of persister cells (Raoof et al., 2019). Such phenotypic changes, traditionally associated with changes in chromatin organization, have also been observed in basal cell carcinoma. Similarly, a quiescent residual tumor population was selected in response to the Hedgehog pathway inhibitor vismodegib (Biehs et al., 2018). Therefore, the phenotypic switches aid the tumor cells in tolerating drug pressure, often achieved through chromatin reorganization. The key players involved and the precise mechanism are still elusive. Fig. 3 schematically illustrates the vulnerabilities of persisters that can be targeted for their adequate clearance.

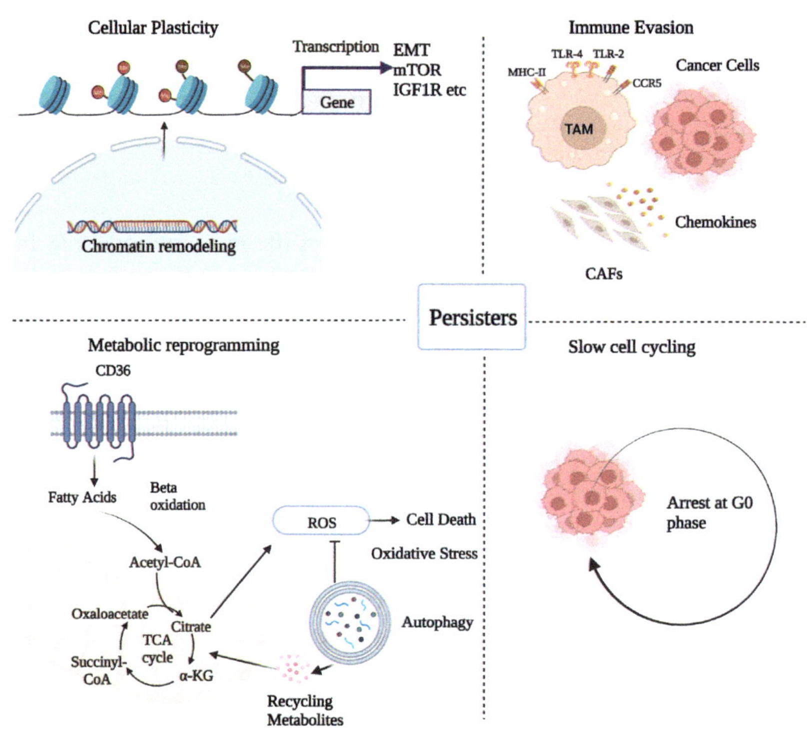

Fig. 3 A schematic diagram illustrating the targetable features of drug-tolerant persisters. Persister cells are characterized by changes in chromatin after drug exposure, leading to the upregulation of genes helping in survival. Cellular plasticity also allows them to avoid getting cleared by immune cells. Persisters, unlike parental cancer cells, utilize oxidative phosphorylation and antioxidant mechanisms. Additionally, they upregulate autophagy and fatty acid oxidation to substitute for nutrients.

4. Epigenetics and associated changes—Key to DTP emergence and survival

Epigenetic modifications lie at the heart of tumor progression, which drives cancer cells to thrive. Therefore, understanding the modifications in the context of "drug tolerance" becomes imperative as it can lead to the development of a transient state of reduced sensitivity to drugs, yielding stable drug-resistant cells (Niveditha, Sharma, Sahu, et al., 2019). In this regard, one of the pioneering studies was performed by Sharma et al. (2010), which showed that non-small cell lung carcinoma (NSCLC) patient-derived cell line PC9 initially responds to TKI drugs; however, a small subpopulation of PC9 cells survive after approximately 9 days of the drug exposure. While although most of the population died after a few days of initial drug exposure, this small population of quiescent cells maintained viability. They termed this as the drug-tolerant persisters (DTP), a highly dynamic state and putatively aided by each cell in the subpopulation. These cells served as the reservoir of cells to initiate proliferation and establish subsequent resistance. Herein, it was hypothesized that epigenetic alterations could mediate the dynamicity of the tolerant cell and their escape from drug pressure. Subsequently, they showed that KDM5A/RBP/JARID1 axis was functioning to maintain these DTPs, and could be selectively targeted using an HDAC inhibitor. This study provided proof of the concept of tolerance. It strongly validated the importance of drug-tolerant cells and that the precise targeting of this transient state can prevent the eventual recurrence and/or resistance. Thus began the exploration of epigenetic plasticity facilitating the survival of these DTPs, which according to some, could be due to dynamic yet transient histone modifications. These epigenetic adaptations are involved with stemness-associated characteristics or with the activation of pathways implicated in the early development of an organism. On a similar note, Hong Yan and colleagues cultured etoposide-treated prostate cancer cell line Du145-VP16 cells and treated them with HDAC inhibitor Trichostatin A or DNMT inhibitor Aza; which resulted in increased expression of epithelial markers like E-cadherin, CD44, and KDM5A, which supports the fact that drug-tolerant tumor cells may involve epigenetic changes (Yan et al., 2011). Given the accumulating evidence and the plethora of adaptive mechanisms relevant to drug tolerance, cancer cells can no longer be classified into only two categories- sensitive or resistant. Still, its forays into a continuum of dynamicity, with DTPs as one of the initially established transient yet essential states. While DTPs are now a bonafide candidate for therapy, there are ongoing investigative studies to understand them better. Herein, we discuss some of those dynamic properties of persisters, with particular emphasis on epigenetics, how they have been targeted and how it improved drug response.

4.1 Altered signaling in DTPs and their targeting

While persisters have long been considered to have originated primarily due to their epigenetic changes, there are speculations that these epigenetic changes might also regulate different intra-cellular signaling and vice versa.

Indeed, epigenetic alterations can often lead to oncogenic signaling in cancer cells, endowing them with capabilities that allow them to cope with the extreme environment (Hanahan & Weinberg, 2011), like tolerating acute drug stress, as described above. Importantly, these cells can turn on stress response pathways instead of the canonical oncogenic signaling to homeostasis (Luo, Solimini, & Elledge, 2009). This adaptive switch to the stress response pathways, putatively mediated by epigenetic reprogramming in DTPs, can explain how they cope with the sudden and acute drug stress. It is, therefore, imperative that manipulating or targeting the epigenetic reprogramming or subsequently linked signaling pathways is an Achilles heel for DTPs. For example, a repression of the DNA damage response (DDR) pathway can induce sensitivity to chemotherapy or radiotherapy in cancer cells (Huang & Zhou, 2020; Huntoon et al., 2013; Prevo et al., 2012). Similarly, autophagy facilitates drug resistance in diverse cancer cells (Kwon, Kim, Jung, Kim, & Jeoung, 2019). Therefore, we hypothesized that if the potential of tumor cells to cope with harsh microenvironment and their perturbations is highly reliant on stress response pathways, it is pretty likely that triggering of such pathways may also be true in DTPs and during their transition to stable resistance. Indeed, several stress response signaling, like induction of autophagy (Vera-Ramirez, Vodnala, Nini, Hunter, & Green, 2018), perturbation of cellular redox balance (Luo et al., 2018), and unfolded protein response (UPR) activation (Ranganathan, Zhang, Adam, & Aguirre-Ghiso, 2006), have been intriguingly found to be associated with the transitory non-proliferative DTPs. However, whether epigenetic modifications induced by acute drug stress stimulate such adaptive signaling pathways or not still remains an open-ended question. Nevertheless, it is understandable that stress-signaling molecules are increasingly considered potential therapeutic targets to prevent the acquisition of full-blown resistance in DTPs. A deeper analysis of the stress pathways revealed specific adaptations by which the growth-arrested DTPs utilize these stress pathways. For instance, autophagy is known to be involved in attenuating oxidative stress through the clearance of damaged mitochondria in DTPs; no wonder the inhibition of autophagy prevents dormancy induces apoptosis and has the potential to sensitize DTPs (Vera-Ramirez et al., 2018). Furthermore, nasopharyngeal carcinoma cells were found to induce MYC-mediated upregulation of CHK1 and CHK2 and subsequent DDR pathways leading to radioresistance, hence ablation of MYC or CHK1/2 reinstated radiosensitivity (Wang et al., 2013). Similarly, protective upstream activation of p38 in dormant human carcinoma cells induced a UPR, facilitating resistance to etoposide or doxorubicin. These data show how the phenotypic plasticity of DTPs enables the rewiring of signaling pathways supporting the sustenance of the slow-proliferating cancer cells under drug stress.

However, though these pathways were implicated in drug tolerance, their association with epigenetics is not fully established. Herein, some initial studies highlighted the probable potential of epigenetics. Using EGFR-driven NSCLC models, Sharma and colleagues showed that an upregulated Insulin Growth Factor-1 Receptor (IGF-1R) signaling helps DTPs to have a repressed chromatin state, which is required for survival in a non-proliferative mode. They emphasized the importance of IGF-1 receptor-mediated signaling in epigenetic transition, and the use of IGF receptor inhibitors (AEW541) could sufficiently prevent the proliferation of the extended persister (DTEP) cells derived from DTPs. Contrarily, adding IGF-1R ligand IGF-1 to the media resulted in a higher population of DTPs. This further suggests that the epigenetic-driven activation of the IGF signaling pathway may have an essential role in drug tolerance, and inhibition of this pathway can forestall resistance. In accordance with the above, several articles have also demonstrated the activation of the IGF pathway in drug-resistant cells. However, differential signaling programs have considerable redundancy or activation in DTPs with different tissue origins or upon exposure to varied drug types. In this context, findings from Risom and colleagues suggest that HCC1143 cells that sustained viability after treatment of PI3K-mTOR inhibitor like BEZ235 showed increased activities of MAPK, Bcl-2, and NFκβ; on the other hand, cells sustaining a MEK inhibitor like Trametinib and AZD6244 treatment showed increased activity of PI3K, integrin, FGF, JNK. Interestingly, both DTP populations showed an upregulated JAK-STAT, Notch, TGF and Ras pathway (Risom et al., 2018). Therefore, it is essential to identify unique pathways in DTPs for future targeting and their probable regulation through the tinkering of the epigenome.

In this regard, signaling pathways that are developmentally relevant or associated with stemness have been consistently found to be triggered in DTPs by multiple studies. For example, Singh and colleagues showed that targeting downstream target genes of Hedgehog pathway might act as an efficient therapeutic strategy against chemo tolerance for Diffuse Large B-Cell Lymphoma (Singh et al., 2011). Similarly, in melanoma cells, an array of receptor tyrosine kinase ligands such as EGF was found to induce MAPK signaling even in the presence of a BRAF inhibitor, thereby yielding a high amount of DTP cells (Lito et al., 2012). No wonder DTP generation was suppressed by the application of a combination of RTK inhibitors with BRAF inhibitors.

4.2 Histone modifications, chromatin remodeling and therapeutic strategies

Adaptive or acquired responses primarily depend on transcriptional changes facilitated by the re-organization of enhancer and promoter landscapes. In this regard, histone modifications promoting adaptive transcriptional changes have been considered a vital non-genetic mediator of drug tolerance. Herein, a plethora of studies shows the involvement of various chromatin modulators, especially histone demethylases, in the survival of DTP cells. For example, H3K4me3 demethylases, KDM5A, and KDM5B were found to play a critical role in response to targeted therapy. Sharma and colleagues found that PC9-derived DTP and DTEP both had an elevated expression of KDM5A, while the DTP cell population showed a reduced level of H3K4Me2/3. Further, they observed that KDM5A knockdown did not significantly alter proliferation in parental PC9 cells but dramatically decreased the amount of DTPs and DTEPs after EGFR TKI therapy (Sharma et al., 2010). In accordance the with above, Liau et al. also observed that the transition of glioblastoma cells to the DTP state is accompanied by a widespread re-distribution of repressive histone methylation marks (Liau et al., 2017). This is also associated with the activation of histone demethylases- KDM6A/B underlining their probable importance in establishing DTPs. Interestingly, it was observed that primitive signaling pathways like Notch are high in primary glioblastomas coupled to a histone demethylase expression even before treatment, thus acting as potential contributing factors to relapse and establishment of resistance. Furthermore, some studies link the involvement of KDM6B to resistance to PI3K/AKT inhibitors in breast cancer. Herein, an important aspect that stamps the involvement of epigenetics is the reversibility of the drug-tolerant phenotype, as cells derived from a drug-tolerant clone when cultured, shun the drug and regains sensitivity. Notably, apart from several studies implicating the importance of KDMs, DTPs have also been found to show increased HDAC activities (Biersack, Nitzsche, & Höpfner, 2022). Accordingly, HDAC inhibitor coupled with ablation of KDMs (or pan KDM5 inhibitor) showed effective results against DTPs. While earlier studies have already revealed the involvement of KDMs as an essential regulatory switch to adaptive transitions and cellular differentiation, their importance in drug tolerance is gradually getting established along with efforts to identify novel small molecules that can inhibit the effect of KDMs to prevent adaptive tolerance.

Apart from alteration in expression pattern and activity of histone demethylases, several studies have also reported coupled alteration in histone methylation pattern with relatively more studies in favor of the drug-induced establishment of repressive histone marks associated with a marked decrease of histone acetylation status. In this regard, a recent study published by Pham et al. suggests that survival of DTPs is highly dependent on histone methyltransferase G9a and EZH2, the latter being a part of the PRC2 repressive complex catalyzing H3K27me3. DTPs were found to harbor enhanced methylation on Jarid2, a PRC member involved in the stabilization of histone repressive marks (Pham et al., 2020). They also observed that EZH2 apart from its canonical function could methylate G9a too, facilitating recruitment of repressive complexes in DTPs. In support of above, other studies also show that DTPs adapt to acute drug stress through the induction of repressive histone marks like H3K9me3 and H3K27me3 to inhibit long-interspersed repeat element-1 (LINE1). Importantly, selective reversion of the repressive chromatin marks over the LINE-1 elements in the persister cells sensitized them. Similarly, in a separate study, it was observed that acquired drug resistance in melanoma cells develops through the induction of multiple signaling cascades and epigenetic reshuffling that might require induction of H3K9me3 and the loss of H3K4me3 and H3K27me3 marks. Integrated genome-wide analysis of the histone tags like H3K9me3, and H3K4me3 in these melanoma cells identified a small group of down-regulated genes, like SPRY4, which is known to be involved in negative regulation of IFN implicating an alleged involvement of IFN signaling in drug tolerance. Interestingly, based on the obtained selective enrichment of interferon and immune signaling across several integrated analyses, it was speculated that the DTPs enable a "viral mimicry" state facilitated by epigenomic regulation in response to stress (Al Emran et al., 2018). The involvement of viral infection-associated signaling, observed by Emran and colleagues, is in accordance with multiple previous findings pointing towards the importance of IFN-related gene signatures in resistance against various modes of anti-cancer therapy. However, it is evident that numerous epigenetic alterations act in cohorts facilitating the DTP state, and it's associated with phenotypic plasticity in a particular cancer type. However, the epigenetic adaptations might vary between cancers of different origins.

In accordance with the above, it has been observed that short-term exposure to MAPK inhibitors may lead to chromatin modification upon upregulated histone demethylation (Menon et al., 2015). In the case of BRAF-mutated colorectal cancer cells, 9–12 days of exposure to MEK inhibitor lead to upregulation of STAT3, following epigenetic remodeling via HDAC (Carson et al., 2015). It has been demonstrated that a contrasting open chromatin architecture and dynamic remodeling can play a significant role in the PI3K-driven DTP state. In addition, Roesch et al. showed that treating melanoma cells with BRAF inhibitor- Vemurafinib, or Cisplatin, results in a high level of H3K4 histone demethylases like JARID1B, KDM5B, and PLU-1, which are in synchrony with previously described findings on the role of histone de-methylases in tolerant cells (Roesch et al., 2013). Importantly, BRAF-mutated melanoma cells generally showed a decreased expression of histone methyltransferase EZH2, thus affecting CpG island methylation in the promoter region. Further, Strub and colleagues identified lysine methyltransferases like SETDB1/2 upregulated in melanoma cells (Strub et al., 2018). In addition, ONECUT2, a well-documented transcription factor, often found to be overexpressed in various cancer cells, modulated the PRC2 complex in lung cancer cells, thus regulating H3K27 methylation, as documented by Guo and colleagues (Guo et al., 2016, 2019). This potentially makes ONECUT2 as a targetable candidate to forestall DTP generation. In summary, the transition of cancer cells to tolerant and henceforth resistant cells may require activation of multiple key signaling pathways driven by diverse epigenomic alterations, w promoting survival under drug stress and facilitating tumor relapse and drug resistance. Thus, modifiers of essential epigenomic alterations can be considered potential targets along with existing therapies to forestall drug resistance in tumors.

4.3 DNA methylation and drug tolerance

DNA methylation is the most well-known epigenetic mechanism at CpG islands and affects drug response by altering gene expression. Typically, gene expression is decreased when the gene promoter is hypermethylated, whereas expression is enhanced when the promoter is hypomethylated. While ample studies delineate aberrant DNA methylation patterns and their association with the de-activation of TSGs or activation of oncogenes in cancer and coupled drug resistance, there are limited studies that analyze global or genomic context-dependent DNA methylation patterns in drug-tolerant cancer cells. In this context, a study published by Emran et al. (2019) reported DNA hyper-methylation in transcriptionally downregulated genes in drug-tolerant melanoma cells (Emran et al., 2019). Deeper integrated transcriptomic and ChIP data analysis revealed that the subset of downregulated genes was marked by repressive histone mark (H3K9me3) at a distal region to the transcription start site (TSS). At the same time, DNA hypermethylation was more prominent in the proximal genomic elements concerning the TSS. Interestingly, they also observed an alteration in the expression of three DNA methyltransferases, DNMT3A, DNMT3B, and DNMT1, in the DTPs, compared to the untreated cells, thus establishing their potential role in re-distribution of DNA methylation in the drug-tolerant state. Notably, the analysis of the global DNA methylation pattern among drug-tolerant cells derived from different cell types like melanoma and breast or colon cancer failed to show any uniformity. While colon cancer-derived DTPs showed considerable global hypo-methylation, lung carcinoma-based DTPs, in contrast, depicted moderate hyper-methylation. Furthermore, analysis of the differentially methylated CpG sites in DTPs of different tissue origins showed that the most affected genomic regions were low CpG density genomic regions termed "Open Sea." Thus, based on the pattern of DNA methylation across different cell-type-derived DTPs, the authors speculated that the distribution in the tolerant cells is not a generalized phenomenon but is cell-type specific in response to prolonged drug pressure. Furthermore, pathway analysis of genes regulated by DNA methylation and H3K9me3 modification in the drug-tolerant cells revealed enrichment of pathways like Wnt and embryonic stem cell-associated signaling TNF, TGF-β receptor, and Toll-like receptor, which can have an apparent connection with the survival of cancer cells under drug pressure. When given a drug holiday, the drug-tolerant melanoma cells regained sensitivity to chemotherapy and targeted therapy, which was associated with altered H3K4me3 and H3K27me3 marks supposedly coupled to a modified DNA methylation pattern as well. A separate study on doxorubicin-tolerant ovarian and breast cancer cell lines also documented methylation changes in genes that contribute to chemoresistance and discovered hypo-methylation of APC, ABCB1, and HIC1 genes concomitant with hyper-methylation of CDH1, BRCA1, SULF2, and DNAJC15 genes suggesting that drug-tolerant cells can acquire gene-specific methylation pattern conducive to their survival under drug stress. Interestingly, Alghamian et al. hypothesized that the development of cisplatin refractoriness is often followed by

EMT-like characteristics. They found that promoter hyper-methylation of epithelial genes like EpCAM and CDH1 leads to down-regulation of these genes and sufficient hypo-methylation of mesenchymal markers such as SNAIL despite having a CpG island proximal to the TSS of SNAIL gene. Furthermore, to verify the role of EMT on DNA methylation, they treated cisplatin-resistant cells with 5-aza-2′-deoxycytidine, which resulted in a partial reversal of the EMT process, suggesting dynamic regulation of the epigenome in response to drug stress (Alghamian, Soukkarieh, Abbady, & Murad, 2022). Thus, the overall analysis of genomic DNA methylation patterns across different DTPs authenticates their putative role in survival under drug stress. However, there is still a scarcity of information in this regard which can be addressed by future research.

4.4 Role of noncoding RNAs in drug tolerance

Non-coding RNAs, including microRNAs (miRNAs) and long non-coding RNAs, constitute an essential arm of epigenetic regulation in cancer. They can play critical roles in regulating tumor initiation and progression via diverse mechanisms, including acting as a sponge for mRNAs, RNA interference, or chromatin regulation. The miRNAs, in particular, have recently been found to regulate tolerance to EGFR tyrosine kinase inhibitor- osimertinib through induction of pseudohypoxia signaling pathways in a miR-147b-dependent manner in human NSCLCs (Zhang, Wells, Chow, et al., 2019). The findings suggest that the NSCLC cells rely on VHL and succinate to tolerate drug pressure. Reducing miR-147b and reactivating the TCA cycle can be an effective strategy against drug tolerance and subsequent tumor relapse. In addition, a comparative miRNAseq experiment between osimertinib tolerant and sensitive parental PC9 and HCC827 cells helped to identify a set of differentially regulated miRNAs that are associated with the development of a transient drug-tolerant state, with the top up-regulated ones being miR181a-2–3p, miR-147b, miR-574–5p, and most down-regulated one being, miR-7641–1, miR-4454 and miR-125b-1–3p (Zhang et al., 2019). Studies on non-small cell lung cancer (Sahu et al., 2016) and osteosarcoma (Sharma, Niveditha, Chowdhury, Mukherjee, & Chowdhury, 2021) have demonstrated that a group of miRNAs such as miR-371-3p, hsa-miR-98-5p, hsa-miR-10a-3p, hsa-miR-365a-3p, hsa-miR-301a-5p, hsa-miR-155-3p and hsa-miR-532-3p is dysregulated in drug-tolerant tumor cells. The miRNAs shown to have an inverse correlation with the expression pattern of their corresponding mRNA partner have been further selected for gene ontology studies, and pathways like TNF signaling, autophagy, and mitophagy have been found to play important roles in reversible drug tolerance in case of osteosarcoma cells (Sharma et al., 2021). Their study also found a set of lncRNA sufficiently dysregulated in osteosarcoma persister and extended persister cells. These lncRNA includes LOC101927575, LINC01194, LINC00636, LINC00311, HAGLR, and LOC100130238. Some of the dysregulated lncRNAs, such as FBXL19-AS1, LINC0024, LINC00865, LINC00674, and LINC00865 have found to have a correlated expression with their corresponding miRNA and mRNA partner as well. Similar studies have also shown that extracellular miR-1246 can boost the radio-resistance of lung cancer cells (Yuan et al., 2016). In addition, functional loss of miR-21 was found to attenuate therapeutic tolerance against EGFR tyrosine kinase inhibitors (Zhang, Skiados, Aftab, et al., 2022). However, the subset of genes or pathways these miRNAs regulate in DTPs is poorly explored and demands further investigation. Until recently, though long noncoding RNAs have been considered another potential key regulator of drug resistance, their role in gene expression via epigenetic modification aiding drug tolerance is poorly elucidated.

4.5 Single-cell analysis and deeper understanding of drug tolerance

One of the long-standing barriers in biology is mapping genotypes to phenotypes, and transcriptome analysis is one way to tackle this problem. Each cell of our body shares the same genotype, but the transcriptome profile is an average of a selected number of genes. While conventional bulk population sequencing can only provide the average expression signal for a group of cells, there is an underlying heterogeneity and cell-to-cell variability, which can now be very well studied using single-cell transcriptomics through technological advancements. Although genomic region coverage of bulk RNAseq is 2–3 folds higher than that of single-cell RNAseq (Lee, Lopez-diaz, et al., 2014), as single-cell analysis covers only those genes which are expressed by the individual cell, single-cell RNAseq can detect low-frequency novel single nucleotide variants (SNV), unlike bulk RNAseq. In this context, Lee and colleagues performed their experiments on MDA-MB-231 cell lines. After one day of treatment with 100 nM paclitaxel, most of the cells died, but a few cells resumed their proliferation and showed DTP-like characteristics. Their single-cell analysis study revealed that drug-tolerant cells had somewhat 30% similar novel SNV with normal untreated cells, unlike stressed cells. Furthermore, Su and colleagues explored single-cell analysis to distinguish between drug-naive and drug-tolerant BRAF mutant melanoma cancer cells based on the heterogeneous expression of markers expressed and identified that phenotypic adaptive nonmutational events like epigenetic reprogramming facilitates distinct transcriptional trajectory in individual cells (Su et al., 2020).

In another study, a well-established preclinical NSCLC (PC9) model that responds to tyrosine kinase inhibitors like erlotinib characterized the different drug-tolerant phenotypes and clustered the cells with similar gene expression (Aissa et al., 2021). They also did a similar analysis after the drug holiday to account for persisters' reversibility. Intriguingly, cells lost expression of genes involved in cholesterol metabolism in response to EGFR TKI. They also discovered a high enrichment in genes pertaining to EMT and tissue development, including epithelium development, drug metabolism, and epigenetic regulation in tolerant cells. This suggests that cellular variability owing to variability in cellular processes may drive tolerance differently. Thus, recent advancement in single-cell analysis offers a new way to profile different types of individual tumor cells and learn more about how they contribute to these processes of drug tolerance. However, given that a single cell line might acquire numerous drug tolerance mechanisms, it is highly unlikely that all cancer cells can be eradicated by simultaneously targeting a single drug tolerance mechanism.

4.6 Drug holiday as a reprogramming and drug sensitivity restoration strategy

One of the interesting facts of the DTPs is their reversible drug sensitivity. Sharma et al. in their initial study, observed that the DTPs, when propagated in drug-free media, resume proliferation and reacquire drug sensitivity (Sharma et al., 2010). The similar rebound of sensitivity was also observed with DTPs isolated from multiple other in vitro models. Significantly, a comparative genome-wide gene expression analysis of parental cells compared to DTPs revealed striking alterations in overall expression profiles and non-random distribution of differentially expressed genes along chromosomes, implicating an epigenetic biases to acquisition of DTP state. Therefore, it is currently hypothesized that an epigenetic re-shuffling is the cause to restoration of sensitivity to drugs, as shown by DTPs upon and drug vacation. Several other studies also reported a re-sensitization upon drug holiday thus establishing and exploration of a probable therapeutic window against DTPs. For example, Becker and colleagues also reported that a period of drug holiday to a group of NSCLC patients showed recovery of "sensitivity" to TKI inhibitor- erlotinib (Becker et al., 2011). Their hypothesis was based on treatment responses from 24 patients with stage IV NSCLC. Initially, they responded to TKI inhibitors, followed by relapse and resistance to erlotinib. But after a certain period of drug holiday, a regain in drug sensitivity was observed, for which Becker and colleagues had suggested the development of non-mutational and reversible resistance to TKI inhibitors instead of EGFR T790M mutation, which was blamed for the initial resistance to EGFR. Cisplatin-resistant osteosarcoma cells have also been shown to regain their sensitivity after a period of drug withdrawal, as reported by Niveditha, Sharma, Majumder, et al. (2019). They cultured untreated osteosarcoma cells (OS), cisplatin-resistant cells (OS-R), non-proliferative persister drug-tolerant cells (OS-P) and the extended persisters (OS-EP). When resistant cells were cultured in drug-free media (OS-DW) for 7 days the cells regained sensitivity to cisplatin, which strongly supports the idea that drug holiday helps resistant cells to revoke their sensitivity to treatment. Further, Niveditha et al. performed a comparative transcriptomic study between OS-DW and OS-R, showing that few transcription factors, such as ATF6, CDX2, and FOSL1 are expressed explicitly in OS-DW cells and not in OS-R cells (Niveditha, Sharma, Majumder, et al., 2019). Importantly, the transcriptomic profile of cells that regained sensitivity to cisplatin was strikingly different from the sensitive parental cells indicating that the resistant cells do not necessarily resemble the parental type upon drug holiday. They acquire their own, unique transcriptomic profile which is different from resistant or parental cells. For example, key genes for OS-R cells like IL6ST, AKT1, RELA are differentially expressed in the case of OS-DW cells, whereas key genes identified for OS-DW included PIK3CA, PIK3R2, BIRC2, etc. Pathway analysis revealed de-regulation of Hedgehog, Wnt, Notch and BMP signaling pathways as key determinants of restoration of sensitivity. Similar results were also obtained on multidrug-resistant (MDR) cancer cells by Zhang, Qin, et al. (2022), where the cells after a long-term "drug holiday" (30 days) re-proliferated and enabled re-acquisition of sensitivity to drug treatment, which was close to parental cells in terms of drug sensitivity. Overall, in spite of research progress, we are still unclear about the transcriptomic or epigenetic signature of the cell populations that emerge when a drug vacation is imposed. Recent single cell-based analysis show that cell subpopulations that emerge early during drug exposure do not pre-exist in the original population; however, there is a striking resemblance in populations emerging after a drug holiday (Aissa et al., 2021). Furthermore, it confirms the non-genetic nature of these transitions providing evidences for future development of epigenetic inhibitors which can be further explored.

5. Conclusion and future perspectives

Overall, several studies have now acknowledged the existence of the slow-cycling drug-tolerant persister cells that not only survive the harsh drug insult but also act as reservoirs to launch the subsequent relapse of cancer or even the more aggressive phase. No doubt, a better understanding of these persisters has immense clinical implications. As presented in this article, diverse strategies are adapted by these small, slow-cycling, dormant subpopulations of cells that are distinctively different from their proliferative counterparts, or cancer cells showing stable resistance. However, these features are also linked to the vulnerabilities of the tolerant cells, which can be utilized to eradicate them. Probably, one of the most universally accepted and widely discussed feature of the DTPs has been their acquisition of non-canonical transcriptional regulation that facilitate a brake on proliferation. DTPs showing such restricted proliferation or "diapause" were often associated with stem cell like features. Literature describes such restricted proliferative state as either 'quiescence' that is often observed in adult stem cells, or as the irreversible 'senescence' that lies at the other end of the spectrum. Importantly, both these cell division restricted states can be induced by chemo or radiotherapy and is often associated with poor cancer prognosis. Given the above, one strategy to sensitize the growth restricted DTPs could be by extending their dormancy and thus allowing extended time for the immune cells to eradicate them. A longer growth arrested state may also allow therapeutic intervention targeting the unique features of the DTPs active during the non-proliferative state. Likewise, blocking transcriptional plasticity; inhibition of stress responsive pathways like MAPK signaling that trigger stem cell like features; or targeting developmentally relevant pathways like Notch or Wnt signaling or IGF-1R signaling has shown promising results. In addition, earlier studies show that drug-induced, senescent tumor cells develop enhanced sensitivity to ABT263 (navitoclax), specific inhibitor of anti-apoptotic proteins. This implies that targeting proteins like, BCL-2 or BCL-XL can be an effective strategy to tame the DTPs.

Herein, several studies point towards an altered chromatin state of the DTPs marked by the presence of heterochromatic foci, overall induction of histone repressive marks, and selective DNA methylation pattern as the underlying mechanism behind the drug-induced, adaptive, restricted growth state of the DTPs. Not surprising, epigenetic modifiers, like the histone demethylases, for example KDM5 has been found to be effective against the DTPs. In addition, the use of histone deacetylase inhibitors (HDACi) has also been able to decrease the number of persister cells in multiple cancer models. Based on the data presented here, it is therefore imperative that a better conceptual and mechanistic understanding of the chromatin dynamics or stress response signaling may deliver ample innovative targeting opportunities to prevent emergence of DTPs or tumor relapse. In this regard, consequent to the dynamic chromatin modification and consequent transcription modulation there is a shift in research focus towards understanding the overall cellular metabolism of the transitory DTPs. The persisters have been found to be addicted to oxidative phosphorylation and cellular antioxidant defense mechanisms to homeostasize under the harsh environment. In accordance to above, an up-regulation of autophagy assisting in metabolite recycling or redox balance through organelle and protein turnover, or a stimulation of unfolded protein response (UPR) have been observed in different DTP models. Autophagy inhibitors are already in the clinical trials while, UPR modulators like tunicamycin have been utilized to trip the metabolic balance of the DTPs leading to their elimination. Finally, chromatin modifications in DTPs were also found to favor a phenotype switch such as the epithelial to mesenchymal transition (EMT) that allows these cells to effectively adapt to the drug stress. During development and wound healing, EMT has been found to be intricately associated with senescence; this is very much compatible with the partial EMT like features observed in DTPs alongside reversible senescence or reduced proliferations state. We speculate that the DTPs activate these distorted programs in a context-dependent manner, which can be conducive to survival and eventual evasion of an immune attack or an apoptotic cell death thus resulting in rebound of cancer post cytotoxic shock. However, further studies are required in this direction to understand the magnitude of adaptive distorted programs enabled in the DTPs preferably at a single cell level.

Truly, a major hindrance to targeting cancer is molecular heterogeneity, which is also an integral feature of DTPs. In accordance with it, variations in adaptive characteristics to drug pressure have been observed even at the single cell level in the persisters, making development of robust therapeutic strategies challenging. In addition, a limited understanding of the precise contribution of major players discussed above, like, altered stress signaling, epigenomic alterations, redox status, autophagy, or metabolic pathways reduces scope for avenues targeting the persisters and subsequent resistance. In addition, the minimal abundance of the tolerant cells and deviations in experimental models utilized to study the molecular dependency of the persister cells also pose a pertinent problem, as they might not faithfully reflect a clinical scenario. Nevertheless, the activation of non-genetic mechanisms and signaling converging on the epigenome regulation has been well-accepted in DTP research. Accordingly, there is a significant drive in identifying novel epigenetic targets through high throughput screening processes and subsequent search for epigenetic modifiers that can restore drug sensitivity. Unfortunately, several studies have also shown that with each sequential treatment failure, the tumor cells acquire and enable increasingly complex and diverse aberrations which is often dependent on cell-type or drug dosing pattern leading to difficulty in targeting. This, therefore, demands multimodality treatment with intense curative intent targeting the DTPs alongside conventional therapy. Alternatively, since acquisition of DTP state occurs in multiple stages they also provide a therapeutic window for efficient targeting. In addition, we believe that comparison of the traits of DTPs with different types of diseases, identifying semblance or common traits and adaptations evading drug stress could further open up new avenues to eliminate the "persisters."

Acknowledgment

NM is a recipient of JRF fellowship from DST_NSM project (Grant DST/NSM/R&D_HPC_Applications/2021/03.24). AD is recipient of SRF fellowship from CSIR, India. This work was supported in part from DST-SERB grant to RC (grant number EMR/2017/004149), CSIR, India grant to SM (grant number 37(1723)/19/EMR II DATED 15.5.2019) ICMR, India, and DST_NSM grants to SC (grant number 5/13/15/2019/NCD-III and DST/NSM/R&D_HPC_Applications/2021/03.24).

Disclosure

Authors declare that they have no conflict of interest, financial or otherwise, regarding the manuscript's content.

References

Aissa, A. F., Islam, A. B. M. M. K., Ariss, M. M., Go, C. C., Rader, A. E., Conrardy, R. D., et al. (2021). Single-cell transcriptional changes associated with drug tolerance and response to combination therapies in cancer. *Nature Communications*, *12*(1), 1628. https://doi.org/10.1038/s41467-021-21884-z.

Al Emran, A., Marzese, D. M., Menon, D. R., Stark, M. S., Torrano, J., Hammerlindl, H., et al. (2018). Distinct histone modifications denote early stress-induced drug tolerance in cancer. *Oncotarget*, *9*(9), 8206–8222. https://doi.org/10.18632/oncotarget.23654.

Alghamian, Y., Soukkarieh, C., Abbady, A. Q., & Murad, H. (2022). Investigation of role of CpG methylation in some epithelial mesenchymal transition gene in a chemoresistant ovarian cancer cell line. *Scientific Reports*, *12*(1), 1–15. https://doi.org/10.1038/s41598-022-11634-6.

Allfrey, G., Faulkner, R., & Mirsky, E. (1964). Acetylation and methylation of histones and their possible role in the regulation of RNA synthesis. *Proceedings of the National Academy of Sciences of the United States of America*, *51*, 786–794. https://doi.org/10.1073/pnas.51.5.786.

Anastasiadou, C., Malousi, A., Maglaveras, N., & Kouidou, S. (2011). Human epigenome data reveal increased CpG methylation in alternatively spliced sites and putative exonic splicing enhancers. *DNA and Cell Biology*, *30*(5), 267–275. https://doi.org/10.1089/dna.2010.1094.

Auclair, G., & Weber, M. (2012). Mechanisms of DNA methylation and demethylation in mammals. *Biochimie*, *94*, 2202–2211.

Baccelli, I., Gareau, Y., Lehnertz, B., Gingras, S., Spinella, J.-F., Corneau, S., et al. (2019). Mubritinib targets the electron transport chain complex I and reveals the landscape of OXPHOS dependency in acute myeloid leukemia. *Cancer Cell*, *36*(1), 84–99. e88. https://doi.org/10.1016/j.ccell.2019.06.003.

Bartel, D. P. (2018). Review metazoan microRNAs. *Cell*, *173*(1), 20–51. https://doi.org/10.1016/j.cell.2018.03.006.

Becker, A., Crombag, L., Heideman, D. A., Thunnissen, F. B., van Wijk, A. W., Postmus, P. E., et al. (2011). Retreatment with erlotinib: Regain of TKI sensitivity following a drug holiday for patients with NSCLC who initially responded to EGFR-TKI treatment. *European Journal of Cancer*, *47*(17), 2603–2606. https://doi.org/10.1016/j.ejca.2011.06.046.

Bejerano, G., Pheasant, M., Makunin, I., Stephen, S., Kent, W. J., Mattick, J. S., et al. (2004). Ultraconserved elements in the human genome. *Science*, *304*(5675), 1321–1325. https://doi.org/10.1126/science.1098119.

Berdasco, M., & Esteller, M. (2010). Aberrant epigenetic landscape in cancer: How cellular identity goes awry. *Developmental Cell*, *19*(5), 698–711. https://doi.org/10.1016/j.devcel.2010.10.005.

Biehs, B., Dijkgraaf, G. J., Piskol, R., Alicke, B., Boumahdi, S., Peale, F., et al. (2018). A cell identity switch allows residual BCC to survive Hedgehog pathway inhibition. *Nature*, *562*(7727), 429–433. https://doi.org/10.1038/s41586-018-0596-y.

Biersack, B., Nitzsche, B., & Höpfner, M. (2022). HDAC inhibitors with potential to overcome drug resistance in castration-resistant prostate cancer. *Cancer Drug Resistance*, *5*, 64–79. http://doi.org/10.20517/cdr.2021.105.

Bigger, J. W. (1944). The bactericidal action of penicillin on Staphylococcus pyogenes. *Irish Journal of Medical Science (1926–1967)*, *19*(11), 553–568. https://doi.org/10.1007/BF02948386.

Brunet, T. D. P., & Doolittle, W. F. (2014). Getting "function " right. *Proceedings of the National Academy of Sciences of the United States of America*, *111*(33), 3365. https://doi.org/10.1073/pnas.1409762111.

Carson, R., Celtikci, B., Fenning, C., Javadi, A., Crawford, N., Carbonell, L. P., et al. (2015). HDAC inhibition overcomes acute resistance to MEK inhibition in BRAF-mutant colorectal cancer by downregulation of c-FLIPL. *Clinical Cancer Research*, *21*(14), 3230–3240. https://doi.org/10.1158/1078-0432.CCR-14-2701.

Chen, J.-H., Hales, C. N., & Ozanne, S. E. (2007). DNA damage, cellular senescence and organismal ageing: Causal or correlative? *Nucleic Acids Research*, *35*(22), 7417–7428. https://doi.org/10.1093/nar/gkm681.

Cheng, Y., He, C., Wang, M., Ma, X., Mo, F., Yang, S., et al. (2019). Targeting epigenetic regulators for cancer therapy: Mechanisms and advances in clinical trials. *Signal Transduction and Targeted Therapy*, *4*(1), 1–39. https://doi.org/10.1038/s41392-019-0095-0.

Cheng, M., Sheng, L., Gao, Q., Xiong, Q., Zhang, H., Wu, M., et al. (2019). The m 6 A methyltransferase METTL3 promotes bladder cancer progression via AFF4/NF-κ B/MYC signaling network. *Oncogene*, *38*(19), 3667–3680. https://doi.org/10.1038/s41388-019-0683-z.

Damante, G., Fabbro, D., Pellizzari, L., Civitareale, D., Guazzi, S., Polycarpou-schwartz, M., et al. (1994). Sequence-specific DNA recognition by the thyroid transcription factor-1 homeodomain. *Nucleic Acids Research*, *22*(15), 3075–3083. https://doi.org/10.1093/nar/22.15.3075.

De Conti, G., Dias, M. H., & Bernards, R. J. C. (2021). Fighting drug resistance through the targeting of drug-tolerant persister cells. *Cancers*, *13*(5), 1118. https://doi.org/10.3390/cancers13051118.

Deniz, Ö., Frost, J. M., & Branco, M. R. (2019). Regulation of transposable elements by DNA modifications. *Nature Reviews Genetics*, *20*(7), 417–431. https://doi.org/10.1038/s41576-019-0106-6.

Djebali, S., Davis, C. A., Merkel, A., Dobin, A., Lassmann, T., Mortazavi, A., et al. (2012). Landscape of transcription in human cells. *Nature*, *489*(7414), 101–108. https://doi.org/10.1038/nature11233.

Dörr, J. R., Yu, Y., Milanovic, M., Beuster, G., Zasada, C., Däbritz, J. H. M., et al. (2013). Synthetic lethal metabolic targeting of cellular senescence in cancer therapy. *Nature*, *501*(7467), 421–425. https://doi.org/10.1038/nature12437.

Dou, Y, & Gorovsky, MA. (2000). Phosphorylation of linker histone H1 regulates gene expression in vivo by creating a charge patch. *Molecular Cell*, *6*(2), 225–231. https://doi.org/10.1016/s1097-2765(00)00024-1.

Du, Z., Sun, T., Hacisuleyman, E., Fei, T., Wang, X., Brown, M., et al. (2016). Integrative analyses reveal a long noncoding RNA-mediated sponge regulatory network in prostate cancer. *Nature Communications*, 7, *10982*. https://doi.org/10.1038/ncomms10982.

Elliott, T. A., Linquist, S., & Gregory, T. R. (2014). Conceptual and empirical challenges of ascribing functions to transposable elements. *American Naturalist*, *184*(1), 14–24. https://doi.org/10.1086/676588.

Emran, A. A., Chatterjee, A., Rodger, E. J., Tiffen, J. C., Gallagher, S. J., Eccles, M. R., et al. (2019). Targeting DNA methylation and EZH2 activity to overcome melanoma resistance to immunotherapy. *Trends in Immunology*, *40*(4), 328–344. https://doi.org/10.1016/j.it.2019.02.004.

Esquela-kerscher, A., & Slack, F. J. (2006). Oncomirs—MicroRNAs with a role in cancer. *Nature Reviews. Cancer*, *6*, 259–269. https://doi.org/10.1038/nrc1840.

Esteller, M. (2007). Cancer epigenomics: DNA methylomes and histone-modification maps. *Nature Reviews Genetics*, *8*(4), 286–298. https://doi.org/10.1038/nrg2005.

Ewald, J. A., Desotelle, J. A., Wilding, G., & Jarrard, D. F. (2010). Therapy-induced senescence in cancer. *Journal of the National Cancer Institute*, *102*(20), 1536–1546. https://doi.org/10.1093/jnci/djq364.

Feingold, E. A., & Pachter, L. (2004). The ENCODE (ENCyclopedia of DNA elements) project. *Science*, *306*(5696), 636–640. https://doi.org/10.1126/science.1105136.

Geertz, M., Shore, D., & Maerkl, S. J. (2012). Massively parallel measurements of molecular interaction kinetics on a microfluidic platform. *Proceedings of the National Academy of Sciences of the United States of America*, *109*(41), 16540–16545. https://doi.org/10.1073/pnas.1206011109.

Giuliani, C., Bacalini, M. G., Sazzini, M., Pirazzini, C., Franceschi, C., Garagnani, P., et al. (2015). The epigenetic side of human adaptation: Hypotheses, evidences and theories. *Annals of Human Biology*, *42*(1), 1–9. https://doi.org/10.3109/03014460.2014.961960.

Goldie, J. H., & Coldman, A. J. (1985). A model for tumor response to chemotherapy: An integration of the stem cell and somatic mutation hypotheses. *Cancer Investigation*, *3*(6), 553–564. https://doi.org/10.3109/07357908509039817.

Guo, H., Ahmed, M., Zhang, F., Yao, C. Q., Li, S., Liang, Y., et al. (2016). Modulation of long noncoding RNAs by risk SNPs underlying genetic predispositions to prostate cancer. *Nature Genetics*, *48*(10), 1142–1150. https://doi.org/10.1038/ng.3637.

Guo, H., Ci, X., Ahmed, M., Hua, J. T., Soares, F., Lin, D., et al. (2019). ONECUT2 is a driver of neuroendocrine prostate cancer. *Nature Communications*, *10*(1), 1–13. https://doi.org/10.1038/s41467-018-08133-6.

Ha, M., Pang, M., Agarwal, V., & Chen, Z. J. (2008). Interspecies regulation of microRNAs and their targets. *Biochimica et Biophysica Acta*, *1779*(11), 735–742. https://doi.org/10.1016/j.bbagrm.2008.03.004.

Hanahan, D., & Weinberg, R. A. (2011). Hallmarks of cancer: The next generation. *Cell*, *144*(5), 646–674. https://doi.org/10.1016/j.cell.2011.02.013.

Hangauer, M. J., Viswanathan, V. S., Ryan, M. J., Bole, D., Eaton, J. K., Matov, A., et al. (2017). Drug-tolerant persister cancer cells are vulnerable to GPX4 inhibition. *Nature*, *551*(7679), 247–250. https://doi.org/10.1038/nature24297.

Harrow, J., Frankish, A., Gonzalez, J. M., Tapanari, E., Diekhans, M., Kokocinski, F., et al. (2012). GENCODE: The reference human genome annotation for the ENCODE project. *Genome Research*, *22*(9), 1760–1774. https://doi.org/10.1101/gr.135350.111.

Hata, A. N., Niederst, M. J., Archibald, H. L., Gomez-Caraballo, M., Siddiqui, F. M., Mulvey, H. E., et al. (2016). Tumor cells can follow distinct evolutionary paths to become resistant to epidermal growth factor receptor inhibition. *Nature Medicine*, *22*(3), 262–269. https://doi.org/10.1038/nm.4040.

Hernando-Herraez, I., Prado-Martinez, J., Garg, P., Fernandez-Callejo, M., Heyn, H., Hvilsom, C., et al. (2013). Dynamics of DNA methylation in recent human and great ape evolution. *PLoS Genetics*, *9*(9), e1003763. https://doi.org/10.1371/journal.pgen.1003763.

Hobby, G. L., Meyer, K., & Chaffee, E. (1942). Observations on the mechanism of action of penicillin. *Experimental Biology and Medicine*, *50*(2), 281–285. https://doi.org/10.3181/00379727-50-13773.

Hosono, Y., Niknafs, Y. S., Prensner, J. R., Iyer, M. K., Dhanasekaran, S. M., Mehra, R., et al. (2017). Oncogenic role of THOR, a conserved cancer/testis long non-coding RNA. *Cell*, *171*(7), 1559–1572.e20. https://doi.org/10.1016/j.cell.2017.11.040.

Huang, R. X., & Zhou, P. K. (2020). DNA damage response signaling pathways and targets for radiotherapy sensitization in cancer. *Signal Transduction and Targeted Therapy*, *5*, 1–27. https://doi.org/10.1038/s41392-020-0150-x.

Huang, Z., Du, Y., Wen, J., Lu, B., & Zhao, Y. (2022). snoRNAs: Functions and mechanisms in biological processes, and roles in tumor pathophysiology. *Cell Death Discovery*, *8*(1), 1–10. https://doi.org/10.1038/s41420-022-01056-8.

Hudson, W. H., Pickard, M. R., De Vera, I. M. S., Kuiper, E. G., Mourt, ada-Maarabouni, M., Conn, G. L., et al. (2014). Conserved sequence-specific lincRNA-steroid receptor interactions drive transcriptional repression and direct cell fate. *Nature Communications*, *5*, *5395*. https://doi.org/10.1038/ncomms6395.

Hung, T., Wang, Y., Lin, M. F., Koegel, A. K., Kotake, Y., Grant, G. D., et al. (2011). Extensive and coordinated transcription of noncoding RNAs within cell-cycle promoters. *Nature Genetics*, *43*(7), 621–629. https://doi.org/10.1038/ng.848.

Hung, C. L., Wang, L. Y., Yu, Y. L., Chen, H. W., Srivastava, S., Petrovics, G., et al. (2014). A long noncoding RNA connects c-Myc to tumor metabolism. *Proceedings of the National Academy of Sciences of the United States of America*, *111*(52), 18697–18702. https://doi.org/10.1073/pnas.1415669112.

Huntoon, C. J., Flatten, K. S., Hendrickson, A. E. W., Huehls, A. M., Sutor, S. L., Kaufmann, S. H., et al. (2013). ATR inhibition broadly sensitizes ovarian cancer cells to chemotherapy independent of BRCA status. *Cancer Research*, *73*, 3683–3691. https://doi.org/10.1158/0008-5472.CAN-13-0110.

Imai-Sumida, M., Dasgupta, P., Kulkarni, P., Shiina, M., Hashimoto, Y., Shahryari, V., et al. (2020). Genistein represses HOTAIR/chromatin remodeling pathways to suppress kidney cancer. *Cellular Physiology and Biochemistry*, *22*, 53–70. https://doi.org/10.33594/000000205.

Jiang, C., & Pugh, B. (2009). Nucleosome positioning and gene regulation: Advances through genomics. *Nature Reviews. Genetics*, *10*, 161–172. https://doi.org/10.1038/nrg2522.

Jones, P. A., & Baylin, S. B. (2007). The epigenomics of cancer. *Cell*, *128*(4), 683–692. https://doi.org/10.1016/j.cell.2007.01.029.

Jones, P. A., & Liang, G. (2009). Rethinking how DNA methylation patterns are maintained. *Nature Reviews. Genetics*, *10*, 805–811. https://doi.org/10.1038/nrg2651.

Ju, H.-Q., Ying, H., Tian, T., Ling, J., Fu, J., Lu, Y., et al. (2017). Mutant Kras-and p16-regulated NOX4 activation overcomes metabolic checkpoints in development of pancreatic ductal adenocarcinoma. *Nature Communications*, *8*(1), 1–14. https://doi.org/10.1038/ncomms14437.

Kino, T., Hurt, D. E., Ichijo, T., Nader, N., & Chrousos, G. P. (2010). Noncoding RNA Gas5 is a growth arrest- and starvation-associated repressor of the glucocorticoid receptor. *Science Signaling*, *3*(107), 1–16. https://doi.org/10.1126/scisignal.2000568.

Kumar, S., Gonzalez, E. A., Rameshwar, P., & Etchegaray, J. P. (2020). Non-coding RNAs as mediators of epigenetic changes in malignancies. *Cancers (Basel)*, *12*(12), 3657. https://doi.org/10.3390/cancers12123657.

Kuntz, E. M., Baquero, P., Michie, A. M., Dunn, K., Tardito, S., Holyoake, T. L., et al. (2017). Targeting mitochondrial oxidative phosphorylation eradicates therapy-resistant chronic myeloid leukemia stem cells. *Nature Medicine*, *23*(10), 1234–1240. https://doi.org/10.1038/nm.4399.

Kurppa, K. J., Liu, Y., To, C., Zhang, T., Fan, M., Vajdi, A., et al. (2020). Treatment-induced tumor dormancy through YAP-mediated transcriptional reprogramming of the apoptotic pathway. *Cancer Cell*, *37*(1), 104–122. e112. https://doi.org/10.1016/j.ccell.2019.12.006.

Kwon, Y., Kim, M., Jung, H. S., Kim, Y., & Jeoung, D. (2019). Targeting autophagy for overcoming resistance to anti-EGFR treatments. *Cancers*, *11*, 1374. https://doi.org/10.3390/cancers11091374.

Landsberg, J., Kohlmeyer, J., Renn, M., Bald, T., Rogava, M., Cron, M., et al. (2012). Melanomas resist T-cell therapy through inflammation-induced reversible dedifferentiation. *Nature*, *490*(7420), 412–416. https://doi.org/10.1038/nature11538.

Lee, J., Inoue, K., Ono, R., Ogonuki, N., Kohda, T., Kaneko-Ishino, T., et al. (2002). Erasing genomic imprinting memory in mouse clone embryos produced from day 11.5 primordial germ cells. *Development*, *129*, 1807–1817. https://doi.org/10.1242/dev.129.8.1807.

Lee, M. W., Lopez-diaz, F. J., Yar, S., Akram, M., Dayn, Y., Joseph, C., et al. (2014). Single-cell analyses of transcriptional heterogeneity during drug tolerance transition in cancer cells by RNA sequencing. *Proceedings of the National Academy of Sciences of the United States of America*, *111*, E4726–E4735. https://doi.org/10.1073/pnas.1404656111.

Lee, H.-J., Zhuang, G., Cao, Y., Du, P., Kim, H.-J., & Settleman, J. J. C. (2014). Drug resistance via feedback activation of Stat3 in oncogene-addicted cancer cells. *Cancer Cell*, *26*(2), 207–221. https://doi.org/10.1016/j.ccr.2014.05.019.

Li, L., Li, C., Mao, H., Du, Z., Chan, W. Y., Murray, P., et al. (2016). Epigenetic inactivation of the CpG demethylase TET1 as a DNA methylation feedback loop in human cancers. *Scientific Reports*, *6*(May), 1–13. https://doi.org/10.1038/srep26591.

Liau, B. B., Sievers, C., Donohue, L. K., Gillespie, S. M., Flavahan, W. A., Miller, T. E., et al. (2017). Adaptive chromatin remodeling drives glioblastoma stem cell plasticity and drug tolerance. *Cell Stem Cell*, *20*(2), 233–246.e7. https://doi.org/10.1016/j.stem.2016.11.003.

Lito, P., Pratilas, C. A., Joseph, E. W., Tadi, M., Halilovic, E., Zubrowski, M., et al. (2012). Relief of profound feedback inhibition of mitogenic signaling by RAF inhibitors attenuates their activity in BRAFV600E melanomas. *Cancer Cell*, *22*(5), 668–682. https://doi.org/10.1016/j.ccr.2012.10.009.

Liu, Y., Liu, Y., Xu, D., & Li, J. (2015). Latanoprost-induced cytokine and chemokine release from human Tenon's capsule fibroblasts: Role of MAPK and NF-κB signaling pathways. *Journal of Glaucoma*, *24*(9), 635–641. https://doi.org/10.1097/IJG.0000000000000140.

Loda, A., & Heard, E. (2019). Xist RNA in action: Past, present, and future. *PLoS Genetics*, *15*(9), e1008333. https://doi.org/10.1371/journal.pgen.1008333.

Lu, A. Q., Lv, B., Qiu, F., Wang, X. Y., & Cao, X. H. (2017). Upregulation of miR-137 reverses sorafenib resistance and cancer-initiating cell phenotypes by degrading ANT2 in hepatocellular carcinoma. *Oncology Reports*, *37*(4), 2071–2078. https://doi.org/10.3892/or.2017.5498.

Luo, M., Shang, L., Brooks, M. D., Jiagge, E., Zhu, Y., Buschhaus, J. M., et al. (2018). Targeting breast cancer stem cell state equilibrium through modulation of redox signaling. *Cell Metabolism*, *28*, 69–86.e6. https://doi.org/10.1016/j.cmet.2018.06.006.

Luo, J., Solimini, N. L., & Elledge, S. J. (2009). Principles of cancer therapy: Oncogene and non-oncogene addiction. *Cell*, *136*, 823–837. https://doi.org/10.1016/j.cell.2009.02.024.

Madakashira, B. P., & Sadler, K. C. (2017). DNA methylation, nuclear organization, and cancer. *Frontiers in Genetics*, *8*, 76. https://doi.org/10.3389/fgene.2017.00076.

Majchrzak-Celińska, A., Warych, A., & Szoszkiewicz, M. (2021). Novel approaches to epigenetic therapies: From drug combinations to epigenetic editing. *Genes*, *12*, 208. https://doi.org/10.3390/genes12020208.

Managadze, D., Rogozin, I. B., Chernikova, D., Shabalina, S. A., & Koonin, E. V. (2011). Negative correlation between expression level and evolutionary rate of long intergenic noncoding RNAs. *Genome Biology and Evolution*, *3*, 1390–1404. https://doi.org/10.1093/gbe/evr116.

Margueron, R., & Reinberg, D. (2010). Chromatin structure and the inheritance of epigenetic information. *Nature Reviews. Genetics*, *11*, 285–296. https://doi.org/10.1038/nrg2752.

Mathelier, A., Fornes, O., Arenillas, D. J., Chen, C. Y., Denay, G., Lee, J., et al. (2016). JASPAR 2016: A major expansion and update of the open-access database of transcription factor binding profiles. *Nucleic Acids Research*, *44*(D1), D110–D115. https://doi.org/10.1093/nar/gkv1176.

Matys, V., Kel-Margoulis, O. V., Fricke, E., Liebich, I., Land, S., Barre-Dirrie, A., et al. (2006). TRANSFAC and its module TRANSCompel: Transcriptional gene regulation in eukaryotes. *Nucleic Acids Research*, *34*(Database issue), 108–110. https://doi.org/10.1093/nar/gkj143.

Meitinger, F., Ohta, M., Lee, K.-Y., Watanabe, S., Davis, R. L., Anzola, J. V., et al. (2020). TRIM37 controls cancer-specific vulnerability to PLK4 inhibition. *Nature*, *585*(7825), 440–446. https://doi.org/10.1038/s41586-020-2710-1.

Meng, F., Li, Z., Zhang, Z., Yang, Z., Kang, Y., Zhao, X., et al. (2018). Theranostics MicroRNA-193b-3p regulates chondrogenesis and chondrocyte metabolism by targeting HDAC3. *Theranostics*, *8*(10), 2862–2883. https://doi.org/10.7150/thno.23547.

Menon, D. R., Das, S., Krepler, C., Vultur, A., Rinner, B., Schauer, S., et al. (2015). Erratum: A stress-induced early innate response causes multidrug tolerance in melanoma (Oncogene (2015) 34 (4448-4459) DOI:10.1038/onc.2014.372). *Oncogene*, *34*(34), 4545. https://doi.org/10.1038/onc.2014.432.

Meseure, D., Vacher, S., Alsibai, K. D., Nicolas, A., Chemlali, W., Caly, M., et al. (2016). Expression of ANRIL-polycomb complexes-CDKN2A/B/ARF genes in breast tumors: Identification of a two-gene (EZH2/CBX7) signature with independent prognostic value. *Molecular Cancer Research*, *14*(7), 623–633. https://doi.org/10.1158/1541-7786.MCR-15-0418.

Mondal, T., Subhash, S., Vaid, R., Enroth, S., Uday, S., Reinius, B., et al. (2015). MEG3 long noncoding RNA regulates the TGF-β pathway genes through formation of RNA-DNA triplex structures. *Nature Communications, 6, 7743.* https://doi.org/10.1038/ncomms8743.

Mukherjee, S., Dash, S., Lohitesh, K., & Chowdhury, R. (2017). The dynamic role of autophagy and MAPK signaling in determining cell fate under cisplatin stress in osteosarcoma cells. *PLoS One, 12*(6), e0179203. https://doi.org/10.1371/journal.pone.0179203.

Ndlovu, M. N., Denis, H., & Fuks, F. (2011). Exposing the DNA methylome iceberg. *Trends in Biochemical Sciences, 36*, 381–387. https://doi.org/10.1016/j.tibs.2011.03.002.

Niveditha, D., Sharma, H., Majumder, S., Mukherjee, S., Chowdhury, R., & Chowdhury, S. (2019). Transcriptomic analysis associated with reversal of cisplatin sensitivity in drug resistant osteosarcoma cells after a drug holiday. *BMC Cancer, 19*(1), 1–14. https://doi.org/10.1186/s12885-019-6300-2.

Niveditha, D., Sharma, H., Sahu, A., Majumder, S., Chowdhury, R., & Chowdhury, S. (2019). Drug tolerant cells: An emerging target with unique transcriptomic features. *Cancer Informatics, 18, 1176935119881633.* https://doi.org/10.1177/1176935119881633.

Palazzo, A. F., & Gregory, T. R. (2014). The case for junk DNA. *PLoS Genetics, 10*(5), e1004351. https://doi.org/10.1371/journal.pgen.1004351.

Palazzo, A. F., & Lee, E. S. (2018). Sequence determinants for nuclear retention and cytoplasmic export of mRNAs and lncRNAs. *Frontiers in Genetics, 9*(OCT), 1–16. https://doi.org/10.3389/fgene.2018.00440.

Pappalardo, X. G., & Barra, V. (2021). Losing DNA methylation at repetitive elements and breaking bad. *Epigenetics & Chromatin, 14*(1), 1–21. https://doi.org/10.1186/s13072-021-00400-z.

Pham, V., Pitti, R., Tindell, C. A., Cheung, T. K., Masselot, A., Stephan, J. P., et al. (2020). Proteomic analyses identify a novel role for EZH2 in the initiation of Cancer cell drug tolerance. *Journal of Proteome Research, 19*(4), 1533–1547. https://doi.org/10.1021/acs.jproteome.

Pontier, D. B., & Gribnau, J. (2011). Xist regulation and function explored. *Human Genetics, 130*(2), 223–236. https://doi.org/10.1007/s00439-011-1008-7.

Prevo, R., Fokas, E., Reaper, P. M., Charlton, P. A., Pollard, J. R., McKenna, W. G., et al. (2012). The novel ATR inhibitor VE-821 increases sensitivity of pancreatic cancer cells to radiation and chemotherapy. *Cancer Biology & Therapy, 13*, 1072–1081. https://doi.org/10.4161/cbt.21093.

Qin, S., Li, B., Ming, H., Nice, E. C., Zou, B., & Huang, C. (2022). Harnessing redox signaling to overcome therapeutic-resistant cancer dormancy. *Biochimica et Biophysica Acta (BBA) - Reviews on Cancer, 1877*, 188749. https://doi.org/10.1016/j.bbcan.2022.188749.

Quinn, J. J., & Chang, H. Y. (2016). Unique features of long non-coding RNA biogenesis and function. *Nature Reviews Genetics, 17*(1), 47–62. https://doi.org/10.1038/nrg.2015.10.

Raha, D., Wilson, T. R., Peng, J., Peterson, D., Yue, P., Evangelista, M., et al. (2014). The cancer stem cell marker aldehyde dehydrogenase is required to maintain a drug-tolerant tumor cell subpopulation. *Cancer Research, 74*(13), 3579–3590. https://doi.org/10.1158/0008-5472.CAN-13-3456.

Rambow, F., Rogiers, A., Marin-Bejar, O., Aibar, S., Femel, J., Dewaele, M., et al. (2018). Toward minimal residual disease-directed therapy in melanoma. *Cell, 174*(4), 843–855. e819. https://doi.org/10.1016/j.cell.2018.06.025.

Ranganathan, A. C., Zhang, L., Adam, A. P., & Aguirre-Ghiso, J. A. (2006). Functional coupling of p38-induced up-regulation of BiP and activation of RNA-dependent protein kinase–like endoplasmic reticulum kinase to drug resistance of dormant carcinoma cells. *Cancer Research, 66*, 1702–1711. https://doi.org/10.1158/0008-5472.CAN-05-3092.

Raoof, S., Mulford, I. J., Frisco-Cabanos, H., Nangia, V., Timonina, D., Labrot, E., et al. (2019). Targeting FGFR overcomes EMT-mediated resistance in EGFR mutant non-small cell lung cancer. *Oncogene*, *38*, 6399–6413. https://doi.org/10.1038/s41388-019-0887-2.

Razin, A., & Cedar, H. (1977). Distribution of 5-methylcytosine in chromatin. *Proceedings of the National Academy of Sciences of the United States of America*, *74*, 2725–2728. https://doi.org/10.1073/pnas.74.7.2725.

Razin, A., & Riggs, A. (1980). DNA methylation and gene function. *Science*, *2010*, 604–610. https://doi.org/10.1126/science.6254144.

Risom, T., Langer, E. M., Chapman, M. P., Rantala, J., Fields, A. J., Boniface, C., et al. (2018). Differentiation-state plasticity is a targetable resistance mechanism in basal-like breast cancer. *Nature Communications*, *9*(1), 1–17. https://doi.org/10.1038/s41467-018-05729-w.

Roesch, A., Vultur, A., Bogeski, I., Wang, H., Zimmermann, K. M., Speicher, D., et al. (2013). Overcoming intrinsic multidrug resistance in melanoma by blocking the mitochondrial respiratory chain of slow-cycling JARID1Bhigh cells. *Cancer Cell*, *23*(6), 811–825. https://doi.org/10.1016/j.ccr.2013.05.003.

Russo, M, Lamba, S, Lorenzato, A, Sogari, A, Corti, G, Rospo, G, et al. (2018). Reliance upon ancestral mutations is maintained in colorectal cancers that heterogeneously evolve during targeted therapies. *Nature Communications*, *9*(1), 2287. https://doi.org/10.1038/s41467-018-04506-z.

Sahu, N., Stephan, J. P., Sahu, N., Dela Cruz, D., Merchant, M., Haley, B., et al. (2016). Functional screening implicates miR-371-3p and peroxiredoxin 6 in reversible tolerance to cancer drugs. *Nature Communications*, 7, 12351. https://doi.org/10.1038/ncomms12351.

Saito, Y., Saito, H., Liang, G., & Friedman, J. M. (2013). Epigenetic alterations and microRNA misexpression in cancer and autoimmune diseases: A critical review. *Clinical Reviews in Allergy and Immunology*, *47*, 128–135. https://doi.org/10.1007/s12016-013-8401-z.

Saudemont, A., & Quesnel, B. J. B. (2004). In a model of tumor dormancy, long-term persistent leukemic cells have increased B7-H1 and B7.1 expression and resist CTL-mediated lysis. *Blood*, *104*, 2124–2133. https://doi.org/10.1182/blood-2004-01-0064.

Sehgal, K., Portell, A., Ivanova, E. V., Lizotte, P. H., Mahadevan, N. R., Greene, J. R., et al. (2021). Dynamic single-cell RNA sequencing identifies immunotherapy persister cells following PD-1 blockade. *The Journal of Clinical Investigation*, *131*(2), e135038. https://doi.org/10.1172/JCI135038.

Shaffer, S. M., Dunagin, M. C., Torborg, S. R., Torre, E. A., Emert, B., Krepler, C., et al. (2017). Rare cell variability and drug-induced reprogramming as a mode of cancer drug resistance. *Nature*, *546*(7658), 431–435. https://doi.org/10.1038/nature22794.

Shahbazian, M. D., & Grunstein, M. (2007). Functions of site-specific histone acetylation and deacetylation. *Annual Review of Biochemistry*, *76*, 75–100. https://doi.org/10.1146/annurev.biochem.76.052705.162114.

Sharma, S. V., Lee, D. Y., Li, B., Quinlan, M. P., Takahashi, F., Maheswaran, S., et al. (2010). A chromatin-mediated reversible drug-tolerant state in Cancer cell subpopulations. *Cell*, *141*(1), 69–80. https://doi.org/10.1016/j.cell.2010.02.027.

Sharma, H., Niveditha, D., Chowdhury, R., Mukherjee, S., & Chowdhury, S. (2021). A genome-wide expression profile of noncoding RNAs in human osteosarcoma cells as they acquire resistance to cisplatin. *Discover Oncology*, *12*(1). https://doi.org/10.1007/s12672-021-00441-6.

Shen, S., Faouzi, S., Souquere, S., Roy, S., Routier, E., Libenciuc, C., et al. (2020). Melanoma persister cells are tolerant to BRAF/MEK inhibitors via ACOX1-mediated fatty acid oxidation. *Cell Reports*, *33*(8), 108421. https://doi.org/10.1016/j.celrep.2020.108421.

Singh, R. R., Kunkalla, K., Qu, C., Schlette, E., Neelapu, S. S., Samaniego, F., et al. (2011). ABCG2 is a direct transcriptional target of hedgehog signaling and involved in stroma-induced drug tolerance in diffuse large B-cell lymphoma. *Oncogene*, *30*(49), 4874–4886. https://doi.org/10.1038/onc.2011.195.

Song, K. A., Hosono, Y., Turner, C., Jacob, S., et al. (2018). Increased synthesis of MCL-1 protein underlies initial survival of EGFR-mutant lung cancer to EGFR inhibitors and provides a novel drug target MCL-1 protects EGFR-mutant cancers from EGFR inhibitors. *Clinical Cancer Research*, *24*(22), 5658–5672. https://doi.org/10.1158/1078-0432.CCR-18-0304.

Soubry, A. (2015). Epigenetic inheritance and evolution: A paternal perspective on dietary influences. *Progress in Biophysics and Molecular Biology*, *118*(1–2), 79–85. https://doi.org/10.1016/j.pbiomolbio.2015.02.008.

Strub, T., Ghiraldini, F. G., Carcamo, S., Li, M., Wroblewska, A., Singh, R., et al. (2018). SIRT6 haploinsufficiency induces BRAFV600E melanoma cell resistance to MAPK inhibitors via IGF signalling. *Nature Communications*, *9*(1), 1–13. https://doi.org/10.1038/s41467-018-05966-z.

Su, Y., Ko, M. E., Cheng, H., Zhu, R., Xue, M., Wang, J., et al. (2020). Multi-omic single-cell snapshots reveal multiple independent trajectories to drug tolerance in a melanoma cell line. *Nature Communications*, *11*(1), 1–12. https://doi.org/10.1038/s41467-020-15956-9.

Subramaniam, D., Thombre, R., Dhar, A., & Anant, S. (2014). DNA methyltransferases: A novel target for prevention and therapy. *Frontiers in Oncology*, *4*, 80. https://doi.org/10.3389/fonc.2014.00080.

Sun, C., Wang, L., Huang, S., Heynen, G. J. J. E., Prahallad, A., Robert, C., et al. (2014). Reversible and adaptive resistance to BRAF(V600E) inhibition in melanoma. *Nature*, *508*(1), 118–122. https://doi.org/10.1038/nature13121.

Sung, H., Ferlay, J., Siegel, R. L., Laversanne, M., Soerjomataram, I., Jemal, A., et al. (2021). Global cancer statistics 2020: GLOBOCAN estimates of incidence and mortality worldwide for 36 cancers in 185 countries. *CA: a Cancer Journal for Clinicians*, *71*(3), 209–249. https://doi.org/10.3322/caac.21660.

Taguchi, Y.-H. (2015). Apparent microRNA-Target-specific histone modification in mammalian spermatogenesis. *Evolutionary Bioinformatics*, *11*(Suppl 1), 13–26. https://doi.org/10.4137/EBO.S21832.

Venkatesh, S., & Workman, J. L. (2015). Histone exchange, chromatin structure and the regulation of transcription. *Nature Reviews Molecular Cell Biology*, *16*(3), 178–189. https://doi.org/10.1038/nrm3941.

Vera-Ramirez, L., Vodnala, S. K., Nini, R., Hunter, K. W., & Green, J. E. (2018). Autophagy promotes the survival of dormant breast cancer cells and metastatic tumour recurrence. *Nature Communications*, *9*, 1–12. https://doi.org/10.1038/s41467-018-04070-6.

Wang, W.-J., Wu, S.-P., Liu, J.-B., Shi, Y.-S., Huang, X., Zhang, Q.-B., et al. (2013). MYC regulation of CHK1 and CHK2 promotes radioresistance in a stem cell-like population of nasopharyngeal carcinoma cells. *Cancer Research*, *73*, 1219–1231. https://doi.org/10.1158/0008-5472.CAN-12-1408.

Weirauch, M. T., & Hughes, T. R. (2010). Dramatic changes in transcription factor binding over evolutionary time. *Genome Biology*, *11*(6), 6–8. https://doi.org/10.1186/gb-2010-11-6-122.

Whyte, W. A., Orlando, D. A., Hnisz, D., Abraham, B. J., Lin, C. Y., Kagey, M. H., et al. (2013). Master transcription factors and mediator establish super-enhancers at key cell identity genes. *Cell*, *153*(2), 307–319. https://doi.org/10.1016/j.cell.2013.03.035.

Wu, H., Yang, L., & Chen, L. L. (2017). The diversity of long noncoding RNAs and their generation. *Trends in Genetics, 33*(8), 540–552. https://doi.org/10.1016/j.tig.2017.05.004.

Xie, B., Ding, Q., Han, H., & Wu, D. (2013). miRCancer: A microRNA-cancer association database constructed by text mining on literature. *Bioinformatics, 29*(5), 638–644. https://doi.org/10.1093/bioinformatics/btt014.

Xie, H., Ma, H., & Zhou, D. (2013). Plasma HULC as a promising novel biomarker for the detection of hepatocellular carcinoma. *BioMed Research International, 2013, 136106.* https://doi.org/10.1155/2013/136106.

Xue, Y., Martelotto, L., Baslan, T., Vides, A., Solomon, M., Mai, T. T., et al. (2017). An approach to suppress the evolution of resistance in BRAFV600E-mutant cancer. *Nature Medicine, 23*(8), 929–937. https://doi.org/10.1038/nm.4369.

Yan, H., Chen, X., Zhang, Q., Qin, J., Li, H., Liu, C., et al. (2011). Drug-tolerant cancer cells show reduced tumor-initiating capacity: Depletion of CD44 + cells and evidence for epigenetic mechanisms. *PLoS One, 6*(9). https://doi.org/10.1371/journal.pone.0024397.

Yoon, J. H., Abdelmohsen, K., Srikantan, S., Yang, X., Martindale, J. L., De, S., et al. (2012). LincRNA-p21 suppresses target mRNA translation. *Molecular Cell, 47*(4), 648–655. https://doi.org/10.1016/j.molcel.2012.06.027.

Yuan, D., Xu, J., Wang, J., Pan, Y., Fu, J., Bai, Y., et al. (2016). Extracellular miR-1246 promotes lung cancer cell proliferation and enhances radioresistance by directly targeting DR5. *Oncotarget*, 7(22), 32707. https://doi.org/10.18632/oncotarget.9017.

Zentner, G. E., & Henikoff, S. (2013). Regulation of nucleosome dynamics by histone modifications. *Nature Structural & Molecular Biology, 20*(3), 259–266. https://doi.org/10.1038/nsmb.2470.

Zhang, E., Han, L., Yin, D., He, X., Hong, L., Si, X., et al. (2017). H3K27 acetylation activated-long non-coding RNA CCAT1 affects cell proliferation and migration by regulating SPRY4 and HOXB13 expression in esophageal squamous cell carcinoma. *Nucleic Acids Research, 45*(6), 3086–3101.

Zhang, Y., Pitchiaya, S., Cieślik, M., Niknafs, Y. S., Tien, J. C. Y., Hosono, Y., et al. (2018). Analysis of the androgen receptor-regulated lncRNA landscape identifies a role for ARLNC1 in prostate cancer progression. *Nature Genetics, 50*(6), 814–824. https://doi.org/10.1038/s41588-018-0120-1.

Zhang, Z., Qin, S., Chen, Y., Zhou, L., Yang, M., Tang, Y., et al. (2022). Inhibition of NPC1L1 disrupts adaptive responses of drug-tolerant persister cells to chemotherapy. *EMBO Molecular Medicine, 14*(2), e14903. https://doi.org/10.15252/emmm.202114903.

Zhang, W. C., Skiados, N., Aftab, F., et al. (2022). MicroRNA-21 guide and passenger strand regulation of adenylosuccinate lyase-mediated purine metabolism promotes transition to an EGFR-TKI-tolerant persister state. *Cancer Gene Therapy, 29*, 1878–1894. https://doi.org/10.1038/s41417-022-00504-y.

Zhang, W. C., Wells, J. M., Chow, K. H., et al. (2019). miR-147b-mediated TCA cycle dysfunction and pseudohypoxia initiate drug tolerance to EGFR inhibitors in lung adenocarcinoma. *Nature Metabolism, 1*, 460–474. https://doi.org/10.1038/s42255-019-0052-9.

Zhou, H. M., Zhang, J. G., Zhang, X., & Li, Q. (2021). Targeting cancer stem cells for reversing therapy resistance: Mechanism, signaling, and prospective agents. *Signal Transduction and Targeted Therapy, 6*(1), 62. https://doi.org/10.1038/s41392-020-00430-1.

CHAPTER EIGHT

The epigenetic regulation of cancer cell recovery from therapy exposure and its implications as a novel therapeutic strategy for preventing disease recurrence

Christiana O. Appiah[a,e,†], Manjulata Singh[a,†], Lauren May[a], Ishita Bakshi[a], Ashish Vaidyanathan[a], Paul Dent[d], Gordon Ginder[c], Steven Grant[a,c,d,f], Harry Bear[b,f], and Joseph Landry[a,*]

[a]Department of Human and Molecular Genetics, VCU Institute of Molecular Medicine, Massey Cancer Center, Virginia Commonwealth University School of Medicine, Richmond, VA, United States
[b]Department of Surgery, Virginia Commonwealth University School of Medicine, Massey Cancer Center, Richmond, VA, United States
[c]Department of Internal Medicine, Massey Cancer Center, Virginia Commonwealth University, Richmond, VA, United States
[d]Department of Biochemistry and Molecular Biology, Massey Cancer Center, Virginia Commonwealth University, Richmond, VA, United States
[e]Wright Center for Clinical and Translational Research, Virginia Commonwealth University, Richmond, VA, United States
[f]Department of Microbiology & Immunology, Virginia Commonwealth University School of Medicine, Massey Cancer Center, Richmond, Richmond, VA, United States
*Corresponding author: e-mail address: joseph.landry@vcuhealth.org

Contents

† Equal contribution.

Advances in Cancer Research, Volume 158
ISSN 0065-230X
https://doi.org/10.1016/bs.acr.2022.11.001

Abstract

The ultimate goal of cancer therapy is the elimination of disease from patients. Most directly, this occurs through therapy-induced cell death. Therapy-induced growth arrest can also be a desirable outcome, if prolonged. Unfortunately, therapy-induced growth arrest is rarely durable and the recovering cell population can contribute to cancer recurrence. Consequently, therapeutic strategies that eliminate residual cancer cells reduce opportunities for recurrence. Recovery can occur through diverse mechanisms including quiescence or diapause, exit from senescence, suppression of apoptosis, cytoprotective autophagy, and reductive divisions resulting from polyploidy. Epigenetic regulation of the genome represents a fundamental regulatory mechanism integral to cancer-specific biology, including the recovery from therapy. Epigenetic pathways are particularly attractive therapeutic targets because they are reversible, without changes in DNA, and are catalyzed by druggable enzymes. Previous use of epigenetic-targeting therapies in combination with cancer therapeutics has not been widely successful because of either unacceptable toxicity or limited efficacy. The use of epigenetic-targeting therapies after a significant interval following initial cancer therapy could potentially reduce the toxicity of combination strategies, and possibly exploit essential epigenetic states following therapy exposure. This review examines the feasibility of targeting epigenetic mechanisms using a sequential approach to eliminate residual therapy-arrested populations, that might possibly prevent recovery and disease recurrence.

Abbreviations

CDK Cyclin-dependent kinase
CHD Chromo and DNA-binding domain
CRCs Chromatin remodeling complexes
CSCs Cancer stem cells
DTP Drug-tolerant persister
ENCODE Encyclopedia of DNA elements
GSCs Glioblastoma stem cells
HATs Histone acetyltransferases
HDACs Histone deacetylases
HKDMs Lysine demethylases

HKMTs	Histone lysine methyltransferases
KDM	Lysine-specific demethylase
MBD	Methyl-cytosine binding domain
MTA	Metastasis associated 1 family member
NSCLC	Non-small cell lung cancer
NURD	Nucleosome remodeling and deacetylase
OIS	Oncogene-induced senescence
PGCCs	Polyploid giant cancer cells
PRC	Polycomb repressive complex
SASP	Senescence-associated secretory phenotype
SWI/SNF	SwItch/sucrose non-fermentable
TET	Ten-eleven translocation
TID	Therapy-induced diapause
TIS	Therapy-induced senescence
CDC	Cell division control

1. Introduction

Cancer treatments, which include chemotherapy, radiotherapy, surgery, and more recently immunotherapy and targeted therapy, have unquestionably improved the quality of life and patient survival (Siegel, Miller, & Jemal, 2020). Within the last several decades, the 5-year survival rate for most cancers has increased, in some cases dramatically (i.e., lung and melanoma) (Siegel et al., 2020). Despite these advances, cancer still remains the second leading cause of death in the United States. Mortality from cancer is due to progressive disease, or in cases of initial complete response, recurrence (National Center for Health Statistics (US), 2017). Consequently, developing novel therapies to improve response in tumors that do not completely respond to traditional therapy and to prevent tumors that initially experience a complete response from reoccurring is currently a major focus of the cancer research community.

The primary goal of cancer therapies (chemotherapy or radiation) is cancer cell death, because cancer cells that die cannot recover and contribute to disease recurrence. Moreover, depending on the mechanism of therapy-induced death, this event can contribute significantly to stimulating anti-tumor immune responses (Galluzzi, Buque, Kepp, Zitvogel, & Kroemer, 2017). Mechanisms of cell death commonly include apoptosis, resulting from excessive damage to cellular components (DNA, mitochondria, plasma membrane, etc.), ferroptosis, a distinct iron-dependent cell death pathway, mitotic catastrophe, which can result from the inhibition of chromosome segregation from microtubule inhibitors, and a cytotoxic form of autophagy that can occur from a combination of endoplasmic reticulum (ER) stress and DNA damage resulting in the immunogenic death of the cell (Fig. 1A) (Galluzzi et al., 2018).

From a practical standpoint, therapies do not result in the death of all tumor cells, as a population of cells are refractory to, or initially respond to but then recover from, exposure to treatment. Recovering populations were reported in some of the earliest studies examining the effects of chemotherapy on cancer cells (Barranco & Flournoy, 1977; Braun & Hahn, 1975). Such incomplete responses, leading to disease recurrence, are due to an inability to achieve lethal doses of therapy in the tumor, as well as a tumor cells intrinsic resistance to therapy. Because of its important role in cancer mortality, the control and prevention of recurrence is critical for improving survival rates. How both local and distant recurrence occur has been investigated intensively for decades (Mahvi, Liu, Grinstaff, Colson, & Raut, 2018; Riggio, Varley, & Welm, 2021). By definition, recurrence occurs when cancer cells exhibit a drug-tolerant persister (DTP) phenotype, which broadly describes a phenotype in which a subpopulation of cells is more resistant to the therapy (Dhanyamraju, Schell, Amin, & Robertson, 2022) (Fig. 1B). Several mechanisms, including the existence of a small preexisting therapy-resistant population, the selection of rare genetic mutations and/or heritable epigenetic changes that confer acquired resistance to therapy, and the escape of cancer cells from sublethal therapy exposure are likely to operate in concert to contribute to disease reoccurrence. Various modifications to therapeutic strategies have been developed to target such subpopulations, most notably the use of combination strategies to forestall or prevent the selection of rare clones resistant to single agents (Fisusi & Akala, 2019; Patel & Minn, 2018). However, in most cases, these approaches are not curative, and as a consequence, continuing to develop novel ways to target these populations will be essential to preventing disease recurrence.

One plausible strategy for eliminating cancer cells recovering from therapy involves targeting epigenetic pathways. This approach has been aggressively investigated over the last few decades, with the primary objective of sensitizing DTP cancer cells prior to or coincident with therapy

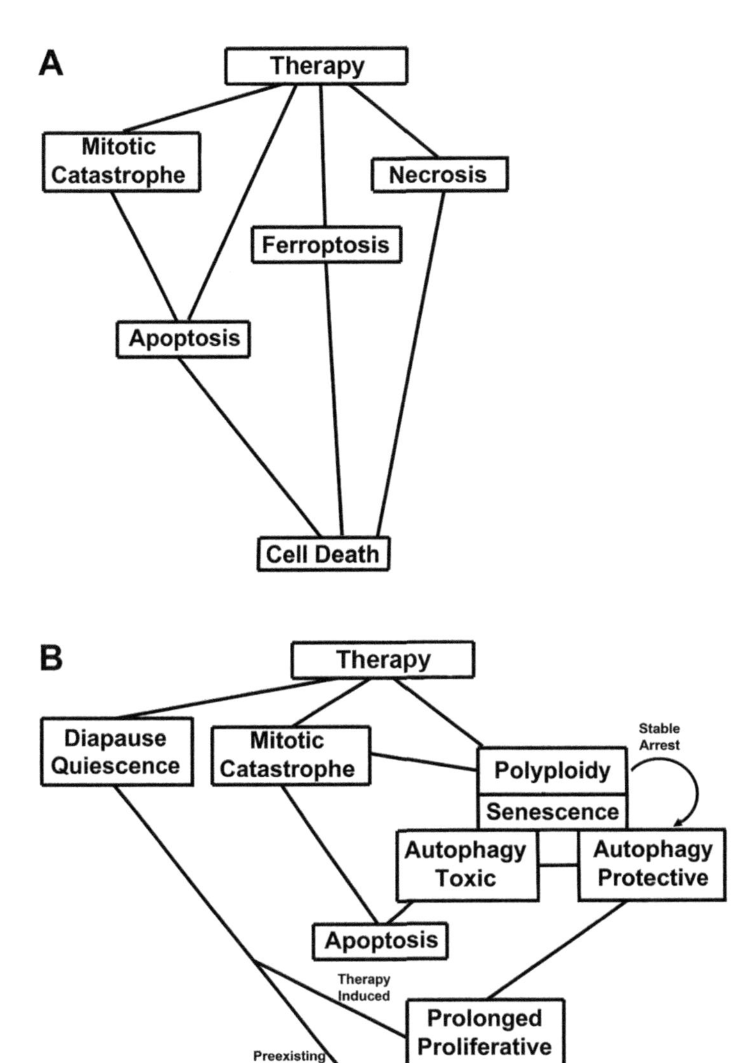

Fig. 1 Autonomous responses of cancer cells to chemotherapy and radiation exposure. (A) Cell death responses of cancer cells from chemotherapy or radiation exposure. Therapy exposure can induce a variety of responses that will eventually lead to the death of the cancer cell. Therapies that target chromosome segregation pathways can lead to a mitotic catastrophe, which when left unresolved lead to polyploidy. Failure to undergo neosis can lead to apoptosis. Therapies of varied mechanisms can lead to apoptosis, ferroptosis, or necrosis if the cellular damage is great enough. (B) In some cases, cancer cells can survive the initial exposure to therapy through a

(Continued)

(Liau et al., 2017; Vinogradova et al., 2016). Such sensitization strategies have as their goal disabling of epigenetic pathways that may be essential for DTP cell survival, and, as a result, the prevention of recurrence due to residual tumor cell elimination. While this strategy has been extensively pursued, it has yielded very limited success in solid tumors in the clinic due to toxicity from combining epigenetic with standard of care therapies, as well as minimal antitumor responses (Morel, Jeffery, Aspeslagh, Almouzni, & Postel-Vinay, 2020).

An alternative and infrequently pursued approach involves a strategy designed to convert cancer cells that have partially responded to initial primary therapy to complete responders using a secondary epigenetic-targeted therapy. This approach is based on the hypothesis that persister cells may be particularly susceptible to either cell death or growth arrest in response to epigenetic agents. This strategy could be particularly beneficial as, for example, cells recovering from therapy-induced senescence exhibit stem cell-like properties and have been shown to be particularly aggressive and tumorigenic (Milanovic et al., 2018). Specifically targeting therapy-arrested senescent cells is currently being actively pursued through the use of senolytics, a class of agents that exhibit selective toxicity to senescent cells (Wang, Lankhorst, & Bernards, 2022; Zhu et al., 2015). However, this phenomenon can also occur by targeting epigenetic pathways essential for maintaining a variety of prosurvival cell stress responses in the residual therapy-exposed cancer cell population. Because epigenetic events play key regulatory roles in diverse cellular stress response pathways (see discussion below), it is plausible that these pathways could be targeted to disable survival-related responses in therapy-exposed cancer cells. Such actions may drive the residual surviving population into one or more forms of cell death. This review summarizes the known epigenetic mechanisms that regulate therapy-induced survival pathways, and provides a rational road map for employing epigenetic-targeting small molecules as a viable strategy for eradicating the residual population of surviving persister cells.

Fig. 1—Cont'd variety of cell stress responses. These responses can result in proliferative arrest of the cell, followed by a recovery. Some cancer cells undergo a state of quiescence prior to therapy exposure, which allows the cells to survive and recover. Another response usually occurs when a cycling cell receives a nonlethal dose of therapy. These responses prominently include senescence, autophagy and polyploidy inducing a prolonged proliferative arrest. A percentage of cells in these cell stress states can recover after repair of the cellular damage. Those that cannot repair the cellular damage enter into apoptosis.

2. A short primer on epigenetics

The field of epigenetics has a complex history (Peixoto, Cartron, Serandour, & Hervouet, 2020) and has been the subject of extensive research efforts over the last several decades (The ENCODE Project Consortium et al., 2020). Epigenetics represents the study of heritable phenotypic changes that do not involve alterations in the DNA sequence (Berger, Kouzarides, Shiekhattar, & Shilatifard, 2009). Epigenetic marks serve as regulatory pathways that act in addition to existing genetic regulatory elements by altering chromatin and its structure through postsynthetic modification of DNA or chromatin proteins. The fundamental unit of chromatin is the nucleosome, which is composed of a histone octamer and approximately 146 base pair (bp) of DNA. The histone octamer is comprised of two copies of four canonical histone proteins (H2A, H2B, H3, and H4) or their variants (Talbert & Henikoff, 2021). Epigenetic regulation occurs through posttranslational modification to histone proteins or DNA and the alteration of nucleosome position relative to the DNA sequence.

DNA and histone modifications are regulated by three main categories of protein complexes: the "writers," "readers," and "erasers" of epigenetic marks (Fig. 2). *Writers* catalyze the addition of posttranslational modifications. Notable writers include DNA methyltransferases (DNMTs), histone acetyltransferases (HATs) and histone lysine methyltransferases (HKMTs). *Erasers* remove posttranslational modifications from the histones or DNA established by the writers and include the ten-eleven translocation (TET) enzymes, histone deacetylases (HDACs), and lysine demethylases (HKDMs). Erasers can return DNA or histones to their original unmodified state. *Readers* are protein complexes that have distinct domains that recognize specific posttranslational modifications established by the writers. The readers recruit structural and enzymatic complexes to chromatin that regulate its structure and function, and ultimately affect access to the DNA sequence. Prominent reader domains include bromodomains, chromodomains, PHD fingers, and the methylcytosine binding domain (MBD) family of proteins.

In addition to posttranslational modifications of DNA and histones, nucleosome position can stably regulate gene expression, and can thus exert epigenetic regulatory activity (Clapier & Cairns, 2009). Nucleosome positions are largely altered by the action of ATP-dependent chromatin remodeling complexes (CRCs). There are four families of CRCs based on homology of the ATPase subunit: SWI/SNF, ISWI, CHD, and INO80.

	Classes	Examples
Writers	DNA Methyltransferases	DNMT
	Histone Methyltransferases	EZH2
	Histone Acetyltransferases	TIP60
Erasers	Ten-eleven translocation	TET1
	Histone Deacetylases	HDAC1
	Histone Demethylases	LSD1
Readers	Bromodomains	BRD4
	Chromodomains	BPTF
	PHD Fingers	MBD2
	Methyl Binding Domains	

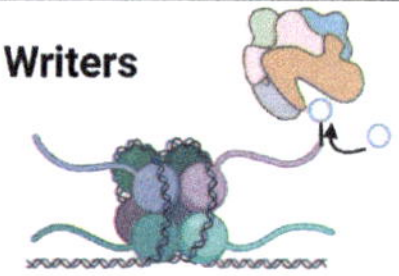

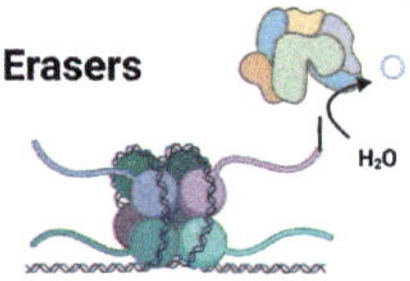

Fig. 2 Summary of epigenetic regulators. Epigenetic regulators have been categorized into three main categories based on their molecular functions. The first are writers which "write" marks, or posttranslational modifications onto proteins, prominently histones, or DNA. Listed are the major classes of writers, and some example enzymes. The second category are the erasers which erase epigenetic marks, or posttranslational modifications, from proteins, predominantly histones, and DNA. The third are the "readers" which are a broad class of proteins which have a functional domain which recognizes and binds to specific posttranslational modifications in proteins or DNA. Readers serve to recruit complexes to specific sites in chromatin with specific epigenetic marks to promote nuclear activities including chromatin remodeling, transcription, DNA replication, DNA-damage repair, to name a few.

The most widely studied CRC is the BRG or BRM ATPase containing SWI/SNF family of complexes, which is composed of more than 30 distinct complexes that vary according to subunit composition (Centore, Sandoval, Soares, Kadoch, & Chan, 2020).

The consequences of DNA and histone posttranslational modifications and chromatin remodeling for gene expression are highly context dependent. We refer the reader to several recent reviews that discuss the functions of epigenetic events in a cancer-relevant context ((Bates, 2020; Esteller, 2008; Villanueva, Alvarez-Errico, & Esteller, 2020)—many others covering specific aspects of cancer biology like metastasis, immune suppression). The functions of these modifications in the context of recovery from therapy exposure is discussed below.

3. The limited success of epigenetic-targeted therapies treating solid tumors

Epigenetics plays a fundamental role in cancer biology, a view that is supported by the elevation of "epigenetic programming" as one of the hallmarks of cancer as recently proposed by Hanahan (Hanahan, 2022). It has been known for decades that epigenetics plays an important role in the ability of hematologic malignancies to respond to, and recover from, exposure to cytotoxic chemotherapy (Kuendgen & Gattermann, 2007; Kuendgen et al., 2004; Schwartsmann et al., 1997). In addition, epigenetic modifying agents such as histone deacetylase inhibitors have been approved for the treatment of certain hematopoietic malignancies e.g., peripheral and cutaneous T-cell lymphoma (Lopez, Bates, & Geskin, 2018). Moreover, a variety of epigenetic pathways have been targeted with small molecules with the goal of improving the effectiveness of cytotoxic chemotherapy in solid tumors (Ganesan, Arimondo, Rots, Jeronimo, & Berdasco, 2019). Such a strategy appears plausible, as epigenetic changes are reversible and are largely catalyzed by enzymes with potentially druggable active sites amenable to targeting with small molecules. Unfortunately, initial studies employing epigenetic-targeting therapies as monotherapies in a variety of solid tumors has been largely unsuccessful (Nervi, De Marinis, & Codacci-Pisanelli, 2015). Subsequently, combinatorial use of epigenetic therapies with chemotherapy, radiation, or targeted therapies has been investigated. However, while such combination approaches have shown some early promise in the clinic to solid tumors, they have largely been unsuccessful due to unacceptable toxicity or limited effectiveness (Morel et al., 2020). Consequently, if epigenetic therapies are to be pursued in the clinic, particularly in the case of solid tumors, new treatment paradigms will be required, to circumvent these limitations.

4. A proposed novel use for epigenetic-targeted therapies

As noted previously, the two major barriers to the success of epigenetic strategies in solid tumors are unacceptable toxicity and limited effectiveness (Morel et al., 2020). The latter has been postulated to stem from poor penetration of epigenetic therapies into solid tumors and the redundancy of epigenetic pathways in cancer cells regulating cancer biology (Gottesman, Lavi, Hall, & Gillet, 2016). The former problem is not unique to epigenetic therapies, and is observed in the case of diverse chemotherapeutic agents (Tannock, Lee, Tunggal, Cowan, & Egorin, 2002). Several novel approaches using nano-delivery methods are being developed to overcome this limitation (Buocikova et al., 2020), and could potentially be extrapolated to epigenetic agents. The redundancy concept can best be explained in relation to the oncogene (i.e., c-Myc) upregulation observed in many cancer types. Oncogenes can be upregulated through multiple mechanisms, including gene amplification, loss of transcriptional repressors, enhancement of transcriptional activators, or modulation of epigenetic pathways (Dhanasekaran et al., 2022). For example, the bromodomain BRD4 has been identified as a key epigenetic factor regulating the transcriptional activation of oncogenes by cancer cell-acquired super enhancers (Loven et al., 2013). The use of BRD4 inhibitors can suppress oncogenes in cases in which their activation is mediated by ectopic super enhancers to achieve beneficial therapeutic responses. However, such a strategy would not be effective in situations in which oncogenes are amplified through mechanisms that do not involve super enhancers (i.e., translocation). These considerations would render epigenetic therapies such as BRD4 inhibitors effective only in a subset of patients whose tumors exhibit superenhancer driven oncogene expression, but not those with alternative mechanisms. However, these limitations may not occur in the case of specific epigenetic pathways that modulate broad aspects of cancer biology. For example, DNA methylation characteristically suppresses tumor suppressors, tumor-specific antigens, and retrotransposons (Chiappinelli et al., 2015; Siebenkas et al., 2017), all of which contribute to tumor growth control. These widespread activities make DNA-demethylating agents (e.g., DNMT1 inhibitors) an effective epigenetic therapy, which is FDA approved for specific cancers e.g., the myelodysplastic syndrome (MDS), and are being used widely in early-stage clinical trials (Mehdipour, Murphy, & De Carvalho, 2020).

In most cases, epigenetic approaches have been employed in conjunction with other therapies to sensitize cancer cells to the actions of the latter. However, this approach may lead to increased toxicity for the combination (Morel et al., 2020). One plausible solution to circumvent such toxicities is to utilize "drug holidays," or pauses in the treatment regimen.

Treatment holidays have had mixed results with respect to patient survival following cytotoxic chemotherapy (Garattini et al., 2021; Labianca et al., 2011; Mittal et al., 2018; Onishi, Sasaki, & Hoshina, 2012), but in the case of targeted therapies, they have generally had a detrimental effect on tumor growth control (Punt, Simkens, & Koopman, 2015). Despite their theoretical disadvantages, drug holidays have been used to provide patients with a respite from the toxic effects of therapy. However, the use of drug holidays could have benefits in the context of epitherapy/chemotherapy combinations. For example, when using the two therapies in sequence (e.g., chemotherapy followed by epitherapy), a drug holiday could reduce the inherent treatment toxicity by allowing clearance of cytotoxic chemotherapy. In addition, there could be unique benefits to the sequential use of epitherapy following cytotoxic chemotherapies. Specifically, such sequential strategy may allow the targeting of unique epigenetic states established postchemotherapy exposure, which may not be present during or immediately after chemotherapy exposure. A similar strategy has been applied for decades in the form of epigenetic priming, a process where cancer cells, or a patients' tumor, are first treated with an epitherapy to sensitize the cancer cells to the effects of the chemotherapy ((Buocikova et al., 2022; Cimino et al., 2006; Ranganathan et al., 2015; Toor et al., 2012; Ye et al., 2016; Duy et al., 2019)—and many others). What we would propose is to first treat with chemotherapy to achieve maximum therapeutic response, thereby allowing novel epigenetic landscapes to form that would be required for viability during therapy-induced arrest and eventual recovery. These modified epigenetic states could then be targeted for disruption using epidrugs, thereby preventing recovery from therapy exposure. This strategy appears plausible as epigenetics plays essential roles in the response of cancer cells to chemotherapy (see discussion below). This strategy would not preclude the option of epigenetic priming, or cotreatments with epidrugs to sensitize cancer cells if combinations of cytotoxic therapies and epidrugs are discovered that are tolerable and effective. Instead, sequential treatment could be used to discover novel uses for epidrugs shown experimentally to be effective in the recovery phase, or alternatively, to salvage epidrugs are too toxic or ineffective when used for epigenic priming or in combination. This concept postulates therapy-induced and delayed epigenetic states exist that are druggable, and when targeted by epidrugs, would prevent the recovery of cancer cells from therapy exposure (Fig. 3).

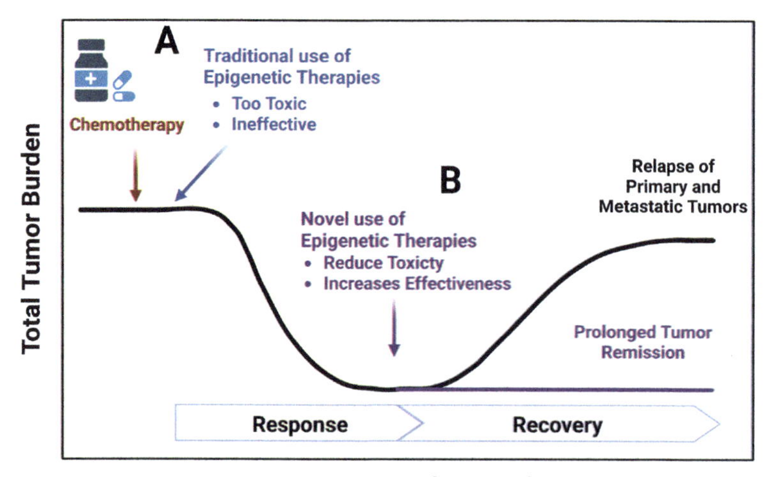

Fig. 3 Proposed strategy for the sequential use of chemotherapies and epigenetic-targeting therapies. (A) Epigenetic therapies have been most commonly used by prepriming, prior to chemotherapy exposure, or in combination with chemotherapies. These strategies served to make the chemotherapy exposure more effective, and therefore more toxic, to the cancer cells and have been largely ineffective to solid tumors. (B) We propose a novel use of epigenetic-targeting therapies where they would be utilized a significant period after chemotherapy exposure to reduce toxicity from the combination, and target novel epigenetic states established as a result of cancer cells entering into a therapy-induced arrested state, which is a significant period of time after chemotherapy exposure.

5. States which promote cancer cell recovery from therapy exposure

5.1 Pretherapy quiescent state

Quiescence is a mechanism of dormancy by which tumors cells persist in a patient for extended periods of time and provides a mechanism by which tumor cells can evade the effects of cytotoxic chemotherapy. Quiescence is also a normal and reversible process of embryonic development and adult tissue homeostasis that is required for both stem and nonstem cells alike to survive stressful microenvironments (Marescal & Cheeseman, 2020). In the context of cancer, quiescence is observed in many tumor types and at several key points in tumor progression (e.g., primary tumor, metastasis, and resistance to therapy), suggesting that quiescence represents a fundamental component of disease progression (Phan & Croucher, 2020; Sistigu,

Musella, Galassi, Vitale, & De Maria, 2020; Sosa, Bragado, & Aguirre-Ghiso, 2014). Quiescent cells can originate in the tumor, exist as circulating cancer cells, or can reside in distant metastatic niches. These environments are usually inhospitable, as they are depleted of oxygen, nutrients, and growth factors, which significantly stresses the cancer cell. The quiescent cell state allows such cells to survive and lie dormant in these inhospitable environments. Because of these adaptations, quiescent cancer cells are also resistant to therapy exposure (Siu, Arooz, & Poon, 1999). The quiescent state is traditionally identified as a G0 arrested cell (i.e., lacking ki67 expression), reduced ATP, glutamate, and amino acid synthesis, reduced intracellular reactive oxygen species (ROS), and diminished apoptosis, all of which render cancer cells resistant to chemotherapies (Cheung & Rando, 2013). There is significant heterogeneity in the quiescent cell state stemming from variability in the signals that induce this state. At a fundamental level, regulation of the G0-arrested quiescent state occurs through $p21^{WAF1}$, $p27^{KIP1}$, $p57^{KIP2}$, the cyclins Cyclin D1, Cyclin E, the cyclin-dependent kinases CDK4/6 and CDK2, and downstream regulators of the p53-$p21^{WAF1}$ and Rb-E2F pathway (Marescal & Cheeseman, 2020). In contrast to therapy-induced senescence, quiescence requires functional p53-$p21^{WAF1}$ and Rb-E2F pathways. Prominent roles for $p27^{KIP1}$ arrest, over $p21^{WAF1}$ arrest, drive the quiescent state due to its ability to arrest the cell cycle in G0 (Terzi, Izmirli, & Gogebakan, 2016). In addition to cell cycle components, signaling pathways such as the Wnt/β-catenin, STAT3, and NF-kB pathways are essential for maintenance of the quiescent state. It is important to note that tumor cells in the quiescent state share some properties with cancer stem cells (CSCs) (e.g., low cell proliferation and altered cellular metabolism).

5.2 Cancer stem cells

CSCs represent a population of pluripotent cancer cells that share stemness traits with their normal stem or progenitor cell counterparts. CSCs have an unlimited proliferation potential, self-renewal capacity, and multilineage differentiation that contributes to tumor progression and heterogeneity (Ayob & Ramasamy, 2018; Calcagno et al., 2010; Chang, 2016). CSCs, like normal stem cells, express phenotypic surface markers like CD133, CD44, CD90, ALDH1 (Beck & Blanpain, 2013; Calcagno et al., 2010; Chen, Dong, Haiech, Kilhoffer, & Zeniou, 2016) as well as pluripotency markers (Sox2 and Oct3/4 expression) (Hu et al., 2012). CSC surface markers are observed in all solid tumors and hematologic malignancies, which include breast, lung, brain, gastric, liver, gastric, colon, prostate, pancreatic, and head and neck cancers, multiple myeloma, leukemia, and melanoma (Chang, 2016; Chen, Huang, & Chen, 2013; Doherty, Smigiel, Junk, & Jackson, 2016; Smith et al., 2008; Takaishi et al., 2009).

CSC attribute to tumor recurrence primarily due to their resistance to traditional cancer therapy such as chemotherapy and radiotherapy (Ayob & Ramasamy, 2018; Desai, Yan, & Gerson, 2019). This situation is further complicated by the fact that standard treatment leads to the eradication of nonstem cancer cells, thereby increasing the relative ratio of CSCs to tumor mass (Doherty et al., 2016), and can convert nonstem cancer cells to CSCs, which further increases the chance for tumor recurrence (Hu et al., 2012; Lagadec, Vlashi, Della Donna, Dekmezian, & Pajonk, 2012; Wang et al., 2014; Xu et al., 2015). CSC resistance to multiple drugs can stem from high expression of genes such as those encoding ATP-binding cassette (ABC) transporters (Begicevic & Falasca, 2017; Zinzi et al., 2014), reduced levels of ROS (Diehn et al., 2009), activation of DNA-damage repair pathways (Diehn et al., 2009), an ability to undergo epithelial-mesenchymal transition (EMT) (Chang, 2016), and quiescent cell phenotypes including slow cell cycle and changes in metabolism (Zeuner, 2015). The quiescent phenotype allows CSC to lie dormant for years, and when revived can lead to recurrence (Chang, 2016).

5.3 Therapy-induced diapause

In addition to cancer cells existing in a dormant state prior to chemotherapy exposure, a therapy-induced diapause (TID) state can result from exposure to chemotherapy (Dhimolea et al., 2021; Duy et al., 2021; Rehman et al., 2021). This state is distinct from that of quiescence wherein the therapy-resistant state is established prior to therapy exposure, but instead occurs as an adaptive mechanism by which the cancer cells survive exposure. Cells in TID can survive in large numbers rather than as rare cells (as is the case with quiescent cells or CSCs) and subsequently enter proliferative recovery. A therapy-induced diapause-like state includes some characteristics of senescence, autophagy, and quiescence. TID exhibits reduced cell cycle progression and reduced metabolic activity, and also expresses markers of senescence and autophagy. Molecular changes associated with a TID include SA-β-galactosidase expression, which provides evidence of a senescent-like state, as well as downregulation of c-Myc and upregulation of the cell cycle inhibitor $p27^{KIP1}$ (coinciding with reduced cell cycle progression). An important point is that the TID state is transient and does not recapitulate with acquired resistance. Cancer cells that survive chemotherapy exposure as a result of TID, once recovered, revert to the same level of sensitivity they had prior to the therapy exposure (Dhimolea et al., 2021; Duy et al., 2021; Rehman et al., 2021).

5.4 Senescence

Senescence is a state of proliferative arrest that can be divided into three main categories based upon the mechanism that induces the senescent state. These include replicative senescence, which occurs when telomeres lose their protective ends; oncogene-induced senescence (OIS), which results from hyperproliferation; and stress-induced premature senescence, which results from ROS or DNA damage and includes therapy-induced senescence (TIS). In many contexts, senescence has been shown to be a barrier to tumor progression. However, several studies have shown it to be an imperfect barrier as senescent cells can escape from the arrest of OIS, stress-induced premature senescence and TIS, and even replicative senescence (Guillon et al., 2019; Li et al., 2020; Medema, 2018; Pacifico et al., 2021, 2022; Saleh et al., 2019; Zampetidis et al., 2021; Zampetidis, Papantonis, & Gorgoulis, 2022). Relevant to this review are both OIS and TIS and each will be discussed in detail below.

5.4.1 Therapy-induced senescence

TIS is widely recognized to be a primary response of human tumors to therapy exposure (Chang, Broude, et al., 1999; Roninson, Broude, & Chang, 2001; te Poele, Okorokov, Jardine, Cummings, & Joel, 2002; Wang et al., 1998). TIS is characterized by a growth arrest, where cells are maintained in the G1 or G2/M stage of the cell cycle, or by enhanced expression of specific cyclin-dependent kinase inhibitors (CDKIs), including $p16^{INK4a}$ (CDKN2A), $p21^{WAF1}$ (CDKN1A, CIP1), and $p27^{KIP1}$ (CDKN1B) (Bringold & Serrano, 2000). p53 activity by senescence-inducing signals (i.e., therapy) activates $p21^{WAF1}$, which further inhibits CDK2/Cyclin E and CDK4 activity and arrests the cell cycle (Dimri, 2005; He et al., 2005; Wierod et al., 2008). $p16^{INK4a}$ is also regulated by senescence-inducing signals (i.e., DNA damage and oncogenes) and when upregulated it inhibits the ability of the CDK4/6/Cyclin D complex to phosphorylate Rb (pRb). The lack of pRb results in a stable G1 arrest and promotes senescence (LaPak & Burd, 2014). $p16^{INK4a}$ is frequently lost in the transformation process, and its loss along with p53, are significant events allowing premalignant cells to bypass and reverse the senescent state (Liggett & Sidransky, 1998; Muller & Vousden, 2013). Cells with functional p53 and Rb are more sensitive to oncogenic stimuli that can induce senescence. TIS can also be induced in a p53-independent manner (Chang, Xuan, et al., 1999; Fang et al., 1999). Telomeric DNA-damage signals involve the p53-p21^{WAF1}-Rb pathway whereas nontelomeric DNA-damage signals induce both the p53-p21^{WAF1}-RB and p16^{INK4a}-RB pathways.

In addition to cell cycle arrest, TIS causes a significant chromatin reorganization and these modifications can serve as therapeutic targets. Cells in TIS create senescence-associated heterochromatin foci (SAHF) in addition to other heterochromatin markers (Adams, 2007). SAHF are rich in heterochromatic markers like hypoacetylated histones, trimethylated histone H3 lysine 9 and 27 (H3K9me3 and H3K27me3), the heterochromatin protein 1 (HP1) family of proteins, and histone variant macroH2A.

Cells in TIS also secrete a senescence-associated secretory phenotype (SASP). SASP consists of several soluble and insoluble factors that induce autocrine and paracrine signaling for tumor progression. The SASP includes various interleukins such as IL-6, IL-8, IL-1, IL-1b, IL-13 and IL-15 (Coppe, Desprez, Krtolica, & Campisi, 2010; Davalos, Coppe, Campisi, & Desprez, 2010), chemokines like MCP-2, MCP-4, MIP-1a, various matrix metalloproteinase like MMP3, MMP6 and MMP9 (Coppe et al., 2010), and inflammatory molecules like TGF-β, GM-CSF, IFN-g, and VEGF (Lopes-Paciencia et al., 2019). These senescent cells, which contribute to therapy-induced tumor control through proliferative arrest and inhibit disease progression in the short term, are likely to serve as a source for cancer relapse at later intervals (Wang, Kohli, & Demaria, 2020). There is evidence that supports the notion that senescent cells and SASP induce a protumoral and procarcinogenic microenvironment (Schosserer, Grillari, & Breitenbach, 2017), induce EMT (Malaquin, Tu, & Rodier, 1896; Was et al., 2018), and contribute to more aggressive tendency to relapse in cancer (Krtolica, Parrinello, Lockett, Desprez, & Campisi, 2001; Saleh et al., 2020).

5.4.2 Oncogene-induced senescence

OIS occurs in cancer cells through aberrant expression of oncogenes such as BRAF, AKT, E2F1, and cyclin E or by inactivation of tumor suppressor genes PTEN, P53, NF1, etc. Oncogenic Ras-induced senescence, which occurs through accumulation of Ras^{V12} or BRAF^{V600E}, p16^{INK4A}, and p19^{ARF}, has been observed in various human cancers (Alcorta et al., 1996; Serrano, Lin, McCurrach, Beach, & Lowe, 1997). OIS is an irreversible phenotype in primary cells (Dimauro & David, 2010). However, the perception of OIS as permanent has been challenged with studies demonstrating escape from OIS (Bellelli et al., 2018; Sadangi et al., 2022).

5.5 Polyploidy

Cells in a polyploid state undergo significant genomic instability and when coupled with a reductive cellular division referred to as neosis (Sundaram, Guernsey, Rajaraman, & Rajaraman, 2004) can yield cells with complex genomes but which are mitotically viable. Daughter cells from these polyploid giant cancer cells (PGCCs) are thought to contribute significantly to the creation of therapy-resistant cells and more aggressive cancers in several solid tumor types (Fei et al., 2015; Mittal et al., 2017; Zhang et al., 2014). Therapy-induced polyploidy is a common occurrence in primary tumors (Zack et al., 2013) and when cancer cells are treated with DNA-damaging or mitosis-interfering chemotherapies (Song, Zhao, Deng, Zhao, & Huang, 2021). In addition to occurring in advanced cancers, PGCCs are therapy-induced, are a common outcome of TIS (Sliwinska et al., 2009), and are believed to represent a step toward recovering from therapy exposure (Wang et al., 2013). Studies involving doxorubicin-induced senescence showed a polyploid population escaping from senescence following cessation of treatment (Mosieniak et al., 2015). In another report, tumor cells ceased mitotic activity but continued DNA replication after cisplatin treatment, resulting in large polyploid cells. Once cisplatin was removed, the cells could form colonies (Puig et al., 2008). Another study demonstrated a multistep process through which ovarian cancer cells become chemotherapy resistant, with one of the steps being the acquisition of polyploidy (Rohnalter et al., 2015). The polyploid cells generated a smaller population of cells that escaped senescence, resumed cell division, and were chemotherapy resistant (Rohnalter et al., 2015).

Polyploid cells are typically cell cycle-arrested and have many of the hallmarks of senescence, including X-gal staining, complex genomes, SASP, and EMT characteristics (Fei et al., 2015; Sikora, Czarnecka-Herok, Bojko, & Sunderland, 2022; Song et al., 2021). Polyploidy can trigger a number of cell cycle checkpoints, such as the spindle and G1 post-mitotic checkpoints, which halt cell cycle and can eventually induce mitotic catastrophe and cell death. Through unknown mechanisms involving neosis, daughter cells are produced after therapy exposure, which are believed to contribute to a recovering population (White-Gilbertson & Voelkel-Johnson, 2020).

5.6 Autophagy

Chemotherapeutics or radiation therapy can activate cytoprotective autophagy or forms of autophagy-related cell death in cancer cells (Graham et al., 2016; Kondo, Kanzawa, Sawaya, & Kondo, 2005; Scarlatti et al., 2004; Shao, Gao, Marks, & Jiang, 2004). In many cases, autophagy is protective in therapy-exposed cancer cells and plays an important role in cancer cell survival posttherapy exposure (Das, Mandal, & Kogel, 2018). Autophagy is a lysosome-dependent, catabolic degradation process that is activated by cell stresses induced by cancer therapeutics (Korhonen, Ylanne, Laitinen, & Virtanen, 1990; Lapierre, Kumsta, Sandri, Ballabio, & Hansen, 2015). Therapy-induced autophagy can be cytoprotective by promoting cell adaptation and survival, or it can be cytotoxic by inducing cell death (Kondo et al., 2005). Cytoprotective autophagy in response to therapy occurs through sequestration and degradation of therapy-damaged macromolecules and organelles in ATG-LC3B mediated autophagosomes to provide nutrients to promote recovery (Gozuacik & Kimchi, 2004; Ogier-Denis & Codogno, 2003). In contrast, cytotoxic autophagy occurs when the tumor undergoes cell death that does not involve necrosis and apoptosis. Cytotoxic autophagy involves both Beclin1-dependent and -independent pathways (Akar et al., 2008; Scarlatti, Maffei, Beau, Codogno, & Ghidoni, 2008). The Beclin1-dependent pathway leads to autophagic cell death after blocking Bcl-2 (Akar et al., 2008). A Beclin1-independent pathway has been observed in MCF7 cells under conditions of a resveratrol-induced autophagic cell death (Scarlatti et al., 2008).

One key regulator of autophagy is a protein complex containing the mTOR kinases: mTORC1 and mTORC2. mTORC1 operates by inhibiting autophagy in response to cellular cues such as nutrition and growth factor availability and phosphorylation and inactivation of ULK/Atg1 and ATG13/Atg13. Both ULK/Atg1 and ATG13/Atg13 are essential to the induction of autophagy (Akkoc, Peker, Akcay, & Gozuacik, 2021; Jung, Ro, Cao, Otto, & Kim, 2010; Lapierre et al., 2015). Molecular components such as p53, p38, Bcl-2, PTEN, and heat shock proteins (HSP70) have also been documented to upregulate or downregulate autophagy (Sui et al., 2013, 2011, 2014). Also, several epigenetic mechanisms such as DNA methylation, chromatin modulation, histone modification, and microRNA are known to regulate autophagy in diverse cancers at several stages. Recently, it has also been shown that methylation of protocadherin 17 in gastric and colorectal cancer exerts antitumor activity through promotion of autophagy as well as induction of apoptosis (Hu et al., 2013; Sui et al., 2012).

Accumulating evidence suggests that autophagy plays a role in cancer recurrence as tumor relapse is accelerated by chemotherapy-induced autophagy (Aqbi et al., 2018). One study reported that autophagy in persistent breast cancer after neoadjuvant chemotherapy is associated with a high risk of tumor relapse (Chen, Jiang, et al., 2013). To augment this, autophagy has also been shown as a mechanism involved in the survival of disseminated dormant cancer cells, which could eventually lead to their proliferation and hence the recurrence of the disease (Vera-Ramirez, Vodnala, Nini, Hunter, & Green, 2018). Additionally, studies have shown that malignant tumor cells that enter a state of nonproliferative but reversible cycle arrest display upregulation of autophagy activity (Lu et al., 2008; Marshall et al., 2012; Vera-Ramirez et al., 2018).

6. Strategies to prevent recovery from therapy exposure

There are several broad strategies for utilizing epigenetic-targeted therapies to eliminate therapy-arrested cells at a significant interval after therapy exposure (Fig. 4). Two leading strategies can be proposed. The first can be thought of as a "lock-in" strategy which attempts to maintain a stable form of therapy-induced proliferative arrest, and the second a "lock-out" strategy which attempts to induce therapy-arrested cells to undergo cell death. Similar strategies have been proposed for eliminating cancer stem cells prior to therapy exposure (Cho et al., 2019). In this context the "lock-in" strategy would seek to target epigenetic factors that prevent the escape from quiescence/diapause and could force cells to maintain a nonproliferative state. The most common therapy-induced cell stress mechanism in actively cycling cells is TIS. "Lock-in" of TIS can be achieved by activating, or reinforcing, epigenetic factors that promote TIS, or alternatively inhibiting epigenetic factors that are required for escaping TIS, thus maintaining a stable TIS state. "Locking-in" TIS is not as desirable as eliminating the TIS population because senescent cells undergo the SASP, which can exert pathologic effects, most prominently reprogramming the tumor microenvironment for increased tumor growth, and is associated with a number of age-related disease pathologies (Krtolica et al., 2001; Saleh et al., 2020; Schosserer et al., 2017). As an example of a "lock-out" strategy, epigenetic factors can be targeted to convert the TIS into apoptosis. This is a more desirable outcome because it eliminates the surviving population, thereby preventing the possibility of recovery, and can also be immunogenic, which can further stimulate the antitumor immune response to residual tumor cells.

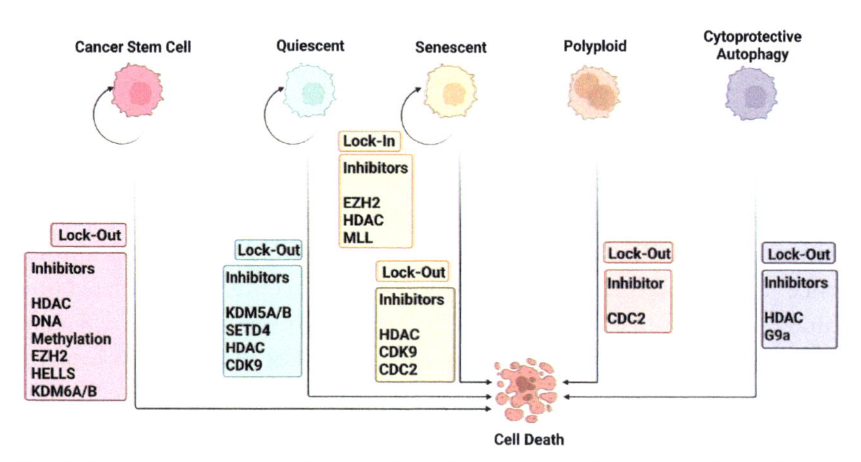

Fig. 4 Suggested strategies for targeting therapy-arrested cancer cell populations. Cancer cells can survive chemotherapy and radiation exposure through several means including quiescence, stem cell state, diapause, senescence, autophagy, and polyploidy. Cancer cell populations can enter into several of these states in response to therapy. Listed are examples of epigenetic regulators that could be targeted to prevent the escape of therapy-exposed cancer cells from quiescence, diapause, or senescence. Therapeutic maintenance of the therapy-arrested state is not desirable because of associated pathologies from senescent cells, and the potential for recovery. Shown below are epigenetic regulators that could be targeted to induce cell death. Cell death is the more desirable outcome because the cells cannot recover and can result in an improved antitumor immune response.

This strategy is being investigated aggressively through the use of novel compounds referred to as senolytics—several classes of compounds including BCL2 pathway inhibitors, Src inhibitors, cardiac glycosides and NF-kB inhibitors that are selectively toxic to senescent cells (Wang et al., 2022; Zhu et al., 2015). In addition to targeting senescent cells, arrested therapy-exposed cells can be eliminated by a "lock-out" strategy through the development of ways to inhibit cytoprotective autophagy or promote cytotoxic autophagy. Inhibiting cytoprotective autophagy could promote apoptosis by suppressing the normally protective functions of autophagy; conversely promoting cytotoxic autophagy could promote an immunogenic process referred to as autophagic cell death (Denton & Kumar, 2019). Autophagic cell death has the direct benefit of eliminating the therapy-exposed cancer cells preventing proliferative recovery, as well as stimulating an antitumor immune response to the residual cells. Polyploid cells are also a significant reservoir of cells entering into proliferative recovery state (Fei et al., 2015; Mittal et al., 2017; Zhang et al., 2014). Developing "lock-out" strategies to eliminate polyploid cells, thus preventing them from undergoing neosis and spinning off mitotically viable cells with complex genomes, could be a valuable strategy to reduce proliferative recovery (White-Gilbertson & Voelkel-Johnson, 2020).

6.1 Targeting cancer stem cells

In most cases a prepriming strategy would be used to target CSC epigenetics prior to therapy exposure to render this therapy-tolerant population more sensitive to the primary therapy (Wainwright & Scaffidi, 2017). However, there have been reports that CSC can be induced as a result of therapy exposure (Hu et al., 2012; Lagadec et al., 2012; Wang et al., 2014; Xu et al., 2015). In such cases there could be benefits in targeting stem cell populations posttherapy exposure as part of a sequential strategy. For instance, HDAC3 and HDAC7 are overexpressed in liver CSCs, and correlates with increased expression of CSC pluripotency factors such as SOX2, OCT4, and NANOG (Liu et al., 2013), and are important for maintaining CSCs (Witt et al., 2017). Consistent with the latter observation, inhibition of HDAC7 in breast CSCs leads to reductions in the CSC phenotype (Caslini, Hong, Ban, Chen, & Ince, 2019). HDAC11 inhibition in nonsmall cell lung cancer (NSCLC) leads to suppression of its CSC self-renewal by reducing the expression of SOX2. This suppression of HDAC epigenetic regulators prevents the growth of NSCLC CSC cells (Bora-Singhal et al., 2020). Consequently, the use of HDAC inhibitors could be an effective strategy to "lock-out" the CSC state by depleting these populations in patents.

In addition to HDACs, DNA methylation leads to increased formation of CSCs and their self-renewal, therapeutic resistance, and metastatic capabilities (Wu & Chu, 2021). In breast cancer, variations of methylation patterns at miRNA promoters enhance breast CSCs. miR-203, a known stemness-inhibiting miRNA, is repressed by DNA methylation thereby abrogating EMT and inducing self-renewal capacities in breast CSCs (Zhang et al., 2011). DNA methylation at the promoter region of downstream targets SOX2 and FOXG1 is observed in glioblastoma CSCs and is responsible for promoting self-renewal and cellular dormancy (Bulstrode et al., 2017; Toh, Lim, & Chow, 2017), including the Wnt/β-catenin pathway which targets *SFRP2*, *SFRP5*, *WIF1*, *DKK*, *WNT5a*, *APC* (Katoh, 2017; Reya & Clevers, 2005). Thus, DNA methyltransferase inhibitors could inhibit the CSC state depleting this population of cells and serve as a "lock-out" strategy.

The methylcytosine binding protein family and its associated NuRD complex could also hold promise as therapeutic targets to inhibit cancer stem cells. Several reports have shown that inhibiting the Nucleosome Remodeling and Deacetylase (NURD) subunit MBD3 promotes cancer stem cells. In one report the inhibition of MBD3, in combination with the ectopic expression of SOX2, OCT4, KLF4 and c-MYC, improves the conversion of liver cancer stem cells (Li et al., 2017). These same functions were observed in pancreatic stem cells where inhibition of MBD3 promotes Hippo signaling and its pro cancer stem cell activities (Wang et al., 2020). Consistent with these results an ATPase CHD4 (Zhang, Lv, Wei, Gong, & Chen, 2022) and MTA3 (Yao et al., 2019) subunits of the NURD complex also has documented roles in promoting SOX2 expression in cancer stem cells.

In addition, increased histone methylation of H3K27me3 in CSCs affects self-renewal pathways (e.g., Wnt and human gonadotropin-releasing hormone [GNRH] pathways), which leads to increased survival, therapy resistance, and drug resistance (Li et al., 2018). EZH2, a histone H3K27 methyltransferase has also been shown to maintain stemness and metastasis in CSCs (Yomtoubian et al., 2020). These reports suggest that using EZH2 inhibitors could suppress CSC, and thus serve as a "lock-out" strategy, which has proven successful in targeting glioma cancer stem cells (Jin et al., 2017). Similar to EZH2, the chromatin remodeler HELLS (Zhang et al., 2019) and the KDM6A/B methyltransferase (Liau et al., 2017) are essential for maintaining glioblastoma stem cells (GSCs) and could be targeted with inhibitors once available.

6.2 Targeting quiescence

Similar to CSC, the preexisting pool of quiescent cancer cells can be targeted using prepriming to alter their epigenetics rendering them more sensitive to therapy exposure. As with CSC there have been several reports that quiescence, in the form of diapause, can be induced with therapy exposure and could be targeted using a sequential strategy (Dhimolea et al., 2021; Duy et al., 2021; Rehman et al., 2021). Targeting this population with the intent of eliminating them from the arrested cell population would represent a "lock-out" strategy.

Histone demethylase KDM5A/B is important for transcriptional heterogeneity in quiescent cancer cell populations, as it promotes resistance to chemotherapies and targeted therapies (Hinohara et al., 2019; Sharma et al., 2010). Similarly, CPI-455 a specific inhibitor of the KDM5 demethylase, reduces survival of DTP cancer cells (Vinogradova et al., 2016).

These data suggest that inhibiting KDM5 could be a viable therapeutic strategy to eliminate cells in a therapy-induced diapause state. SETD4 has been shown to be essential for the maintenance of a quiescent state and resistance to chemotherapy through facilitating heterochromatin and H4K20me3 modification (Ye et al., 2019). In addition, HDAC and CDK9 inhibitors can sensitize $BRAF^{V600}$ mutant melanoma cells to BRAF inhibitors in quiescent cancer cell populations (Madorsky Rowdo et al., 2020). These inhibitors could be used to eliminate $BRAF^{V600}$ mutant therapy-induced quiescent cells in melanoma patients receiving BRAF inhibitors.

6.3 Elimination of therapy-induced polyploid cells

Polyploid, as well as senescent cells, can escape from senescence by overexpressing CDC2 and upregulating its downstream target survivin (Wang et al., 2013). These studies suggest that targeting the epigenetic regulation of CDC2 could be a viable strategy to prevent recovery from therapy exposure and serve as an example of a "lock-out" strategy. CDK1 could be targeted directly using a number of CDK1 inhibitors (Mou, Chen, & Deng, 2020) or alternatively by targeting regulators of its expression. CDK1 is regulated by NF-kB transcription factors (Voce et al., 2021), promoter methylation status (Li et al., 2021) through the DEAD-box RNA helicase 21 (DDX21) (Lu et al., 2022) each of which could be leveraged to regulate its activity.

6.4 Promoting the senescent state

Recently, various epigenetic mechanisms have been investigated for induction of senescence in cancers (Cheng et al., 2019; Ewald, Desotelle, Wilding, & Jarrard, 2010; Kumar, Paul, Rameshwar, & Pillai, 2019). Growing evidence suggests that targeting epigenetic regulators of senescence, and in some cases specifically TIS, with the intent of prolonging the senescent state, in essence a "lock-in" strategy, could advance therapeutic strategies for cancer and reduce cancer relapse.

The homeostasis of p16 gene expression and regulation is maintained by the polycomb group complexes (PcGs) polycomb repressor complex 1 and 2 (PRC1 and PRC2). In the nucleus, H3K27me3 is recognized by a chromodomain-containing chromobox (CBX) protein family member associated with PRC1. PRC1 is recruited to p16, where it catalyzes the ubiquitination of histone H2A lysine 119 (H2AK119ub), resulting in further chromatin compaction and gene silencing (Bracken et al., 2007). Jmjd3, a H3K27me3 demethylase, has been reported as a critical factor in senescence.

The knockdown of Jmjd3 sustained the repression of p16 by H3K27me3, promoting the escape from senescence (Agger et al., 2009; Barradas et al., 2009). Similarly, polycomb proteins BMI1, EZH2, and SUZ12 are instrumental in repressing p16 expression through H3K27me3 (Bracken et al., 2007), suggesting that EZH2 inhibitors could be used in a "lock-in" strategy.

p53, Rb, and p16 are regulated by the multivalent epigenetic factor CTCF (De La Rosa-Velazquez, Rincon-Arano, Benitez-Bribiesca, & Recillas-Targa, 2007; Witcher & Emerson, 2009), which binds to promoter sequences and regulates gene expression by promoting various histone modifications including H3K9me3, H3K27me3, and H4K20me3 (Soto-Reyes & Recillas-Targa, 2010). In breast cancer cells, the epigenetic silencing of the $p16^{INK4a}$ tumor suppressor is associated with loss of CTCF binding and a chromosome boundary (Witcher & Emerson, 2009). Some studies have suggested a possible role for the epigenetic modulator EZH2—a member of PRC2. One study in both a breast cancer cell line (MCF7) and a colorectal cancer cell line (LS174T) demonstrated that the upregulation of cyclin D1 in the early stages of TIS led to later upregulation of EZH2 in those cells that managed to escape from senescence (Le Duff et al., 2018). Downregulation of EZH2 promotes senescence through two distinct mechanisms. First, depletion of EZH2 in proliferating cells rapidly initiates a DNA-damage response prior to a reduction in the levels of H3K27me3 marks. Second, the eventual loss of H3K27me3 induces p16 (CDKN2A) gene expression independent of DNA damage and potentially activates SASP genes. PRC2 has methyl transferase-like activity and methylates H3K27. PRC2s, like EZH2, also plays an important role in senescence through EZH2/TET1 (Ito, Teo, Evans, Neretti, & Sedivy, 2018; Yu et al., 2019). SUZ12 is another histone methyltransferase that mediates senescence through $p16^{INK4A}$-coupled miRNA pathways (Overhoff et al., 2014). SUZ12 levels are reduced in response to the reduction of the polycomb protein CBX7. Reducing SUZ12 levels increases expression of $p15^{INK4B}$ and $p16^{INK4A}$. Long noncoding RNA ANRIL is required for the PRC2 recruitment and silencing of the p15 (CDKN2B) tumor suppressor gene (Kotake et al., 2011). These reports in aggregate would suggest inhibiting EZH2 with small molecules could be a viable "lock-in" strategy.

Various epigenetic regulators that contribute to $p21^{WAF1}$ regulation are writers such as DNMT, TIP60, PCAF, CTIF2, EZH2, DOT1L, and epigenetic erasers such as TET, KDM1A/LSD1, and HDAC (Ocker, Bitar, Monteiro, Gali-Muhtasib, & Schneider-Stock, 2019). Several histone deacetylases have been implicated in the recovery of senescent cells. p53 and

HDAC1 are antagonistic regulators of the p21 gene, which is a critical regulator of senescence. HDAC1 inhibition results in the upregulation of p21^{WAF1}, which favors cell cycle arrest (Lagger et al., 2003). Among all of the HDACs, sirtuins have been found to play an important role in senescence. Overexpression of SIRT1 may block senescence by direct deacetylation of p53 (Ota et al., 2009). Treating human squamous cell carcinoma (SCC) cells with the anticancer compounds resveratrol and doxorubicin triggered p53-independent premature senescence by mTOR-dependent phosphorylation of SIRT1. The pharmacologic and genetic inhibition of mTOR dephosphorylates S47A and sensitizes prematurely senescent SCC cells to apoptosis (Back et al., 2011). Sirtinol induced senescence-like growth arrest characterized by induction of SA-β-galactosidase activity and increased expression of plasminogen activator inhibitor-1 in human breast cancer MCF7 cells and lung cancer H1299 cells (Ota et al., 2006). As in the case of EZH2 inhibitors, HDAC inhibitors could be used in "lock-in" strategies.

Mixed-lineage leukemia 1 (MLL1) is a H3K4 histone methyl transferase. MLL1 inhibition may prevent the negative outcomes of SASP as well as prevent recovery from senescence. For example, MLL1 is a positive regulator of recovery-related proteins such as aurora kinases, survivin, PLK1, and FOXMI, as MLL1 inhibition lowers the expression of these genes (Capell et al., 2016). Another histone methyltransferase, SUV39H1, specifically methylates H3K9 and helps maintain genomic stability and proliferative potential (Sidler, Woycicki, et al., 2014). Decreased SUV39H1 expression increases genomic instability and cell cycle arrest in cells exposed to ionizing radiation (Sidler, Li, Wang, Kovalchuk, & Kovalchuk, 2014). SUV39H1 promotes senescence by inhibiting E2F1 target genes (Rao, Pal, & Taneja, 2017; Tamgue & Lei, 2017). p53 downregulates SUV39H1 expression through p21^{WAF1} suppression of E2F during the DNA-damage response. p53 causes an increase of p21^{WAF1}, which triggers indirect inhibition of SUV39H1 activity (Sidler, Li, et al., 2014). Decreased SUV39H1 decreases H3K9me3 and thereby reinstates the DNA-damage response and delays senescence. Inhibiting either of these methyltransferases with could serve as a "lock-in" strategy.

6.5 Promoting apoptosis of therapy-arrested cells

Evasion of cell death is one of the hallmarks of cancer (Hanahan & Weinberg, 2011) and poses a challenge to effective treatment (Adams & Cory, 2007; Pfeffer & Singh, 2018). As apoptosis has been observed in multiple tumor cells that have epigenetic aberrations, it can be postulated that destruction of epigenetic homeostasis can induce apoptosis of tumor cells (Yu et al., 2021). In this strategy the targeting of epigenetic regulators would result in a "lock-out" strategy where the therapy-exposed cell is ejected from stable therapy arrest into a state of cell death.

HDAC inhibitors are represented prominently in this category because of their ability to stimulate apoptosis in tumor cells as the hypoacetylated state of chromatin is linked to the downregulation of antiapoptotic genes (Emanuele, Lauricella, & Tesoriere, 2008; Peart et al., 2005). This was shown in studies where suberoylanilide hydroxamic acid (SAHA; vorinostat) induced apoptosis in several tumor cells as opposed to normal cells, which appear to be resistant (Garber, 2007; Marks & Dokmanovic, 2005). Similarly, sodium butyrate is another potential anticancer chemotherapeutic drug that has been shown to induce apoptosis in human gastric cancer cells (Shin, Lee, & Lee, 2012). In this study, sodium butyrate-induced apoptosis increased expression of caspase-3 and DAPK1/2 genes but decreased expression of Bcl-2. Sodium valproate can promote apoptosis and inhibit cell proliferation by upregulation of WWOX and p27 (CDKN1B) genes (Yan & Zhang, 2012). HDAC inhibitors, such as SAHA and MGCD0103, and the CDK9 inhibitor CDKI-73 induced apoptosis in melanoma cells that underwent senescence after PLX4032 (vemurafenib) treatment (Madorsky Rowdo et al., 2020). Panobinostat has been tested as a senolytic for its role in selectively targeting NSCLC cells that have been induced into senescence by paclitaxel or cisplatin through downregulating the antiapoptotic factor BclXL (Samaraweera, Adomako, Rodriguez-Gabin, & McDaid, 2017).

Several plant-based drugs with HDAC inhibitory activity have been repurposed as chemotherapeutic agents (extensively reviewed by Montalvo-Casimiro and colleagues (Montalvo-Casimiro et al., 2020)). Artemisinin, first identified as an antimalarial drug, has been shown to inhibit HDAC1, HDAC2, HDAC6 and is reported to induce apoptosis in breast cancer cells (Kumari, Keshari, Sengupta, Sabat, & Mishra, 2017; Mao et al., 2013). In addition, HC toxin, which is an analog of artemisinin, inhibited HDAC activity leading to the induction of cell cycle arrest and apoptosis in both breast cancer and neuroblastoma cells (Deubzer et al., 2008; Joung, Kim, & Sheen, 2004). In vitro studies show that hydralazine induced demethylation and reexpression of the tumor suppressor adenomatous polyposis (APC) in human cervical cancer cells (HeLa, CaSki, and ECV304) is associated with an increase in apoptosis of the cells as well as cell cycle arrest (Song & Zhang, 2009). Hydralazine can be potentially used for the clinical treatment of human cervical cancer but could be limited due to its significant toxicity (PMID: 31643895). Also, disrupting SAHF with a HDAC inhibitor can activate apoptosis (Di Micco et al., 2011).

Targeting the ATPase and NURD subunit CHD4 could also be a means to promote apoptosis. Inhibiting CHD4 relaxes chromatin of AML blasts which renders them more sensitive to DNA-damaging therapeutic agents promoting apoptosis (Sperlazza et al., 2015). With this strategy inhibiting CHD4 would occur through a prepriming or coincident treatment strategy to sensitize the chromatin for the damaging agent.

A recurring cell cycle regulator that has been implicated in the recovery from TIS is CDC2. Expression of CDC2 is required for the progression of the cell cycle from G1 phase into S phase, and when it forms a complex with another cyclin CDC13, where it facilitates the initiation of mitosis. CDC2 is overexpressed in both lung and breast cancer cells that have recovered from chemotherapy exposure (Elmore, Di, Dumur, Holt, & Gewirtz, 2005; Roberson, Kussick, Vallieres, Chen, & Wu, 2005). These recovering cells have a higher expression of proliferative cell cycle regulators, including CDC2. In the case of breast cancer cells treated with doxorubicin the recovering population was stably resistant indicating that their recovery could have been from a rare genetic mutation resulting in acquired resistance (Elmore et al., 2005). Conversely, in the lung cancer study the recovering population reacquired sensitivity to camptothecin suggesting a reversable recovery state, similar to what is observed with DTP phenotypes (Roberson et al., 2005). Pharmacologic inhibition of CDC2 resulted in cell death of recovered populations, and suppressed the ability of naive cells to recover, suggesting that inhibition of CDC2 through epigenetic strategies could be a viable approach to inducing cell death of therapy-treated cancer cells to prevent recovery (see polyploid section above) (Roberson et al., 2005).

6.6 Inhibiting cytoprotective autophagy or promoting cytotoxic autophagy

Epigenetic factors associated with autophagy, especially those that when inhibited induce autophagic-dependent cell death, or those that are essential for protective autophagy, can be targeted to prevent recovery (Bhol et al., 2020). Autophagy is known to be essential to CSCs (Nazio, Bordi, Cianfanelli, Locatelli, & Cecconi, 2019). Inhibition of autophagy in combination with therapy exposure can selectively target CSCs (Angeletti et al., 2016). This strategy, as with that of inducing apoptosis of therapy-arrested cells would represent a "lock-out" strategy. Ideally, inhibition of the epigenic factor should induce an immunogenic autophagy dependent cell death to enhance the antitumor immune response (Denton & Kumar, 2019).

Epigenetic regulators that promote cancer cell death in part through mechanisms involving autophagy are prominently HDAC inhibitors (De et al., 2018; Di Fazio et al., 2016; Han et al., 2017; Mrakovcic, Bohner, Hanisch, & Frohlich, 2018; Shao et al., 2004). HDAC6 is well known to promote autophagy in cancer cells, and when autophagy is cytoprotective, inhibition of HDAC6 could promote the death of therapy-exposed cancer cells (Passaro et al., 2021). Similarly, HDAC10 promotes a protective form of autophagy in therapy-exposed neuroblastoma cancers and, when inhibited, promotes cell death (Oehme et al., 2013). Conversely, inhibition of the histone methylase G9a results in increased cytotoxic autophagy through the upregulation of Beclin 1 (Park et al., 2016). Valproic acid (VPA) cotreatment with doxorubicin improved the drugs internalization leading to enhanced autophagy and ROS in HepG2 cells that led to increased cell death (Saha et al., 2017).

7. Future directions

In this review, we propose a new strategy for the use of epigenetic-targeting therapies in the prevention of recurrent disease under conditions of partial or complete response. We hypothesize that the use epigenetic-targeting therapies in sequence, a significant interval after the primary cytotoxic therapy, could reduce the toxicity from cotreatments and could target novel epigenetic states that have been established posttherapy exposure.

In the past, epigenetic-targeting therapies have been used to improve cytotoxic therapies in the form of epigenetic priming (an epigenetic-targeting therapy followed by a chemotherapeutic approach) or a cotreatment strategy (epigenetic-targeting therapy with chemotherapy). While these strategies have been effective for hematologic malignancies, neither of these approaches have been widely successful for solid tumors. Combination strategies have resulted in increased toxicity and overall low effectiveness in solid tumors (Morel et al., 2020). Epigenetic priming has been used in a number of preclinical studies in solid tumors with low effectiveness when combined in sequence with cytotoxic chemotherapies (Oza et al., 2020). One key difference between a combination approach and a priming approach is that the priming approach has been shown to be well tolerated with lower toxicity than combination therapies (Lee et al., 2018; Schneider et al., 2017).

This suggests that sequential use of an epidrug followed by chemotherapy could result in reduced and tolerable toxicity, and as a corollary, suggests that the reverse sequence (chemotherapy then epidrug) could similarly result in reduced toxicity.

A reduction in toxicity from a sequential chemotherapy followed by epitherapy treatment strategy is supported by the use of "drug holidays." Drug holidays are commonly used in the clinic to alleviate the toxic effects of therapy to patients. As such, our proposed sequential use of chemotherapy followed by epitherapy, with the major objective being to elevate toxicity from use of multiple therapies and to kill recovering cells, could be successful. While drug holidays do provide reductions in toxicity, they are in most cases neutral, or less effective in controlling disease than an uninterrupted course of therapy (Garattini et al., 2021; Labianca et al., 2011; Mittal et al., 2018; Onishi et al., 2012). While the objective of exposing tumor cells to the toxic effects of the combined epidrug and chemotherapy combinations is desirable, we propose that these combinations are not required, and are actually counterproductive due to toxicity. We propose that a sequential strategy using a cytotoxic therapy followed by an epitherapy could target epigenetic states established a significant time after the initial therapy which are essential for recovery. This possibility would result in benefits for a sequential strategy, and an actual requirement for a delay between the cytotoxic chemotherapy and the epitherapy, allowing enough time for novel epigenetic states to be established, which when targeted by epidrugs could eliminate therapy-arrested cancer cells. Conceptually, targeting epigenetic states posttherapy is similar to targeting epigenetic states pretherapy using a priming approach, the main difference being that the former intends to inhibit epigenetic states that promote recovery, whereas the latter intends to inhibit epigenetic states that promote survival.

Novel epigenetic states that are possibly exploitable are those that promote recovery posttherapy exposure. Several theoretical possibilities include converting cytoprotective autophagy to a cytotoxic autophagy to promote cancer cell death or converting senescent cells into a state of cell death. However, promoting a stable quiescent or senescent cell state by preventing the recovery of therapy-exposed cancer cells could also be pursued. Strategies with the highest priority would be those that result in the death of the therapy-exposed tumor cells; thereby preventing them from exhibiting the potential to recover and contribute to disease recurrence. Promoting the death of cancer cells in TIS has been aggressively pursued by the cancer research community, most recently through the use of senolytic agents (Kirkland & Tchkonia, 2020). Epitherapy-induced death could have the benefit of promoting an immunogenic cell death, which can promote antitumor immune responses. Sustaining senescence is less desirable largely because it is accompanied by SASP. In many cases, SASP is protumorigenic and it can have significant proinflammatory activities (Tchkonia, Zhu, van Deursen, Campisi, & Kirkland, 2013). These deleterious effects are not seen in all cases, and SASP is very context dependent, with its makeup varying from tumor to tumor. However, the possibility of prolonging a SASP, which is associated with protumor activities, would make it less desirable of an outcome and could theoretically lead to recurrence (Krtolica et al., 2001; Saleh et al., 2020).

To investigate the hypothesis that a sequential chemotherapy then epigenetic therapy could prevent recovery, efforts need to be spent in determining which epigenetic factors regulate recovery. Initial efforts need to focus on developing tractable models of recovery. The simplest systems use in vitro cell culture, of which there have been many used to study recovery from therapy exposure. High-dose therapy treatments could monitor the recovery of quiescent cells and CSCs, which would be enriched in the surviving pool of cells. Lower doses could monitor cells recovering from sublethal exposure and allow the study of the recovery from polyploidy, senescence, and cytoprotective autophagy. Ideally in vivo studies should be performed, which could involve transplanting in vitro therapy-exposed cells into mice to allow recovery in vivo. This approach would ensure that the cancer cells are exposed to a uniform amount of therapy that provides a uniform population of exposed cells to study. More technically difficult screens could be designed in vivo using tumor bearing mice. However, numerous difficulties will need to be overcome, which include controlling for variable exposure of tumor cells to chemotherapies dosed into mice and controlling the effects of variabilities in tumor microenvironments.

Priorities could be spent on functional studies using either loss-of-function (LOF), or gain-of-function (GOF) approaches posttherapy exposure (Joung et al., 2017). Genetic CRISPR/Cas9 or shRNA KD LOF screening could be done posttherapy exposure to identify epigenetic factors important for recovery from therapy exposure. GOF screens are possibly equally informative since epigenetic factors could be overexpressed using CAS9 activator (CAS9a) methodology, which would allow overexpression of endogenous genes or ligand-inducible transgene expression systems. An advantage of these systems is that all epigenetic factors could be identified regardless of whether if it is druggable or not. Therapeutically relevant studies could screen small molecule libraries with a focus on epigenetic targets in similar in vitro screens to identify therapeutically usable small molecules. These studies could be adapted for in vivo studies but would be vulnerable to the difficulties of doing in vivo treatment and recovery experiments as described above.

Another promising area of research is in understanding and cataloging changes in epigenetic marks, which occur at various time points after therapy exposure. There are numerous studies of changes in epigenetics including histone modification, miRNAs, chromatin remodeling, DNA methylation pre- and posttherapy exposure. However, we do not know the kinetics of these changes and whether epigenetics is fundamentally different between the early stages posttherapy exposure, when the cells are responding initially to therapy exposure, and the late stages, when cells have entered into stable therapy-induced growth arrest. These studies could mirror the objectives of the ENCODE project with the objective of cataloging the epigenetic state of regulatory elements, and associated gene expression patterns (The ENCODE Project Consortium et al., 2020). Along these lines, histone codes have been proposed for the entry and exit of cells from the quiescent state (Bonitto, Sarathy, Atai, Mitra, & Coller, 2021). These efforts could lead to the rational identification of druggable histone-modifying enzymes, which when inhibited could prevent exit from the quiescent state. Another study documented differences in the DNA methylation profile of cancer cells in peroxide-induced premature senescence vs replicative senescence (Zhang et al., 2008). This study shows that there are differences in the epigenetic states for the different types of senescence. Similar efforts would need to be performed for different forms of therapy-induced cell stress (senescence, autophagy, polyploidy), over a time course from exposure to full response, and then recovery. While discovering these changes does not imply functionality, it would provide advancements in our understanding of how epigenetics changes after therapy exposure and how epigenic changes can influence gene expression, and could provide an epigenic and regulatory road map to interpret any LOF and GOF approaches as described above.

Acknowledgments

The authors would like to acknowledge our funding sources that contributed to this work. These include a grant from the DoD BCRP W81XWH1910489 to J.L., a Closing the Gap pilot grant from the Virginia Commonwealth University (VCU) Massey Cancer Center to J.L. and H.B. In addition, we received internal funds and support from the VCU School of Medicine, the VCU Wright Center for Clinical and Translational Research. Biorender was used to create figures. We would like to acknowledge Heidi Sankala for expert editing assistance.

References

Adams, P. D. (2007). Remodeling of chromatin structure in senescent cells and its potential impact on tumor suppression and aging. *Gene*, *397*(1–2), 84–93. Epub 2007/06/05. https://doi.org/10.1016/j.gene.2007.04.020. 17544228. PMC2755200.

Adams, J. M., & Cory, S. (2007). The Bcl-2 apoptotic switch in cancer development and therapy. *Oncogene*, *26*(9), 1324–1337. Epub 2007/02/27. https://doi.org/10.1038/sj.onc.1210220. 17322918. PMC2930981.

Agger, K., Cloos, P. A., Rudkjaer, L., Williams, K., Andersen, G., Christensen, J., et al. (2009). The H3K27me3 demethylase JMJD3 contributes to the activation of the INK4A-ARF locus in response to oncogene- and stress-induced senescence. *Genes & Development*, *23*(10), 1171–1176. Epub 2009/05/20. https://doi.org/10.1101/gad.510809. 19451217. PMC2685535.

Akar, U., Chaves-Reyez, A., Barria, M., Tari, A., Sanguino, A., Kondo, Y., et al. (2008). Silencing of Bcl-2 expression by small interfering RNA induces autophagic cell death in MCF-7 breast cancer cells. *Autophagy*, *4*(5), 669–679. Epub 2008/04/22. https://doi.org/10.4161/auto.6083. 18424910.

Akkoc, Y., Peker, N., Akcay, A., & Gozuacik, D. (2021). Autophagy and cancer dormancy. *Frontiers in Oncology*, *11*, 627023. Epub 2021/04/06. https://doi.org/10.3389/fonc.2021.627023. 33816262. PMC8017298.

Alcorta, D. A., Xiong, Y., Phelps, D., Hannon, G., Beach, D., & Barrett, J. C. (1996). Involvement of the cyclin-dependent kinase inhibitor p16 (INK4a) in replicative senescence of normal human fibroblasts. *Proceedings of the National Academy of Sciences of the United States of America*, *93*(24), 13742–13747. Epub 1996/11/26. https://doi.org/10.1073/pnas.93.24.13742. 8943005. PMC19411.

Angeletti, F., Fossati, G., Pattarozzi, A., Wurth, R., Solari, A., Daga, A., et al. (2016). Inhibition of the autophagy pathway synergistically potentiates the cytotoxic activity of givinostat (ITF2357) on human glioblastoma cancer stem cells. *Frontiers in Molecular Neuroscience*, *9*, 107. Epub 2016/11/12. https://doi.org/10.3389/fnmol.2016.00107. 27833530. PMC5081386.

Aqbi, H. F., Tyutyunyk-Massey, L., Keim, R. C., Butler, S. E., Thekkudan, T., Joshi, S., et al. (2018). Autophagy-deficient breast cancer shows early tumor recurrence and escape from dormancy. *Oncotarget*, *9*(31), 22113–22122. Epub 2018/05/19. 10.18632/oncotarget.25197. 29774126. PMC5955162.

Ayob, A. Z., & Ramasamy, T. S. (2018). Cancer stem cells as key drivers of tumour progression. *Journal of Biomedical Science*, *25*(1), 20. Epub 2018/03/07. https://doi.org/10.1186/s12929-018-0426-4. 29506506. PMC5838954.

Back, J. H., Rezvani, H. R., Zhu, Y., Guyonnet-Duperat, V., Athar, M., Ratner, D., et al. (2011). Cancer cell survival following DNA damage-mediated premature senescence is regulated by mammalian target of rapamycin (mTOR)-dependent inhibition of sirtuin 1. *The Journal of Biological Chemistry*, *286*(21), 19100–19108. Epub 2011/04/08. https://doi.org/10.1074/jbc.M111.240598. 21471201. PMC3099723.

Barradas, M., Anderton, E., Acosta, J. C., Li, S., Banito, A., Rodriguez-Niedenfuhr, M., et al. (2009). Histone demethylase JMJD3 contributes to epigenetic control of INK4a/ARF by oncogenic RAS. *Genes & Development*, *23*(10), 1177–1182. Epub 2009/05/20. https://doi.org/10.1101/gad.511109. 19451218. PMC2685533.

Barranco, S. C., & Flournoy, D. R. (1977). Cell killing, kinetics, and recovery responses induced by 1,2:5,6-dianhydrogalactitol in dividing and nondividing cells in vitro. *Journal of the National Cancer Institute*, *58*(3), 657–663. Epub 1977/03/01. https://doi.org/10.1093/jnci/58.3.657. 839561.

Bates, S. E. (2020). Epigenetic therapies for cancer. *The New England Journal of Medicine*, *383*(7), 650–663. Epub 2020/08/14. https://doi.org/10.1056/NEJMra1805035. 32786190.

Beck, B., & Blanpain, C. (2013). Unravelling cancer stem cell potential. *Nature Reviews. Cancer, 13*(10), 727–738. Epub 2013/09/26. https://doi.org/10.1038/nrc3597. 24060864.

Begicevic, R. R., & Falasca, M. (2017). ABC transporters in cancer stem cells: Beyond chemoresistance. *International Journal of Molecular Sciences, 18*(11). https://doi.org/10.3390/ijms18112362. Epub 2017/11/09. 29117122. PMC5713331.

Bellelli, R., Vitagliano, D., Federico, G., Marotta, P., Tamburrino, A., Salerno, P., et al. (2018). Oncogene-induced senescence and its evasion in a mouse model of thyroid neoplasia. *Molecular and Cellular Endocrinology, 460*, 24–35. Epub 2017/06/28. https://doi.org/10.1016/j.mce.2017.06.023. 28652169. PMC5741508.

Berger, S. L., Kouzarides, T., Shiekhattar, R., & Shilatifard, A. (2009). An operational definition of epigenetics. *Genes & Development, 23*(7), 781–783. Epub 2009/04/03. https://doi.org/10.1101/gad.1787609. 19339683. PMC3959995.

Bhol, C. S., Panigrahi, D. P., Praharaj, P. P., Mahapatra, K. K., Patra, S., Mishra, S. R., et al. (2020). Epigenetic modifications of autophagy in cancer and cancer therapeutics. *Seminars in Cancer Biology, 66*, 22–33. Epub 2019/06/04. https://doi.org/10.1016/j.semcancer.2019.05.020. 31158463.

Bonitto, K., Sarathy, K., Atai, K., Mitra, M., & Coller, H. A. (2021). Is there a histone code for cellular quiescence? *Frontiers in Cell and Development Biology, 9*, 739780. Epub 2021/11/16. https://doi.org/10.3389/fcell.2021.739780. 34778253. PMC8586460.

Bora-Singhal, N., Mohankumar, D., Saha, B., Colin, C. M., Lee, J. Y., Martin, M. W., et al. (2020). Novel HDAC11 inhibitors suppress lung adenocarcinoma stem cell self-renewal and overcome drug resistance by suppressing Sox2. *Scientific Reports, 10*(1), 4722. Epub 2020/03/15. https://doi.org/10.1038/s41598-020-61295-6. 32170113. PMC7069992.

Bracken, A. P., Kleine-Kohlbrecher, D., Dietrich, N., Pasini, D., Gargiulo, G., Beekman, C., et al. (2007). The polycomb group proteins bind throughout the INK4A-ARF locus and are disassociated in senescent cells. *Genes & Development, 21*(5), 525–530. Epub 2007/03/09. https://doi.org/10.1101/gad.415507. 17344414. PMC1820894.

Braun, J., & Hahn, G. M. (1975). Enhanced cell killing by bleomycin and 43 degrees hyperthermia and the inhibition of recovery from potentially lethal damage. *Cancer Research, 35*(11 Pt. 1), 2921–2927. Epub 1975/11/01. 52401.

Bringold, F., & Serrano, M. (2000). Tumor suppressors and oncogenes in cellular senescence. *Experimental Gerontology, 35*(3), 317–329. Epub 2000/06/01. https://doi.org/10.1016/s0531-5565(00)00083-8. 10832053.

Bulstrode, H., Johnstone, E., Marques-Torrejon, M. A., Ferguson, K. M., Bressan, R. B., Blin, C., et al. (2017). Elevated FOXG1 and SOX2 in glioblastoma enforces neural stem cell identity through transcriptional control of cell cycle and epigenetic regulators. *Genes & Development, 31*(8), 757–773. Epub 2017/05/04. https://doi.org/10.1101/gad.293027.116. 28465359. PMC5435889.

Buocikova, V., Rios-Mondragon, I., Pilalis, E., Chatziioannou, A., Miklikova, S., Mego, M., et al. (2020). Epigenetics in breast cancer therapy—New strategies and future nanomedicine perspectives. *Cancers, 12*(12). https://doi.org/10.3390/cancers12123622. Epub 2020/12/09. 33287297. PMC7761669.

Buocikova, V., Longhin, E. M., Pilalis, E., Mastrokalou, C., Miklikova, S., Cihova, M., et al. (2022). Decitabine potentiates efficacy of doxorubicin in a preclinical trastuzumab-resistant HER2-positive breast cancer models. *Biomedicine & Pharmacotherapy, 147*, 112662. Epub 2022/01/30. https://doi.org/10.1016/j.biopha.2022.112662. 35091237.

Calcagno, A. M., Salcido, C. D., Gillet, J. P., Wu, C. P., Fostel, J. M., Mumau, M. D., et al. (2010). Prolonged drug selection of breast cancer cells and enrichment of cancer stem cell characteristics. *Journal of the National Cancer Institute, 102*(21), 1637–1652. Epub 2010/10/12. https://doi.org/10.1093/jnci/djq361. 20935265. PMC2970576.

Capell, B. C., Drake, A. M., Zhu, J., Shah, P. P., Dou, Z., Dorsey, J., et al. (2016). MLL1 is essential for the senescence-associated secretory phenotype. *Genes & Development*, *30*(3), 321–336. Epub 2016/02/03. https://doi.org/10.1101/gad.271882.115. 26833731. PMC4743061.

Caslini, C., Hong, S., Ban, Y. J., Chen, X. S., & Ince, T. A. (2019). HDAC7 regulates histone 3 lysine 27 acetylation and transcriptional activity at super-enhancer-associated genes in breast cancer stem cells. *Oncogene*, *38*(39), 6599–6614. Epub 2019/08/04. https://doi.org/10.1038/s41388-019-0897-0. 31375747.

Centore, R. C., Sandoval, G. J., Soares, L. M. M., Kadoch, C., & Chan, H. M. (2020). Mammalian SWI/SNF chromatin remodeling complexes: Emerging mechanisms and therapeutic strategies. *Trends in Genetics*, *36*(12), 936–950. Epub 2020/09/03. https://doi.org/10.1016/j.tig.2020.07.011. 32873422.

Chang, J. C. (2016). Cancer stem cells: Role in tumor growth, recurrence, metastasis, and treatment resistance. *Medicine (Baltimore)*, *95*(1 Suppl. 1), S20–S25. Epub 2016/09/10. https://doi.org/10.1097/MD.0000000000004766. 27611935. PMC5599212.

Chang, B. D., Broude, E. V., Dokmanovic, M., Zhu, H., Ruth, A., Xuan, Y., et al. (1999). A senescence-like phenotype distinguishes tumor cells that undergo terminal proliferation arrest after exposure to anticancer agents. *Cancer Research*, *59*(15), 3761–3767. Epub 1999/08/14. 10446993.

Chang, B. D., Xuan, Y., Broude, E. V., Zhu, H., Schott, B., Fang, J., et al. (1999). Role of p53 and p21waf1/cip1 in senescence-like terminal proliferation arrest induced in human tumor cells by chemotherapeutic drugs. *Oncogene*, *18*(34), 4808–4818. Epub 1999/09/22. https://doi.org/10.1038/sj.onc.1203078. 10490814.

Chen, S., Jiang, Y. Z., Huang, L., Zhou, R. J., Yu, K. D., Liu, Y., et al. (2013). The residual tumor autophagy marker LC3B serves as a prognostic marker in local advanced breast cancer after neoadjuvant chemotherapy. *Clinical Cancer Research*, *19*(24), 6853–6862. Epub 2013/10/22. https://doi.org/10.1158/1078-0432.CCR-13-1617. 24141623.

Chen, W., Dong, J., Haiech, J., Kilhoffer, M. C., & Zeniou, M. (2016). Cancer stem cell quiescence and plasticity as major challenges in cancer therapy. *Stem Cells International*, *2016*, 1740936. Epub 2016/07/16. https://doi.org/10.1155/2016/1740936. 27418931. PMC4932171.

Chen, K., Huang, Y. H., & Chen, J. L. (2013). Understanding and targeting cancer stem cells: Therapeutic implications and challenges. *Acta Pharmacologica Sinica*, *34*(6), 732–740. Epub 2013/05/21. https://doi.org/10.1038/aps.2013.27. 23685952. PMC3674516.

Cheng, Y., He, C., Wang, M., Ma, X., Mo, F., Yang, S., et al. (2019). Targeting epigenetic regulators for cancer therapy: Mechanisms and advances in clinical trials. *Signal Transduction and Targeted Therapy*, *4*, 62. Epub 2019/12/25. https://doi.org/10.1038/s41392-019-0095-0. 31871779. PMC6915746.

Cheung, T. H., & Rando, T. A. (2013). Molecular regulation of stem cell quiescence. *Nature Reviews. Molecular Cell Biology*, *14*(6), 329–340. Epub 2013/05/24. https://doi.org/10.1038/nrm3591. 23698583. PMC3808888.

Chiappinelli, K. B., Strissel, P. L., Desrichard, A., Li, H., Henke, C., Akman, B., et al. (2015). Inhibiting DNA methylation causes an interferon response in cancer via dsRNA including endogenous retroviruses. *Cell*, *162*(5), 974–986. Epub 2015/09/01. https://doi.org/10.1016/j.cell.2015.07.011. 26317466. PMC4556003.

Cho, I. J., Lui, P. P., Obajdin, J., Riccio, F., Stroukov, W., Willis, T. L., et al. (2019). Mechanisms, hallmarks, and implications of stem cell quiescence. *Stem Cell Reports*, *12*(6), 1190–1200. Epub 2019/06/13. https://doi.org/10.1016/j.stemcr.2019.05.012. 31189093. PMC6565921.

Cimino, G., Lo-Coco, F., Fenu, S., Travaglini, L., Finolezzi, E., Mancini, M., et al. (2006). Sequential valproic acid/all-trans retinoic acid treatment reprograms differentiation in refractory and high-risk acute myeloid leukemia. *Cancer Research*, *66*(17), 8903–8911. Epub 2006/09/05. https://doi.org/10.1158/0008-5472.CAN-05-2726. 16951208.

Clapier, C. R., & Cairns, B. R. (2009). The biology of chromatin remodeling complexes. *Annual Review of Biochemistry*, *78*, 273–304. Epub 2009/04/10. https://doi.org/10.1146/annurev.biochem.77.062706.153223. 19355820.

Coppe, J. P., Desprez, P. Y., Krtolica, A., & Campisi, J. (2010). The senescence-associated secretory phenotype: The dark side of tumor suppression. *Annual Review of Pathology*, *5*, 99–118. Epub 2010/01/19. https://doi.org/10.1146/annurev-pathol-121808-102144. 20078217. PMC4166495.

Das, C. K., Mandal, M., & Kogel, D. (2018). Pro-survival autophagy and cancer cell resistance to therapy. *Cancer Metastasis Reviews*, *37*(4), 749–766. Epub 2018/03/15. https://doi.org/10.1007/s10555-018-9727-z. 29536228.

Davalos, A. R., Coppe, J. P., Campisi, J., & Desprez, P. Y. (2010). Senescent cells as a source of inflammatory factors for tumor progression. *Cancer Metastasis Reviews*, *29*(2), 273–283. Epub 2010/04/15. https://doi.org/10.1007/s10555-010-9220-9. 20390322. PMC2865636.

De, U., Son, J. Y., Sachan, R., Park, Y. J., Kang, D., Yoon, K., et al. (2018). A new synthetic histone deacetylase inhibitor, MHY2256, induces apoptosis and autophagy cell death in endometrial cancer cells via p53 acetylation. *International Journal of Molecular Sciences*, *19*(9). https://doi.org/10.3390/ijms19092743. Epub 2018/09/16. 30217020. PMC6164480.

De La Rosa-Velazquez, I. A., Rincon-Arano, H., Benitez-Bribiesca, L., & Recillas-Targa, F. (2007). Epigenetic regulation of the human retinoblastoma tumor suppressor gene promoter by CTCF. *Cancer Research*, *67*(6), 2577–2585. Epub 2007/03/17. https://doi.org/10.1158/0008-5472.CAN-06-2024. 17363576.

Denton, D., & Kumar, S. (2019). Autophagy-dependent cell death. *Cell Death and Differentiation*, *26*(4), 605–616. Epub 2018/12/21. https://doi.org/10.1038/s41418-018-0252-y. 30568239. PMC6460387.

Desai, A., Yan, Y., & Gerson, S. L. (2019). Concise reviews: Cancer stem cell targeted therapies: Toward clinical success. *Stem Cells Translational Medicine*, *8*(1), 75–81. Epub 2018/10/18. https://doi.org/10.1002/sctm.18-0123. 30328686. PMC6312440.

Deubzer, H. E., Ehemann, V., Westermann, F., Heinrich, R., Mechtersheimer, G., Kulozik, A. E., et al. (2008). Histone deacetylase inhibitor Helminthosporium carbonum (HC)-toxin suppresses the malignant phenotype of neuroblastoma cells. *International Journal of Cancer*, *122*(8), 1891–1900. Epub 2007/12/13. https://doi.org/10.1002/ijc.23295. 18074352.

Dhanasekaran, R., Deutzmann, A., Mahauad-Fernandez, W. D., Hansen, A. S., Gouw, A. M., & Felsher, D. W. (2022). The MYC oncogene—The grand orchestrator of cancer growth and immune evasion. *Nature Reviews. Clinical Oncology*, *19*(1), 23–36. Epub 2021/09/12. https://doi.org/10.1038/s41571-021-00549-2. 34508258. PMC9083341.

Dhanyamraju, P. K., Schell, T. D., Amin, S., & Robertson, G. P. (2022). Drug-tolerant persister cells in cancer therapy resistance. *Cancer Research*, *82*(14), 2503–2514. https://doi.org/10.1158/0008-5472.CAN-21-3844. Epub 2022/05/19. 35584245. PMC9296591.

Dhimolea, E., de Matos, S. R., Kansara, D., Al'Khafaji, A., Bouyssou, J., Weng, X., et al. (2021). An embryonic diapause-like adaptation with suppressed Myc activity enables tumor treatment persistence. *Cancer Cell*, *39*(2). https://doi.org/10.1016/j.ccell.2020.12.002. 240–256.e11. Epub 2021/01/09. 33417832. PMC8670073.

Di Fazio, P., Waldegger, P., Jabari, S., Lingelbach, S., Montalbano, R., Ocker, M., et al. (2016). Autophagy-related cell death by pan-histone deacetylase inhibition in liver cancer. *Oncotarget*, 7(20), 28998–29010. Epub 2016/04/09. 10.18632/oncotarget.8585. 27058414. PMC5045373.

Di Micco, R., Sulli, G., Dobreva, M., Liontos, M., Botrugno, O. A., Gargiulo, G., et al. (2011). Interplay between oncogene-induced DNA damage response and heterochromatin in senescence and cancer. *Nature Cell Biology*, *13*(3), 292–302. Epub 2011/02/22. https://doi.org/10.1038/ncb2170. 21336312. PMC3918344.

Diehn, M., Cho, R. W., Lobo, N. A., Kalisky, T., Dorie, M. J., Kulp, A. N., et al. (2009). Association of reactive oxygen species levels and radioresistance in cancer stem cells. *Nature*, *458*(7239), 780–783. Epub 2009/02/06. https://doi.org/10.1038/nature07733. 19194462. PMC2778612.

Dimauro, T., & David, G. (2010). Ras-induced senescence and its physiological relevance in cancer. *Current Cancer Drug Targets*, *10*(8), 869–876. Epub 2010/08/20. https://doi.org/10.2174/156800910793357998. 20718709. PMC4023163.

Dimri, G. P. (2005). What has senescence got to do with cancer? *Cancer Cell*, *7*(6), 505–512. Epub 2005/06/14. https://doi.org/10.1016/j.ccr.2005.05.025. 15950900. PMC1769521.

Doherty, M. R., Smigiel, J. M., Junk, D. J., & Jackson, M. W. (2016). Cancer stem cell plasticity drives therapeutic resistance. *Cancers*, *8*(1). https://doi.org/10.3390/cancers8010008. Epub 2016/01/08. 26742077. PMC4728455.

Duy, C., Teater, M., Garrett-Bakelman, F. E., Lee, T. C., Meydan, C., Glass, J. L., et al. (2019). Rational targeting of cooperating layers of the epigenome yields enhanced therapeutic efficacy against AML. *Cancer Discovery*, *9*(7), 872–889. https://doi.org/10.1158/2159-8290.CD-19-0106.

Duy, C., Li, M., Teater, M., Meydan, C., Garrett-Bakelman, F. E., Lee, T. C., et al. (2021). Chemotherapy induces senescence-like resilient cells capable of initiating AML recurrence. *Cancer Discovery*, *11*(6), 1542–1561. Epub 2021/01/28. https://doi.org/10.1158/2159-8290.CD-20-1375. 33500244. PMC8178167.

Elmore, L. W., Di, X., Dumur, C., Holt, S. E., & Gewirtz, D. A. (2005). Evasion of a single-step, chemotherapy-induced senescence in breast cancer cells: Implications for treatment response. *Clinical Cancer Research*, *11*(7), 2637–2643. Epub 2005/04/09. https://doi.org/10.1158/1078-0432.CCR-04-1462. 15814644.

Emanuele, S., Lauricella, M., & Tesoriere, G. (2008). Histone deacetylase inhibitors: Apoptotic effects and clinical implications (review). *International Journal of Oncology*, *33*(4), 637–646. Epub 2008/09/25. 18813776.

Esteller, M. (2008). Epigenetics in cancer. *The New England Journal of Medicine*, *358*(11), 1148–1159. Epub 2008/03/14. https://doi.org/10.1056/NEJMra072067. 18337604.

Ewald, J. A., Desotelle, J. A., Wilding, G., & Jarrard, D. F. (2010). Therapy-induced senescence in cancer. *Journal of the National Cancer Institute*, *102*(20), 1536–1546. Epub 2010/09/23. https://doi.org/10.1093/jnci/djq364. 20858887. PMC2957429.

Fang, L., Igarashi, M., Leung, J., Sugrue, M. M., Lee, S. W., & Aaronson, S. A. (1999). p21Waf1/Cip1/Sdi1 induces permanent growth arrest with markers of replicative senescence in human tumor cells lacking functional p53. *Oncogene*, *18*(18), 2789–2797. Epub 1999/06/11. https://doi.org/10.1038/sj.onc.1202615. 10362249.

Fei, F., Zhang, D., Yang, Z., Wang, S., Wang, X., Wu, Z., et al. (2015). The number of polyploid giant cancer cells and epithelial-mesenchymal transition-related proteins are associated with invasion and metastasis in human breast cancer. *Journal of Experimental & Clinical Cancer Research*, *34*, 158. Epub 2015/12/26. https://doi.org/10.1186/s13046-015-0277-8. 26702618. PMC4690326.

Fisusi, F. A., & Akala, E. O. (2019). Drug combinations in breast cancer therapy. *Pharmaceutical Nanotechnology*, 7(1), 3–23. Epub 2019/01/23. https://doi.org/10.2174/2211738507666190122111224. 30666921. PMC6691849.

Galluzzi, L., Vitale, I., Aaronson, S. A., Abrams, J. M., Adam, D., Agostinis, P., et al. (2018). Molecular mechanisms of cell death: Recommendations of the nomenclature committee on cell death 2018. *Cell Death and Differentiation*, *25*(3), 486–541. Epub 2018/01/25. https://doi.org/10.1038/s41418-017-0012-4. 29362479. PMC5864239.

Galluzzi, L., Buque, A., Kepp, O., Zitvogel, L., & Kroemer, G. (2017). Immunogenic cell death in cancer and infectious disease. *Nature Reviews. Immunology*, *17*(2), 97–111. Epub 2016/11/01. https://doi.org/10.1038/nri.2016.107. 27748397.

Ganesan, A., Arimondo, P. B., Rots, M. G., Jeronimo, C., & Berdasco, M. (2019). The timeline of epigenetic drug discovery: From reality to dreams. *Clinical Epigenetics*, *11*(1), 174. Epub 2019/12/04. https://doi.org/10.1186/s13148-019-0776-0. 31791394. PMC6888921.

Garattini, S. K., Basile, D., Bonotto, M., Ongaro, E., Porcu, L., Corvaja, C., et al. (2021). Drug holidays and overall survival of patients with metastatic colorectal cancer. *Cancers*, *13*(14). https://doi.org/10.3390/cancers13143504. Epub 2021/07/25. 34298718. PMC8304309.

Garber, K. (2007). HDAC inhibitors overcome first hurdle. *Nature Biotechnology*, *25*(1), 17–19. Epub 2007/01/11. https://doi.org/10.1038/nbt0107-17. 17211382.

Gottesman, M. M., Lavi, O., Hall, M. D., & Gillet, J. P. (2016). Toward a better understanding of the complexity of cancer drug resistance. *Annual Review of Pharmacology and Toxicology*, *56*, 85–102. https://doi.org/10.1146/annurev-pharmtox-010715-103111.

Gozuacik, D., & Kimchi, A. (2004). Autophagy as a cell death and tumor suppressor mechanism. *Oncogene*, *23*(16), 2891–2906. Epub 2004/04/13. https://doi.org/10.1038/sj.onc.1207521. 15077152.

Graham, C. D., Kaza, N., Klocke, B. J., Gillespie, G. Y., Shevde, L. A., Carroll, S. L., et al. (2016). Tamoxifen induces cytotoxic autophagy in glioblastoma. *Journal of Neuropathology and Experimental Neurology*, *75*(10), 946–954. Epub 2016/08/16. https://doi.org/10.1093/jnen/nlw071. 27516117. PMC5029439.

Guillon, J., Petit, C., Moreau, M., Toutain, B., Henry, C., Roche, H., et al. (2019). Regulation of senescence escape by TSP1 and CD47 following chemotherapy treatment. *Cell Death & Disease*, *10*(3), 199. Epub 2019/03/01. https://doi.org/10.1038/s41419-019-1406-7. 30814491. PMC6393582.

Han, H., Li, J., Feng, X., Zhou, H., Guo, S., & Zhou, W. (2017). Autophagy-related genes are induced by histone deacetylase inhibitor suberoylanilide hydroxamic acid via the activation of cathepsin B in human breast cancer cells. *Oncotarget*, *8*(32), 53352–53365. Epub 2017/09/09. 10.18632/oncotarget.18410. 28881816. PMC5581115.

Hanahan, D. (2022). Hallmarks of cancer: New dimensions. *Cancer Discovery*, *12*(1), 31–46. Epub 2022/01/14. https://doi.org/10.1158/2159-8290.CD-21-1059. 35022204.

Hanahan, D., & Weinberg, R. A. (2011). Hallmarks of cancer: The next generation. *Cell*, *144*(5), 646–674. Epub 2011/03/08. https://doi.org/10.1016/j.cell.2011.02.013. 21376230.

He, G., Siddik, Z. H., Huang, Z., Wang, R., Koomen, J., Kobayashi, R., et al. (2005). Induction of p21 by p53 following DNA damage inhibits both Cdk4 and Cdk2 activities. *Oncogene*, *24*(18), 2929–2943. Epub 2005/03/01. https://doi.org/10.1038/sj.onc.1208474. 15735718.

Hinohara, K., Wu, H. J., Sebastien, V., McDonald TO, Igarashi, K. J., Yamamoto, K. N., et al. (2019). KDM5 histone demethylase activity links cellular transcriptomic heterogeneity to therapeutic resistance. *Cancer Cell*, *35*(2), 330–332. Epub 2019/02/13. https://doi.org/10.1016/j.ccell.2019.01.012. 30753830. PMC6428693.

Hu, X., Ghisolfi, L., Keates, A. C., Zhang, J., Xiang, S., Lee, D. K., et al. (2012). Induction of cancer cell stemness by chemotherapy. *Cell Cycle*, *11*(14), 2691–2698. Epub 2012/06/27. https://doi.org/10.4161/cc.21021. 22732500.

Hu, X., Sui, X., Li, L., Huang, X., Rong, R., Su, X., et al. (2013). Protocadherin 17 acts as a tumour suppressor inducing tumour cell apoptosis and autophagy, and is frequently methylated in gastric and colorectal cancers. *The Journal of Pathology*, *229*(1), 62–73. Epub 2012/08/29. https://doi.org/10.1002/path.4093. 22926751.

Ito, T., Teo, Y. V., Evans, S. A., Neretti, N., & Sedivy, J. M. (2018). Regulation of cellular senescence by polycomb chromatin modifiers through distinct DNA damage- and histone methylation-dependent pathways. *Cell Reports*, *22*(13), 3480–3492. Epub 2018/03/29. https://doi.org/10.1016/j.celrep.2018.03.002. 29590617. PMC5915310.

Jin, X., Kim, L. J. Y., Wu, Q., Wallace, L. C., Prager, B. C., Sanvoranart, T., et al. (2017). Targeting glioma stem cells through combined BMI1 and EZH2 inhibition. *Nature Medicine, 23*(11), 1352–1361. Epub 2017/10/17. https://doi.org/10.1038/nm.4415. 29035367. PMC5679732.

Joung, J., Konermann, S., Gootenberg, J. S., Abudayyeh, O. O., Platt, R. J., Brigham, M. D., et al. (2017). Genome-scale CRISPR-Cas9 knockout and transcriptional activation screening. *Nature Protocols, 12*(4), 828–863. Epub 2017/03/24. https://doi.org/10.1038/nprot.2017.016. 28333914. PMC5526071.

Joung, K. E., Kim, D. K., & Sheen, Y. Y. (2004). Antiproliferative effect of trichostatin A and HC-toxin in T47D human breast cancer cells. *Archives of Pharmacal Research, 27*(6), 640–645. Epub 2004/07/31. https://doi.org/10.1007/BF02980164. 15283467.

Jung, C. H., Ro, S. H., Cao, J., Otto, N. M., & Kim, D. H. (2010). mTOR regulation of autophagy. *FEBS Letters, 584*(7), 1287–1295. Epub 2010/01/20. https://doi.org/10.1016/j.febslet.2010.01.017. 20083114. PMC2846630.

Katoh, M. (2017). Canonical and non-canonical WNT signaling in cancer stem cells and their niches: Cellular heterogeneity, omics reprogramming, targeted therapy and tumor plasticity (review). *International Journal of Oncology, 51*(5), 1357–1369. Epub 2017/10/20. https://doi.org/10.3892/ijo.2017.4129. 29048660. PMC5642388.

Kirkland, J. L., & Tchkonia, T. (2020). Senolytic drugs: From discovery to translation. *Journal of Internal Medicine, 288*(5), 518–536. Epub 2020/07/21. https://doi.org/10.1111/joim.13141. 32686219. PMC7405395.

Kondo, Y., Kanzawa, T., Sawaya, R., & Kondo, S. (2005). The role of autophagy in cancer development and response to therapy. *Nature Reviews. Cancer, 5*(9), 726–734. Epub 2005/09/09. https://doi.org/10.1038/nrc1692. 16148885.

Korhonen, M., Ylanne, J., Laitinen, L., & Virtanen, I. (1990). The alpha 1-alpha 6 subunits of integrins are characteristically expressed in distinct segments of developing and adult human nephron. *The Journal of Cell Biology, 111*(3), 1245–1254. Epub 1990/09/01. https://doi.org/10.1083/jcb.111.3.1245. 2144000. PMC2116295.

Kotake, Y., Nakagawa, T., Kitagawa, K., Suzuki, S., Liu, N., Kitagawa, M., et al. (2011). Long non-coding RNA ANRIL is required for the PRC2 recruitment to and silencing of p15(INK4B) tumor suppressor gene. *Oncogene, 30*(16), 1956–1962. Epub 2010/12/15. https://doi.org/10.1038/onc.2010.568. 21151178. PMC3230933.

Krtolica, A., Parrinello, S., Lockett, S., Desprez, P. Y., & Campisi, J. (2001). Senescent fibroblasts promote epithelial cell growth and tumorigenesis: A link between cancer and aging. *Proceedings of the National Academy of Sciences of the United States of America, 98*(21), 12072–12077. Epub 2001/10/11. https://doi.org/10.1073/pnas.211053698. 11593017. PMC59769.

Kuendgen, A., & Gattermann, N. (2007). Valproic acid for the treatment of myeloid malignancies. *Cancer, 110*(5), 943–954. Epub 2007/07/25. https://doi.org/10.1002/cncr.22891. 17647267.

Kuendgen, A., Strupp, C., Aivado, M., Bernhardt, A., Hildebrandt, B., Haas, R., et al. (2004). Treatment of myelodysplastic syndromes with valproic acid alone or in combination with all-trans retinoic acid. *Blood, 104*(5), 1266–1269. Epub 2004/05/25. https://doi.org/10.1182/blood-2003-12-4333. 15155466.

Kumar, R., Paul, A. M., Rameshwar, P., & Pillai, M. R. (2019). Epigenetic dysregulation at the crossroad of women's cancer. *Cancers, 11*(8). https://doi.org/10.3390/cancers11081193. Epub 2019/08/21. 31426393. PMC6721458.

Kumari, K., Keshari, S., Sengupta, D., Sabat, S. C., & Mishra, S. K. (2017). Transcriptome analysis of genes associated with breast cancer cell motility in response to artemisinin treatment. *BMC Cancer, 17*(1), 858. Epub 2017/12/17. https://doi.org/10.1186/s12885-017-3863-7. 29246124. PMC5732364.

Labianca, R., Sobrero, A., Isa, L., Cortesi, E., Barni, S., Nicolella, D., et al. (2011). Intermittent versus continuous chemotherapy in advanced colorectal cancer: A randomised 'GISCAD' trial. *Annals of Oncology*, *22*(5), 1236–1242. Epub 2010/11/17. https://doi.org/10.1093/annonc/mdq580. 21078826.

Lagadec, C., Vlashi, E., Della Donna, L., Dekmezian, C., & Pajonk, F. (2012). Radiation-induced reprogramming of breast cancer cells. *Stem Cells*, *30*(5), 833–844. Epub 2012/04/11. https://doi.org/10.1002/stem.1058. 22489015. PMC3413333.

Lagger, G., Doetzlhofer, A., Schuettengruber, B., Haidweger, E., Simboeck, E., Tischler, J., et al. (2003). The tumor suppressor p53 and histone deacetylase 1 are antagonistic regulators of the cyclin-dependent kinase inhibitor p21/WAF1/CIP1 gene. *Molecular and Cellular Biology*, *23*(8), 2669–2679. Epub 2003/04/01. https://doi.org/10.1128/MCB.23.8.2669-2679.2003. 12665570. PMC152549.

LaPak, K. M., & Burd, C. E. (2014). The molecular balancing act of p16(INK4a) in cancer and aging. *Molecular Cancer Research*, *12*(2), 167–183. Epub 2013/10/19. https://doi.org/10.1158/1541-7786.MCR-13-0350. 24136988. PMC3944093.

Lapierre, L. R., Kumsta, C., Sandri, M., Ballabio, A., & Hansen, M. (2015). Transcriptional and epigenetic regulation of autophagy in aging. *Autophagy*, *11*(6), 867–880. Epub 2015/04/04. https://doi.org/10.1080/15548627.2015.1034410. 25836756. PMC4502732.

Le Duff, M., Gouju, J., Jonchere, B., Guillon, J., Toutain, B., Boissard, A., et al. (2018). Regulation of senescence escape by the cdk4-EZH2-AP2M1 pathway in response to chemotherapy. *Cell Death & Disease*, *9*(2), 199. Epub 2018/02/09. https://doi.org/10.1038/s41419-017-0209-y. 29415991. PMC5833455.

Lee, V., Wang, J., Zahurak, M., Gootjes, E., Verheul, H. M., Parkinson, R., et al. (2018). A phase I trial of a guadecitabine (SGI-110) and irinotecan in metastatic colorectal cancer patients previously exposed to irinotecan. *Clinical Cancer Research*, *24*(24), 6160–6167. Epub 2018/08/12. https://doi.org/10.1158/1078-0432.CCR-18-0421. 30097434.

Li, R., He, Q., Han, S., Zhang, M., Liu, J., Su, M., et al. (2017). MBD3 inhibits formation of liver cancer stem cells. *Oncotarget*, *8*(4), 6067–6078. https://doi.org/10.18632/oncotarget.13496.

Li, G., Wang, D., Ma, W., An, K., Liu, Z., Wang, X., et al. (2018). Transcriptomic and epigenetic analysis of breast cancer stem cells. *Epigenomics*, *10*(6), 765–783. Epub 2018/02/27. https://doi.org/10.2217/epi-2018-0008. 29480027.

Li, G., Zhang, R., Zhang, X., Shao, S., Hu, F., & Feng, Y. (2020). Human colorectal cancer derived-MSCs promote tumor cells escape from senescence via P53/P21 pathway. *Clinical & Translational Oncology*, *22*(4), 503–511. Epub 2019/06/21. https://doi.org/10.1007/s12094-019-02152-5. 31218648.

Li, S., Li, H., Cao, Y., Geng, H., Ren, F., Li, K., et al. (2021). Integrated bioinformatics analysis reveals CDK1 and PLK1 as potential therapeutic targets of lung adenocarcinoma. *Medicine (Baltimore)*, *100*(32), e26474. https://doi.org/10.1097/MD.0000000000026474.

Liau, B. B., Sievers, C., Donohue, L. K., Gillespie, S. M., Flavahan, W. A., Miller, T. E., et al. (2017). Adaptive chromatin remodeling drives glioblastoma stem cell plasticity and drug tolerance. *Cell Stem Cell*, *20*(2). https://doi.org/10.1016/j.stem.2016.11.003. 233–246.e7. Epub 2016/12/19. 27989769. PMC5291795.

Liggett, W. H., Jr., & Sidransky, D. (1998). Role of the p16 tumor suppressor gene in cancer. *Journal of Clinical Oncology*, *16*(3), 1197–1206. Epub 1998/03/21. https://doi.org/10.1200/JCO.1998.16.3.1197. 9508208.

Liu, C., Liu, L., Shan, J., Shen, J., Xu, Y., Zhang, Q., et al. (2013). Histone deacetylase 3 participates in self-renewal of liver cancer stem cells through histone modification. *Cancer Letters*, *339*(1), 60–69. Epub 2013/07/25. https://doi.org/10.1016/j.canlet.2013.07.022. 23879963.

Lopes-Paciencia, S., Saint-Germain, E., Rowell, M. C., Ruiz, A. F., Kalegari, P., & Ferbeyre, G. (2019). The senescence-associated secretory phenotype and its regulation. *Cytokine*, *117*, 15–22. Epub 2019/02/19. https://doi.org/10.1016/j.cyto.2019.01.013. 30776684.

Lopez, A. T., Bates, S., & Geskin, L. (2018). Current status of HDAC inhibitors in cutaneous T-cell lymphoma. *American Journal of Clinical Dermatology*, *19*(6), 805–819. https://doi.org/10.1007/s40257-018-0380-7.

Loven, J., Hoke, H. A., Lin, C. Y., Lau, A., Orlando, D. A., Vakoc, C. R., et al. (2013). Selective inhibition of tumor oncogenes by disruption of super-enhancers. *Cell*, *153*(2), 320–334. Epub 2013/04/16. https://doi.org/10.1016/j.cell.2013.03.036. 23582323. PMC3760967.

Lu, Z., Luo, R. Z., Lu, Y., Zhang, X., Yu, Q., Khare, S., et al. (2008). The tumor suppressor gene ARHI regulates autophagy and tumor dormancy in human ovarian cancer cells. *The Journal of Clinical Investigation*, *118*(12), 3917–3929. Epub 2008/11/27. https://doi.org/10.1172/JCI35512. 19033662. PMC2582930.

Lu, P., Yu, Z., Wang, K., Zhai, Y., Chen, B., Liu, M., et al. (2022). DDX21 interacts with WDR5 to promote colorectal cancer cell proliferation by activating CDK1 expression. *Journal of Cancer*, *13*(5), 1530–1539. https://doi.org/10.7150/jca.69216.

Madorsky Rowdo, F. P., Baron, A., Gallagher, S. J., Hersey, P., Emran, A. A., Von Euw, E. M., et al. (2020). Epigenetic inhibitors eliminate senescent melanoma BRAFV600E cells that survive long-term BRAF inhibition. *International Journal of Oncology*, *56*(6), 1429–1441. Epub 2020/04/03. https://doi.org/10.3892/ijo.2020.5031. 32236593. PMC7170042.

Mahvi, D. A., Liu, R., Grinstaff, M. W., Colson, Y. L., & Raut, C. P. (2018). Local cancer recurrence: The realities, challenges, and opportunities for new therapies. *CA: A Cancer Journal for Clinicians*, *68*(6), 488–505. Epub 2018/10/18. https://doi.org/10.3322/caac.21498. 30328620. PMC6239861.

Malaquin, N., Tu, V., & Rodier, F. (1896). Assessing functional roles of the senescence-associated secretory phenotype (SASP). *Methods in Molecular Biology*, *2019*, 45–55. Epub 2018/11/27. https://doi.org/10.1007/978-1-4939-8931-7_6. 30474839.

Mao, H., Gu, H., Qu, X., Sun, J., Song, B., Gao, W., et al. (2013). Involvement of the mitochondrial pathway and Bim/Bcl-2 balance in dihydroartemisinin-induced apoptosis in human breast cancer in vitro. *International Journal of Molecular Medicine*, *31*(1), 213–218. Epub 2012/11/10. https://doi.org/10.3892/ijmm.2012.1176. 23138847.

Marescal, O., & Cheeseman, I. M. (2020). Cellular mechanisms and regulation of quiescence. *Developmental Cell*, *55*(3), 259–271. Epub 2020/11/11. https://doi.org/10.1016/j.devcel.2020.09.029. 33171109. PMC7665062.

Marks, P. A., & Dokmanovic, M. (2005). Histone deacetylase inhibitors: Discovery and development as anticancer agents. *Expert Opinion on Investigational Drugs*, *14*(12), 1497–1511. Epub 2005/11/26. https://doi.org/10.1517/13543784.14.12.1497. 16307490.

Marshall, J. C., Collins, J. W., Nakayama, J., Horak, C. E., Liewehr, D. J., Steinberg, S. M., et al. (2012). Effect of inhibition of the lysophosphatidic acid receptor 1 on metastasis and metastatic dormancy in breast cancer. *Journal of the National Cancer Institute*, *104*(17), 1306–1319. Epub 2012/08/23. https://doi.org/10.1093/jnci/djs319. 22911670. PMC3611817.

Medema, J. P. (2018). Escape from senescence boosts tumour growth. *Nature*, *553*(7686), 37–38. Epub 2018/01/01. https://doi.org/10.1038/d41586-017-08652-0. 32086500.

Mehdipour, P., Murphy, T., & De Carvalho, D. D. (2020). The role of DNA-demethylating agents in cancer therapy. *Pharmacology & Therapeutics*, *205*, 107416. Epub 2019/10/19. https://doi.org/10.1016/j.pharmthera.2019.107416. 31626871.

Milanovic, M., Fan, D. N. Y., Belenki, D., Dabritz, J. H. M., Zhao, Z., Yu, Y., et al. (2018). Senescence-associated reprogramming promotes cancer stemness. *Nature*, *553*(7686), 96–100. Epub 2017/12/21. https://doi.org/10.1038/nature25167. 29258294.

Mittal, K., Donthamsetty, S., Kaur, R., Yang, C., Gupta, M. V., Reid, M. D., et al. (2017). Multinucleated polyploidy drives resistance to docetaxel chemotherapy in prostate cancer. *British Journal of Cancer, 116*(9), 1186–1194. Epub 2017/03/24. https://doi.org/10.1038/bjc.2017.78. 28334734. PMC5418452.

Mittal, K., Derosa, L., Albiges, L., Wood, L., Elson, P., Gilligan, T., et al. (2018). Drug holiday in metastatic renal-cell carcinoma patients treated with vascular endothelial growth factor receptor inhibitors. *Clinical Genitourinary Cancer, 16*(3), e663–e667. Epub 2018/02/13. https://doi.org/10.1016/j.clgc.2017.12.014. 29428404.

Montalvo-Casimiro, M., Gonzalez-Barrios, R., Meraz-Rodriguez, M. A., Juarez-Gonzalez, V. T., Arriaga-Canon, C., & Herrera, L. A. (2020). Epidrug repurposing: Discovering new faces of old acquaintances in cancer therapy. *Frontiers in Oncology, 10*, 605386. Epub 2020/12/15. https://doi.org/10.3389/fonc.2020.605386. 33312959. PMC7708379.

Morel, D., Jeffery, D., Aspeslagh, S., Almouzni, G., & Postel-Vinay, S. (2020). Combining epigenetic drugs with other therapies for solid tumours—Past lessons and future promise. *Nature Reviews. Clinical Oncology, 17*(2), 91–107. Epub 2019/10/02. https://doi.org/10.1038/s41571-019-0267-4. 31570827.

Mosieniak, G., Sliwinska, M. A., Alster, O., Strzeszewska, A., Sunderland, P., Piechota, M., et al. (2015). Polyploidy formation in doxorubicin-treated cancer cells can favor escape from senescence. *Neoplasia, 17*(12), 882–893. Epub 2015/12/24. https://doi.org/10.1016/j.neo.2015.11.008. 26696370. PMC4688565.

Mou, J., Chen, D., & Deng, Y. (2020). Inhibitors of cyclin-dependent kinase 1/2 for anticancer treatment. *Journal of Medicinal Chemistry, 16*(3), 307–325. https://doi.org/10.2174/1573406415666190626113900.

Mrakovcic, M., Bohner, L., Hanisch, M., & Frohlich, L. F. (2018). Epigenetic targeting of autophagy via HDAC inhibition in tumor cells: Role of p53. *International Journal of Molecular Sciences, 19*(12). https://doi.org/10.3390/ijms19123952. Epub 2018/12/14. 30544838. PMC6321134.

Muller, P. A., & Vousden, K. H. (2013). p53 mutations in cancer. *Nature Cell Biology, 15*(1), 2–8. Epub 2012/12/25. https://doi.org/10.1038/ncb2641. 23263379.

National Center for Health Statistics (US). (2017). *Health, United States, 2016: With chartbook on long-term trends in health*. Hyattsville, MD.

Nazio, F., Bordi, M., Cianfanelli, V., Locatelli, F., & Cecconi, F. (2019). Autophagy and cancer stem cells: Molecular mechanisms and therapeutic applications. *Cell Death and Differentiation, 26*(4), 690–702. Epub 2019/02/08. https://doi.org/10.1038/s41418-019-0292-y. 30728463. PMC6460398.

Nervi, C., De Marinis, E., & Codacci-Pisanelli, G. (2015). Epigenetic treatment of solid tumours: A review of clinical trials. *Clinical Epigenetics*, 7, 127. Epub 2015/12/23. https://doi.org/10.1186/s13148-015-0157-2. 26692909. PMC4676165.

Ocker, M., Bitar, S. A., Monteiro, A. C., Gali-Muhtasib, H., & Schneider-Stock, R. (2019). Epigenetic regulation of p21(cip1/waf1) in human cancer. *Cancers, 11*(9). https://doi.org/10.3390/cancers11091343. Epub 2019/09/14. 31514410. PMC6769618.

Oehme, I., Linke, J. P., Bock, B. C., Milde, T., Lodrini, M., Hartenstein, B., et al. (2013). Histone deacetylase 10 promotes autophagy-mediated cell survival. *Proceedings of the National Academy of Sciences of the United States of America, 110*(28), E2592–E2601. Epub 2013/06/27. https://doi.org/10.1073/pnas.1300113110. 23801752. PMC3710791.

Ogier-Denis, E., & Codogno, P. (2003). Autophagy: A barrier or an adaptive response to cancer. *Biochimica et Biophysica Acta, 1603*(2), 113–128. Epub 2003/03/06. https://doi.org/10.1016/s0304-419x(03)00004-0. 12618311.

Onishi, T., Sasaki, T., & Hoshina, A. (2012). Intermittent chemotherapy is a treatment choice for advanced urothelial cancer. *Oncology, 83*(1), 50–56. Epub 2012/06/23. https://doi.org/10.1159/000338770. 22722700.

Ota, H., Tokunaga, E., Chang, K., Hikasa, M., Iijima, K., Eto, M., et al. (2006). Sirt1 inhibitor, Sirtinol, induces senescence-like growth arrest with attenuated Ras-MAPK signaling in human cancer cells. *Oncogene*, *25*(2), 176–185. Epub 2005/09/20. https://doi.org/10.1038/sj.onc.1209049. 16170353.

Ota, H., Eto, M., Ako, J., Ogawa, S., Iijima, K., Akishita, M., et al. (2009). Sirolimus and everolimus induce endothelial cellular senescence via sirtuin 1 down-regulation: Therapeutic implication of cilostazol after drug-eluting stent implantation. *Journal of the American College of Cardiology*, *53*(24), 2298–2305. Epub 2009/06/13. https://doi.org/10.1016/j.jacc.2009.01.072. 19520256.

Overhoff, M. G., Garbe, J. C., Koh, J., Stampfer, M. R., Beach, D. H., & Bishop, C. L. (2014). Cellular senescence mediated by p16INK4A-coupled miRNA pathways. *Nucleic Acids Research*, *42*(3), 1606–1618. Epub 2013/11/13. https://doi.org/10.1093/nar/gkt1096. 24217920. PMC3919591.

Oza, A. M., Matulonis, U. A., Alvarez Secord, A., Nemunaitis, J., Roman, L. D., Blagden, S. P., et al. (2020). A randomized phase II trial of epigenetic priming with guadecitabine and carboplatin in platinum-resistant, recurrent ovarian cancer. *Clinical Cancer Research*, *26*(5), 1009–1016. Epub 2019/12/14. https://doi.org/10.1158/1078-0432.CCR-19-1638. 31831561. PMC7056559.

Pacifico, F., Badolati, N., Mellone, S., Stornaiuolo, M., Leonardi, A., & Crescenzi, E. (2021). Glutamine promotes escape from therapy-induced senescence in tumor cells. *Aging (Albany NY)*, *13*(17), 20962–20991. Epub 2021/09/08. 10.18632/aging.203495. 34492636. PMC8457561.

Pacifico, F., Mellone, S., D'Incalci, M., Stornaiuolo, M., Leonardi, A., & Crescenzi, E. (2022). Trabectedin suppresses escape from therapy-induced senescence in tumor cells by interfering with glutamine metabolism. *Biochemical Pharmacology*, *202*, 115159. Epub 2022/07/06. https://doi.org/10.1016/j.bcp.2022.115159. 35780827.

Park, S. E., Yi, H. J., Suh, N., Park, Y. Y., Koh, J. Y., Jeong, S. Y., et al. (2016). Inhibition of EHMT2/G9a epigenetically increases the transcription of Beclin-1 via an increase in ROS and activation of NF-kappaB. *Oncotarget*, 7(26), 39796–39808. Epub 2016/10/26. 10.18632/oncotarget.9290. 27174920. PMC5129971.

Passaro, E., Papulino, C., Chianese, U., Toraldo, A., Congi, R., Del Gaudio, N., et al. (2021). HDAC6 inhibition extinguishes autophagy in cancer: Recent insights. *Cancers*, *13*(24). https://doi.org/10.3390/cancers13246280. Epub 2021/12/25. 34944907. PMC8699196.

Patel, S. A., & Minn, A. J. (2018). Combination cancer therapy with immune checkpoint blockade: Mechanisms and strategies. *Immunity*, *48*(3), 417–433. Epub 2018/03/22. https://doi.org/10.1016/j.immuni.2018.03.007. 29562193. PMC6948191.

Peart, M. J., Smyth, G. K., van Laar, R. K., Bowtell, D. D., Richon, V. M., Marks, P. A., et al. (2005). Identification and functional significance of genes regulated by structurally different histone deacetylase inhibitors. *Proceedings of the National Academy of Sciences of the United States of America*, *102*(10), 3697–3702. Epub 2005/03/02. https://doi.org/10.1073/pnas.0500369102. 15738394. PMC552783.

Peixoto, P., Cartron, P. F., Serandour, A. A., & Hervouet, E. (2020). From 1957 to nowadays: A brief history of epigenetics. *International Journal of Molecular Sciences*, *21*(20). https://doi.org/10.3390/ijms21207571. Epub 2020/10/18. 33066397. PMC7588895.

Pfeffer, C. M., & Singh, A. T. K. (2018). Apoptosis: A target for anticancer therapy. *International Journal of Molecular Sciences*, *19*(2). https://doi.org/10.3390/ijms19020448. Epub 2018/02/03. 29393886. PMC5855670.

Phan, T. G., & Croucher, P. I. (2020). The dormant cancer cell life cycle. *Nature Reviews. Cancer*, *20*(7), 398–411. Epub 2020/06/04. https://doi.org/10.1038/s41568-020-0263-0. 32488200.

Puig, P. E., Guilly, M. N., Bouchot, A., Droin, N., Cathelin, D., Bouyer, F., et al. (2008). Tumor cells can escape DNA-damaging cisplatin through DNA endoreduplication and reversible polyploidy. *Cell Biology International*, *32*(9), 1031–1043. Epub 2008/06/14. https://doi.org/10.1016/j.cellbi.2008.04.021. 18550395.

Punt, C. J., Simkens, L. H., & Koopman, M. (2015). Systemic treatment: Maintenance compared with holiday. *American Society of Clinical Oncology Educational Book*, 85–90. Epub 2015/05/21. 10.14694/EdBook_AM.2015.35.85. 25993146.

Ranganathan, P., Yu, X., Santhanam, R., Hofstetter, J., Walker, A., Walsh, K., et al. (2015). Decitabine priming enhances the antileukemic effects of exportin 1 (XPO1) selective inhibitor selinexor in acute myeloid leukemia. *Blood*, *125*(17), 2689–2692. Epub 2015/02/27. https://doi.org/10.1182/blood-2014-10-607648. 25716206. PMC4408293.

Rao, V. K., Pal, A., & Taneja, R. (2017). A drive in SUVs: From development to disease. *Epigenetics*, *12*(3), 177–186. Epub 2017/01/21. https://doi.org/10.1080/15592294.2017.1281502. 28106510. PMC5406210.

Rehman, S. K., Haynes, J., Collignon, E., Brown, K. R., Wang, Y., Nixon, A. M. L., et al. (2021). Colorectal cancer cells enter a diapause-like DTP state to survive chemotherapy. *Cell*, *184*(1). https://doi.org/10.1016/j.cell.2020.11.018. 226–242.e21. Epub 2021/01/09. 33417860. PMC8437243.

Reya, T., & Clevers, H. (2005). Wnt signalling in stem cells and cancer. *Nature*, *434*(7035), 843–850. Epub 2005/04/15. https://doi.org/10.1038/nature03319. 15829953.

Riggio, A. I., Varley, K. E., & Welm, A. L. (2021). The lingering mysteries of metastatic recurrence in breast cancer. *British Journal of Cancer*, *124*(1), 13–26. Epub 2020/11/27. https://doi.org/10.1038/s41416-020-01161-4. 33239679. PMC7782773.

Roberson, R. S., Kussick, S. J., Vallieres, E., Chen, S. Y., & Wu, D. Y. (2005). Escape from therapy-induced accelerated cellular senescence in p53-null lung cancer cells and in human lung cancers. *Cancer Research*, *65*(7), 2795–2803. Epub 2005/04/05. https://doi.org/10.1158/0008-5472.CAN-04-1270. 15805280.

Rohnalter, V., Roth, K., Finkernagel, F., Adhikary, T., Obert, J., Dorzweiler, K., et al. (2015). A multi-stage process including transient polyploidization and EMT precedes the emergence of chemoresistent ovarian carcinoma cells with a dedifferentiated and pro-inflammatory secretory phenotype. *Oncotarget*, *6*(37), 40005–40025. Epub 2015/10/28. 10.18632/oncotarget.5552. 26503466. PMC4741876.

Roninson, I. B., Broude, E. V., & Chang, B. D. (2001). If not apoptosis, then what? Treatment-induced senescence and mitotic catastrophe in tumor cells. *Drug Resistance Updates*, *4*(5), 303–313. Epub 2002/05/07. https://doi.org/10.1054/drup.2001.0213. 11991684.

Sadangi, S., Milosavljevic, K., Castro-Perez, E., Lares, M., Singh, M., Altameemi, S., et al. (2022). Role of the skin microenvironment in melanomagenesis: Epidermal keratinocytes and dermal fibroblasts promote BRAF oncogene-induced senescence escape in melanocytes. *Cancers*, *14*(5). https://doi.org/10.3390/cancers14051233. Epub 2022/03/11. 35267541. PMC8909265.

Saha, S. K., Yin, Y., Kim, K., Yang, G. M., Dayem, A. A., Choi, H. Y., et al. (2017). Valproic acid induces endocytosis-mediated doxorubicin internalization and shows synergistic cytotoxic effects in hepatocellular carcinoma cells. *International Journal of Molecular Sciences*, *18*(5). https://doi.org/10.3390/ijms18051048. Epub 2017/05/13. 28498322. PMC5454960.

Saleh, T., Tyutyunyk-Massey, L., Murray, G. F., Alotaibi, M. R., Kawale, A. S., Elsayed, Z., et al. (2019). Tumor cell escape from therapy-induced senescence. *Biochemical Pharmacology*, *162*, 202–212. Epub 2018/12/24. https://doi.org/10.1016/j.bcp.2018.12.013. 30576620.

Saleh, T., Bloukh, S., Carpenter, V. J., Alwohoush, E., Bakeer, J., Darwish, S., et al. (2020). Therapy-induced senescence: An "old" friend becomes the enemy. *Cancers*, *12*(4). https://doi.org/10.3390/cancers12040822. Epub 2020/04/03. 32235364. PMC7226427.

Samaraweera, L., Adomako, A., Rodriguez-Gabin, A., & McDaid, H. M. (2017). A novel indication for panobinostat as a senolytic drug in NSCLC and HNSCC. *Scientific Reports*, 7(1), 1900. Epub 2017/05/17. https://doi.org/10.1038/s41598-017-01964-1. 28507307. PMC5432488.

Scarlatti, F., Bauvy, C., Ventruti, A., Sala, G., Cluzeaud, F., Vandewalle, A., et al. (2004). Ceramide-mediated macroautophagy involves inhibition of protein kinase B and up-regulation of beclin 1. *The Journal of Biological Chemistry*, *279*(18), 18384–18391. Epub 2004/02/19. https://doi.org/10.1074/jbc.M313561200. 14970205.

Scarlatti, F., Maffei, R., Beau, I., Codogno, P., & Ghidoni, R. (2008). Role of non-canonical Beclin 1-independent autophagy in cell death induced by resveratrol in human breast cancer cells. *Cell Death and Differentiation*, *15*(8), 1318–1329. Epub 2008/04/19. https://doi.org/10.1038/cdd.2008.51. 18421301.

Schneider, B. J., Shah, M. A., Klute, K., Ocean, A., Popa, E., Altorki, N., et al. (2017). Phase I study of epigenetic priming with azacitidine prior to standard neoadjuvant chemotherapy for patients with resectable gastric and esophageal adenocarcinoma: Evidence of tumor hypomethylation as an indicator of major histopathologic response. *Clinical Cancer Research*, *23*(11), 2673–2680. Epub 2016/11/12. https://doi.org/10.1158/1078-0432.CCR-16-1896. 27836862. PMC5425331.

Schosserer, M., Grillari, J., & Breitenbach, M. (2017). The dual role of cellular senescence in developing tumors and their response to cancer therapy. *Frontiers in Oncology*, *7*, 278. Epub 2017/12/09. https://doi.org/10.3389/fonc.2017.00278. 29218300. PMC5703792.

Schwartsmann, G., Fernandes, M. S., Schaan, M. D., Moschen, M., Gerhardt, L. M., Di Leone, L., et al. (1997). Decitabine (5-aza-2′-deoxycytidine; DAC) plus daunorubicin as a first line treatment in patients with acute myeloid leukemia: Preliminary observations. *Leukemia*, *11*(Suppl. 1), S28–S31. Epub 1997/03/01. 9130689.

Serrano, M., Lin, A. W., McCurrach, M. E., Beach, D., & Lowe, S. W. (1997). Oncogenic ras provokes premature cell senescence associated with accumulation of p53 and p16INK4a. *Cell*, *88*(5), 593–602. Epub 1997/03/07. https://doi.org/10.1016/s0092-8674(00)81902-9. 9054499.

Shao, Y., Gao, Z., Marks, P. A., & Jiang, X. (2004). Apoptotic and autophagic cell death induced by histone deacetylase inhibitors. *Proceedings of the National Academy of Sciences of the United States of America*, *101*(52), 18030–18035. Epub 2004/12/15. https://doi.org/10.1073/pnas.0408345102. 15596714. PMC539807.

Sharma, S. V., Lee, D. Y., Li, B., Quinlan, M. P., Takahashi, F., Maheswaran, S., et al. (2010). A chromatin-mediated reversible drug-tolerant state in cancer cell subpopulations. *Cell*, *141*(1), 69–80. Epub 2010/04/08. https://doi.org/10.1016/j.cell.2010.02.027. 20371346. PMC2851638.

Shin, H., Lee, Y. S., & Lee, Y. C. (2012). Sodium butyrate-induced DAPK-mediated apoptosis in human gastric cancer cells. *Oncology Reports*, *27*(4), 1111–1115. Epub 2011/12/14. https://doi.org/10.3892/or.2011.1585. 22160140. PMC3583600.

Sidler, C., Woycicki, R., Li, D., Wang, B., Kovalchuk, I., & Kovalchuk, O. (2014). A role for SUV39H1-mediated H3K9 trimethylation in the control of genome stability and senescence in WI38 human diploid lung fibroblasts. *Aging (Albany NY)*, *6*(7), 545–563. Epub 2014/07/27. 10.18632/aging.100678. 25063769. PMC4153622.

Sidler, C., Li, D., Wang, B., Kovalchuk, I., & Kovalchuk, O. (2014). SUV39H1 downregulation induces deheterochromatinization of satellite regions and senescence after exposure to ionizing radiation. *Frontiers in Genetics*, *5*, 411. Epub 2014/12/09. https://doi.org/10.3389/fgene.2014.00411. 25484892. PMC4240170.

Siebenkas, C., Chiappinelli, K. B., Guzzetta, A. A., Sharma, A., Jeschke, J., Vatapalli, R., et al. (2017). Inhibiting DNA methylation activates cancer testis antigens and expression of the antigen processing and presentation machinery in colon and ovarian cancer cells. *PLoS One*, *12*(6), e0179501. Epub 2017/06/18. https://doi.org/10.1371/journal.pone.0179501. 28622390. PMC5473589.

Siegel, R. L., Miller, K. D., & Jemal, A. (2020). Cancer statistics, 2020. *CA: A Cancer Journal for Clinicians*, *70*(1), 7–30. Epub 2020/01/09. https://doi.org/10.3322/caac.21590. 31912902.

Sikora, E., Czarnecka-Herok, J., Bojko, A., & Sunderland, P. (2022). Therapy-induced polyploidization and senescence: Coincidence or interconnection? *Seminars in Cancer Biology*, *81*, 83–95. Epub 2020/12/04. https://doi.org/10.1016/j.semcancer.2020.11.015. 33271316.

Sistigu, A., Musella, M., Galassi, C., Vitale, I., & De Maria, R. (2020). Tuning cancer fate: Tumor microenvironment's role in cancer stem cell quiescence and reawakening. *Frontiers in Immunology*, *11*, 2166. Epub 2020/11/17. https://doi.org/10.3389/fimmu.2020.02166. 33193295. PMC7609361.

Siu, W. Y., Arooz, T., & Poon, R. Y. (1999). Differential responses of proliferating versus quiescent cells to adriamycin. *Experimental Cell Research*, *250*(1), 131–141. Epub 1999/07/02. https://doi.org/10.1006/excr.1999.4551. 10388527.

Sliwinska, M. A., Mosieniak, G., Wolanin, K., Babik, A., Piwocka, K., Magalska, A., et al. (2009). Induction of senescence with doxorubicin leads to increased genomic instability of HCT116 cells. *Mechanisms of Ageing and Development*, *130*(1–2), 24–32. Epub 2008/06/10. https://doi.org/10.1016/j.mad.2008.04.011. 18538372.

Smith, L. M., Nesterova, A., Ryan, M. C., Duniho, S., Jonas, M., Anderson, M., et al. (2008). CD133/prominin-1 is a potential therapeutic target for antibody-drug conjugates in hepatocellular and gastric cancers. *British Journal of Cancer*, *99*(1), 100–109. Epub 2008/06/11. https://doi.org/10.1038/sj.bjc.6604437. 18542072. PMC2453027.

Song, Y., & Zhang, C. (2009). Hydralazine inhibits human cervical cancer cell growth in vitro in association with APC demethylation and re-expression. *Cancer Chemotherapy and Pharmacology*, *63*(4), 605–613. Epub 2008/06/04. https://doi.org/10.1007/s00280-008-0773-z. 18521605.

Song, Y., Zhao, Y., Deng, Z., Zhao, R., & Huang, Q. (2021). Stress-induced polyploid giant cancer cells: Unique way of formation and non-negligible characteristics. *Frontiers in Oncology*, *11*, 724781. Epub 2021/09/17. https://doi.org/10.3389/fonc.2021.724781. 34527590. PMC8435787.

Sosa, M. S., Bragado, P., & Aguirre-Ghiso, J. A. (2014). Mechanisms of disseminated cancer cell dormancy: An awakening field. *Nature Reviews. Cancer*, *14*(9), 611–622. Epub 2014/08/15. https://doi.org/10.1038/nrc3793. 25118602. PMC4230700.

Soto-Reyes, E., & Recillas-Targa, F. (2010). Epigenetic regulation of the human p53 gene promoter by the CTCF transcription factor in transformed cell lines. *Oncogene*, *29*(15), 2217–2227. Epub 2010/01/27. https://doi.org/10.1038/onc.2009.509. 20101205.

Sperlazza, J., Rahmani, M., Beckta, J., Aust, M., Hawkins, E., Wang, S. Z., et al. (2015). Depletion of the chromatin remodeler CHD4 sensitizes AML blasts to genotoxic agents and reduces tumor formation. *Blood*, *126*(12), 1462–1472. https://doi.org/10.1182/blood-2015-03-631606.

Sui, X., Jin, L., Huang, X., Geng, S., He, C., & Hu, X. (2011). p53 signaling and autophagy in cancer: A revolutionary strategy could be developed for cancer treatment. *Autophagy*, 7(6), 565–571. Epub 2010/11/26. https://doi.org/10.4161/auto.7.6.14073. 21099252.

Sui, X., Wang, D., Geng, S., Zhou, G., He, C., & Hu, X. (2012). Methylated promoters of genes encoding protocadherins as a new cancer biomarker family. *Molecular Biology Reports*, *39*(2), 1105–1111. Epub 2011/05/21. https://doi.org/10.1007/s11033-011-0837-8. 21598112.

Sui, X., Chen, R., Wang, Z., Huang, Z., Kong, N., Zhang, M., et al. (2013). Autophagy and chemotherapy resistance: A promising therapeutic target for cancer treatment. *Cell Death & Disease*, *4*, e838. Epub 2013/10/12. https://doi.org/10.1038/cddis.2013.350. 24113172. PMC3824660.

Sui, X., Kong, N., Ye, L., Han, W., Zhou, J., Zhang, Q., et al. (2014). p38 and JNK MAPK pathways control the balance of apoptosis and autophagy in response to chemotherapeutic agents. *Cancer Letters*, *344*(2), 174–179. Epub 2013/12/18. https://doi.org/10.1016/j.canlet.2013.11.019. 24333738.

Sundaram, M., Guernsey, D. L., Rajaraman, M. M., & Rajaraman, R. (2004). Neosis: A novel type of cell division in cancer. *Cancer Biology & Therapy*, *3*(2), 207–218. Epub 2004/01/17. https://doi.org/10.4161/cbt.3.2.663. 14726689.

Takaishi, S., Okumura, T., Tu, S., Wang, S. S., Shibata, W., Vigneshwaran, R., et al. (2009). Identification of gastric cancer stem cells using the cell surface marker CD44. *Stem Cells*, *27*(5), 1006–1020. Epub 2009/05/06. https://doi.org/10.1002/stem.30. 19415765. PMC2746367.

Talbert, P. B., & Henikoff, S. (2021). Histone variants at a glance. *Journal of Cell Science*, *134*(6). https://doi.org/10.1242/jcs.244749. Epub 2021/03/28. 33771851. PMC8015243.

Tamgue, O., & Lei, M. (2017). Triptolide promotes senescence of prostate cancer cells through histone methylation and heterochromatin formation. *Asian Pacific Journal of Cancer Prevention*, *18*(9), 2519–2526. Epub 2017/09/28. 10.22034/APJCP.2017.18.9. 2519. 28952292. PMC5720660.

Tannock, I. F., Lee, C. M., Tunggal, J. K., Cowan, D. S., & Egorin, M. J. (2002). Limited penetration of anticancer drugs through tumor tissue: A potential cause of resistance of solid tumors to chemotherapy. *Clinical Cancer Research*, *8*(3), 878–884.

Tchkonia, T., Zhu, Y., van Deursen, J., Campisi, J., & Kirkland, J. L. (2013). Cellular senescence and the senescent secretory phenotype: Therapeutic opportunities. *The Journal of Clinical Investigation*, *123*(3), 966–972. Epub 2013/03/05. https://doi.org/10.1172/JCI64098. 23454759. PMC3582125.

te Poele, R. H., Okorokov, A. L., Jardine, L., Cummings, J., & Joel, S. P. (2002). DNA damage is able to induce senescence in tumor cells in vitro and in vivo. *Cancer Research*, *62*(6), 1876–1883. Epub 2002/03/26. 11912168.

Terzi, M. Y., Izmirli, M., & Gogebakan, B. (2016). The cell fate: Senescence or quiescence. *Molecular Biology Reports*, *43*(11), 1213–1220. Epub 2016/10/19. https://doi.org/10.1007/s11033-016-4065-0. 27558094.

The ENCODE Project Consortium, Snyder, M. P., Gingeras, T. R., Moore, J. E., Weng, Z., Gerstein, M. B., et al. (2020). Perspectives on ENCODE. *Nature*, *583*(7818), 693–698. Epub 2020/07/31. https://doi.org/10.1038/s41586-020-2449-8. 32728248. PMC7410827.

Toh, T. B., Lim, J. J., & Chow, E. K. (2017). Epigenetics in cancer stem cells. *Molecular Cancer*, *16*(1), 29. Epub 2017/02/06. https://doi.org/10.1186/s12943-017-0596-9. 28148257. PMC5286794.

Toor, A. A., Payne, K. K., Chung, H. M., Sabo, R. T., Hazlett, A. F., Kmieciak, M., et al. (2012). Epigenetic induction of adaptive immune response in multiple myeloma: Sequential azacitidine and lenalidomide generate cancer testis antigen-specific cellular immunity. *British Journal of Haematology*, *158*(6), 700–711. Epub 2012/07/24. https://doi.org/10.1111/j.1365-2141.2012.09225.x. 22816680. PMC4968567.

Vera-Ramirez, L., Vodnala, S. K., Nini, R., Hunter, K. W., & Green, J. E. (2018). Autophagy promotes the survival of dormant breast cancer cells and metastatic tumour recurrence. *Nature Communications*, *9*(1), 1944. Epub 2018/05/24. https://doi.org/10.1038/s41467-018-04070-6. 29789598. PMC5964069.

Villanueva, L., Alvarez-Errico, D., & Esteller, M. (2020). The contribution of epigenetics to cancer immunotherapy. *Trends in Immunology*, *41*(8), 676–691. Epub 2020/07/06. https://doi.org/10.1016/j.it.2020.06.002. 32622854.

Vinogradova, M., Gehling, V. S., Gustafson, A., Arora, S., Tindell, C. A., Wilson, C., et al. (2016). An inhibitor of KDM5 demethylases reduces survival of drug-tolerant cancer cells. *Nature Chemical Biology*, *12*(7), 531–538. Epub 2016/05/24. https://doi.org/10.1038/nchembio.2085. 27214401.

Voce, D. J., Bernal, G. M., Cahill, K. E., Wu, L., Mansour, N., Crawley, C. D., et al. (2021). CDK1 is up-regulated by temozolomide in an NF-κB dependent manner in glioblastoma. *Scientific Reports*, *11*(1), 5665. https://doi.org/10.1038/s41598-021-84912-4.

Wainwright, E. N., & Scaffidi, P. (2017). Epigenetics and cancer stem cells: Unleashing, hijacking, and restricting cellular plasticity. *Trends in Cancer*, *3*(5), 372–386. Epub 2017/07/19. https://doi.org/10.1016/j.trecan.2017.04.004. 28718414. PMC5506260.

Wang, X., Wong, S. C., Pan, J., Tsao, S. W., Fung, K. H., Kwong, D. L., et al. (1998). Evidence of cisplatin-induced senescent-like growth arrest in nasopharyngeal carcinoma cells. *Cancer Research*, *58*(22), 5019–5022. Epub 1998/11/21. 9823301.

Wang, Q., Wu, P. C., Dong, D. Z., Ivanova, I., Chu, E., Zeliadt, S., et al. (2013). Polyploidy road to therapy-induced cellular senescence and escape. *International Journal of Cancer*, *132*(7), 1505–1515. Epub 2012/09/05. https://doi.org/10.1002/ijc.27810. 22945332.

Wang, Y., Li, W., Patel, S. S., Cong, J., Zhang, N., Sabbatino, F., et al. (2014). Blocking the formation of radiation-induced breast cancer stem cells. *Oncotarget*, *5*(11), 3743–3755. Epub 2014/07/09. 10.18632/oncotarget.1992. 25003837. PMC4116517.

Wang, H., Feng, W., Chen, W., He, J., Min, J., Liu, Y., et al. (2020). Methyl-CpG-binding domain 3 inhibits stemness of pancreatic cancer cells via Hippo signaling. *Experimental Cell Research*, *393*(1), 112091. https://doi.org/10.1016/j.yexcr.2020.112091.

Wang, B., Kohli, J., & Demaria, M. (2020). Senescent cells in cancer therapy: Friends or foes? *Trends in Cancer*, *6*(10), 838–857. Epub 2020/06/03. https://doi.org/10.1016/j.trecan.2020.05.004. 32482536.

Wang, L., Lankhorst, L., & Bernards, R. (2022). Exploiting senescence for the treatment of cancer. *Nature Reviews. Cancer*, *22*(6), 340–355. Epub 2022/03/05. https://doi.org/10.1038/s41568-022-00450-9. 35241831.

Was, H., Czarnecka, J., Kominek, A., Barszcz, K., Bernas, T., Piwocka, K., et al. (2018). Some chemotherapeutics-treated colon cancer cells display a specific phenotype being a combination of stem-like and senescent cell features. *Cancer Biology & Therapy*, *19*(1), 63–75. Epub 2017/10/21. https://doi.org/10.1080/15384047.2017.1385675. 29053388. PMC5790359.

White-Gilbertson, S., & Voelkel-Johnson, C. (2020). Giants and monsters: Unexpected characters in the story of cancer recurrence. *Advances in Cancer Research*, *148*, 201–232. Epub 2020/07/30. https://doi.org/10.1016/bs.acr.2020.03.001. 32723564. PMC7731118.

Wierod, L., Rosseland, C. M., Lindeman, B., Oksvold, M. P., Grosvik, H., Skarpen, E., et al. (2008). Activation of the p53-p21(Cip1) pathway is required for CDK2 activation and S-phase entry in primary rat hepatocytes. *Oncogene*, *27*(19), 2763–2771. Epub 2007/11/21. https://doi.org/10.1038/sj.onc.1210937. 18026139.

Witcher, M., & Emerson, B. M. (2009). Epigenetic silencing of the p16(INK4a) tumor suppressor is associated with loss of CTCF binding and a chromatin boundary. *Molecular Cell*, *34*(3), 271–284. Epub 2009/05/20. https://doi.org/10.1016/j.molcel.2009.04.001. 19450526. PMC2723750.

Witt, A. E., Lee, C. W., Lee, T. I., Azzam, D. J., Wang, B., Caslini, C., et al. (2017). Identification of a cancer stem cell-specific function for the histone deacetylases, HDAC1 and HDAC7, in breast and ovarian cancer. *Oncogene*, *36*(12), 1707–1720. Epub 2016/10/04. https://doi.org/10.1038/onc.2016.337. 27694895. PMC5364039.

Wu, H. J., & Chu, P. Y. (2021). Epigenetic regulation of breast cancer stem cells contributing to carcinogenesis and therapeutic implications. *International Journal of Molecular Sciences*, *22*(15). https://doi.org/10.3390/ijms22158113. Epub 2021/08/08. 34360879. PMC8348144.

Xu, Z. Y., Tang, J. N., Xie, H. X., Du, Y. A., Huang, L., Yu, P. F., et al. (2015). 5-Fluorouracil chemotherapy of gastric cancer generates residual cells with properties of cancer stem cells. *International Journal of Biological Sciences*, *11*(3), 284–294. Epub 2015/02/14. https://doi.org/10.7150/ijbs.10248. 25678847. PMC4323368.

Yan, H. C., & Zhang, J. (2012). Effects of sodium valproate on the growth of human ovarian cancer cell line HO8910. *Asian Pacific Journal of Cancer Prevention*, *13*(12), 6429–6433. Epub 2012/01/01. https://doi.org/10.7314/apjcp.2012.13.12.6429. 23464470.

Yao, Z., Du, L., Xu, M., Li, K., Guo, H., Ye, G., et al. (2019). MTA3-SOX2 module regulates cancer stemness and contributes to clinical outcomes of tongue carcinoma. *Frontiers in Oncology, 9*, 816. https://doi.org/10.3389/fonc.2019.00816.

Ye, X. N., Zhou, X. P., Wei, J. Y., Xu, G. X., Li, Y., Mao, L. P., et al. (2016). Epigenetic priming with decitabine followed by low-dose idarubicin/cytarabine has an increased anti-leukemic effect compared to traditional chemotherapy in high-risk myeloid neoplasms. *Leukemia & Lymphoma, 57*(6), 1311–1318. Epub 2015/09/16. https://doi.org/10.3109/10428194.2015.1091931. 26372888.

Ye, S., Ding, Y. F., Jia, W. H., Liu, X. L., Feng, J. Y., Zhu, Q., et al. (2019). SET domain-containing protein 4 epigenetically controls breast cancer stem cell quiescence. *Cancer Research, 79*(18), 4729–4743. Epub 2019/07/17. https://doi.org/10.1158/0008-5472.CAN-19-1084. 31308046.

Yomtoubian, S., Lee, S. B., Verma, A., Izzo, F., Markowitz, G., Choi, H., et al. (2020). Inhibition of EZH2 catalytic activity selectively targets a metastatic subpopulation in triple-negative breast cancer. *Cell Reports, 30*(3), 755–770. e6. Epub 2020/01/23. https://doi.org/10.1016/j.celrep.2019.12.056. 31968251.

Yu, Y., Qi, J., Xiong, J., Jiang, L., Cui, D., He, J., et al. (2019). Epigenetic co-deregulation of EZH2/TET1 is a senescence-countering, actionable vulnerability in triple-negative breast cancer. *Theranostics, 9*(3), 761–777. Epub 2019/02/28. https://doi.org/10.7150/thno.29520. 30809307. PMC6376470.

Yu, X., Li, M., Guo, C., Wu, Y., Zhao, L., Shi, Q., et al. (2021). Therapeutic targeting of cancer: Epigenetic homeostasis. *Frontiers in Oncology, 11*, 747022. Epub 2021/11/13. https://doi.org/10.3389/fonc.2021.747022. 34765551. PMC8576334.

Zack, T. I., Schumacher, S. E., Carter, S. L., Cherniack, A. D., Saksena, G., Tabak, B., et al. (2013). Pan-cancer patterns of somatic copy number alteration. *Nature Genetics, 45*(10), 1134–1140. Epub 2013/09/28. https://doi.org/10.1038/ng.2760. 24071852. PMC3966983.

Zampetidis, C. P., Galanos, P., Angelopoulou, A., Zhu, Y., Polyzou, A., Karamitros, T., et al. (2021). A recurrent chromosomal inversion suffices for driving escape from oncogene-induced senescence via subTAD reorganization. *Molecular Cell, 81*(23). https://doi.org/10.1016/j.molcel.2021.10.017. 4907–4923.e8. Epub 2021/11/19. 34793711.

Zampetidis, C. P., Papantonis, A., & Gorgoulis, V. G. (2022). Escape from senescence: Revisiting cancer therapeutic strategies. *Molecular & Cellular Oncology, 9*(1), 2030158. Epub 2022/03/08. https://doi.org/10.1080/23723556.2022.2030158. 35252554. PMC8890391.

Zeuner, A. (2015). The secret life of quiescent cancer stem cells. *Molecular & Cellular Oncology, 2*(1), e968067. Epub 2015/01/01. https://doi.org/10.4161/23723548.2014.968067. 27308385. PMC4905231.

Zhang, J., Lv, X., Wei, B., Gong, X., & Chen, L. (2022). CHD4 mediates SOX2 transcription through TRPS1 in luminal breast cancer. *Cellular Signalling, 100*, 110464. https://doi.org/10.1016/j.cellsig.2022.110464.

Zhang, W., Ji, W., Yang, J., Yang, L., Chen, W., & Zhuang, Z. (2008). Comparison of global DNA methylation profiles in replicative versus premature senescence. *Life Sciences, 83*(13–14), 475–480. Epub 2008/08/30. https://doi.org/10.1016/j.lfs.2008.07.015. 18723031.

Zhang, Z., Zhang, B., Li, W., Fu, L., Fu, L., Zhu, Z., et al. (2011). Epigenetic silencing of miR-203 upregulates SNAI2 and contributes to the invasiveness of malignant breast cancer cells. *Genes & Cancer, 2*(8), 782–791. Epub 2012/03/07. https://doi.org/10.1177/1947601911429743. 22393463. PMC3278899.

Zhang, S., Mercado-Uribe, I., Xing, Z., Sun, B., Kuang, J., & Liu, J. (2014). Generation of cancer stem-like cells through the formation of polyploid giant cancer cells. *Oncogene*, *33*(1), 116–128. Epub 2013/03/26. https://doi.org/10.1038/onc.2013.96. 23524583. PMC3844126.

Zhang, G., Dong, Z., Prager, B. C., Kim, L. J., Wu, Q., Gimple, R. C., et al. (2019). Chromatin remodeler HELLS maintains glioma stem cells through E2F3 and MYC. *JCI Insight*, *4*(7). https://doi.org/10.1172/jci.insight.126140. Epub 2019/02/20. 30779712. PMC6483649.

Zhu, Y., Tchkonia, T., Pirtskhalava, T., Gower, A. C., Ding, H., Giorgadze, N., et al. (2015). The Achilles' heel of senescent cells: From transcriptome to senolytic drugs. *Aging Cell*, *14*(4), 644–658. Epub 2015/03/11. https://doi.org/10.1111/acel.12344. 25754370. PMC4531078.

Zinzi, L., Contino, M., Cantore, M., Capparelli, E., Leopoldo, M., & Colabufo, N. A. (2014). ABC transporters in CSCs membranes as a novel target for treating tumor relapse. *Frontiers in Pharmacology*, *5*, 163. Epub 2014/07/30. https://doi.org/10.3389/fphar.2014.00163. 25071581. PMC4091306.

CHAPTER NINE

Targeting the super elongation complex for oncogenic transcription driven tumor malignancies: Progress in structure, mechanisms and small molecular inhibitor discovery

Xinyu Wu[a,b,†], Yanqiu Xie[a,b,†], Kehao Zhao[a,*], and Jing Lu[a,*]

[a]School of Pharmacy, Key Laboratory of Molecular Pharmacology and Drug Evaluation (Yantai University), Ministry of Education, Collaborative Innovation Center of Advanced Drug Delivery System and Biotech Drugs in Universities of Shandong, Yantai University, Yantai, China

[b]Drug Discovery and Design Center, State Key Laboratory of Drug Research, Shanghai Institute of Materia Medica, Chinese Academy of Sciences, Shanghai, China

*Corresponding authors: e-mail address: kehaozhao@gmail.com; lujing_ytu@126.com

Contents

† Those two authors contributed equally to the manuscript.

Advances in Cancer Research, Volume 158
ISSN 0065-230X
https://doi.org/10.1016/bs.acr.2022.12.007

Abstract

Oncogenic transcription activation is associated with tumor development and resistance derived from chemotherapy or target therapy. The super elongation complex (SEC) is an important complex regulating gene transcription and expression in metazoans closely related to physiological activities. In normal transcriptional regulation, SEC can trigger promoter escape, limit proteolytic degradation of transcription elongation factors and increase the synthesis of RNA polymerase II (POL II), and regulate many normal human genes to stimulate RNA elongation. Dysregulation of SEC accompanied by multiple transcription factors in cancer promotes rapid transcription of oncogenes and induce cancer development. In this review, we summarized recent progress in understanding the mechanisms of SEC in regulating normal transcription, and importantly its roles in cancer development. We also highlighted the discovery of SEC complex target related inhibitors and their potential applications in cancer treatment.

1. Introduction of SEC structure and function

Super elongation complex (SEC) is an important complex regulates gene transcription in metazoans closely related to physiological activities such as stress response and heat shock response (Liang et al., 2018). SEC includes different combinations of non-coexisting homologous proteins derived from different families, which increases gene control options and regulatory diversity during SEC-intervened transcriptional activation of different genes. SEC is composed of positive transcription elongation factor b (P-TEFb), the RNA polymerase II (Pol II) elongation factors 11–19 Lys-rich leukemia (ELL) proteins, the ELL enhancers EAF1/EAF2 and several frequent mixed lineage leukemia (MLL) translocation partners such as AFF1/AFF4 and ENL/AF9 (Fig. 1A). Among them, AFF1/AFF4 is considered as the backbone of SEC and has direct interactions with

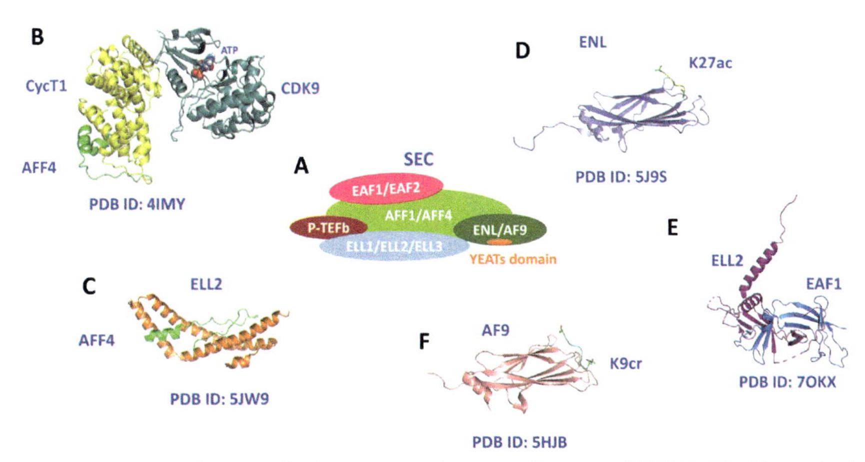

Fig. 1 Structures of SEC and subunits. (A) Schematic diagram of SEC. (B) The N-terminal helix in CycT1 (yellow cartoon) forms interactions to CDK9 (*dark cyan*), but no connection of the phospho-threonine site to the cyclin (PDB ID: 4IMY) (Schulze-Gahmen et al., 2013). The AFF4 peptide and ATP in CDK9 are shown as *green cartoon* and *red sphere*, respectively. (C) The residues 301–351 in AFF4 (*green*) folds up into an elbow joint that forms extensive contacts with the occludin domain of ELL2 (*orange*, PDB ID: 5JW9) (Qi et al., 2017). (D) The YEATS domain of ENL (*blue*) folds up into an eight-stranded β-sandwich with the H3K27ac peptide (*yellow*, PDB ID: 5J9S) (Wan et al., 2017). (E) The ELL2-EAF1 dimerization adopts a triple-barrel fold formed by eight β strands and two α helices in ELL2 (*magenta*) and eight β strands in EAF1 (marine) (PDB ID: 7OKX) (Chen et al., 2021). (F) The YEATS domain of AF9 (*purple*) recognizes H3K9cr peptide (*cyan*) using the same Kac-binding aromatic sandwich cage (PDB ID: 5HJB) (Li et al., 2016).

P-TEFb, ELL1/ELL2, and ENL/AF9 (Fig. 1B and C) (Qi et al., 2017; Schulze-Gahmen et al., 2013). P-TEFb synergizes with ELLs to inhibit POL II transcriptional pausing and trigger full length mRNA transcription (Chen, Smith, & Shilatifard, 2018). ELL2-EAF1 forms a heterodimer and is important for the elongation of RNA polymerase complex II (Fig. 1E) (Chen, Vos, et al., 2021). ENL and AF9 are recognized by acetylated and crotonoylated histones to regulate oncogenic transcriptional programs in multiple cancer types (Fig. 1D and F) (Zhang et al., 2016).

1.1 AFF1/AFF4

The AFF1 and AFF4 of AF4/FRMs family members are homologous proteins encoded by different genes with single amino acid variation. However, AFF1 and AFF4 do not co-exist in the same SEC (Lu et al., 2015). AFF1 and

AFF4 also have distinct interaction with partners. While AFF1 is associated with HIV transactivation, AFF4 is shown to sense Hsp70 upon heat shock (Luo, Lin, & Shilatifard, 2012).

AFF1/AFF4 is the scaffold protein consisting of an intrinsically largely disordered N-terminal region (1–1210 for AFF1 and 1–1163 for AFF4) which harbors the dispersed binding sites for P-TEFb, ELL1/ELL2, and ENL/AF9, and a highly conserved C-terminal homology domain (CHD) among AF4/FMR2 family proteins responsible for AFF1/AFF4 proteins homodimer or heterodimer formation (Fig. 2A). P-TEFb is folded into two small helices (Fig. 1B) (Qi et al., 2017) with its hydrophobic residues binding to residues 1–70 of AFF4. ELL1/ELL2 anchors to AFF4 residues 301–350, and the ELL1/ELL2 closure protein domain forms an interactive cavity with AFF4, namely the ELL binding pocket (Fig. 1C). ENL/AF9 binds to AFF4 residues 761–774. In MLL translocation leukemia, the ANC1 homology domain of AF9 was found to interact with AFF4 to form an intermolecular triple-stranded β-sheet (Chen & Cramer, 2019).

Recent structure studies on the C-terminal of AFF1/AFF4 indicated that a C-terminal homology domain (CHD) in AFF4 (residues 970–1163) or AFF1 (residues 906–1210) has tetratrico peptide repeat (TPR)-like properties (Chen & Cramer, 2019; Tang et al., 2020). AFF4-CHD dimerization is mainly intervened by its two hydrophobic cores involving two central residues Phe1014 and Tyr1096 (Gao et al., 2020; Schulze-Gahmen et al., 2016). The F1014L mutation of AFF4 was found to occur in bladder urothelial carcinoma and lung cancer, indicating that the dimerization of AFF4 could be important in tumor development (Chen & Cramer, 2019; Tang et al., 2020). In addition to self-dimerization of AFF1/AFF4, AFF1/AFF4 can form heterodimers through CHD which is supported by the evidence that deletion of CHD in AFF1 completely interrupted AFF4 binding to AFF1 (Chen & Cramer, 2019). Interestingly, SEC assembly does not require AFF1/AFF4 heterodimers (Chen & Cramer, 2019; Lu, Li, et al., 2015). CHD-mediated dimerization also connects the fusion of AFF1/AFF4 to mixed lineage leukemia (MLL) protein, thereby inducing the transcriptional upregulation of HOXa9 and increasing oncogenic potential, particularly in MLL-fusion driven acute myeloid leukemia (AML) (Leach et al., 2013).

1.2 P-TEFb

P-TEFb originally purified from the extracts of *Drosophila melanogaster* cells, consists of the catalytic subunit CDK9 and the regulatory subunit Cyclin T1

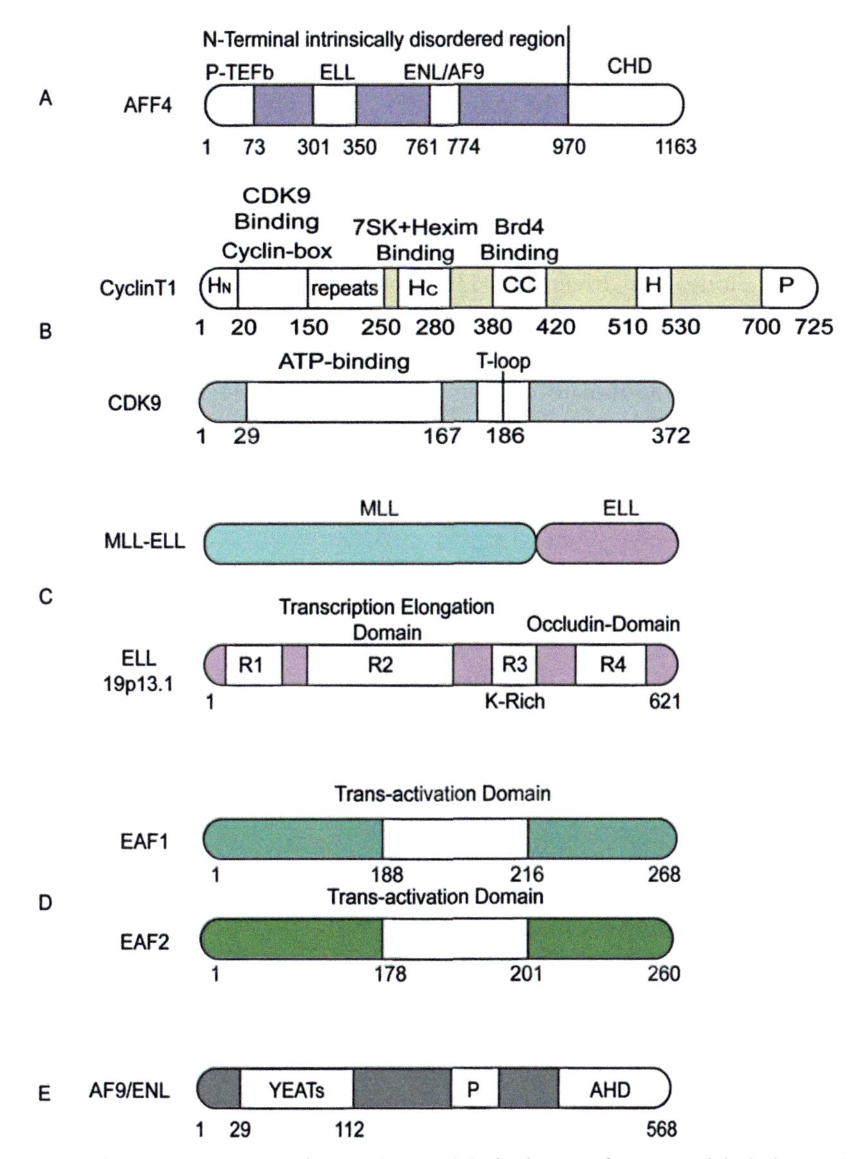

Fig. 2 Schematic structures of SEC subunits. (A) The boxes of AFF4 are labeled as CHD and binding sites with its partners including P-TEFb, ELL2, ENL/AF9. (B) CycT1 contains two canonical cyclin region repeats and accessory N- and C-terminal helices. The less conserved central region including coiled coil fragments (CC) and histidine-rich fragments (H). There are 20 prolines at the C-terminal region. CDK9 includes ATP-binding domain with multiple phosphorylation sites and T-loop. T-loop can be phosphorylated to interact with HEXIM1, integrating CDK9 into 7SK snRNP storing P-TEFb. (C) ELL includes four domains (R1–R4). The translocation *t*(11;19) (q23;p13.1) links the amino-terminal region of MLL to the R2-R4 domain of ELL. Normal ELL and ELL mutants have transcription elongation (elongated), lysine-rich domain (K-rich), and occludin-like domains. Stimulation of EAF1-dependent ELL elongation requires the ELL N-terminal and transcription elongation domains. (D) EAF1 and EAF2 have a transactivation domain for activating transcriptional elongation. (E) ENL/AF9 includes YEATs, ANC1 homology domain (AHD), and polyproline region (P).

(CycT1) or T2a/2b. As the most active P-TEFb-containing complex, P-TEFb and ELL2 co-stimulates transcriptional elongation of POL II and promotes elongation-coupled mRNA 3′ processing to generate full-length polyadenylated mRNA transcripts, which is necessary for rapid induction of gene expression (Chen et al., 2018). This activity is independent of transcription elongation driven by H3K79 methylation by DOTL1 histone methyltransferase, but there are some co-regulations between SEC and DOT1L (Chen et al., 2018; Mohan et al., 2010). In addition, P-TEFb is also found in bromodomain-containing protein 4 (BRD4) involved in gene transcription and 7sk small nuclear ribonucleoprotein (7sk snRNP) (Ji, Lu, Zhou, & Luo, 2014). The 7sk snRNP can sequester the activity of P-TEFb, and activation of P-TEFb requires the release of CDK9/CycT1 from 7sk RNP (Michels et al., 2004).

The CDK family is a proline-directed serine/threonine protein kinase family positioned in chromatin 9934.1 (Fig. 2B). CDK9 has two isoforms, long CDK9 with 55-KDa (CDK9–55) and short CDK9 with 42-KDa (CDK9–42), which share the same gene code and exons (Ahmed et al., 2022; Bacon & D'Orso, 2019). CDK9–55 has an additional N-terminal extension of 117 amino acids compared to CDK9–42. Both isoforms are expressed in cells with the short one being more abundant. Interestingly, the majority of the CDK9–42s are distributed in the nucleoplasm, while CDK9–55 is mainly distributed in the nucleolus (Bacon & D'Orso, 2019; Franco, Morales, Boffo, & Giordano, 2018). Ectopic expression of the corresponding epitope markers demonstrated that CDK9–55 was involved in the regulation of apoptosis and DNA repair, while CDK9–42 was involved in cellular transcriptional regulation (Bacon & D'Orso, 2019). Dysregulation of CDK9-related pathways indicated that CDK9 overexpression promoted cell proliferation and the synthesis of anti-apoptotic factors such as MCL-1, BCL-2, and XLAP, which are key factors of carcinogenesis (Boffo, Damato, Alfano, & Giordano, 2018).

CycT1 acts as an adaptor for transcription factors and substrates, provides the substrate for CDKs, and its distal end interacts with the disordered protein AFF1/AFF4 (Meinhart, Kamenski, Hoeppner, Baumli, & Cramer, 2005). The N terminal (HN) and C-terminal (HC) in CycT1 are two characteristic α-helical repeats that are arranged into a unique structure containing the binding pockets of CDK9 (Figs. 1B and 2B) (Meinhart et al., 2005). The phosphorylation of CDK9's T186 in T-loop induces its binding with CycT1, which stimulates CDK9's interaction with HEXIM and integration into the 7sksnRNP (Bacon & D'Orso, 2019; Franco et al., 2018; Meinhart et al., 2005; Zhou, Li, & Price, 2012). The central region of CycT1 includes a coiled helix (CC) responsible for the binding with BRD4, a highly conserved histidine-rich fragment, and proline-rich fragments, respectively (Anand, Schulte, Fujinaga, Scheffzek, & Geyer, 2007). In transcriptional activation, histidine-rich extension strands in the C-terminal of CycT1/2 can be regarded as substrate recognition motifs for P-TEFb to interact with the CTD of POL II.

1.3 ELL1/2/3

The ELL family, first purified from rat liver extract, is known as the lysine-rich leukemia protein family, and also called the extended stimulation family. ELLs inhibit POL II transcriptional pausing, synergize with P-TEFb to align mRNA to the catalytic site at the 3′ end of nascent RNA and prevent RNA backtracking, and improve RNA transcription efficiency (He et al., 2011). The stability of the ELLs is affected by the interaction between AFF4 and P-TEFb (Chen et al., 2018).

The ELL family includes three members: ELL1, ELL2, and ELL3. Homology analyses of the ELL family showed that ELLs generally share POL II inhibitory domain (R1), transcription elongation domain (R2), lysine-rich domain (R3) and carboxy-terminal domain (R4), with the except of ELL2 having no R1 domain (Anand et al., 2007). The R2-R4 domains of ELLs tend to be fused to the MLL to form the MLL-ELL fusion proteins, making ELLs become the common translocation partners of MLL (Fig. 2C) (Anand et al., 2007). The highly conserved R4 domain (Occludin-like domain) exhibits robust transcriptional activation properties, and is required for the immortalization associated with MLL-ELL fusion protein (DiMartino et al., 2000; Mozzetta, Boyarchuk, Pontis, & Ait-Si-Ali, 2015).

ELL1 is a component of the so-called small elongation complex involved in snRNA gene expression (Liu, Chen, Li, Huang, & Xue, 2020). In transcription elongation, human ELL1 is a POL II elongation factor being capable of improving the catalytic efficiency of elongation by reducing the transient pause of POL II at multiple sites on the DNA (Lu, Li, et al., 2015). ELL2 has a chemical dose-limiting and unique regulatory role in stabilizing protein stability, such as stabilizing the SEC by interacting with AFF4 (Lu, Li, et al., 2015). ELL3 appears to specifically initiate gene expression by labeling enhancers in embryonic stem cells (Lu, Li, et al., 2015).

Overexpression of the ELL family proteins results in c-MYC decline/degradation, and addition of any of the three respective ELLs results in decreased c-MYC levels (Ghobrial, Flick, Daly, Hoffman, & Milcarek, 2019).

1.4 EAF1/EAF2

Liu et al. first identified EAF1 and EAF2 from a yeast two-hybrid screening by using the ELL as a bait (Liu et al., 2020). EAF1 and EAF2 are homologous proteins with 60% sequence identity and 75% similarity. Both C-terminals of EAF1 (residues 188–216) and EAF2 (residues 178–201) contain a transactivation domain rich in serine, glutamate, and aspartate residues (Fig. 2D), which are highly homologous to the transcriptional activation domains of several MLL translocation chaperones such as AF4, LAF4, and AF5q31 (Simone et al., 2001).

The binding of EAF1/EAF2 to ELLs promotes the transcriptional elongation of ELLs, which is attributed to the region of ELLs required for POL II elongation (residues 60–373) which is the same of EAF1/EAF2 stably bound to ELL (Liu et al., 2020). Therefore, EAF1/2 is a positive modulator of ELL elongation activity (Liu et al., 2020). Interestingly, the N-terminal of ELL (residues 1–90) binds to EAF1 and EAF2, while the C-terminal of ELL (residues 508–621) only binds to EAF1 (Simone, Luo, Polak, Kaberlein, & Thirman, 2003). Moreover, ELL and EAF1 are components of the Cajal body, which contains many critical factors for RNA processing and gene expression (Polak, Simone, Kaberlein, Luo, & Thirman, 2003). EAF2 as an apoptosis-inducing agent inhibits in vivo prostate tumor growth, and the consumption of EAF2 enhanced cell proliferation and greatly increased cancer risk, suggesting that EAF2 may act as a potential tumor suppressor (Liu et al., 2020).

EAF2-ELL2 complex has shown allosteric regulation of RNA Pol II and significantly enhances RNA Pol II extension far from the active site (Fig. 2D) (Chen, Vos, et al., 2021). The first EAF1-ELL2 complex resolved by CRYO-EM indicated that two modules (ELL2-EAF1 dimer and the linker connected to RNA Pol II) are responsible for this allosteric regulation (Chen, Vos, et al., 2021). Moreover, the EAF2-ELL2 module of SEC simulates the extension of RNA POL II, and affects the stability of ELL2 occludin-like domain (Chen, Vos, et al., 2021).

1.5 ENL/AF9

ENL as a fusion target of leukemia is an intrinsically disordered transcriptional regulator that recruits multiple partners to assemble SECs, which are the results of protein-protein interactions of ENL with AFF1/AFF4, and the N-terminal coupling folding of ENL (Chen et al., 2018; Leach et al., 2013). AF9 at Chromosome 9 is known as the MLL1 fusion gene. The C-terminal of ENL/AF9 is associated with recruiting chromatin. Both AF9 and ENL contain YEATs domains with an extended acyl-binding repertoire, proline-rich region P (Biswas et al., 2011; Kuntimaddi et al., 2015; Li et al., 2014), and an AHD domain which binds to SECs as well as DOT1L (Fig. 2E) (Cao et al., 2020). The YEATs domain of ENL/AF9 recognizes H3K9ac, H3K18ac, H3K27ac through hydrogen bonds and c-π bonds (Chen et al., 2018). Both ENL and AF9 showed higher binding affinity to crotonylated H3K9 and H3K18 than acetylated histones (Li et al., 2016). In the absence of sequence-specific recruitment factors TAT and MLL, the YEAT domain in the N-terminal of ENL/AF9 also interacts with the human polymerase-associated factor 1 complex (PAF1C) which is in charge of targeting chromatin (Leach et al., 2013).

Although AF9 and ENL are homologous proteins with similar structures, they mediate different SEC-dependent transcriptional functions. AF9 and ENL exclusively bind to the same region of AFF4 (residues 601–900), indicating that they cannot simultaneously exist in the same SEC (Lu, Li, et al., 2015). AF9 plays a dominant role in physiological conditions. However, ENL has a higher priority over AF9 in MLL, and AF9 can unidirectionally compensate for the loss of ENL (Wan et al., 2017).

2. Relationship between SEC and transcriptional regulation

2.1 Gene transcription regulation

Transcription is an essential biological process to transcribe DNA genetic information into proteins in eukaryotes. Transcriptional regulation consists of six stages including preinitiation complexes (PIC) formation (Fig. 3A), promoter clearance (Fig. 3B), mRNA's 5′ capping (Fig. 3C), transcription pause (Fig. 3D), productive elongation (Fig. 3E), splicing (Fig. 3F), and polyadenylation (Fig. 3G). There are several published excellent reviews on transcriptional regulation (Chen et al., 2018; Li et al., 2022; Meinhart et al., 2005). This review will highlight the mechanisms of SEC in productive elongation and splicing & polyadenylation in transcription process.

2.2 POL II regulation before productive elongation

Prior to the involvement of SEC complex in transcription elongation, the PIC is assembled at the promoter which includes RNA POL II, general

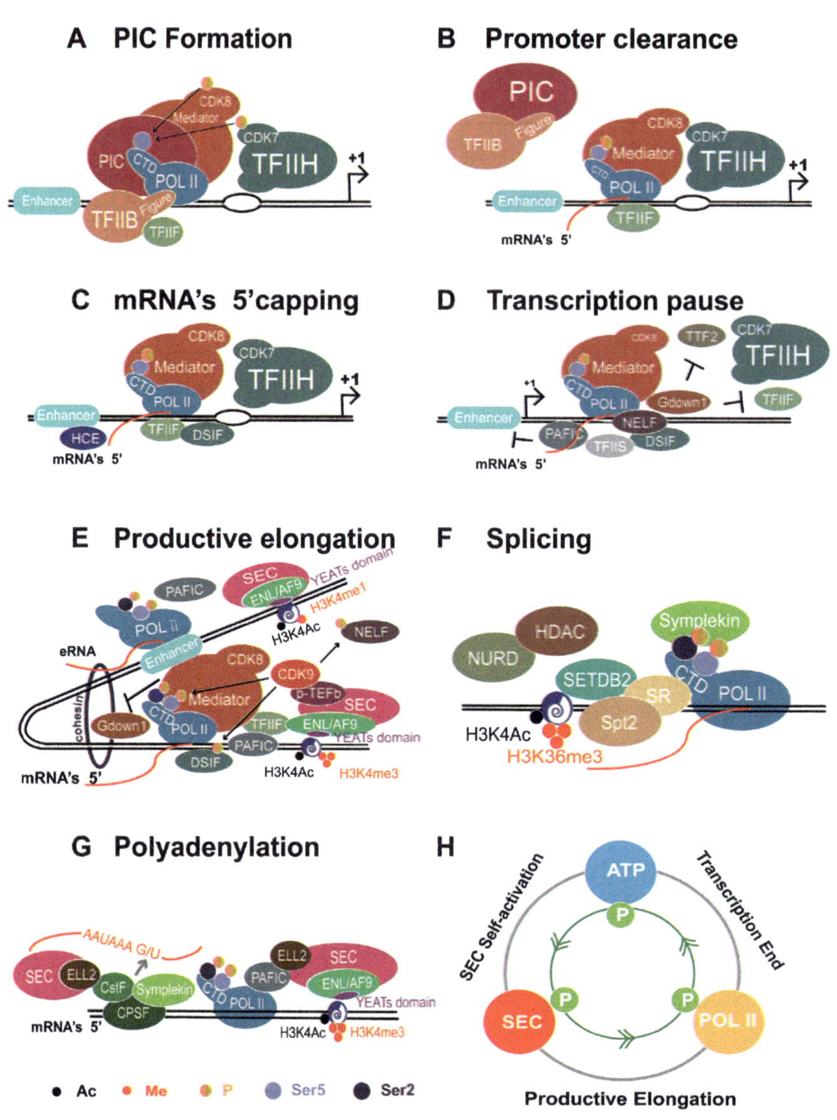

Fig. 3 Crosstalk between SEC and biological processes in vivo. (A) POL II recruits transcription factors to assemble the PIC at the active promoter. (B) Then PIC and TFIIB is ejected and promoter clearance starts. (C) HCE is recruited to POL II for capping the 5′ terminal of mRNAs. (D) Gdown1, DSIF and NELF are recruited to POL II for maintaining the pausing of POL II. (E) SEC is recruited to the POL II reversal pause state at the active enhancer. (F) Splicing factors such as SR, STDB2 and ST are recruited by NURD, HDAC to splice the newly generated mRNA. (G) SEC subunit ELL2 polyadenylates the 3′ end of mRNA by directly recruiting CstF, CPSF, and symbolekin or using PAF1C to specifically identify G/U sequences. (H) The cycle of phosphorylation is a process including self-activation via ATP to SEC, a leap from SEC to POL II Ser2, and a return from POL II to ATP via transcriptional termination.

transcription factors (GTFs) (including TFFIIF, TFIIH and TFIIB) and a variety of co-regulatory factors (including mediator) (Chen et al., 2018). This step requires Ser5 phosphorylation by subunits CDK7 of TFIIH (Chen et al., 2018; Compe & Egly, 2012) and CDK8 of mediator (Meinhart et al., 2005; Takahashi et al., 2020) and deep insertion of finger of TFIIB (Saunders, Core, & Lis, 2006). TFIIF can act as an adaptor to bring POL II into the PIC and thus play an active role in transcriptional elongation (DiMartino et al., 2000). PIC and TFIIB are then ejected, representing the start of promoter clearance (Saunders et al., 2006). Promoter clearance is the step that POL II must undergo to become an mRNA in eukaryotic transcription (Chen et al., 2018). When transcripts reach a certain level, phosphorylated Ser5 recruits and stimulates human capping enzyme (HCE) (Baumli, Hole, Wang, Noble, & Endicott, 2012; Rahl et al., 2010; Saunders et al., 2006; Zhou et al., 2012), which adds methylated guanine to the 5′ end of the RNA to increase RNA stability (Chen et al., 2018; Saunders et al., 2006). DRB sensitivity-inducing factor (DSIF) assists in recruiting HCE to facilitate transcriptional elongation (Gilmour & Lis, 1986; Yamada et al., 2006). In most eukaryotes, the vast majority of transcription factors exhibit proximal promoter pausing in the region after the TSS (nt: 20–120), as the pause maintains the active state of the gene (Chen et al., 2018; Jonkers & Lis, 2015; Yamaguchi et al., 1999). NELF prevents premature termination of the proximal promoter to stabilize paused POL II and causes DSIF to change from a positive elongation factor to a negative elongation factor (Core et al., 2012; Gilchrist et al., 2008; Henriques et al., 2013; Yamaguchi et al., 1999). TFIIH is uncoupled from the DNA double strand and the negative transcription factor Gdown1 is recruited to the pause site, maintaining transcriptional pause by preventing the recruitment of TFIIF and antagonizing transcription termination factor 2 (TTF2) (Cheng et al., 2012; DeLaney & Luse, 2016; Ishibashi et al., 2014; Mullen Davis, Guo, Price, & Luse, 2014; Tan, Conaway, & Conaway, 1995; Wu et al., 2012). PAF1C is recruited to prevent enhancer activity and the release of POL II from the proximal promoter pause. Retrograde POL II is cut off by THIIS and the retrograde site of POL II is rearranged (Chen et al., 2018).

2.3 SEC in productive elongation

To ensure the transcription continues, RNA needs to antagonize this state by recruiting transcription factors (Fish & Kane, 2002; Zhou et al., 2012). Co-activators SEC, BRD4 and Mediator are attracted, which may lead to enhancer-promoter interaction with non-coding RNA (NcRNA) and the activation of p-TEFb, allowing transcription into productive elongation (Chen et al., 2018). For SEC and BRD4, another viewpoint indicated that protein phosphatases are recruited by transcription factors for dephosphorylation of the T-loop in P-TEFb, blocking the reassociation of P-TEFb-Hexim1 onto the 7SK RNA, and promoting the release of P-TEFb and some other productive elongation release factors including DDX21 and SF2 (Chen et al., 2008; McNamara, McCann, Gudipaty, & D'Orso, 2013; Wang et al., 2008).

Productive elongation of POL II affects chromatin structure through nucleosome turnover and co-transcriptional modification of proteins and DNA (Chen et al., 2018; Lai & Pugh, 2017; Talbert & Henikoff, 2017; Venkatesh & Workman, 2015). When the synthesized length of RNA achieves 20–40 nt, PAF1C is depleted, and the positive elongation factor SEC is recruited to the promoter (Chen et al., 2018; Ji et al., 2013). SEC is localized in the acetylated H3 nucleosomes through the YEAT domain of AF9/ENL, and subsequently the CTD of POL II, NELF and DSIF are phosphorylated by CDK9 of SEC (Erb et al., 2017). The process that SEC phosphorylates Ser2 on the CTD is considered as a marker of RNA transcription into productive elongation. Spt5 of DSIF is also phosphorylated by SEC, switching DSIF from a negative elongation factor to a positive one. DSIF interacts with the elongated POL II, enhances the closed conformation of the POL II clamp, and facilitates template DNA through the central cleft of the polymerase to improve productive elongation (Doamekpor, Sanchez, Schwer, Shuman, & Lima, 2014; Klein et al., 2011; Shetty et al., 2017). The subunit E of NELF (NELF-E) is phosphorylated by SEC and is shed from the promoter. Mediator inhibits POL II pause-dependent stabilization of Gdown1, resulting in detachment of Gdown1, increased TFIIF recruitment and continued RNA elongation (Chen et al., 2018; DeLaney & Luse, 2016; Mullen Davis et al., 2014; Tan et al., 1995; Wu et al., 2012).

Transcription elongation can be directly regulated by the interaction between promoter and active enhancer. In this phase, transcriptional efficiency is mainly dependent on the association of active enhancers and transcriptional pausing sites linked by cohesin (Fay et al., 2011; Izumi et al., 2015; Kagey et al., 2010; Schaaf et al., 2013). The depletion of cohesin affects the release of paused POL II, but not the establishment of pauses (Schaaf et al., 2013). The phenotypic similarities between loss of adhesin function and enhanced SEC function increase the activity of adhesin-dependent enhancer and regulation of SEC on transcription elongation. Genes close to less active enhancers typically have higher pausing rates (Fay et al., 2011). SEC can facilitate enhancer activation by recognizing acetylated nucleosomes close to active enhancers (Chen et al., 2018). Active enhancers are rich in H3K4me1 and H3K27ac, which are mainly catalyzed by MLL3 (also known as KMT2C), MLL4 (KMT2D), p300 and CREB-binding protein (CBP), respectively (Chen et al., 2018; Herz et al., 2012; Hu et al., 2013). P300, CBP, and H3 acetyltransferases are recruited to enhancers by transcription factors and catalyze histone acetylation at enhancers and promoters (Creyghton et al., 2010; Shen et al., 2012; Shilatifard, 2012). Enhancer RNAs (eRNAs) can bind to transcription factors and cofactors to maintain enhancer dynamics (Bradner, Hnisz, & Young, 2017). eRNAs engage with promoters in order to initiate the release of stalled POL II and facilitate NELF clearance from promoters. This creates a clear connection between enhancer activation and transcription elongation (Bradner et al., 2017). Moreover, the mediator complex subunit MED26 interaction with EAF1/EAF2 leads to MED26 recruiting SECs to target genes, indicating SEC and MED26 may synergistically activate genes of super-enhancers (Ji et al., 2014; Wang et al., 2013).

2.4 SEC in splicing and polyadenylation

Prior to polyadenylation, the nascent RNA needs to be sheared. Histone lysine *N*-methyltransferase SETD2 binds to the CTD of phosphorylated POL II, and precipitates H3K36me3 in the gene body, affecting RNA splicing and preventing cryptic transcription (Ebmeier et al., 2017; Mozzetta et al., 2015). This regulation is mediated by a variety of transcription factors that recognize H3K36me3, such as histone deacetylases (HDACs) and nucleosome remodels (NURDs) (Carrozza et al., 2005; Smolle et al., 2012). Phosphorylation of Ser2 is shown to interact with splicing factors SR and Spt6, mRNA on the CTD is spliced by splicing factors (Ji et al., 2013). The splicing factors not only use the POL II elongation complex to successfully delete introns from pre-mRNA, but also promote elongation, ensuring that pre-mRNA is not produced until the processing machinery is correctly formed and localized (Mozzetta et al., 2015; Zhou et al., 2012).

3′ end processing mainly depends on polyadenylation factors such as CstF, CPSF, and symbolekin, which are recruited to POL II by the synergistic effects of phosphorylated Ser2 on the CTD, the ELL2 subunit of the

SEC as well as PAF1C (Chen et al., 2018). These cleavage factors track POL II. Once the cleaved poly (A) signal (G/U-rich sequence after AAUAAA) appears in the nascent mRNA, these cleavage factors specifically recognize this sequence, inducing efficient polyadenylation (Zhou et al., 2012). In addition, ELL2 contributes to 3′ end processing of mRNA and POL II elongation by linking POL II and the gene enhancer CstF encoding the immunoglobulin heavy chain complex in plasma cells (Martincic, Alkan, Cheatle, Borghesi, & Milcarek, 2009; Nagaike et al., 2011; Zhou et al., 2012). ELL2 accelerates the use of a weak promoter-proximal secretory-specific poly (A) site, and enhances exon skipping of the non-consensus splicing signal involving in direct competition at the POL II site (Martincic et al., 2009; Nagaike et al., 2011; Zhou et al., 2012).

2.5 Phosphorylation and transcriptional elongation

The process of POL II transcriptional elongation cannot proceed smoothly without the involvement of phosphorylation. Phosphorylated CTD increases the affinity of the elongation complex to HCE, and serves as a scaffold for the serine/arginine-rich splicing factor and polyadenylation machinery (Meinhart et al., 2005; Rahl et al., 2010). The cycle of CDK9 phosphorylation includes three steps (Fig. 3H): in the first phase, T186 autophosphorylation occurs on the conserved T-loop of CDK9, and SEC is thus self-activated (Franco et al., 2018). In the phase of productive elongation, CDK9 phosphorylates Ser2 on the CTD of POL II to antagonize transcriptional pausing. In the final phase, POL II is dephosphorylated, and the phosphate group is returned to ATP, transcription ends.

3. SEC dysregulation and cancer

As described above, the SEC complex has been shown to be a tissue-specific transcription factor in the physiological state, which regulates the transcriptional elongation checkpoint control (TECC) phase (Core & Adelman, 2019). SEC allows rapid activation of genes in response to environmental signals and appropriate transitions in chromatin state, or prevents epigenetic repression of genes (Luo et al., 2012). This stage is crucial in the development of the human body, and dysregulation of this stage can lead to the development of cancers.

In cancer, the SEC complex is recruited to mutated oncogenes involved in the rapid transcription of oncogenes. For example, AFF4 can promote melanoma cell invasion and migration by mediating epithelial-to-mesenchymal filtration (EMT) (Hu et al., 2021), and AFF4 protein expression is closely related to the prognosis of melanoma patients (Deng et al., 2018). SEC-mediated signaling promotes the clonogenic potential of tumor cells and self-renewal of tumor stem cells (Yu, Liu, Yang, & Zhou, 2018). The self-renewal and differentiation of cancer stem cells (CSC) is controlled by multiple critical and highly regulated transcription factors (TFs) (Huang et al., 2020). The biological phenotypes of CSCs are also regulated by the integrated transcriptional, post-transcriptional, metabolic, and epigenetic regulatory networks. CSCs contribute to tumor progression, and disease recurrence through their sustained proliferation, invasion into normal tissue, promotion of angiogenesis, evasion of the immune system, and resistance to conventional anticancer therapies (Yang et al., 2020). For example, in triple-negative breast cancer, c-MYC as a key transcriptional factors of CSCs transcriptionally regulates MTDH expression and then stimulates Twist1 signaling to maintain an EMT signature and CSC-like state (Luo et al., 2018). A marker in many CSCs of aldehyde dehydrogenase (ALDH) (Ginestier et al., 2007), eliminates oxidative stress and enhances resistance to chemotherapeutic drugs, such as oxazolidine, taxanes, and platinum drugs (Yang et al., 2020). Given the roles of SEC complex in regulating oncogenic transcriptions, its inhibitory modulation could be an attractive strategy in alleviating CSC differentiation and self-renewal associated acquired tumor resistance derived from traditional chemotherapy, target therapy or even immunotherapy.

3.1 Role of MLL and DOTL1 methylation in cancer

In human, the MLL gene is involved in a large number of chromosomal translocations by fusing the N terminus of MLL to several proteins with minimal sequence similarity (Meyer et al., 2013), resulting in the production of fusion proteins ultimately causing the development of MLL translocation leukemia, which accounts for 10% of the incidence of acute leukemia (AML) and acute lymphoblastic leukemia (T-AML) (Kuntimaddi et al., 2015). The SEC subunit AFF1, AFF4, AF9, ENL, and ELL families are common translocated proteins in MLL, suggesting that a major mechanism of leukemogenesis may be the dysregulation of the transcriptional elongation phase (Lin et al., 2010).

Wild-type MLL with histone methyltransferase activity is able to methylate H3K4, and recruit CBP/P300 to regulate the initiation of transcription factors (Fig. 4A) (Liang et al., 2017). However, in the leukemia with MLL and SEC fusion's protein, the regulation of transcriptional elongation is disrupted. When the N-terminal of MLL fuses with the C-terminal

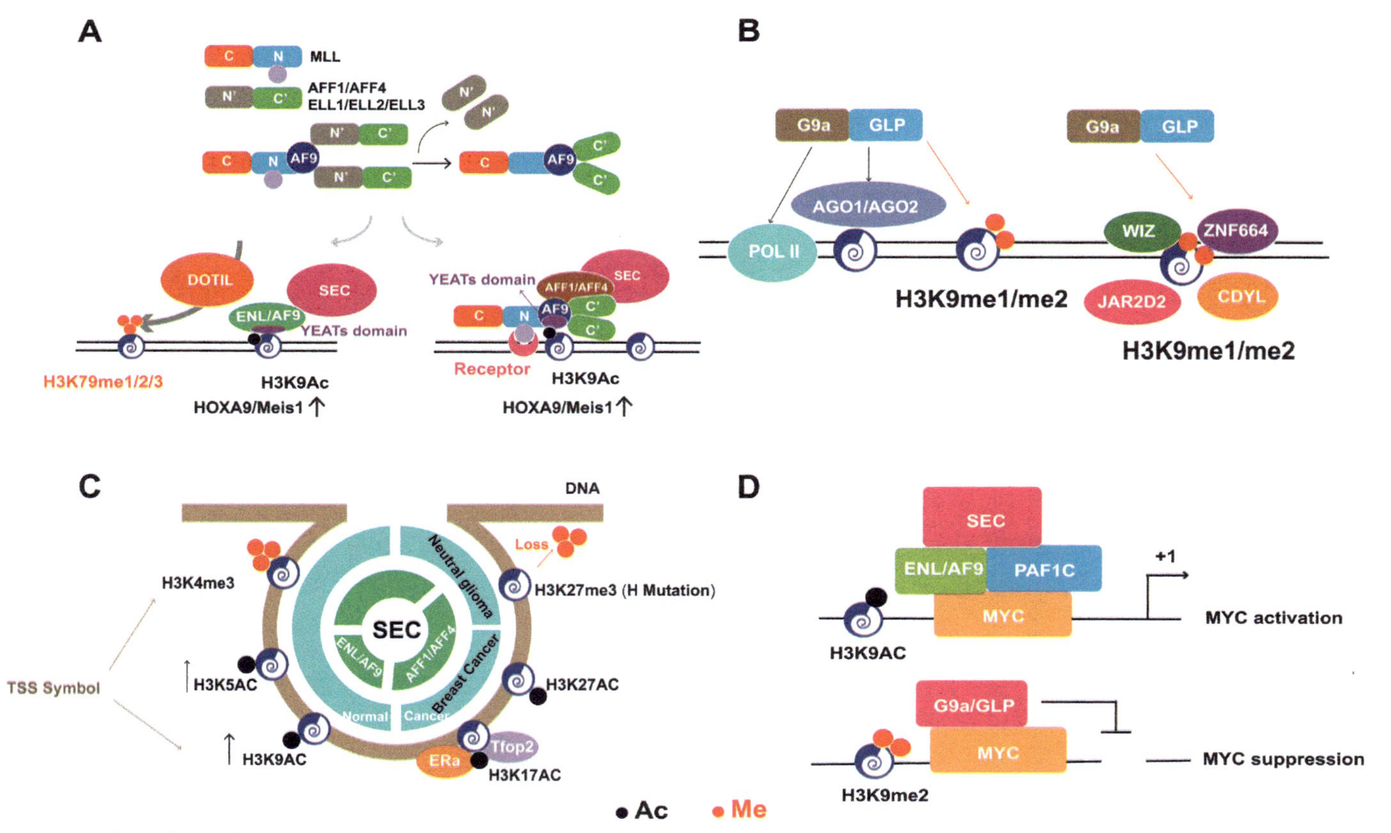

Fig. 4 See figure legend on opposite page.

of AF9, aberrant transcription is produced by co-recruiting the N-terminal structural domain of MLL, Menin and other proteins, instead of only recruiting SEC (Kuntimaddi et al., 2015). On the other hand, these fusion proteins may also be recruited to the normal target genes of MLL, leading to premature activation of transcriptional elongation and overexpression of the MLL target genes HOXA9 and Meis1, resulting in the development of leukemia (Smith, Lin, & Shilatifard, 2011).

Another epigenetically regulated fusion protein involved in MLL is DOT1L, the only known methyltransferase of H3K79, which can methylate H3K79 responsible for the repair of cellular DNA damage and cell cycle (Cao et al., 2020; Kuntimaddi et al., 2015). DOT1L is involved in the differentiation and development of stem cells and tissue systems, and abnormal expression of DOT1L is also closely associated with tumorigenesis and prognosis (Vatapalli et al., 2020). Erb et al. show that the ENL protein is a key factor affecting the viability of MLL-rearranged cells in leukemia (Erb et al., 2017). ENL contains a YEATS structural domain that recognizes a specific acetyl group (Ac) on histone H3 (Li et al., 2016), which binds to unfused SEC/DOT1L proteins or fused SEC/DOT1L protein, and then in cells the SEC complex forms a fusion protein with the MLL gene (Andrews et al., 2016). The major cause of hematological tumors is a chromosomal translocation that produces a MLL fusion protein which binds to SEC

Fig. 4 SEC dysregulation and cancer. (A) Role of SEC and DOT1L methylation in MLL. In MLL translocation leukemia, the fusion of the N-terminal chromatin-targeting portion of MLL and MLL ligand (AFF1/AFF4, ELL1/2/3) results in partial loss of the N-terminus of the MLL ligand, and retention of the C-terminal structural domain in AF9 named the ANC1 homolog structure (AHD). There are two pathways for AHD to upregulate the HOXA9 gene and the meis1 gene. One is the recruitment of DOT1L, where DOTIL is specifically deposited on H3K79me1/me2/me3, leading to gene upregulation; the other is the direct recruitment of SEC by AHD, where the YEATs domain recognizes H3K9Ac, inducing MLL binding to receptors on DNA and gene upregulation (Kaithoju, 2014; Kuntimaddi et al., 2015; Leach et al., 2013). (B) Role of G9a methylation in SEC dysregulation. G9a-GLP needs to be recruited to the promoter by specific transcription factors to function. The examples of transcription factors include the IFP widely spaced zinc finger protein (WIZ), ZNF664, JAR2D2, and chromatin Y-like protein (CDYL) (Mozzetta et al., 2015). (C) Role of acetylation in SEC dysregulation. Under normal conditions, H3K4me3 acts as an active promoter, recruiting SEC to H3K9AC and H3K5AC. In breast cancer, SEC is recruited to H3K17AC and H3K27AC for rapid transcription (Dahl et al., 2020; Gao, Chen, et al., 2020). (D) MYC and SEC in cancer development SEC recruited by MYC is deposited on H3K9AC, which rapidly induced MYC gene transcriptional expression. The expression of MYC is blocked when G9a-GLP was deposited on MYC promoter H3K9me2 (Tu et al., 2018).

complex or DOT1L protein triggering abnormal transcription activation or histone methylation associated gene expression (Liang et al., 2017).

3.2 Role of G9a methylation in SEC dysregulation

G9a is a histone methylation transferase with a classical SET domain that anchors protein repeats and participates in protein-protein interaction. G9a is mainly responsible for methylation of histone H3K9 and H3K27 in euchromatin (Poulard, Noureddine, Pruvost, & Le Romancer, 2021). G9a regulates gene transcription by two different mechanisms: (i) G9a can promote histone methylation or DNA methylation in the promoter region of genes, thereby inhibiting gene transcription and (ii) G9a can act as a scaffold protein to recruit transcriptional activators and thus activate gene transcription (Haebe, Bergin, Sandouka, & Benoit, 2021). AGO1 and AGO2 are recruited to chromatin by non-coding RNAs, and then co-recruit SUV39H1/H2 and SET domain bifurcation 1 (SETDB1) (Fig. 4B), which leads to H3K9me3-labeled deposition and subsequently heterologic recruitment of H3K9me3 reader (Mozzetta et al., 2015). The inhibiting environment of chromatin slows down POL II progression, promotes recruitment of the spliceosome, and regulates splicing (Mozzetta et al., 2015).

G9a plays an important role in regulating the autophagic response and cell differentiation (Tachibana et al., 2002), and G9a upregulation is relevant in a variety of human cancers, including bladder tumors and acute myeloid leukemia (Gouda, Zidane, Abdelhady, & Hassan, 2022; Segovia et al., 2019). In colon cancer, Forehead box O (FOXO) family members are multifunctional transcription factors that regulate the transcription of target genes which is methylated by G9a (Salih & Brunet, 2008). As a result, the interaction between FOXO1 and a specific E3 ligase for FOXO1 promotes the development of colon cancer (Chae et al., 2019). Moreover, G9a drives cancer progression through molecular networks maintaining self-renewal and tumorigenicity, and knock out of G9a can slow down AML development (Lehnertz et al., 2014; Tachibana et al., 2002).

3.3 Role of methylation and acetylation in SEC dysregulation

SEC has a preference for acetylation due to an extended acetyl binding profile of the YEATs domain (Schulze, Wang, & Kobor, 2010). H3K9ac recognition by AF9 is essential for the chromatin recruitment of DOT1L and the subsequent deposition of H3K79 methylation on target genes promotes active transcription (Li et al., 2014). Some mutations occurring in the YEATS domain have been found in Wilms tumors, in which the mutations weaken the binding of YEATS domain to H3K9ac or H3K27ac, resulting in increased MYC gene expression and HOX dysregulation (Zhao, Li, Xiong, Chen, & Li, 2017). In ER-positive breast cancer cells, depletion of AFF4 by CRISPR-Cas9 reduces the recruitment of transcriptional elongation factors and RNA polymerase II to the ESR1 transcription start sites, and hence inhibits the growth of ER-positive breast cancer cells (Gao, Chen, et al., 2020). Moreover, mutations in the H3 gene are drivers of diffuse midline gliomas, and these recurrent oncogenic histones result in the total loss of inhibitory H3K27me3 residues and extensive epigenetic dysregulation (Fig. 4C) (Dahl et al., 2020).

3.4 MYC and SEC in cancer

MYC is an oncogenic driver that regulates transcriptional activation and repression, which can promote cell proliferation, increase the sensitivity of cells to apoptosis, and prevent differentiation (Fig. 4D) (Kanazawa, Soucek, Evan, Okamoto, & Peterlin, 2003). The oncogenic mechanism of MYC is to make mRNA cleavage uncontrolled through gaining genetic copy number, abnormally activating upstream signaling and altering protein stability (Zlotorynski, 2018). The burden of appropriate mRNA processing by the core spliceosome is increased by MYC hyperactivation, which causes transcriptional amplification and increases mRNA synthesis, suggesting that mRNA splicing represents a therapeutic gap in MYC-driven cancer (Liang et al., 2018). The co-localization ratio of MYC and SEC subunits on chromatin is higher in cell lines with high MYC expression than those with low MYC expression (Zlotorynski, 2018). Deletion of MYC reduces chromatin occupancy of SEC subunits and decreases elongation of POL II, suggesting that SEC-dependent transcription elongation is an effector of MYC (Liang et al., 2018). Cells can compensate for the loss of CDK9 by maintaining the production of many mRNAs through MYC, including MYC itself, and therefore treatment of cancer requires inhibition of both CDK9 and MYC activation (Lu et al., 2015). As a critical TF of CSCs, c-MYC is perhaps the most frequently dysregulated protein in human tumorigenesis among these MYC members (c-MYC, N-MYC, and L-MYC), serving as a promising therapeutic target for cancer treatments (Kress, Sabò, & Amati, 2015). C-MYC amplification in patient-derived CSCs generates sensitivity to PARP inhibition through c-MYC-mediated transcriptional repression of CDK18, CCNE2, and CDKN1A (Ning et al., 2019).

In human cancer cells, G9a and associated proteins interact with MYC to catalyze gene restrictive demethylation of histone H3K9me2 (Tu et al., 2018). The oncogenic function of MYC is influenced by both SEC-mediated gene activation and G9a-mediated gene expression. G9a and GLP can inhibit MYC expression by deposition of H3K9me3 onto MYC (Liang et al., 2018; Zlotorynski, 2018). MYC is also found to induce the gene encoding G9a, and its depletion leads to reduction of G9a and H3K9me2 levels (Liang et al., 2018).

In acute myeloid leukemia, SEC and PAF1C complex are recruited to the MYC (or heat shock protein 70 (HSP70)) locus by BRD4, and SEC catalyzes MYC to rapidly transcribe RNA to produce cancer cell proteins (Luo et al., 2012; Mandal, Becker, & Strebhardt, 2021). Hepatocellular carcinoma cells exhibit MYC overexpression, maintain malignancy by increasing P-TEFb activity of specific promoters responsible for malignant cell growth, and alter the ratio of CDK9–42/CDK9–55 (Huang et al., 2014). Sengupta et al. describes that the transcriptional overexpression of MYC gene is also involved in the upstream molecular events of estrogen-independent proliferating breast cancer cells, and CDK9 was identified as a potential drug target for the treatment of endocrine-resistant breast cancer (Sengupta, Biarnes, & Jordan, 2014).

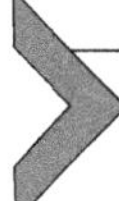

4. Inhibitors targeting SEC or regulating SEC components

Many small molecular inhibitors either directly targeting SEC subunits or its associated proteins have been reported. Here we summarize some of the most noticeable or advanced inhibitors.

4.1 CDK9 inhibitors

AZD-4573 (Originator: AstraZeneca) with the amidopyridine scaffold (Fig. 5A) is a potent CDK9 inhibitor ($IC_{50} < 0.004\,\mu M$) with high selectivity against other CDK family kinases (>5.8–10-fold) (Barlaam et al., 2020). AZD-4573 induces cell death in hematologic cancer cells by decreasing pSer2-RNAP2 and Mcl-1, and activating caspase 3/7 and cell apoptosis (Barlaam et al., 2020). The crystal structure of AZD-4573 and CDK9/ Cyclin T1 (PDB ID: 6Z45 (Barlaam et al., 2020)) showed that AZD-4573 made a bidentate hydrogen-bond with Cys106 in the ATP binding site of CDK9 (Fig. 5A) but has no direct interactions with Cyclin T1. Phase I/II clinical trials involving AZD-4573 for hematologic malignancies have been

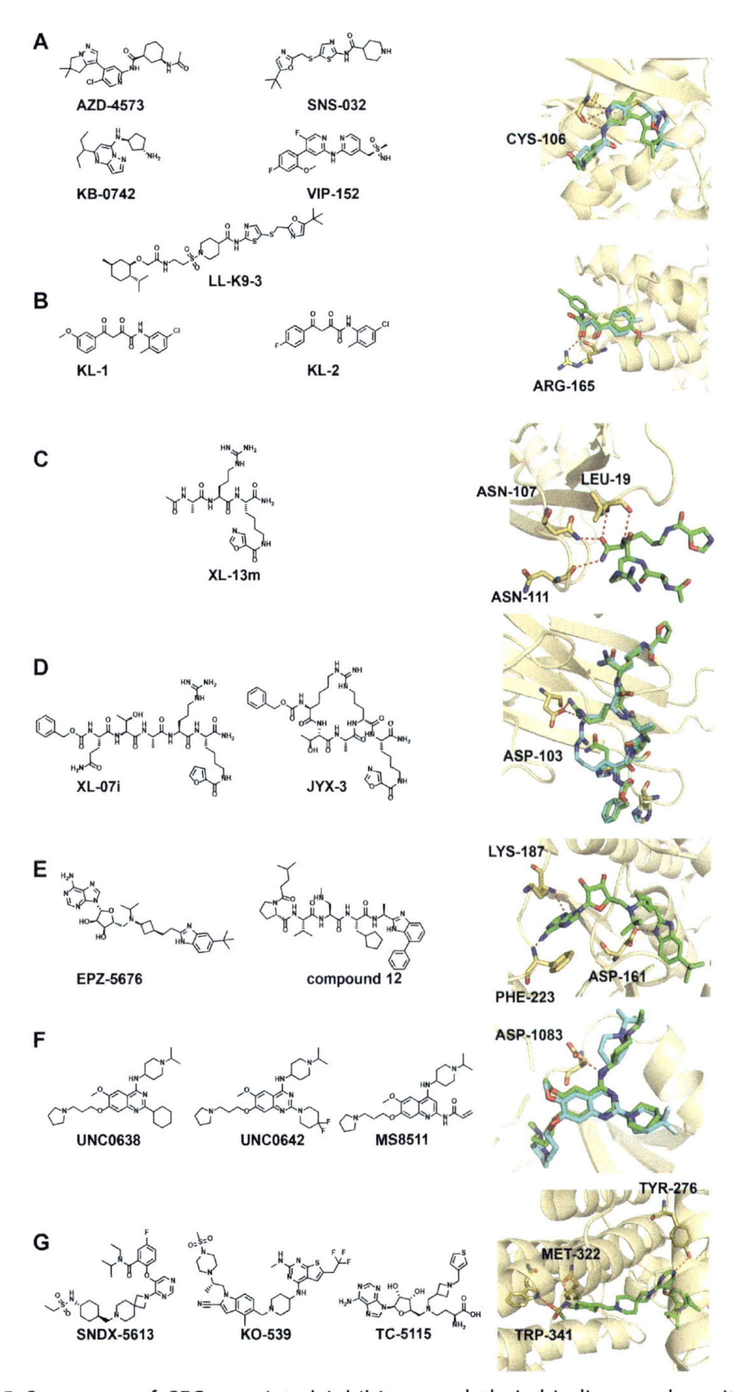

Fig. 5 Structures of SEC-associated inhibitors and their binding modes with the corresponding targets. (A) CDK9 inhibitors: AZD-4573 (*green*, PDB ID: 6Z45), SNS-032 (*cyan*, docking mode), LL-K9–3, KB-0742, and VIP152. (B) CycT1 inhibitors: KL-1

(Continued)

conducted (NCT04630756 and NCT03263637). SNS-032 (Originator: Bristol-Myers Squibb) with the amidothiazole scaffold is a selective inhibitor of CDKs 2, 7, and 9, showing similar binding mode of AZD-4573 to CDK9 by molecular docking analysis (Fig. 5A). However, the phase I results of SNS-032 were not satisfactory due to lack of clinical efficacy and adverse effects (Heath, Bible, Martell, Adelman, & Lorusso, 2008; Tong et al., 2010). Moreover, multiple selective CDK9 inhibitors, such as KB-0742 and VIP152, are being evaluated in the clinic (NCT04718675, NCT02635672, and NCT04978779) (Day et al., 2022; Frigault et al., 2022). Recently, the first small-molecule CDK9 degrader (LL-K9–3) was reported. It employed a hydrophobic tagging for SNS032 to promote degradation of CDK9-Cyc T1 heterodimer (Li et al., 2022). In comparison to the parental SNS032, LL-K9–3 demonstrated enhanced anti-proliferative and pro-apoptotic activities, as well as inhibitory effects on a number of AR-targeted genes.

4.2 CycT1-AFF4 inhibitors

The AFF4 pentapeptide LFAEP (leucine-phenylalanine-alanine-glutamic acid-proline)-mimicking inhibitors KL-1 and KL-2 against CycT1 were initially identified by in silico screening (PDB ID: 4IMY, Fig. 5B) (Liang et al., 2018). KL-1 and KL-2 showed dose-dependent inhibitory effects on the CycT1-AFF4 interaction with K_i of 3.48 and 1.50 μM, respectively, whose mechanisms are distinct from CDK9 and BET domain inhibitors. The inhibition of KL-1 resulted in abnormal release of POL II from promoter-proximal pause sites and slowed POL II elongation rate, attenuated heat-shock response (Katagi et al., 2021). Furthermore, KL-1 inhibition reduced expression of MYC and RNA splicing-related genes, delayed tumor progression in MYC-driven models of DIPG and breast cancer (Katagi et al., 2021; Liang et al., 2018), indicating that CycT1 can be used as a potential target for the therapy of MYC-induced cancer. Given the low activity of KL-1 and KL-2, optimization is required to improve their activities in order to further understand their anti-tumor activities in vivo, and importantly advancement in clinical setting.

Fig. 5—Cont'd (*green*, docking mode) and KL-2 (*cyan*, docking mode). (C) ENL inhibitor: XL-13 m (*green*, docking mode). (D) AF9 inhibitors: XL-07i (*green*, PDB ID: 5YYF) and JYX-3 (*cyan*, PDB ID: 6L5Z). (E) DOT1L inhibitor: EPZ-5676 (*green*, PDB ID: 4HRA) and compound 12. (F) G9a inhibitors: UNC0638 (*green*, PDB ID: 3RJW), UNC0642 (*cyan*, docking mode) and MS8511. (G) Menin/MLL inhibitors: SNDX-5613 (*green*, PDB ID: 6PKC), KO-539 and TC-5115. Key residues in targets and hydrogen bonds with the inhibitors are shown as *golden sticks* and *red dashed lines*, respectively.

4.3 ENL/AF9 inhibitors

Guided by a specific π-π-π stacking interaction of the YEATS-H3K9cr complex, a linear tripeptide XL-13m (Fig. 5C) was designed to show strong inhibition for YEATS-Kcr interaction of ENL but not AF9. The docking analysis showed that XL-13m formed a hydrogen-bond network with key residues Leu19, Asn107, and Asn111 in ENL (PDB ID: 5J9S) (Wan et al., 2017). XL-13m synergizes with DOT1L inhibitorEPZ-5676 to reduce *MYC* transcription in MLL-rearranged (MLL-r) acute leukemia (Li et al., 2018). In contrast, a longer linear pentapeptide XL-07i (Fig. 5D) showed a higher potency to AF9 than ENL.XL-07i targets both the acyl lysine-binding pocket and its proximal site which can differentiate AF9 YEATS (H107 and H111, PDB ID: 5YYF) from ENL YEATS (N107 and N111) (Jiang et al., 2020; Li et al., 2018). Furthermore, an cyclized analogue of XL-07i (JYX-3, Fig. 5D) was designed to rigidify the inhibitor into the AF9 YEATS-favored conformation (PDB ID: 6L5Z), resulting in producing higher selectivity than XL-07i for AF9 YEATS over ENL YEATS (38-fold vs 3-fold) (Jiang et al., 2020).

4.4 DOT1L inhibitors

EPZ-5676 (Pinometostat, Fig. 5E) developed by Epizyme is an aminonucleoside-based DOT1L inhibitor that has completed Phase I/II clinical studies for the treatment of refractory/relapsed patients with MLL-r leukemia (NCT03701295, NCT02141828, and NCT01684150). EPZ-5676 showed inhibitory activity against DOT1L by occupying the *S*-adenosyl methionine (SAM) binding site, and opened a hydrophobic pocket not seen in the DOT1L-SAM complex, forming a hydrogen bond network with the key residues of DOT1L (PDB ID: 4HRA, Fig. 5E) (Daigle et al., 2013). EPZ-5676 selectively killed the cell lines bearing MLL-AF9 and MLL-ENL fusions with IC_{50} of 0.028 and 0.17 μM, respectively. However, more than 3 weeks of continued treatment of EPZ-5676 induced the resistance in MLL-ENL fusion (KOPN-8) cells due to increased expression of the drug efflux transporter ABCB1, which is different from the resistance in MLL-AF9 fusion (NOMO-1) cells independent of ABCB1 expression (Campbell et al., 2017). Moreover, EPZ-5676 dramatically suppressed the development of tumor in TNBC patient-derived xenograft models, suggesting the potential of EPZ-5676 against stem cell-enriched TNBC (Kurani et al., 2022). Yuan et al. designed a DOT1L peptide mimetic showing comparable anticancer activity with EPZ5676 by inhibiting the protein-protein interactions between DOT1L and MLL fusion proteins, which suggested a new strategy for the treatment of MLL-r leukemia (Yuan et al., 2022).

4.5 G9a inhibitors

The development of G9a inhibitor EZM8266 is discontinued by Epizyme due to a preclinical toxicology finding (Cortellis, n.d.). Based on the template of 2,4-diamino-6,7-dimethoxyquinazoline, a chemical probe (UNC0638, Fig. 5F) was developed to selectively inhibit G9a and GLP activity in cells. This inhibitor is non-competitive with SAM and competitive with the peptide substrate (PDB ID: 3RJW) (Vedadi et al., 2011). However, UNC0638 has high clearance, short half-life, and low exposure levels. 4,4-Difluoropiperidine substitution (UNC0642, Fig. 5F) for the cyclohexane of UNC0638 improved inhibitory activities against G9a and in vivo PK properties compared with UNC0638 (Liu et al., 2013). Moreover, UNC0642 treatment decreased H3K9me1/2 (two G9a-regulated marks) and MYC binding at MYC-repressed gene promoters, but not affected MYC-bound activated genes (Tu et al., 2018). Furthermore, Park et al. designed a first-in-class G9a covalent inhibitor (MS8511, Fig. 5F) using an acrylamide group substituted for the cyclohexane in UNC0638, which made a covalent bond with Cys1098 in the substrate binding site of G9a (Park et al., 2022). In cellular experiments, the irreversible inhibitor has shown greater potency than UNC0642 in lowering the level of H3K9me2 and enhancing anti-proliferation ability. It also demonstrated kinetic advantage by covalently binding to G9a rather than GLP.

4.6 Menin-MLL inhibitors

Menin is a scaffold protein that tethers MLL1 to chromatin, and the wild type menin-MLL1 complex is one key oncogenic drivers of MEIS1, FLT3, and HOXA9 for promoting self-renewal of mutant leukemic blasts (Fiskus et al., 2021). KO-539 (Ziftomenib, Fig. 5G) is a small molecule inhibitor interfering with the Menin-MLL interaction for the treatment of patients with relapsed or refractory AML (NCT04067336) in clinical I/IIA trials. Another inhibitor is SNDX-5613 (revumenib fumarate, Fig. 5G) developed by SyndaxPharmaceuticals. SNDX-5613 monotherapy demonstrated high objective remissions in the phase I/II clinical trial for patients with previously treated MLL-r AML (Fiskus et al., 2021). SNDX-5613 (PDB ID: 6PKC) (Krivtsov et al., 2019) bridged the MLL binding pocket to establish two H-bonds with Trp341 as well as Tyr276 (Fig. 5G), and showed downregulation of multiple MLL-targeted genes responsible for MLL-r leukemia proliferation. Chern et al. designed a SAM-analogue (TC-5115) with IC_{50} of 15.2 nM for MLL1 by locking the MLL SET domain in an open and inactive conformation (Chern et al., 2020), which may develop a new therapy for treating MLL-r leukemia.

5. Future perspective

This review first systematically summarizes the structure of SEC and its subunits, and the role of SEC in normal transcription and the coordination of SEC dysregulation with various types of transcription factors mainly in multiple cancers, which provides a theoretical reference for subsequent study of each subunit. However, the coincidence of inhibition of DOT1L and SEC is still controversial (Cao et al., 2020; Leach et al., 2013), and the complete mechanism needs to be studies through more experiments. Understanding the association of the SEC complex and its component in cancer and other disease progression is an involving process. The advancement will be sure to provide further insights on targeting SEC for treatment cancers such as AML and DISP. Importantly, significant progression has been made in developing inhibitors targeting SEC complex or its critical components including CDK9, G9a and the other associated proteins. CDK9 inhibitors have been in clinical studies for cancer and sickle cell indications. Although there are no clinical results for those inhibitors, it is predicted that the clinical benefits would rely on patient stratification in selecting right population based on precision biomarker in order to maximize the clinical outcomes.

Cancer stem cells (CSCs) have been a key area of research in cancer mechanisms since their discovery in 1994 (Yang et al., 2020). The high division potential and low differentiation properties of CSCs make them be associated with recurrence, metastasis, heterogeneity, multidrug resistance, and radiation resistance in a variety of cancers (Huang et al., 2020). Recently SEC has been found to be closely related to cancer stem cells. For example, Liu et al. demonstrated that SEC acted as an intrinsic amplifier within tumor stem cells to regulate cell fate in Drosophila neuroblastoma cells (Liu et al., 2017). SEC is highly expressed in neuroblastoma cells, and promotes self-renewal through physical binding to the Notch transcriptional activation complex enhancing HES (hair and E(SPL)) transcription. HES in turn upregulates SEC activity, forming an unexpected self-reinforcing feedback loop. Gao et al. demonstrated that the SEC subunit AFF4 exerts oncogenic effects in bladder cancer stem cells (BCSCs) by influencing downstream oncogene transcription through the mutually reinforced modifications of

METTL3-AFF4-SOX2/MYC axis and m6A (Gao et al., 2020). This suggests that SEC, as a highly active target in cancer stem cells, may be involved in the regulation of chemoresistance and cellular heterogeneity in cancer stem cells, which may bring more novel insights into the subsequent development of SEC inhibitors.

With the rapid advancement of artificial intelligent technologies, the development of more potent and specific SEC inhibitors has become a reality through de novo searching and designing for inhibitors by targeting complex machinery (Chen, Min, Parthasarathy, & Ning, 2021; Jayatunga, Xie, Ruder, Schulze, & Meier, 2022). For example, an algorithm called molecular contrastive learning of representations via graph neural networks built a self-supervised learning framework that leverages large unlabeled data for efficient molecular property prediction and drug discovery (Wang, Wang, Cao, & Barati Farimani, 2022). A geometric-based deep learning algorithm was developed to predict the binding conformations of ligands and targets by learning a statistical potential derived from the distance likelihood of the ligand-target pair coupled with global optimization algorithms, which performed similarly or better than well-established docking scoring functions (Méndez-Lucio, Ahmad, del Rio-Chanona, & Wegner, 2021). More recently, one deep docking protocol significantly accelerated the exploration of beyond a billion molecules by the structure-based virtual screening via random sampling of the chemical library, model training and inference, and the residual docking (Gentile et al., 2022). These AI methods can be utilized to support the ligand-based and structure-based drug design targeting SEC or regulating SEC components. Moreover, the development of constrained peptides based on protein–protein interactions is a hot research field, which shows significant benefits including enhanced affinity, stability and cellular penetration (Kelly et al., 2021; Wang et al., 2021). In particular, the undruggable protein–protein interaction interface now can be targeted for therapeutic intervention. For example, we can develop peptidomimetic inhibitors mimicking key interactions between CycT1-AFF4 or AF9-ENL, or other interfaces that could better block CDK9 phosphorylation activities to limit the transcription elongation activities in disease setting. Such novel, potentially allosteric and selective inhibitors would be valuable as chemical probes for further exploring the disease biology of SEC complex.

In conclusion, here we have summarized the structure and function of the SEC complex in transcription elongation. SEC plays important roles in increasing the synthesis of RNA polymerase II (POL II) and regulating many normal human genes to stimulate RNA elongation in normal physiological conditions. However, the dysfunction of SEC complex is associated with oncogenic transcription activation such as C-Myc which is commonly found in cancer development and resistance derived from chemotherapy or target therapy. We highlighted the progress in developing therapeutics inhibitors targeting key components of SEC, extended partners, or SEC complex itself. Among them, several inhibitors of CDK9, Dot1L and MLL-Menin have been advanced to clinical trials. Such inhibitors could become important therapies in treating cancers related to resistance associated with oncogenic transcriptional upregulation from chemotherapy or target therapy.

Acknowledgments

This work is supported by the National Natural Science Foundation of China (21877095 and 81603024), Natural Science Foundation of Shandong Province (ZR2022MH291), and Taishan Scholar Project.

Conflict of interest

The authors declare no potential conflicts of interest.

References

Ahmed, O., Affar, M., Masclef, L., Echbicheb, M., Gushul-Leclaire, M., Estavoyer, B., et al. (2022). Immunoprecipitation and Western blot-based detection of protein O-GlcNAcylation in cells. *STAR Protocols*, *3*(1), 101108.

Anand, K., Schulte, A., Fujinaga, K., Scheffzek, K., & Geyer, M. (2007). Cyclin box structure of the P-TEFb subunit cyclin T1 derived from a fusion complex with EIAV tat. *Journal of Molecular Biology*, *370*(5), 826–836.

Andrews, F. H., Shinsky, S. A., Shanle, E. K., Bridgers, J. B., Gest, A., Tsun, I. K., et al. (2016). The Taf14 YEATS domain is a reader of histone crotonylation. *Nature Chemical Biology*, *12*(6), 396–398.

Bacon, C. W., & D'Orso, I. (2019). CDK9: A signaling hub for transcriptional control. *Transcription*, *10*(2), 57–75.

Barlaam, B., Casella, R., Cidado, J., Cook, C., De Savi, C., Dishington, A., et al. (2020). Discovery of AZD4573, a potent and selective inhibitor of CDK9 that enables short duration of target engagement for the treatment of hematological malignancies. *Journal of Medicinal Chemistry*, *63*(24), 15564–15590.

Baumli, S., Hole, A. J., Wang, L. Z., Noble, M. E., & Endicott, J. A. (2012). The CDK9 tail determines the reaction pathway of positive transcription elongation factor b. *Structure*, *20*(10), 1788–1795.

Biswas, D., Milne, T. A., Basrur, V., Kim, J., Elenitoba-Johnson, K. S., Allis, C. D., et al. (2011). Function of leukemogenic mixed lineage leukemia 1 (MLL) fusion proteins through distinct partner protein complexes. *Proceedings of the National Academy of Sciences of the United States of America*, *108*(38), 15751–15756.

Boffo, S., Damato, A., Alfano, L., & Giordano, A. (2018). CDK9 inhibitors in acute myeloid leukemia. *Journal of Experimental & Clinical Cancer Research*, *37*(1), 36.

Bradner, J. E., Hnisz, D., & Young, R. A. (2017). Transcriptional addiction in cancer. *Cell, 168*(4), 629–643.
Campbell, C. T., Haladyna, J. N., Drubin, D. A., Thomson, T. M., Maria, M. J., Yamauchi, T., et al. (2017). Mechanisms of Pinometostat (EPZ-5676) treatment-emergent resistance in MLL-rearranged leukemia. *Molecular Cancer Therapeutics, 16*(8), 1669–1679.
Cao, K., Ugarenko, M., Ozark, P. A., Wang, J., Marshall, S. A., Rendleman, E. J., et al. (2020). DOT1L-controlled cell-fate determination and transcription elongation are independent of H3K79 methylation. *Proceedings of the National Academy of Sciences of the United States of America, 117*(44), 27365–27373.
Carrozza, M. J., Li, B., Florens, L., Suganuma, T., Swanson, S. K., Lee, K. K., et al. (2005). Histone H3 methylation by Set2 directs deacetylation of coding regions by Rpd3S to suppress spurious intragenic transcription. *Cell, 123*(4), 581–592.
Chae, Y. C., Kim, J. Y., Park, J. W., Kim, K. B., Oh, H., Lee, K. H., et al. (2019). FOXO1 degradation via G9a-mediated methylation promotes cell proliferation in colon cancer. *Nucleic Acids Research, 47*(4), 1692–1705.
Chen, Y., & Cramer, P. (2019). Structure of the super-elongation complex subunit AFF4 C-terminal homology domain reveals requirements for AFF homo- and heterodimerization. *The Journal of Biological Chemistry, 294*(27), 10663–10673.
Chen, R., Liu, M., Li, H., Xue, Y., Ramey, W. N., He, N., et al. (2008). PP2B and PP1alpha cooperatively disrupt 7SK snRNP to release P-TEFb for transcription in response to Ca2+ signaling. *Genes & Development, 22*(10), 1356–1368.
Chen, Z., Min, M. R., Parthasarathy, S., & Ning, X. (2021). A deep generative model for molecule optimization via one fragment modification. *Nature Machine Intelligence, 3*(12), 1040–1049.
Chen, F. X., Smith, E. R., & Shilatifard, A. (2018). Born to run: Control of transcription elongation by RNA polymerase II. *Nature Reviews. Molecular Cell Biology, 19*(7), 464–478.
Chen, Y., Vos, S. M., Dienemann, C., Ninov, M., Urlaub, H., & Cramer, P. (2021). Allosteric transcription stimulation by RNA polymerase II super elongation complex. *Molecular Cell, 81*(16), 3386–3399.e10.
Cheng, B., Li, T., Rahl, P. B., Adamson, T. E., Loudas, N. B., Guo, J., et al. (2012). Functional association of Gdown1 with RNA polymerase II poised on human genes. *Molecular Cell, 45*(1), 38–50.
Chern, T. R., Liu, L., Petrunak, E., Stuckey, J. A., Wang, M., Bernard, D., et al. (2020). Discovery of potent small-molecule inhibitors of MLL methyltransferase. *ACS Medicinal Chemistry Letters, 11*(6), 1348–1352.
Compe, E., & Egly, J. M. (2012). TFIIH: When transcription met DNA repair. *Nature Reviews. Molecular Cell Biology, 13*(6), 343–354.
Core, L., & Adelman, K. (2019). Promoter-proximal pausing of RNA polymerase II: A nexus of gene regulation. *Genes & Development, 33*(15–16), 960–982.
Core, L. J., Waterfall, J. J., Gilchrist, D. A., Fargo, D. C., Kwak, H., Adelman, K., et al. (2012). Defining the status of RNA polymerase at promoters. *Cell Reports, 2*(4), 1025–1035.
Cortellis. https://www.cortellis.com/drugdiscovery/entity/drug/1017792/product?ent=L8QWsxYl.
Creyghton, M. P., Cheng, A. W., Welstead, G. G., Kooistra, T., Carey, B. W., Steine, E. J., et al. (2010). Histone H3K27ac separates active from poised enhancers and predicts developmental state. *Proceedings of the National Academy of Sciences of the United States of America, 107*(50), 21931–21936.
Dahl, N. A., Danis, E., Balakrishnan, I., Wang, D., Pierce, A., Walker, F. M., et al. (2020). Super elongation complex as a targetable dependency in diffuse midline glioma. *Cell Reports, 31*(1), 107485.

Daigle, S. R., Olhava, E. J., Therkelsen, C. A., Basavapathruni, A., Jin, L., Boriack-Sjodin, P. A., et al. (2013). Potent inhibition of DOT1L as treatment of MLL-fusion leukemia. *Blood*, *122*(6), 1017–1025.

Day, M. A., Saffran, D. C., Hood, T., Obholzer, N., Pandey, A., Lin, C. Y., et al. (2022). Abstract 2565: CDK9 inhibitor KB-0742 is active in preclinical models of small-cell lung cancer. *Cancer Research*, *82*(12_Supplement), 2565.

DeLaney, E., & Luse, D. S. (2016). Gdown1 associates efficiently with RNA polymerase II after promoter clearance and displaces TFIIF during transcript elongation. *PLoS One*, *11*(10), e0163649.

Deng, P., Wang, J., Zhang, X., Wu, X., Ji, N., Li, J., et al. (2018). AFF4 promotes tumorigenesis and tumor-initiation capacity of head and neck squamous cell carcinoma cells by regulating SOX2. *Carcinogenesis*, *39*(7), 937–947.

DiMartino, J. F., Millar, T., Ayton, P. M., Landewe, T., Hess, J. L., Cleary, M. L., et al. (2000). A carboxy-terminal domain of ELL is required and sufficient for immortalization of myeloid progenitors by MLL-ELL. *Blood*, *96*, 3887–3893.

Doamekpor, S. K., Sanchez, A. M., Schwer, B., Shuman, S., & Lima, C. D. (2014). How an mRNA capping enzyme reads distinct RNA polymerase II and Spt5 CTD phosphorylation codes. *Genes & Development*, *28*(12), 1323–1336.

Ebmeier, C. C., Erickson, B., Allen, B. L., Allen, M. A., Kim, H., Fong, N., et al. (2017). Human TFIIH kinase CDK7 regulates transcription-associated chromatin modifications. *Cell Reports*, *20*(5), 1173–1186.

Erb, M. A., Scott, T. G., Li, B. E., Xie, H., Paulk, J., Seo, H.-S., et al. (2017). Transcription control by the ENL YEATS domain in acute leukaemia. *Nature*, *543*(7644), 270–274.

Fay, A., Misulovin, Z., Li, J., Schaaf, C. A., Gause, M., Gilmour, D. S., et al. (2011). Cohesin selectively binds and regulates genes with paused RNA polymerase. *Current Biology*, *21*(19), 1624–1634.

Fish, R. N., & Kane, C. M. (2002). Promoting elongation with transcript cleavage stimulatory factors. *Biochimica et Biophysica Acta*, *1577*(2), 287–307.

Fiskus, W. C., Mill, C. P., Birdwell, C., Davis, J. A., Salazar, A., Philip, K., et al. (2021). Preclinically effective menin inhibitor SNDX-50469 and SNDX-5613-based combinations against MLL1-rearranged (MLL-r) or NPM1-mutant AML models. *Blood*, *138*(Supplement 1), 3340.

Franco, L. C., Morales, F., Boffo, S., & Giordano, A. (2018). CDK9: A key player in cancer and other diseases. *Journal of Cellular Biochemistry*, *119*(2), 1273–1284.

Frigault, M. M., Garban, H., Greer, J. M., Hwang, S., Izumi, R., Johnson, A. J., et al. (2022). Abstract 1859: VIP152, a selective CDK9 inhibitor, demonstrates sensitivity in gynecologic cell lines that are cisplatin sensitive or resistant and delivers in vivo antitumor efficacy. *Cancer Research*, *82*(12_Supplement), 1859.

Gao, Y., Chen, L., Han, Y., Wu, F., Yang, W. S., Zhang, Z., et al. (2020). Acetylation of histone H3K27 signals the transcriptional elongation for estrogen receptor alpha. *Communications Biology*, *3*(1), 165.

Gao, Q., Zheng, J., Ni, Z., Sun, P., Yang, C., Cheng, M., et al. (2020). The m(6)a methylation-regulated AFF4 promotes self-renewal of bladder cancer stem cells. *Stem Cells International*, *2020*, 8849218.

Gentile, F., Yaacoub, J. C., Gleave, J., Fernandez, M., Ton, A. T., Ban, F., et al. (2022). Artificial intelligence-enabled virtual screening of ultra-large chemical libraries with deep docking. *Nature Protocols*, *17*(3), 672–697.

Ghobrial, A., Flick, N., Daly, R., Hoffman, M., & Milcarek, C. (2019). ELL2 influences transcription elongation, splicing, Ig secretion and growth. *Journal of Mucosal Immunology Research*, *3*(1).

Gilchrist, D. A., Nechaev, S., Lee, C., Ghosh, S. K., Collins, J. B., Li, L., et al. (2008). NELF-mediated stalling of Pol II can enhance gene expression by blocking promoter-proximal nucleosome assembly. *Genes & Development*, *22*(14), 1921–1933.

Gilmour, D. S., & Lis, J. T. (1986). RNA polymerase II interacts with the promoter region of the noninduced hsp70 gene in Drosophila melanogaster cells. *Molecular and Cellular Biology*, *6*(11), 3984–3989.

Ginestier, C., Hur, M. H., Charafe-Jauffret, E., Monville, F., Dutcher, J., Brown, M., et al. (2007). ALDH1 is a marker of normal and malignant human mammary stem cells and a predictor of poor clinical outcome. *Cell Stem Cell*, *1*(5), 555–567.

Gouda, M. B. Y., Zidane, M. A., Abdelhady, A. S., & Hassan, N. M. (2022). Expression and prognostic significance of chromatin modulators EHMT2/G9a and KDM2b in acute myeloid leukemia. *Journal of Cellular Biochemistry*, *123*(8), 1340–1355.

Haebe, J. R., Bergin, C. J., Sandouka, T., & Benoit, Y. D. (2021). Emerging role of G9a in cancer stemness and promises as a therapeutic target. *Oncogene*, *10*(11), 76.

He, N., Chan, C. K., Sobhian, B., Chou, S., Xue, Y., Liu, M., et al. (2011). Human polymerase-associated factor complex (PAFc) connects the super elongation complex (SEC) to RNA polymerase II on chromatin. *Proceedings of the National Academy of Sciences of the United States of America*, *108*(36), E636–E645.

Heath, E. I., Bible, K., Martell, R. E., Adelman, D. C., & Lorusso, P. M. (2008). A phase 1 study of SNS-032 (formerly BMS-387032), a potent inhibitor of cyclin-dependent kinases 2, 7 and 9 administered as a single oral dose and weekly infusion in patients with metastatic refractory solid tumors. *Investigational New Drugs*, *26*(1), 59–65.

Henriques, T., Gilchrist, D. A., Nechaev, S., Bern, M., Muse, G. W., Burkholder, A., et al. (2013). Stable pausing by RNA polymerase II provides an opportunity to target and integrate regulatory signals. *Molecular Cell*, *52*(4), 517–528.

Herz, H. M., Mohan, M., Garruss, A. S., Liang, K., Takahashi, Y. H., Mickey, K., et al. (2012). Enhancer-associated H3K4 monomethylation by Trithorax-related, the Drosophila homolog of mammalian Mll3/Mll4. *Genes & Development*, *26*(23), 2604–2620.

Hu, D., Gao, X., Morgan, M. A., Herz, H. M., Smith, E. R., & Shilatifard, A. (2013). The MLL3/MLL4 branches of the COMPASS family function as major histone H3K4 monomethylases at enhancers. *Molecular and Cellular Biology*, *33*(23), 4745–4754.

Hu, H. Y., Zhang, Y., Zhao, L. F., Zhao, W. T., Wang, X. X., Ye, E., et al. (2021). AFF4 facilitates melanoma cell progression by regulating c-Jun activity. *Experimental Cell Research*, *399*(2).

Huang, C. H., Lujambio, A., Zuber, J., Tschaharganeh, D. F., Doran, M. G., Evans, M. J., et al. (2014). CDK9-mediated transcription elongation is required for MYC addiction in hepatocellular carcinoma. *Genes & Development*, *28*(16), 1800–1814.

Huang, T., Song, X., Xu, D., Tiek, D., Goenka, A., Wu, B., et al. (2020). Stem cell programs in cancer initiation, progression, and therapy resistance. *Theranostics*, *10*(19), 8721–8743.

Ishibashi, T., Dangkulwanich, M., Coello, Y., Lionberger, T. A., Lubkowska, L., Ponticelli, A. S., et al. (2014). Transcription factors IIS and IIF enhance transcription efficiency by differentially modifying RNA polymerase pausing dynamics. *Proceedings of the National Academy of Sciences of the United States of America*, *111*(9), 3419–3424.

Izumi, K., Nakato, R., Zhang, Z., Edmondson, A. C., Noon, S., Dulik, M. C., et al. (2015). Germline gain-of-function mutations in AFF4 cause a developmental syndrome functionally linking the super elongation complex and cohesin. *Nature Genetics*, *47*(4), 338–344.

Jayatunga, M. K. P., Xie, W., Ruder, L., Schulze, U., & Meier, C. (2022). AI in small-molecule drug discovery: A coming wave? *Nature Reviews. Drug Discovery*, *21*(3), 175–176.

Ji, X., Lu, H., Zhou, Q., & Luo, K. (2014). LARP7 suppresses P-TEFb activity to inhibit breast cancer progression and metastasis. *eLife*, *3*, e02907.

Ji, X., Zhou, Y., Pandit, S., Huang, J., Li, H., Lin, C. Y., et al. (2013). SR proteins collaborate with 7SK and promoter-associated nascent RNA to release paused polymerase. *Cell*, *153*(4), 855–868.

Jiang, Y., Chen, G., Li, X. M., Liu, S., Tian, G., Li, Y., et al. (2020). Selective targeting of AF9 YEATS domain by cyclopeptide inhibitors with preorganized conformation. *Journal of the American Chemical Society*, *142*(51), 21450–21459.

Jonkers, I., & Lis, J. T. (2015). Getting up to speed with transcription elongation by RNA polymerase II. *Nature Reviews. Molecular Cell Biology*, *16*(3), 167–177.

Kagey, M. H., Newman, J. J., Bilodeau, S., Zhan, Y., Orlando, D. A., van Berkum, N. L., et al. (2010). Mediator and cohesin connect gene expression and chromatin architecture. *Nature*, *467*(7314), 430–435.

Kaithoju, S. (2014). Epigenetics and cancer therapy. *Journal of Cancer Biology & Research*, *1502*, 2.

Kanazawa, S., Soucek, L., Evan, G., Okamoto, T., & Peterlin, B. M. (2003). c-Myc recruits P-TEFb for transcription, cellular proliferation and apoptosis. *Oncogene*, *22*(36), 5707–5711.

Katagi, H., Takata, N., Aoi, Y., Zhang, Y., Rendleman, E. J., Blyth, G. T., et al. (2021). Therapeutic targeting of transcriptional elongation in diffuse intrinsic pontine glioma. *Neuro-Oncology*, *23*(8), 1348–1359.

Kelly, C. N., Townsend, C. E., Jain, A. N., Naylor, M. R., Pye, C. R., Schwochert, J., et al. (2021). Geometrically diverse lariat peptide scaffolds reveal an untapped chemical space of high membrane permeability. *Journal of the American Chemical Society*, *143*(2), 705–714.

Klein, B. J., Bose, D., Baker, K. J., Yusoff, Z. M., Zhang, X., & Murakami, K. S. (2011). RNA polymerase and transcription elongation factor Spt4/5 complex structure. *Proceedings of the National Academy of Sciences of the United States of America*, *108*(2), 546–550.

Kress, T. R., Sabò, A., & Amati, B. (2015). MYC: Connecting selective transcriptional control to global RNA production. *Nature Reviews. Cancer*, *15*(10), 593–607.

Krivtsov, A. V., Evans, K., Gadrey, J. Y., Eschle, B. K., Hatton, C., Uckelmann, H. J., et al. (2019). A menin-MLL inhibitor induces specific chromatin changes and eradicates disease in models of MLL-rearranged leukemia. *Cancer Cell*, *36*(6), 660–673.e11.

Kuntimaddi, A., Achille, N. J., Thorpe, J., Lokken, A. A., Singh, R., Hemenway, C. S., et al. (2015). Degree of recruitment of DOT1L to MLL-AF9 defines level of H3K79 di- and tri-methylation on target genes and transformation potential. *Cell Reports*, *11*(5), 808–820.

Kurani, H., Razavipour, S. F., Harikumar, K. B., Dunworth, M., Ewald, A. J., Nasir, A., et al. (2022). DOT1L is a novel cancer stem cell target for triple-negative breast cancer. *Clinical Cancer Research*, *28*(9), 1948–1965.

Lai, W. K. M., & Pugh, B. F. (2017). Understanding nucleosome dynamics and their links to gene expression and DNA replication. *Nature Reviews. Molecular Cell Biology*, *18*(9), 548–562.

Leach, B. I., Kuntimaddi, A., Schmidt, C. R., Cierpicki, T., Johnson, S. A., & Bushweller, J. H. (2013). Leukemia fusion target AF9 is an intrinsically disordered transcriptional regulator that recruits multiple partners via coupled folding and binding. *Structure*, *21*(1), 176–183.

Lehnertz, B., Pabst, C., Su, L., Miller, M., Liu, F., Yi, L., et al. (2014). The methyltransferase G9a regulates HoxA9-dependent transcription in AML. *Genes & Development*, *28*(4), 317–327.

Li, X., Li, X. M., Jiang, Y., Liu, Z., Cui, Y., Fung, K. Y., et al. (2018). Structure-guided development of YEATS domain inhibitors by targeting pi-pi-pi stacking. *Nature Chemical Biology*, *14*(12), 1140–1149.

Li, J., Liu, T., Song, Y., Wang, M., Liu, L., Zhu, H., et al. (2022). Discovery of small-molecule degraders of the CDK9-cyclin T1 complex for targeting transcriptional addiction in prostate cancer. *Journal of Medicinal Chemistry*, *65*(16), 11034–11057.

Li, Y., Sabari, B. R., Panchenko, T., Wen, H., Zhao, D., Guan, H., et al. (2016). Molecular coupling of histone crotonylation and active transcription by AF9 YEATS domain. *Molecular Cell*, *62*(2), 181–193.

Li, Y., Wen, H., Xi, Y., Tanaka, K., Wang, H., Peng, D., et al. (2014). AF9 YEATS domain links histone acetylation to DOT1L-mediated H3K79 methylation. *Cell*, *159*(3), 558–571.

Liang, K., Smith, E. R., Aoi, Y., Stoltz, K. L., Katagi, H., Woodfin, A. R., et al. (2018). Targeting processive transcription elongation via SEC disruption for MYC-induced cancer therapy. *Cell*, *175*(3), 766–779.e17.

Liang, K., Volk, A. G., Haug, J. S., Marshall, S. A., Woodfin, A. R., Bartom, E. T., et al. (2017). Therapeutic targeting of MLL degradation pathways in MLL-rearranged leukemia. *Cell*, *168*(1–2), 59–72.e13.

Lin, C., Smith, E. R., Takahashi, H., Lai, K. C., Martin-Brown, S., Florens, L., et al. (2010). AFF4, a component of the ELL/P-TEFb elongation complex and a shared subunit of MLL chimeras, can link transcription elongation to leukemia. *Molecular Cell*, *37*(3), 429–437.

Liu, R., Chen, C., Li, Y., Huang, Q., & Xue, Y. (2020). ELL-associated factors EAF1/2 negatively regulate HIV-1 transcription through inhibition of super elongation complex formation. *Biochimica et Biophysica Acta, Gene Regulatory Mechanisms*, *1863*(5), 194508.

Liu, K., Shen, D., Shen, J., Gao, S. M., Li, B., Wong, C., et al. (2017). The super elongation complex drives neural stem cell fate commitment. *Developmental Cell*, *40*(6), 537–551.e6.

Liu, J. X., Zhang, D., Xie, X., Ouyang, G., Liu, X., Sun, Y., et al. (2013). Eaf1 and Eaf2 negatively regulate canonical Wnt/beta-catenin signaling. *Development*, *140*(5), 1067–1078.

Lu, H., Li, Z., Zhang, W., Schulze-Gahmen, U., Xue, Y., & Zhou, Q. (2015). Gene target specificity of the super elongation complex (SEC) family: How HIV-1 tat employs selected SEC members to activate viral transcription. *Nucleic Acids Research*, *43*(12), 5868–5879.

Lu, H., Xue, Y., Yu, G. K., Arias, C., Lin, J., Fong, S., et al. (2015). Compensatory induction of MYC expression by sustained CDK9 inhibition via a BRD4-dependent mechanism. *eLife*, *4*, e06535.

Luo, Z., Lin, C., & Shilatifard, A. (2012). The super elongation complex (SEC) family in transcriptional control. *Nature Reviews. Molecular Cell Biology*, *13*(9), 543–547.

Luo, L., Tang, H., Ling, L., Li, N., Jia, X., Zhang, Z., et al. (2018). LINC01638 lncRNA activates MTDH-Twist1 signaling by preventing SPOP-mediated c-Myc degradation in triple-negative breast cancer. *Oncogene*, *37*(47), 6166–6179.

Mandal, R., Becker, S., & Strebhardt, K. (2021). Targeting CDK9 for anti-cancer therapeutics. *Cancers (Basel)*, *13*(9).

Martincic, K., Alkan, S. A., Cheatle, A., Borghesi, L., & Milcarek, C. (2009). Transcription elongation factor ELL2 directs immunoglobulin secretion in plasma cells by stimulating altered RNA processing. *Nature Immunology*, *10*(10), 1102–1109.

McNamara, R. P., McCann, J. L., Gudipaty, S. A., & D'Orso, I. (2013). Transcription factors mediate the enzymatic disassembly of promoter-bound 7SK snRNP to locally recruit P-TEFb for transcription elongation. *Cell Reports*, *5*(5), 1256–1268.

Meinhart, A., Kamenski, T., Hoeppner, S., Baumli, S., & Cramer, P. (2005). A structural perspective of CTD function. *Genes & Development*, *19*(12), 1401–1415.

Méndez-Lucio, O., Ahmad, M., del Rio-Chanona, E. A., & Wegner, J. K. (2021). A geometric deep learning approach to predict binding conformations of bioactive molecules. *Nature Machine Intelligence*, *3*(12), 1033–1039.

Meyer, C., Hofmann, J., Burmeister, T., Groger, D., Park, T. S., Emerenciano, M., et al. (2013). The MLL recombinome of acute leukemias in 2013. *Leukemia*, *27*(11), 2165–2176.

Michels, A. A., Fraldi, A., Li, Q., Adamson, T. E., Bonnet, F., Nguyen, V. T., et al. (2004). Binding of the 7SK snRNA turns the HEXIM1 protein into a P-TEFb (CDK9/cyclin T) inhibitor. *The EMBO Journal, 23*(13), 2608–2619.

Mohan, M., Herz, H. M., Takahashi, Y. H., Lin, C., Lai, K. C., Zhang, Y., et al. (2010). Linking H3K79 trimethylation to Wnt signaling through a novel Dot1-containing complex (DotCom). *Genes & Development, 24*(6), 574–589.

Mozzetta, C., Boyarchuk, E., Pontis, J., & Ait-Si-Ali, S. (2015). Sound of silence: The properties and functions of repressive Lys methyltransferases. *Nature Reviews. Molecular Cell Biology, 16*(8), 499–513.

Mullen Davis, M. A., Guo, J., Price, D. H., & Luse, D. S. (2014). Functional interactions of the RNA polymerase II-interacting proteins Gdown1 and TFIIF. *The Journal of Biological Chemistry, 289*(16), 11143–11152.

Nagaike, T., Logan, C., Hotta, I., Rozenblatt-Rosen, O., Meyerson, M., & Manley, J. L. (2011). Transcriptional activators enhance polyadenylation of mRNA precursors. *Molecular Cell, 41*(4), 409–418.

Ning, J. F., Stanciu, M., Humphrey, M. R., Gorham, J., Wakimoto, H., Nishihara, R., et al. (2019). Myc targeted CDK18 promotes ATR and homologous recombination to mediate PARP inhibitor resistance in glioblastoma. *Nature Communications, 10*(1), 2910.

Park, K. S., Xiong, Y., Yim, H., Velez, J., Babault, N., Kumar, P., et al. (2022). Discovery of the first-in-class G9a/GLP covalent inhibitors. *Journal of Medicinal Chemistry, 65*(15), 10506–10522.

Polak, P. E., Simone, F., Kaberlein, J. J., Luo, R. T., & Thirman, M. J. (2003). ELL and EAF1 are Cajal body components that are disrupted in MLL-ELL leukemia. *Molecular Biology of the Cell, 14*(4), 1517–1528.

Poulard, C., Noureddine, L. M., Pruvost, L., & Le Romancer, M. (2021). Structure, activity, and function of the protein lysine methyltransferase G9a. *Life (Basel), 11*(10).

Qi, S., Li, Z., Schulze-Gahmen, U., Stjepanovic, G., Zhou, Q., & Hurley, J. H. (2017). Structural basis for ELL2 and AFF4 activation of HIV-1 proviral transcription. *Nature Communications, 8*, 14076.

Rahl, P. B., Lin, C. Y., Seila, A. C., Flynn, R. A., McCuine, S., Burge, C. B., et al. (2010). c-Myc regulates transcriptional pause release. *Cell, 141*(3), 432–445.

Salih, D. A., & Brunet, A. (2008). FoxO transcription factors in the maintenance of cellular homeostasis during aging. *Current Opinion in Cell Biology, 20*(2), 126–136.

Saunders, A., Core, L. J., & Lis, J. T. (2006). Breaking barriers to transcription elongation. *Nature Reviews. Molecular Cell Biology, 7*(8), 557–567.

Schaaf, C. A., Kwak, H., Koenig, A., Misulovin, Z., Gohara, D. W., Watson, A., et al. (2013). Genome-wide control of RNA polymerase II activity by cohesin. *PLoS Genetics, 9*(3), e1003382.

Schulze, J. M., Wang, A. Y., & Kobor, M. S. (2010). Reading chromatin: Insights from yeast into YEATS domain structure and function. *Epigenetics, 5*(7), 573–577.

Schulze-Gahmen, U., Echeverria, I., Stjepanovic, G., Bai, Y., Lu, H., Schneidman-Duhovny, D., et al. (2016). Insights into HIV-1 proviral transcription from integrative structure and dynamics of the tat:AFF4:P-TEFb:TAR complex. *eLife*, 5.

Schulze-Gahmen, U., Upton, H., Birnberg, A., Bao, K., Chou, S., Krogan, N. J., et al. (2013). The AFF4 scaffold binds human P-TEFb adjacent to HIV tat. *eLife, 2*, e00327.

Segovia, C., San Jose-Eneriz, E., Munera-Maravilla, E., Martinez-Fernandez, M., Garate, L., Miranda, E., et al. (2019). Inhibition of a G9a/DNMT network triggers immune-mediated bladder cancer regression. *Nature Medicine, 25*(7), 1073–1081.

Sengupta, S., Biarnes, M. C., & Jordan, V. C. (2014). Cyclin dependent kinase-9 mediated transcriptional de-regulation of cMYC as a critical determinant of endocrine-therapy resistance in breast cancers. *Breast Cancer Research and Treatment, 143*(1), 113–124.

Shen, Y., Yue, F., McCleary, D. F., Ye, Z., Edsall, L., Kuan, S., et al. (2012). A map of the cis-regulatory sequences in the mouse genome. *Nature, 488*(7409), 116–120.

Shetty, A., Kallgren, S. P., Demel, C., Maier, K. C., Spatt, D., Alver, B. H., et al. (2017). Spt5 plays vital roles in the control of sense and antisense transcription elongation. *Molecular Cell, 66*(1), 77–88.e5.

Shilatifard, A. (2012). The COMPASS family of histone H3K4 methylases: Mechanisms of regulation in development and disease pathogenesis. *Annual Review of Biochemistry, 81*, 65–95.

Simone, F., Luo, R. T., Polak, P. E., Kaberlein, J. J., & Thirman, M. J. (2003). ELL-associated factor 2 (EAF2), a functional homolog of EAF1 with alternative ELL binding properties. *Blood, 101*(6), 2355–2362.

Simone, F., Polak, P. E., Kaberlein, J. J., Luo, R. T., Levitan, D. A., & Thirman, M. J. (2001). EAF1, a novel ELL-associated factor that is delocalized by expression of the MLL-ELL fusion protein. *Blood, 98*(1), 201–209.

Smith, E., Lin, C., & Shilatifard, A. (2011). The super elongation complex (SEC) and MLL in development and disease. *Genes & Development, 25*(7), 661–672.

Smolle, M., Venkatesh, S., Gogol, M. M., Li, H., Zhang, Y., Florens, L., et al. (2012). Chromatin remodelers Isw1 and Chd1 maintain chromatin structure during transcription by preventing histone exchange. *Nature Structural & Molecular Biology, 19*(9), 884–892.

Tachibana, M., Sugimoto, K., Nozaki, M., Ueda, J., Ohta, T., Ohki, M., et al. (2002). G9a histone methyltransferase plays a dominant role in euchromatic histone H3 lysine 9 methylation and is essential for early embryogenesis. *Genes & Development, 16*(14), 1779–1791.

Takahashi, H., Ranjan, A., Chen, S., Suzuki, H., Shibata, M., Hirose, T., et al. (2020). The role of mediator and little elongation complex in transcription termination. *Nature Communications, 11*(1), 1063.

Talbert, P. B., & Henikoff, S. (2017). Histone variants on the move: Substrates for chromatin dynamics. *Nature Reviews. Molecular Cell Biology, 18*(2), 115–126.

Tan, S., Conaway, R. C., & Conaway, J. W. (1995). Dissection of transcription factor TFIIF functional domains required for initiation and elongation. *Proceedings of the National Academy of Sciences of the United States of America, 92*(13), 6042–6046.

Tang, D., Chen, C. J., Liao, G., Liu, J. M., Liao, B. H., Huang, Q. Q., et al. (2020). Structural and functional insight into the effect of AFF4 dimerization on activation of HIV-1 proviral transcription. *Cell Discovery, 6*(1), 11.

Tong, W. G., Chen, R., Plunkett, W., Siegel, D., Sinha, R., Harvey, R. D., et al. (2010). Phase I and pharmacologic study of SNS-032, a potent and selective Cdk2, 7, and 9 inhibitor, in patients with advanced chronic lymphocytic leukemia and multiple myeloma. *Journal of Clinical Oncology, 28*(18), 3015–3022.

Tu, W. B., Shiah, Y. J., Lourenco, C., Mullen, P. J., Dingar, D., Redel, C., et al. (2018). MYC interacts with the G9a histone methyltransferase to drive transcriptional repression and tumorigenesis. *Cancer Cell, 34*(4), 579–595.e8.

Vatapalli, R., Sagar, V., Rodriguez, Y., Zhao, J. C., Unno, K., Pamarthy, S., et al. (2020). Histone methyltransferase DOT1L coordinates AR and MYC stability in prostate cancer. *Nature Communications, 11*(1), 4153.

Vedadi, M., Barsyte-Lovejoy, D., Liu, F., Rival-Gervier, S., Allali-Hassani, A., Labrie, V., et al. (2011). A chemical probe selectively inhibits G9a and GLP methyltransferase activity in cells. *Nature Chemical Biology, 7*(8), 566–574.

Venkatesh, S., & Workman, J. L. (2015). Histone exchange, chromatin structure and the regulation of transcription. *Nature Reviews. Molecular Cell Biology, 16*(3), 178–189.

Wan, L., Wen, H., Li, Y., Lyu, J., Xi, Y., Hoshii, T., et al. (2017). ENL links histone acetylation to oncogenic gene expression in acute myeloid leukaemia. *Nature, 543*(7644), 265–269.

Wang, H., Dawber, R. S., Zhang, P., Walko, M., Wilson, A. J., & Wang, X. (2021). Peptide-based inhibitors of protein-protein interactions: Biophysical, structural and cellular consequences of introducing a constraint. *Chemical Science, 12*(17), 5977–5993.

Wang, Y., Dow, E. C., Liang, Y. Y., Ramakrishnan, R., Liu, H., Sung, T. L., et al. (2008). Phosphatase PPM1A regulates phosphorylation of Thr-186 in the Cdk9 T-loop. *The Journal of Biological Chemistry, 283*(48), 33578–33584.

Wang, Y., Wang, J., Cao, Z., & Barati Farimani, A. (2022). Molecular contrastive learning of representations via graph neural networks. *Nature Machine Intelligence, 4*(3), 279–287.

Wang, W., Yao, X., Huang, Y., Hu, X., Liu, R., Hou, D., et al. (2013). Mediator MED23 regulates basal transcription in vivo via an interaction with P-TEFb. *Transcription, 4*(1), 39–51.

Wu, Y. M., Chang, J. W., Wang, C. H., Lin, Y. C., Wu, P. L., Huang, S. H., et al. (2012). Regulation of mammalian transcription by Gdown1 through a novel steric crosstalk revealed by cryo-EM. *The EMBO Journal, 31*(17), 3575–3587.

Yamada, T., Yamaguchi, Y., Inukai, N., Okamoto, S., Mura, T., & Handa, H. (2006). P-TEFb-mediated phosphorylation of hSpt5 C-terminal repeats is critical for processive transcription elongation. *Molecular Cell, 21*(2), 227–237.

Yamaguchi, Y., Takagi, T., Wada, T., Yano, K., Furuya, A., Sugimoto, S., et al. (1999). NELF, a multisubunit complex containing RD, cooperates with DSIF to repress RNA polymerase II elongation. *Cell, 97*(1), 41–51.

Yang, L., Shi, P., Zhao, G., Xu, J., Peng, W., Zhang, J., et al. (2020). Targeting cancer stem cell pathways for cancer therapy. *Signal Transduction and Targeted Therapy, 5*(1), 8.

Yu, D., Liu, R., Yang, G., & Zhou, Q. (2018). The PARP1-Siah1 Axis controls HIV-1 transcription and expression of Siah1 substrates. *Cell Reports, 23*(13), 3741–3749.

Yuan, Y., Du, L., Tan, R., Yu, Y., Jiang, J., Yao, A., et al. (2022). Design, synthesis, and biological evaluations of DOT1L peptide mimetics targeting the protein-protein interactions between DOT1L and MLL-AF9/MLL-ENL. *Journal of Medicinal Chemistry, 65*(11), 7770–7785.

Zhang, Q., Zeng, L., Zhao, C., Ju, Y., Konuma, T., & Zhou, M. M. (2016). Structural insights into histone crotonyl-lysine recognition by the AF9 YEATS domain. *Structure, 24*(9), 1606–1612.

Zhao, D., Li, Y., Xiong, X., Chen, Z., & Li, H. (2017). YEATS domain—A histone acylation reader in health and disease. *Journal of Molecular Biology, 429*(13), 1994–2002.

Zhou, Q., Li, T., & Price, D. H. (2012). RNA polymerase II elongation control. *Annual Review of Biochemistry, 81*, 119–143.

Zlotorynski, E. (2018). Myc in elongation and repression. *Cancer Biology, 19*, 751.

9780443194184